T0186380

TRANSCRANIAL
BRAIN STIMULATION

FRONTIERS IN NEUROSCIENCE

Series Editor
Sidney A. Simon, Ph.D.

Published Titles

Apoptosis in Neurobiology
Yusuf A. Hannun, M.D., Professor of Biomedical Research and Chairman, Department of
   Biochemistry and Molecular Biology, Medical University of South Carolina, Charleston,
   South Carolina
Rose-Mary Boustany, M.D., tenured Associate Professor of Pediatrics and Neurobiology, Duke
   University Medical Center, Durham, North Carolina

Neural Prostheses for Restoration of Sensory and Motor Function
John K. Chapin, Ph.D., Professor of Physiology and Pharmacology, State University of New York
   Health Science Center, Brooklyn, New York
Karen A. Moxon, Ph.D., Assistant Professor, School of Biomedical Engineering, Science, and
   Health Systems, Drexel University, Philadelphia, Pennsylvania

Computational Neuroscience: Realistic Modeling for Experimentalists
Eric DeSchutter, M.D., Ph.D., Professor, Department of Medicine, University of Antwerp,
   Antwerp, Belgium

Methods in Pain Research
Lawrence Kruger, Ph.D., Professor of Neurobiology (Emeritus), UCLA School of Medicine and
   Brain Research Institute, Los Angeles, California

Motor Neurobiology of the Spinal Cord
Timothy C. Cope, Ph.D., Professor of Physiology, Wright State University, Dayton, Ohio

Nicotinic Receptors in the Nervous System
Edward D. Levin, Ph.D., Associate Professor, Department of Psychiatry and Pharmacology and
   Molecular Cancer Biology and Department of Psychiatry and Behavioral Sciences, Duke
   University School of Medicine, Durham, North Carolina

Methods in Genomic Neuroscience
Helmin R. Chin, Ph.D., Genetics Research Branch, NIMH, NIH, Bethesda, Maryland
Steven O. Moldin, Ph.D., University of Southern California, Washington, D.C.

Methods in Chemosensory Research
Sidney A. Simon, Ph.D., Professor of Neurobiology, Biomedical Engineering, and Anesthesiology,
   Duke University, Durham, North Carolina
Miguel A.L. Nicolelis, M.D., Ph.D., Professor of Neurobiology and Biomedical Engineering,
   Duke University, Durham, North Carolina

The Somatosensory System: Deciphering the Brain's Own Body Image
Randall J. Nelson, Ph.D., Professor of Anatomy and Neurobiology,
   University of Tennessee Health Sciences Center, Memphis, Tennessee

The Superior Colliculus: New Approaches for Studying Sensorimotor Integration
William C. Hall, Ph.D., Department of Neuroscience, Duke University, Durham, North Carolina
Adonis Moschovakis, Ph.D., Department of Basic Sciences, University of Crete, Heraklion, Greece

New Concepts in Cerebral Ischemia
Rick C.S. Lin, Ph.D., Professor of Anatomy, University of Mississippi Medical Center, Jackson, Mississippi

DNA Arrays: Technologies and Experimental Strategies
Elena Grigorenko, Ph.D., Technology Development Group, Millennium Pharmaceuticals, Cambridge, Massachusetts

Methods for Alcohol-Related Neuroscience Research
Yuan Liu, Ph.D., National Institute of Neurological Disorders and Stroke, National Institutes of Health, Bethesda, Maryland
David M. Lovinger, Ph.D., Laboratory of Integrative Neuroscience, NIAAA, Nashville, Tennessee

Primate Audition: Behavior and Neurobiology
Asif A. Ghazanfar, Ph.D., Princeton University, Princeton, New Jersey

Methods in Drug Abuse Research: Cellular and Circuit Level Analyses
Barry D. Waterhouse, Ph.D., MCP-Hahnemann University, Philadelphia, Pennsylvania

Functional and Neural Mechanisms of Interval Timing
Warren H. Meck, Ph.D., Professor of Psychology, Duke University, Durham, North Carolina

Biomedical Imaging in Experimental Neuroscience
Nick Van Bruggen, Ph.D., Department of Neuroscience Genentech, Inc.
Timothy P.L. Roberts, Ph.D., Associate Professor, University of Toronto, Canada

The Primate Visual System
John H. Kaas, Department of Psychology, Vanderbilt University, Nashville, Tennessee
Christine Collins, Department of Psychology, Vanderbilt University, Nashville, Tennessee

Neurosteroid Effects in the Central Nervous System
Sheryl S. Smith, Ph.D., Department of Physiology, SUNY Health Science Center, Brooklyn, New York

Modern Neurosurgery: Clinical Translation of Neuroscience Advances
Dennis A. Turner, Department of Surgery, Division of Neurosurgery, Duke University Medical Center, Durham, North Carolina

Sleep: Circuits and Functions
Pierre-Hervé Luppi, Université Claude Bernard, Lyon, France

Methods in Insect Sensory Neuroscience
Thomas A. Christensen, Arizona Research Laboratories, Division of Neurobiology, University of Arizona, Tuscon, Arizona

Motor Cortex in Voluntary Movements
Alexa Riehle, INCM-CNRS, Marseille, France
Eilon Vaadia, The Hebrew University, Jerusalem, Israel

Neural Plasticity in Adult Somatic Sensory-Motor Systems
Ford F. Ebner, Vanderbilt University, Nashville, Tennessee

Advances in Vagal Afferent Neurobiology
Bradley J. Undem, Johns Hopkins Asthma Center, Baltimore, Maryland
Daniel Weinreich, University of Maryland, Baltimore, Maryland

The Dynamic Synapse: Molecular Methods in Ionotropic Receptor Biology
Josef T. Kittler, University College, London, England
Stephen J. Moss, University College, London, England

Animal Models of Cognitive Impairment
Edward D. Levin, Duke University Medical Center, Durham, North Carolina
Jerry J. Buccafusco, Medical College of Georgia, Augusta, Georgia

The Role of the Nucleus of the Solitary Tract in Gustatory Processing
Robert M. Bradley, University of Michigan, Ann Arbor, Michigan

Brain Aging: Models, Methods, and Mechanisms
David R. Riddle, Wake Forest University, Winston-Salem, North Carolina

Neural Plasticity and Memory: From Genes to Brain Imaging
Frederico Bermudez-Rattoni, National University of Mexico, Mexico City, Mexico

Serotonin Receptors in Neurobiology
Amitabha Chattopadhyay, Center for Cellular and Molecular Biology, Hyderabad, India

TRP Ion Channel Function in Sensory Transduction and Cellular Signaling Cascades
Wolfgang B. Liedtke, M.D., Ph.D., Duke University Medical Center, Durham, North Carolina
Stefan Heller, Ph.D., Stanford University School of Medicine, Stanford, California

Methods for Neural Ensemble Recordings, Second Edition
Miguel A.L. Nicolelis, M.D., Ph.D., Professor of Neurobiology and Biomedical Engineering,
    Duke University Medical Center, Durham, North Carolina

Biology of the NMDA Receptor
Antonius M. VanDongen, Duke University Medical Center, Durham, North Carolina

Methods of Behavioral Analysis in Neuroscience
Jerry J. Buccafusco, Ph.D., Alzheimer's Research Center, Professor of Pharmacology and Toxicology,
    Professor of Psychiatry and Health Behavior, Medical College of Georgia, Augusta, Georgia

In Vivo Optical Imaging of Brain Function, Second Edition
Ron Frostig, Ph.D., Professor, Department of Neurobiology, University of California,
Irvine, California

Fat Detection: Taste, Texture, and Post Ingestive Effects
Jean-Pierre Montmayeur, Ph.D., Centre National de la Recherche Scientifique, Dijon, France
Johannes le Coutre, Ph.D., Nestlé Research Center, Lausanne, Switzerland

The Neurobiology of Olfaction
Anna Menini, Ph.D., Neurobiology Sector International School for Advanced Studies, (S.I.S.S.A.),
    Trieste, Italy

Neuroproteomics
Oscar Alzate, Ph.D., Department of Cell and Developmental Biology, University of North
    Carolina, Chapel Hill, North Carolina

Translational Pain Research: From Mouse to Man
Lawrence Kruger, Ph.D., Department of Neurobiology, UCLA School of Medicine, Los Angeles,
    California
Alan R. Light, Ph.D., Department of Anesthesiology, University of Utah, Salt Lake City, Utah

Advances in the Neuroscience of Addiction
Cynthia M. Kuhn, Duke University Medical Center, Durham, North Carolina
George F. Koob, The Scripps Research Institute, La Jolla, California

Neurobiology of Huntington's Disease: Applications to Drug Discovery
Donald C. Lo, Duke University Medical Center, Durham, North Carolina
Robert E. Hughes, Buck Institute for Age Research, Novato, California

Neurobiology of Sensation and Reward
Jay A. Gottfried, Northwestern University, Chicago, Illinois

The Neural Bases of Multisensory Processes
Micah M. Murray, CIBM, Lausanne, Switzerland
Mark T. Wallace, Vanderbilt Brain Institute, Nashville, Tennessee

Neurobiology of Depression
Francisco López-Muñoz, University of Alcalá, Madrid, Spain
Cecilio Álamo, University of Alcalá, Madrid, Spain

Astrocytes: Wiring the Brain
Eliana Scemes, Albert Einstein College of Medicine, Bronx, New York
David C. Spray, Albert Einstein College of Medicine, Bronx, New York

Dopamine–Glutamate Interactions in the Basal Ganglia
Susan Jones, University of Cambridge, United Kingdom

Alzheimer's Disease: Targets for New Clinical Diagnostic and Therapeutic Strategies
Renee D. Wegrzyn, Booz Allen Hamilton, Arlington, Virginia
Alan S. Rudolph, Duke Center for Neuroengineering, Potomac, Maryland

The Neurobiological Basis of Suicide
Yogesh Dwivedi, University of Illinois at Chicago

Transcranial Brain Stimulation
Carlo Miniussi, University of Brescia, Italy
Walter Paulus, Georg-August University Medical Center, Göttingen, Germany
Paolo M. Rossini, Institute of Neurology, Catholic University of Rome, Italy

TRANSCRANIAL BRAIN STIMULATION

Edited by
Carlo Miniussi
Department of Clinical and Experimental Sciences
University of Brescia
and
IRCCS The Saint John of God Clinical Research Centre
Brescia, Italy

Walter Paulus
Department of Clinical Neurophysiology
Georg-August University Medical Center
Göttingen, Germany

Paolo M. Rossini
Institute of Neurology
Catholic University of Rome
Rome, Italy

CRC Press
Taylor & Francis Group
Boca Raton London New York

CRC Press is an imprint of the
Taylor & Francis Group, an **informa** business

CRC Press
Taylor & Francis Group
6000 Broken Sound Parkway NW, Suite 300
Boca Raton, FL 33487-2742

First issued in paperback 2019

© 2013 by Taylor & Francis Group, LLC
CRC Press is an imprint of Taylor & Francis Group, an Informa business

No claim to original U.S. Government works

ISBN-13: 978-1-4398-7570-4 (hbk)
ISBN-13: 978-0-367-38057-1 (pbk)

This book contains information obtained from authentic and highly regarded sources. Reasonable efforts have been made to publish reliable data and information, but the author and publisher cannot assume responsibility for the validity of all materials or the consequences of their use. The authors and publishers have attempted to trace the copyright holders of all material reproduced in this publication and apologize to copyright holders if permission to publish in this form has not been obtained. If any copyright material has not been acknowledged please write and let us know so we may rectify in any future reprint.

Except as permitted under U.S. Copyright Law, no part of this book may be reprinted, reproduced, transmitted, or utilized in any form by any electronic, mechanical, or other means, now known or hereafter invented, including photocopying, microfilming, and recording, or in any information storage or retrieval system, without written permission from the publishers.

For permission to photocopy or use material electronically from this work, please access www.copyright.com (http://www.copyright.com/) or contact the Copyright Clearance Center, Inc. (CCC), 222 Rosewood Drive, Danvers, MA 01923, 978-750-8400. CCC is a not-for-profit organization that provides licenses and registration for a variety of users. For organizations that have been granted a photocopy license by the CCC, a separate system of payment has been arranged.

Trademark Notice: Product or corporate names may be trademarks or registered trademarks, and are used only for identification and explanation without intent to infringe.

Library of Congress Cataloging-in-Publication Data

Transcranial brain stimulation / editors, Carlo Miniussi, Walter Paulus, and Paolo M. Rossini.
    p. ; cm. -- (Frontiers in neuroscience)
  Includes bibliographical references and index.
  ISBN 978-1-4398-7570-4 (hardcover : alk. paper)
  I. Miniussi, Carlo. II. Paulus, Walter. III. Rossini, Paolo M. IV. Series: Frontiers in neuroscience (Boca Raton, Fla.)
  [DNLM: 1. Transcranial Magnetic Stimulation--methods. 2. Brain--physiology. 3. Transcutaneous Electric Nerve Stimulation--methods. WL 141.5.T7]

612.8'26--dc23   2012023192

Visit the Taylor & Francis Web site at
http://www.taylorandfrancis.com

and the CRC Press Web site at
http://www.crcpress.com

Contents

PART I Basic Aspects

PART II   Biology

PART III   Imaging-Brain Mapping

PART IV   Perception and Cognition

PART V   *Therapeutic Applications*

PART VI   *Safety*

Series Preface

The Frontiers in Neuroscience Series presents the insights of experts on emerging experimental technologies and theoretical concepts that are, or will be, at the vanguard of neuroscience.

The books cover new and exciting multidisciplinary areas of brain research and describe breakthroughs in fields like visual, gustatory, auditory, olfactory neuroscience as well as aging biomedical imaging. Recent books cover the rapidly evolving fields of multisensory processing, depression, and different aspects of reward.

Each book is edited by experts and consists of chapters written by leaders in a particular field. The books have been richly illustrated and contain comprehensive bibliographies. The chapters provide substantial background material relevant to the particular subject.

The goal is for these books to be the references every neuroscientist uses in order to acquaint themselves with new information and methodologies in brain research. I view my task as series editor to produce outstanding products that contribute to the broad field of neuroscience. Now that the chapters are available online, the effort put in by us—the publisher, the book editors, and individual authors—will contribute to the further development of brain research. To the extent that you learn from these books, we will have succeeded.

Sidney A. Simon, PhD
Series Editor

Preface

Recent years have seen the emergence of exciting new techniques for the understanding of the human brain. An important contribution has come from the introduction of noninvasive brain stimulation (NIBS). NIBS techniques include transcranial magnetic stimulation (TMS) and transcranial electric stimulation (tES). Since the discovery of TMS and tES, these techniques have been used to investigate the state of cortical excitability, the excitability of the corticocortical and corticospinal pathways, the role of a given brain region in a particular cognitive function, and the timing of its activity as well as the pathophysiology of various disorders. TMS and tES are also relevant to clinical neuroscience as a means to improve plasticity and therefore deficits of several functions in individuals with neurological or psychiatric complaints. This book reviews recent advances made in the field of brain stimulation techniques. Moreover, NIBS techniques exert their effects on the neuronal state through different mechanisms at cellular and functional levels. These mechanisms are discussed and recent results are reported.

In this context, the principal goal of this book is to present new knowledge about these innovative approaches by presenting and discussing basic and applied research topics. It brings together leading international experts in the field and provides an authoritative review of the scientific and technical background required to understand transcranial stimulation techniques.

The book offers an overview of basic principles, methodological aspects, and safety implications of transcranial brain stimulation. It also gives practical suggestions and reports the results of numerous TMS and tES studies on biological and behavioral effects. The future of TMS and tES as a multimodal brain imaging approach is discussed. The book also presents a wide range of possible brain stimulation applications and discusses what new information can be gained on the dynamics of brain functions, hierarchical organization, and effective connectivity by using this technique. Implications of recent findings related to therapeutic application are also discussed, presenting what the new hopes for patients and clinicians are.

Finally, the book gives an overview of how NIBS can make a substantial impact on many areas of clinical and basic neuroscience, becoming a fundamental tool in the armamentarium of a neuroscientist.

The development of NIBS techniques to study the central nervous system and also to induce neuroplasticity constitutes a significant breakthrough in our understanding of the brain at work.

Contributors

Giovanni Abbruzzese
Department of Neurosciences,
Ophthalmology and Genetics
University of Genoa
Genoa, Italy

Géza Gergely Ambrus
Department of Clinical Neurophysiology
Georg-August University of Göttingen
Göttingen, Germany

Andrea Antal
Department of Clinical Neurophysiology
Georg-August University Medical Center
Göttingen, Germany

Sven Bestmann
Sobell Department of Motor
Neuroscience and Movement
Disorders
UCL Institute of Neurology
London, United Kingdom

Marom Bikson
Department of Biomedical Engineering
The City College of New York
The City University of New York
New York, New York

Luisella Bocchio-Chiavetto
Neuropsychopharmacology Unit
IRCCS The Saint John of God Clinical
Research Centre
Brescia, Italy

Binith Cheeran
Department of Neurology
The John Radcliffe Hospital
Oxford, United Kingdom

Massimo Cincotta
Unit of Neurology
Florence Health Authority
Florence, Italy

Joseph Classen
Department of Neurology
University of Leipzig
Leipzig, Germany

Vincenzo Di Lazzaro
Division of Neurology
Università Campus Bio–Medico
Rome, Italy

Felipe Fregni
Laboratory of Neuromodulation
Department of Physical Medicine
and Rehabilitation
Spaulding Rehabilitation Hospital
and
Massachusetts General Hospital
and
Department of Continuing Education
Harvard Medical School
Boston, Massachusetts

Stefan Jun Groiss
Department of Neurology
and
Institute of Clinical Neuroscience and
Medical Physiology
Heinrich–Heine University Düsseldorf
Düsseldorf, Germany

Ying-Zu Huang
Department of Neurology
Chang Gung Memorial Hospital
and
Chang Gung University College
    of Medicine
Taoyuan, Taiwan, Republic of China

Friedhelm C. Hummel
Department of Neurology
University Medical Center
    Hamburg-Eppendorf
Hamburg, Germany

Risto J. Ilmoniemi
Department of Biomedical Engineering
    and Computational Science
Aalto University School of Science
Espoo, Finland

Anke Karabanov
Danish Research Centre for Magnetic
    Resonance
Copenhagen University
    Hospital-Hvidovre
Hvidovre, Denmark

Anli Liu
Berenson-Allen Center for Noninvasive
    Brain Stimulation
Division of Cognitive Neurology
Beth Israel Deaconess Medical Center
Harvard Medical School
Boston, Massachusetts

Hanna Mäki
Department of Biomedical Engineering
    and Computational Science
Aalto University School of Science
Espoo, Finland

Carlo Miniussi
Department of Clinical and
    Experimental Sciences
University of Brescia
and
IRCCS The Saint John of God Clinical
    Research Centre
Brescia, Italy

Pedro Cavaleiro Miranda
Faculty of Science
Institute of Biophysics and Biomedical
    Engineering
University of Lisbon
Lisbon, Portugal

Michael A. Nitsche
Department of Clinical
    Neurophysiology
Georg-August University Medical Center
Göttingen, Germany

Jacinta O'Shea
Nuffield Department of Clinical
    Neurosciences
and
Oxford Centre for Functional Magnetic
    Resonance Imaging of the Brain
University of Oxford
Oxford, United Kingdom

Alvaro Pascual-Leone
Berenson-Allen Center for Noninvasive
    Brain Stimulation
Beth Israel Deaconess Medical Center
Harvard-Catalyst Clinical Research
    Center
Harvard Medical School
Boston, Massachusetts

Walter Paulus
Department of Clinical
    Neurophysiology
Georg-August University Medical Center
Göttingen, Germany

Maria Concetta Pellicciari
Cognitive Neuroscience Section
IRCCS The Saint John of God Clinical
    Research Centre
Brescia, Italy

Angelo Quartarone
Department of Neuroscience
Psychiatry and Anaesthesiological
    Sciences
University of Messina
Messina, Italy

Asif Rahman
Department of Biomedical
    Engineering
The City College of New York
The City University of New York
New York, New York

Davide Reato
Department of Biomedical
    Engineering
The City College of New York
The City University of New York
New York, New York

Michael Charles Ridding
School of Paediatrics and Reproductive
    Health
and
Robinson Institute
University of Adelaide
Adelaide, South Australia, Australia

Simone Rossi
Dipartimento di Neuroscienze
Sezione Neurologia e Neurofisiologia
Clinica Azienda Ospedaliero-
    Universitaria Senese
Policlinico Le Scotte
Siena, Italy

Paolo M. Rossini
Institute of Neurology
Catholic University of Rome
Rome, Italy

Yiftach Roth
Department of Life Sciences
Ben-Gurion University
Beer-Sheva, Israel

John C. Rothwell
UCL Institute of Neurology
London, United Kingdom

Hartwig Roman Siebner
Danish Research Centre for Magnetic
    Resonance
Copenhagen University
    Hospital-Hvidovre
Hvidovre, Denmark

Martin Sommer
Department of Clinical
    Neurophysiology
Georg-August University Medical Center
Göttingen, Germany

James T.H. Teo
Sobell Department of Motor
    Neuroscience
UCL Institute of Neurology
London, United Kingdom

Gregor Thut
Institute of Neuroscience
and
Psychology Centre for Cognitive
    Neuroimaging
University of Glasgow
Glasgow, United Kingdom

Yoshikazu Ugawa
Department of Neurology
Fukushima Medical University
School of Medicine
Fukushima, Japan

Timothy A. Wagner
Berenson-Allen Center for Noninvasive
    Brain Stimulation
Beth Israel Deaconess Medical Center
Harvard Medical School
and
Highland Instruments
and
MIT Division of Health Sciences
    and Technology
Boston, Massachusetts

Vincent Walsh
Institute of Cognitive Neuroscience
and
Department of Psychology
University College London
London, United Kingdom

Abraham Zangen
Department of Life Sciences
Ben-Gurion University
Beer-Sheva, Israel

Ulf Ziemann
Department of Neurology
Goethe-University of Frankfurt
Frankfurt am Main, Germany

Part I

Basic Aspects

Part I

Basic Aspects

1 Basic Principles and Methodological Aspects of Transcranial Magnetic Stimulation

Yiftach Roth and Abraham Zangen*

CONTENTS

---

* Drs. Roth and Zangen are key inventors of the H-coils and serve as consultants for and have financial interest in Brainsway, Inc.

1.1   HISTORICAL BACKGROUND

The discovery of electricity in the nervous system dates back to the end of the eighteenth century. Luigi Galvani discovered the role of electricity in the contractions of a frog's muscle (Galvani, 1791). Moreover, Galvani concluded, following a series of experiments, that the convulsions were caused by an intrinsic source of electricity stored in the frog's muscles. Curiously, Alessandro Volta disagreed with Galvani and claimed that the contractions resulted from an outside source of electricity and, along his efforts to prove this thesis, invented the electric battery (1800). This debate was unequivocally resolved only in the 1840s, when Du Bois-Reymond has demonstrated that a peripheral nerve activity was accompanied by an electric potential change (Du Bois-Reymond, 1848, 1849).

Several scientists have promoted the understanding of animal electricity (termed "Galvanism") through a series of experiments, which included electrical stimulation of animals' brain regions (For a review, see Parent, 2004). Galvani's nephew, Giovanni Aldini, was the first to stimulate the human cerebral cortex of one hemisphere and to obtain facial muscle contractions on the contralateral side (Aldini, 1804). Aldini also made a first demonstration of electroconvulsive therapy by treating some patients with mental disorders by the administration of electric current to the skull with a partial clinical success.

Following these early achievements, there were practically no advances for a long period, until the late decades of the nineteenth century when various attempts were made to electrically stimulate several regions of the brain in animals. Fritsch and Hitzig (1870) and Ferrier (1873, 1876) stimulated the animal motor cortex and elicited evoked motor responses on contralateral limb muscles. In 1874, Bartholow (1874) stimulated electrically the exposed cerebral cortex in a subject with cranial fracture and obtained motor responses on contralateral upper and lower limb muscles. In 1882, Sciammanna (1882) applied electrical stimulation to the exposed dura mater in a patient with a skull breach in the parietal cortex. Sciammanna obtained motor responses of various facial and upper limb muscles and practically drew an initial partial map of motor representations. Yet it is highly probable that he actually stimulated the somatosensory cortex rather than the motor cortex. A few years later, Alberti (1884, 1886) did a series of electrical stimulation experiments in a patient with an exposed superior frontal and parietal lobes due to tumor erodation. Alberti elicited motor responses from upper and lower limbs as well as shoulder and facial muscles.

Several improvements in electrical stimulation methods and electrodes design were introduced in the first part of the twentieth century. Penfield and Jasper (Penfield and Boldrey, 1937; Penfield and Jasper, 1954) explored the human brain systematically with electrical stimulation during surgery and drew the well-known homunculus, in which the motor representations of various body parts are arranged in an orderly fashion.

Gualtierotti and Paterson (1954) tried to stimulate the unexposed cerebral cortex. They applied electrical stimulation through the intact scalp and obtained motor responses. Yet this procedure was extremely painful and inefficient, since most of the current flowed through the scalp rather than into the brain.

The first clinically applicable method of nonsurgical transcranial electrical stimulation was performed by Merton and Morton (1980). They used a single high-voltage capacitive discharge that allowed penetration of the current into the brain structure with less pain. With this device, they produced contralateral motor responses and phosphenes, by stimulating the motor and the visual cortex, respectively. Still with this type of stimulator, however, the pain accompanying stimulation was strong enough to keep it from wide clinical or research use.

In parallel with the development of direct electrical neurostimulation via implanted or skin-surface electrodes, the notion of noninvasive and indirect brain stimulation using a time-varying magnetic field has gradually evolved during the nineteenth century, following the discovery of the electromagnetic induction in 1831 by Faraday (1831, 1839). In 1855, Foucault discovered the eddy currents, which are induced in a conductive medium when exposed to a time-varying magnetic field. It was soon realized that this principle might be utilized to stimulate neuronal tissue.

The first known attempt to induce magnetic brain stimulation was by d'Arsonval (1896), who applied an alternating current to a coil surrounding the head, and induced phosphenes, vertigo, and syncope.

In the following years, several researchers (Beer, 1902; Thompson, 1910; Dunlap, 1911; Magnuson and Stevens, 1911, 1914) induced visual sensations by alternating currents at various frequencies in large coils located near the head.

In 1959, Kolin et al. (1959) gave the first demonstration of a magnetic stimulation of a nerve, when they stimulated a frog sciatic nerve and induced muscle contractions.

In 1965, Bickford and Freming induced magnetic stimulation of peripheral nerves in animals and human subjects (Bickford and Freming, 1965).

The birth date of transcranial magnetic stimulation (TMS) method was in 1985, when Barker and colleagues achieved noninvasive and painless stimulation of human motor cortex, using a stimulator consisting of a capacitor discharging into a stimulating coil placed on the scalp (Barker et al., 1985a,b). The TMS technique introduced a novel research tool for studying the functionality, morphology, and connectivity of various cortical regions, especially the motor cortex (Terao and Ugawa, 2002).

By the early 1990s, further development of magnetic stimulators expanded the range of stimulus frequency, allowing rapid-rate repetitive TMS (rTMS) at frequencies of up to 30 Hz (Dhuna et al., 1991; Pascual-Leone et al., 1991, 1994). Several studies demonstrated that application of rTMS to the motor cortex can produce minutes of increased (Pascual-Leone et al., 1994) or decreased (Wassermann et al., 1996a,b; Chen et al., 1997) corticospinal excitability.

Generally, two principal rTMS modalities have been applied in previous intervention studies: low-frequency rTMS (<3 Hz), which is proposed to reduce cortical excitability; and high-frequency rTMS ($\geq$5 Hz), which is proposed to increase cortical excitability (Padberg et al., 2007).

Since the 1990s, there is a rapidly growing interest and research addressing the potential of rTMS to modulate excitability of various brain regions and to treat various neurological and psychiatric disorders (Wassermann and Lisanby, 2001; Ridding and Rothwell, 2007; Rossini and Rossi, 2007).

1.2 PHYSICAL PRINCIPLES OF TMS

TMS is a technique for noninvasive stimulation of neuronal structures. Magnetic pulses are administered by passing a strong current through an electromagnetic coil placed upon the scalp that induces an electric field and therefore current in the underlying cortical tissue.

The TMS technique is based on the law of electromagnetic induction discovered by Michael Faraday back in 1831. Faraday's set of experiments constituted a key milestone of electromagnetism and also a remarkable example of a masterpiece of empirical science in general.

Faraday was surprised and disappointed to discover that a constant current in a conductor had no effect on a nearby conductor. Yet he noticed that upon switching on or off the primary current in the first conductor, there was an induced current in the secondary wire. He pursued this line of research and discovered that an *alternating*—but not a constant—electric current in one conductor induces current in the opposite direction in nearby conductors.

This law of induction was naturally termed Faraday's law and was generalized as one of the four fundamental equations of classical electromagnetism, the Maxwell equations

$$\vec{\nabla} \times \vec{E} = -\frac{d\vec{B}}{dt} \tag{1.1}$$

where
  $\vec{E}$ is the induced primary electric field
  $\vec{B}$ is the magnetic field
  t is the time

Basically, this law states that a changing magnetic field produces an electric field. Any electric current in a conductor induces a magnetic field around it. The magnetic field $\vec{B}$ is proportional to the coil current $\vec{I}$

$$\vec{B}(t) \propto \vec{I}(t) \tag{1.2}$$

Hence the induced electric field is proportional to the rate of change of the current

$$\vec{E}(t) \propto \frac{\partial \vec{I}}{\partial t} \tag{1.3}$$

The general effect of the current I at any point in space r is expressed by the vector potential $\vec{A}(r)$, defined by

$$\vec{A}(r) = \frac{\mu_0 I}{4\pi} \int \frac{d\vec{l'}}{|\vec{r} - \vec{r'}|} \tag{1.4}$$

where

$\mu_0 = 4\pi * 10^{-7}$ Tm/A is the magnetic permeability of free space

the integral of dl' is over the whole wire path of the coil

$\vec{r}$ is a vector indicating the position of the relevant point

$\vec{r}'$ is a vector indicating the position of the wire element

The magnetic field is related to the vector potential by

$$\vec{B} = \vec{\nabla} \times \vec{A} \tag{1.5}$$

Any electric field in a conductive medium generates electric currents. Hence, when an alternating current is passed in a coil, current (called eddy currents) will be induced in a nearby conductive medium.

If the electromagnetic coil is placed near a human head, an electric field is induced in the brain tissue.

The general expression for the electric field is

$$\vec{E} = -\frac{\partial \vec{A}}{\partial t} - \nabla \phi \tag{1.6}$$

The first term is the primary electric field induced due to the coil alternating current. Accumulation of charges in a medium with an inhomogeneous conductivity distribution (caused by, for example, the head surface at the macroscopic level and cellular membranes at the cellular level) produces a secondary electrostatic field, which is denoted by the second term on the right side of Equation 1.6. The effective electric field induced in the brain tissue is therefore the sum of the induced primary field and the secondary field.

At high enough intensity, the electric field can be sufficient to cause membrane depolarization in neuronal structures, initiation of action potential, and hence neuronal activation.

The basics of the neuronal response to the TMS pulse are described in the next sections.

1.3 NEURONAL ACTIVATION

1.3.1 MECHANISMS THAT ACTIVATE A NEURON WHEN STIMULATED

Electromagnetic fields can induce excitation of neurons without the need for mechanical contact. The basic mechanism of neural activation induced by implanted electrodes in the brain or by direct electrical or magnetic stimulation relies on forcing free charges in intra- and extracellular volumes to move coherently by an electric field. Depolarization or hyperpolarization is induced in cell membranes that interrupt current progress, and eventually neural action potential is triggered by depolarization of the axon membrane (Ruohonen, 2003).

It should be noted that TMS, and especially deep TMS, does not activate solely the target area, but also tissues around and above it and, indirectly, distant interconnected sites in the brain (Keck, 2007).

In general, several excitation mechanisms may be involved in the process of neuronal activation by TMS, where different mechanisms are dominant in different cases. Straight long axons are stimulated at the strongest point of the electric field gradient along the axon (Katz, 1939; Rattay, 1986; Reilly, 1989; Roth and Basser, 1990). This seems to be the dominant mechanism in long peripheral nerves (Maccabee et al., 1993). By contrast, in cortical excitation, it was found that it is the peak of the macroscopic applied electric field rather than its first spatial derivative that effectively controls the location of excitation (Thielscher and Kammer, 2002). This apparent contradiction is resolved when accounting for the finding that curved axons are preferably stimulated at the bends, where effective electric field gradient is maximal (Tranchina and Nicholson, 1986; Reilly, 1989; Amassian et al., 1992; Maccabee et al., 1993; Nagarajan et al., 1993). Short axons are most easily stimulated at their ends (Ruohonen, 2003). For a nerve with a series of bends, a complex pattern of hyperpolarization and depolarization is expected (Iles, 2005).

Indeed, most neural structures in the brain have complex geometry including bend points, terminations, and branches. Recent modeling studies imply that cortical excitation is predominantly induced by TMS at the bends of corticocortical or corticospinal fibers, at nerve endings, or at constrictions near the surface of the brain (Silva et al., 2008; Opitz et al., 2011; Salvador et al., 2011; Thielscher et al., 2011), although the complex shapes of neurons make predictions of precise excitation sites difficult. In all situations, membrane current must be outward for excitation to occur.

Volume-conductor inhomogeneities introduce another complexity to the equation. The most sharp discontinuity occurs at the brain–bone boundaries, where charge accumulation leads to a reduction in the amplitude of the electric field induced in the brain and changes its spatial distribution (Tofts, 1990; Maccabee et al., 1991; Roth et al., 1991a,b; Tofts and Branston, 1991; Eaton, 1992). Other important boundaries are the CSF–gray matter and the gray matter–white matter (GM–WM) interfaces. Thereby, a distinction must be made between electric field's primary and secondary components (Nagarajan and Durand, 1996), the latter arising from abrupt changes in the electric field at brain boundaries. Recent modeling studies indicate that the brain tissue heterogeneity and anisotropy can affect significantly the electric field distribution, the location of stimulation sites (Kobayashi et al., 1997; Liu and Ueno, 2000; Miranda et al., 2003, 2007; Silva et al., 2008; Opitz et al., 2011; Thielscher et al., 2011), and the threshold of neuronal stimulation (Salvador et al., 2011). Moreover, the jump in the electric field component normal to the boundary at a CSF–brain tissue or a GM–WM interface may introduce an independent mechanism for membrane depolarization and action potential (Miranda et al., 2003, 2007; Silva et al., 2008). Hence, we can speak about three distinct mechanisms by which the TMS pulse may lead to membrane depolarization in the brain: (1) the peak electric field induced at axons terminations, bend points, and branching points, (2) the jump in the electric field at a tissue interface, and (3) the gradient of the electric field along the fiber.

Membrane space constant, λ, is another important parameter that governs excitation site (Basser and Roth, 1991; Roth, 1994). The axon length with respect to λ and the coil (Nagarajan et al., 1993), as well as the coil position, orientation, and current polarity with respect to a bend or termination site dictate the primary source of influence and hence excitation site (Nagarajan et al., 1997; Silva et al., 2008; Opitz et al., 2011; Salvador et al., 2011; Thielscher et al., 2011). Thus in the cortex, stimulation may preferably occur where the induced electric field is perpendicular to CSF–GM or GM–WM interfaces (Opitz et al., 2011; Salvador et al., 2011; Thielscher et al., 2011).

TMS of the motor cortex has been extensively studied and characterized. When stimulation intensity of the motor cortex is high enough to produce activation of peripheral muscles, electromyography (EMG) measures can indicate both D-waves, representing direct stimulation of the corticospinal axon, and indirect I-waves that arise from trans-synaptic activation of various types of cortical neurons (Rothwell et al., 1991; Ziemann and Rothwell, 2000; Terao and Ugawa, 2002; Esser et al., 2005), which may include interneurons and collaterals of pyramidal tract neurons (Salvador et al., 2011). The resulting wave pattern is sensitive to coil orientation and to current polarity. Using a computer model of the coil and a heterogeneous, isotropic cortical sulcus, the complexity of the situation was demonstrated with observation of an array of potential excitation sites that depended on the interaction of coil orientation with the underlying tissue geometry and heterogeneity (Silva et al., 2008).

1.3.2 Cable Equation

In order to gain a basic understanding of the interaction between the TMS pulse and the neural tissue, we will use a simplistic model of an axon.

Let us consider a long straight axon stimulated by a figure-8 TMS coil (Figure 1.1). A changing current $\vec{I}(t)$ is passed through the coil. The axon is located beneath the central segment of the coil. A longitudinal current $\vec{i}_l(t)$ is induced in the axon in an opposite direction. At this stage, we will assume that the axon is straight and very long compared with the coil dimensions. This assumption is much more realistic for peripheral nerves than for cortical neuronal structures. Yet for the sake of clarity, we will first deal with this simplistic case and, in the next two sections, describe cases that are more relevant for cortical stimulation.

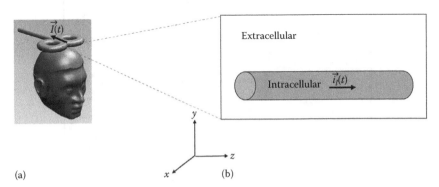

(a) (b)

FIGURE 1.1 A sketch of a figure-8 TMS coil (a) activating a long straight axon (b). The axon is beneath the central segment of the coil. A changing current $\vec{I}(t)$ is passed through the coil with $\partial \vec{I}/\partial t$ in the $-z$ direction, inducing an electric field in the underlying axon. This induces a longitudinal current $\vec{i}_l(t)$ in the opposite direction ($+z$) in the axon.

1.3.2.1 Transmembrane Potential

The neural parameter that is most relevant for the initiation of an action potential is the transmembrane potential. In neurons, there is an inherent difference in the electric potential between the intracellular and the extracellular media (Figure 1.2). This results from differences in ion concentrations between the two media and from the presence of macromolecules (such as proteins) in the intracellular space, which are partially charged. Thus, at a baseline state, there is an excess of positive sodium (Na$^+$) and negative chlorine (Cl$^-$) ions in the extracellular medium. On the other hand, there is an excess of potassium (K$^+$) ions in the intracellular medium. The overall effect of all these concentrations is that at baseline the intracellular potential is more negative than the extracellular one, and hence the transmembrane potential V_m is about −70 mV.

An action potential occurs when the transmembrane potential is depolarized below a threshold value. The main stages of an action potential are shown in Figure 1.3.

A detailed description of the action potential stages is beyond the scope of this chapter. We will only focus on the main condition for action potential initiation. When V_m is depolarized (i.e., becomes less negative) above a certain critical value (−60 mV in the example of Figure 1.3), an action potential is initiated.

In order to understand how V_m is affected, we will use the passive cable model for an axon (Hodgkin and Rushton, 1946; de Lorente, 1947; Roth and Basser, 1990). The model is shown in Figure 1.4.

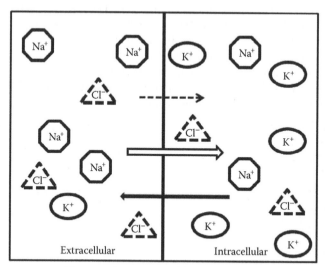

Transmembrane potential ΔV_m
= Intracellular − Extracellular potentia
= −70 mV

FIGURE 1.2 An illustration of the differences in ion concentrations between the intracellular cytoplasm of a neuron or a nerve fiber, and the extracellular space.

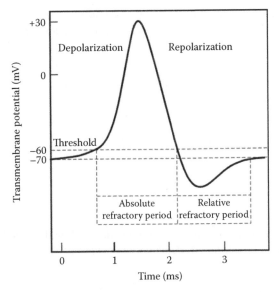

FIGURE 1.3 An illustration of the action potential. The transmembrane potential V_m is plotted as a function of time in milliseconds. When V_m is depolarized below a critical value (−60 mV in this example), a self-amplified process of ion channels is opening and current influx is initiated, and an action potential is produced. The action potential then propagates from the initiation point in both directions along the axon.

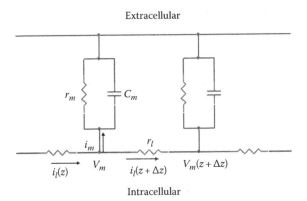

FIGURE 1.4 A passive cable model for an axon. The membrane segment is represented by a resistance r_m and a capacitance C_m. The axon is modeled by a longitudinal resistance per unit length r_1 in the intracellular medium. The membrane current per unit length is $\vec{i}_m(z)$ and the longitudinal intracellular current is $\vec{i}_l(z)$, where z is the spatial variable along the axon longitudinal axis. The transmembrane potential is $V_m(z)$.

The axon is modeled by a longitudinal resistance per unit length r_1 in the intracellular medium. It can be seen from Figure 1.4 that the longitudinal current $i_i(z)$ is given by

$$\vec{i_i}(z) = -\frac{\left(V_m(z+\Delta z) - V_m(z)\right)}{r_1} \tag{1.7}$$

At the limit when Δz approaches 0, we obtain

$$\vec{i_i}(z) = -\frac{1}{r_1}\frac{\partial V_m}{\partial z} \tag{1.8}$$

The membrane current per unit length $\vec{i}_m(z)$ is dictated by the membrane resistance r_m and capacitance C_m and the transmembrane potential $V_m(z)$

$$\vec{i}_m(z) = C_m\frac{\partial V_m}{\partial t} + \frac{V_m}{r_m}. \tag{1.9}$$

The two currents are related by Kirchhoff's law of current conservation

$$\vec{i}_m(z) = -\left(\vec{i_i}(z+\Delta z) - \vec{i_i}(z)\right) = -\frac{\partial \vec{i_i}}{\partial z} \tag{1.10}$$

where in the last equality, we made Δz approach 0.

Equations 1.8 through 1.10 lead to the cable equation for V_m

$$\lambda^2\frac{\partial^2 V_m}{\partial z^2} = \tau\frac{\partial V_m}{\partial t} + V_m \tag{1.11}$$

where

$$\lambda^2 = \frac{r_m}{r_1} \tag{1.12}$$

and

$$\tau = C_m r_m \tag{1.13}$$

The parameter τ is the time constant of the membrane, which indicates the rate of leakage of an excess potential on it. The zero-frequency length constant λ describes the decay of the potential along z when DC currents are applied. In realistic axons,

λ depends on axonal internal and external diameters and on axonal electric properties such as conductivities of the axoplasm and of the myelin sheath surrounding the axon (Fitzhugh, 1969; Basser and Roth, 1991; Silva et al., 2008).

1.3.2.2 Addition of the TMS Effect to the Cable Equation: Long Straight Axons

We now consider the effect of applying a TMS pulse over the axon. The magnetic pulse induces an electric field E_z along the axon. In this case, Equation 1.8 should be modified to

$$\vec{i}_l(z) = -\frac{1}{r_l}\left(\frac{\partial V_m}{\partial z} - \vec{E}_z\right) \tag{1.14}$$

The addition of the TMS-induced field leads to a modified cable equation

$$\lambda^2 \frac{\partial^2 V_m}{\partial z^2} = \tau \frac{\partial V_m}{\partial t} + V_m + \lambda^2 \frac{\partial \vec{E}_z}{\partial z} \tag{1.15}$$

The last term in Equation 1.15 is the source for the electromagnetic induction effect on V_m. It can be seen that it is the gradient of the induced electric field along the axon (z axis in our case) that is the crucial factor in the modulation of V_m in case of long straight axons.

The effect of the electromagnetic induction on the axon is illustrated in Figure 1.5.

At point **a** in Figure 1.5, there is a positive gradient of $\vec{E}_z$, and the current equation is

$$\vec{i}_m(z)\big|_a = \vec{i}_l(z + \Delta z) - \vec{i}_l(z) \tag{1.16}$$

resulting in a membrane current $\vec{i}_m$ going inward into the intracellular space. This current leads to membrane hyperpolarization.

By contrast, at point **b** in Figure 1.5, the membrane current $\vec{i}_m$ is going outward and membrane depolarization occurs. Hence, in this case, an action potential may be initiated at point **b**.

In case the coil is turned to the opposite direction (in the +z direction in Figure 1.5), the sites of hyperpolarization and depolarization will be exchanged. Hence, in that case, depolarization and possible action potential initiation may occur at point **a**.

Thus far we discussed only the spatial properties of the induced electric field. We looked at the situation at a certain point in time. In Section 1.3.2, we will deal with some temporal issues such as the pulse shape and will see that the depolarization–hyperpolarization scheme of each site may vary with time.

The main conclusion of this derivation is that the most important parameter for initiation of action potential in a case of a long straight nerve fiber is the gradient of the induced electric field along the nerve axis. In the next sections, we will discuss some different neuronal structures.

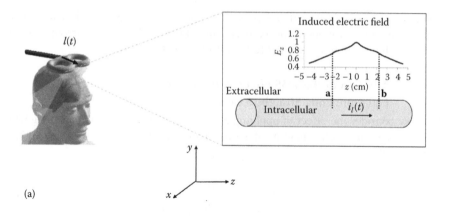

(a)

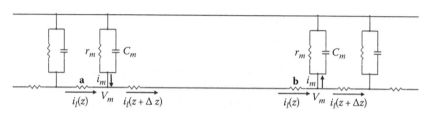

(b)

FIGURE 1.5 An illustration of the effect of a TMS pulse on a long straight axon. (a) The normalized electric field induced along the axon. At points **a** and **b**, the electric field gradient along the z axis is maximally positive and negative, respectively. (b) The passive cable model for this case. At point **a**, the membrane current $\vec{i}_m$ flows inward leading to membrane hyperpolarization, while at point **b**, $\vec{i}_m$ flows outward leading to membrane depolarization and initiation of an action potential.

1.3.3 LONG AND CURVED AXONS

The effect of the electromagnetic induction on a long curved axon is illustrated in Figure 1.6. Due to the axon curvature, the spatial derivative of the effective electric field along the axon is maximal at this point, and this is the key factor affecting the neural response.

The current equation at point **c** is now

$$\vec{i}_m(r)\big|_c = \vec{i}_{ax}(r + \Delta r) - \vec{i}_{ax}(r) \tag{1.17}$$

where
i_{ax} is the intracellular axial current along the axon axis
r is the coordinate along the axon axis

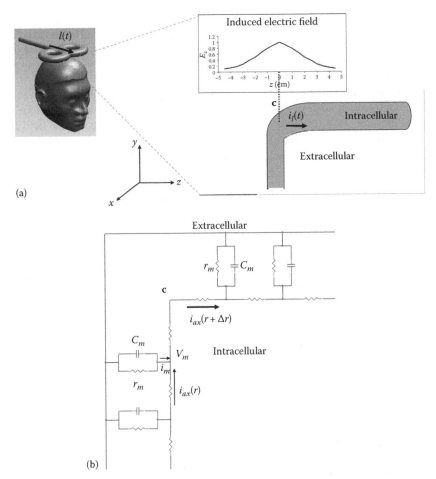

FIGURE 1.6 An illustration of the effect of a TMS pulse on a long curved axon. (a) The normalized electric field induced along the z axis. (b) The passive cable model for this case. At point **c**, the membrane current $\vec{i}_m$ flows inward leading to membrane hyperpolarization.

In this case, the intracellular axial current going upward is much smaller than the current going to the right, resulting in a membrane current $\vec{i}_m$ flowing inward into the intracellular space. This current leads to membrane hyperpolarization.

In case the coil current is in the opposite direction, the induced intracellular current $\vec{i}_{ax}$ will flow to the left, hence a membrane current $\vec{i}_m$ will flow outward and a membrane depolarization will occur.

We can conclude that, in the case of a curved fiber, the most important parameters for initiation of action potential are the intensity and the direction of the induced electric field itself, and not its gradient, at the bent point. At bent point, the electric field gradient *along the fiber tract* is maximal. Hence bent points are especially prone to neural stimulation by TMS.

This conclusion will be generalized in the next section, where we will describe the practical case of brain stimulation by TMS.

1.3.4 NERVE TERMINALS AND CONSTRICTIONS

As demonstrated in the previous section, a location of maximal effective electric field derivative will exist for an induced electric field along the axon at points where the axon terminates or bends away from the field. This appears to be a common situation in the cortex (Thielscher and Kammer, 2002; Silva et al., 2008; Opitz et al., 2011; Salvador et al., 2011; Thielscher et al., 2011). In the brain, the neurons are short compared with the spatial extent of the field and are usually bent and convoluted. Hence there are numerous points of bend, termination, or branching of nerve fibers. These points are the most likely sites for stimulation.

Hence when TMS of brain neuronal structures is discussed, as opposed to long peripheral nerves, the following assertions may be made:

1. The intensity of the induced electric field itself, and not its derivative, is the key factor for stimulation.
2. The electric field orientation relative to the neuronal structure is important. The lowest threshold for activation occurs where the induced field is parallel to the neuronal structure.
3. The direction along the nerve axis is crucial. As was demonstrated earlier, for a certain direction, a membrane depolarization will occur at the bend point, which above a critical value may lead to neural stimulation. On the other hand, the opposite direction will lead to membrane hyperpolarization and will reduce the chance for stimulation.

1.4 TMS ELECTRONICS

1.4.1 TMS CIRCUIT DESIGN

The goal of the TMS circuit is to create a brief current pulse in a stimulating coil. This current pulse can induce an electric field in an adjacent tissue, thus leading to neuronal activation. The TMS stimulation circuit consists of a high-voltage power supply that charges a capacitor or a bank of capacitors, which are then rapidly discharged via a fast electronic switch into the TMS coil, to create the briefly changing magnetic field pulse. A typical circuit is shown in Figure 1.7, where low-voltage AC is transformed into high-voltage DC, which charges the capacitor. A crucial component is the fast switch, which is usually a thyristor or an insulated-gate bipolar transistor and has to traverse very high current of very short duration of 50–250 μs.

The stimulator must include a control unit that operates the switch and enable the operator to program and determine the operation parameters.

The discharge circuit is basically an RCL circuit, characterized by R—the resistance, C—the capacitance, and L—the inductance.

The capacitance C is a characteristic of the capacitor, which reflects how much electrical energy it can store in the form of an electric charge. The inductance L is a

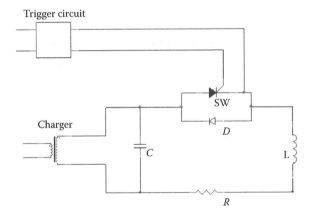

FIGURE 1.7 A schematic TMS circuit for biphasic pulses. An AC–DC transformer charges the capacitor C to a certain voltage V_C. A controlled switch SW enables the capacitor to discharge via the coil L. The switch gate is controlled by a trigger circuit. A diode D is connected in parallel with the switch in order to enable current flow in the opposite direction and capacitor recharging. The total resistance in the discharge circuit is R.

property that determines how much voltage is required to change the current in the circuit. The inductance is defined by

$$\frac{\partial I}{\partial t} = \frac{V_L}{L} \tag{1.18}$$

where
  V_L is the voltage
  I is the current
  $\partial I/\partial t$ is the rate of change of the current

In the TMS circuit, the main contributor to the inductance is the stimulating coil, although there may be additional inductance of the leads and other components.

The resistance R determines the amount of energy dissipated as heat and is mainly determined by the TMS coil.

During a pulse, the TMS circuit must sustain high peak currents and voltages for short cycle times. Typical ranges of circuit parameters and pulse widths are shown in Table 1.1.

The first TMS stimulators produced a monophasic pulse of electric current, where the current flows in a single direction. Yet the most widely used TMS stimulators nowadays produce a biphasic current, where the current flows in both directions and the pulse is terminated after a single sinusoidal cycle.

Representative monophasic and biphasic pulses are shown in Figure 1.8.

A scheme of a basic TMS circuit for biphasic pulses is shown in Figure 1.7.

With biphasic pulses, a significant portion of the energy returns to the capacitor at the end of the cycle, thus shortening the time for recharging and enabling to save energy. By contrast, for monophasic pulses, the energy is dissipated as heat, and at

TABLE 1.1

Typical Ranges of Parameters in a TMS Circuit

Property	Typical Range
Peak voltage	0.5–3 kV
Peak current	2–10 kA
Pulse width	60–1000 μs
Inductance L	10–30 μH
Capacitance C	10–250 μF
Resistance R	20–80 mΩ

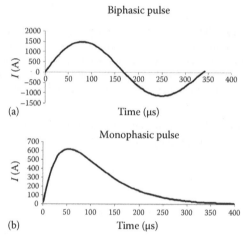

FIGURE 1.8   (a) A typical biphasic current pulse. The current is in amperes, and the duration is in microseconds. The current swings between two polarities; at the end of the second phase, the switch is closed and the pulse is terminated. (b) A monophasic pulse. The current flows in a single direction; after the peak, it is dissipated and does not return to the stimulator for recharging.

the end of the pulse, the capacitor voltage is zero. Hence stimulators intended for rTMS are designed to produce biphasic pulses.

In the following discussion, we will focus on a biphasic stimulator.

The equation in an RCL circuit is

$$L\frac{\partial^2 I}{\partial t^2} + R\frac{\partial I}{\partial t} + \frac{I}{C} = 0 \tag{1.19}$$

At time $t = 0$, before the switch is connected, the current I is 0, all the energy W is stored in the capacitor, and is given by

$$W = \frac{1}{2}CV_C^2 \tag{1.20}$$

where V_C is the capacitor voltage.

Equation 1.19 can be solved with the earlier initial conditions, and the resulting current I is

$$I(t) = \frac{V_C}{\omega L} \exp(-\alpha t) \sin(\omega t) \tag{1.21}$$

where $\alpha = R/2L$ and the circuit resonance frequency is

$$\omega = \sqrt{\left(\frac{1}{LC} - \frac{R^2}{4L^2}\right)}$$

The duration T of one pulse cycle is

$$T = \frac{2\pi}{\omega} \approx 2\pi\sqrt{LC} \tag{1.22}$$

where in the last equality, we employed the condition

$$R \ll \sqrt{\frac{L}{C}} \tag{1.23}$$

where R is measured in Ohm, L in Henri, and C in Farad. In a TMS circuit, the resistance has to be minimal and is usually on the order of 50 mΩ. Hence condition (1.23) is normally obeyed, and the pulse duration is governed by the L and C.

What are the most favorable values of L, C, and the pulse duration?

This question leads us to discuss the temporal characteristics of the neuronal response to the TMS pulse.

The neuronal response depends not only on the electric field magnitude but also on the pulse duration. As the pulse duration is extended, the electric field required to reach neuronal threshold E_{thr} becomes smaller. The dependence of E_{thr} on pulse duration is given by a strength–duration curve (Bourland et al., 1996) of the form

$$E_{thr} = \beta\left(1 + \frac{\gamma}{T}\right) \tag{1.24}$$

The biologic parameters determining neural response are the threshold at infinite duration, termed the rheobase (β, measured in Volt per meter), and the duration at which the threshold is twice the rheobase, termed the chronaxie (γ, in microseconds), which is related to the time constant of the neuronal membrane. The chronaxie and rheobase depend on many biologic and experimental factors, such as whether the nerves are myelinated or not (hence peripheral and cortical parameters are different) and other factors.

It can be shown that the strength–duration curve is equivalent to the requirement that the transmembrane potential V_m is depolarized to the threshold value.

Figure 1.9 shows an example of a strength–duration curve (Roth et al., 2007).

As can be seen from Equation 1.22, the duration of a TMS pulse can be extended in two ways: by increasing the capacitance C or by increasing the coil inductance L.

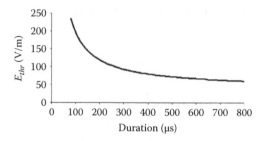

FIGURE 1.9 A strength–duration curve of the minimal electric field required to reach the threshold for neuronal activation, E_{thr}, in units of volt per meter, as a function of the pulse duration in microseconds. (From Roth, Y. et al., Transcranial magnetic stimulation of deep brain regions: Principles and methods, In *Transcranial Stimulation as Treatment in Mental Disorders*, Marcolin, M. and Padberg, F., Eds., Advances in Biological Psychiatry, Vol. 23, Karger Publishers, Zürich, Switzerland, pp. 204–224, 2007.)

The electric field induced by the TMS pulse is in general given by

$$E = N \frac{\partial I}{\partial t} F(r) \tag{1.25}$$

where

  N is the number of windings in the coil
  $F(r)$ depends on coil geometry, configuration, and orientation relative to the brain, which we will not discuss here

In Equation 1.25, the quasi-static approximation (Plonsey and Heppner, 1967) is used, where all the effects of electromagnetic wave propagation are neglected. The frequencies of TMS pulses are on the order of kilohertz. This means that the wavelength of the TMS pulse electromagnetic waves is on the order of kilometers, far above the size of the head. The quasi-static approximation allows the separation of the temporal and the spatial dependencies of the electric field, as is done in Equation 1.25. In this section, we discuss only the temporal characteristics of the electric field and the neuronal response to the TMS pulse.

The maximal induced electric field occurs at time $t = 0$ when the fast switch connects the capacitor to the coil. At this instant, the rate of current change is maximal and obeys

$$\frac{\partial I}{\partial t} = \frac{V_C}{L} \tag{1.26}$$

For a simple circular coil, the number of windings N is related to the inductance L by

$$N \propto \sqrt{L} \tag{1.27}$$

Usually there is a parasitic inductance L_{par} due to the leads and other components, and relation 1.27 should be replaced by

$$N \propto \sqrt{(L - L_{par})} \qquad (1.28)$$

Combining Equations 1.20 and 1.24 through 1.28, we can see how the total energy W^{thr} and the capacitor voltage V_C^{thr}, required at the threshold for neuronal activation, depend on the circuit parameters C and L.

For V_C^{thr}, we obtain the following dependence:

$$V_C^{thr} = a_1 \frac{L}{\sqrt{L - L_{par}}} + a_2 \sqrt{\frac{L}{(L - L_{par})C}} \qquad (1.29)$$

where a_1 and a_2 were introduced in order to emphasize the dependence on L and C.

For the energy W^{thr}, we obtain

$$W^{thr} = \frac{L}{L - L_{par}} \left(a_3 LC + a_4 \sqrt{LC} + a_5 \right) \qquad (1.30)$$

It can be seen that increasing the capacitance C leads to increased energy consumption, on one hand, but to lower required capacitor voltage, on the other hand. From Equation 1.26, it can be seen that lower voltage leads to lower required $\partial I/\partial t$. Both of these effects enable to use cheaper circuit elements and to simplify the circuit design. Hence, all these conflicting considerations must be accounted for when choosing the optimized circuit capacitor.

Regarding the inductance L, we can see that both W^{thr} and V_C^{thr} increase with increasing L. In addition, increased inductance is usually related to increased number of windings and to increased coil resistance. Both the resistance and the pulse duration are associated with the energy dissipation and heating rate in the stimulating coil (these issues will be discussed in Sections 1.4.6 and 1.4.7).

All these reasons point to the need to minimize the circuit inductance. On the other hand, very small L values lead to very high $\partial I/\partial t$ in the circuit, which may require expensive and bulky circuit elements such as the switch. Hence, optimization must be done. In practice, most TMS coil inductances are in the range of between 10 and 20 µH.

1.4.2 Pulse Waveforms and Sequences of Pulses

As discussed earlier, the most widely used pulse shapes in TMS are monophasic and biphasic pulses. Another possibility is a polyphasic pulse, where, unlike a biphasic pulse, the oscillation is not terminated after a single cycle, but the signal alternates for many cycles until it is almost zero. This waveform is less favorable since the energy is dissipated completely and does not return to the capacitor, and the second and later cycles have lower amplitude and hence are less effective than the first cycle. Hence polyphasic pulses are rarely used.

A recent development demonstrated a novel TMS stimulator design that enables one to control the pulse width, to terminate the current at a desired time point, and thus to reduce energy consumption and heating losses (Peterchev et al., 2008). Various other pulse shapes may be developed in future TMS stimulators.

The characteristics of typical biphasic and monophasic pulses are shown in Figure 1.10, for $C = 180 \ \mu F$, $L = 16 \ \mu H$, and $R = 50 \ m\Omega$.

Biphasic pulses are used for rTMS, since recharging time is much shorter and a train of pulses with short intervals is much more feasible.

As can be seen in Figure 1.10e, in each neuronal site, both depolarization and hyperpolarization occur during a biphasic pulse. In the example of Figure 1.10e, depolarization occurs at first, followed by subsequent hyperpolarization. Neuronal activation may occur only at depolarization. Since the change in V_m is larger during the second phase, the threshold for neuronal activation would be lower if the current polarity is reversed in this case. This demonstrates the current polarity dependence of TMS effect on motor activation, as seen in many studies.

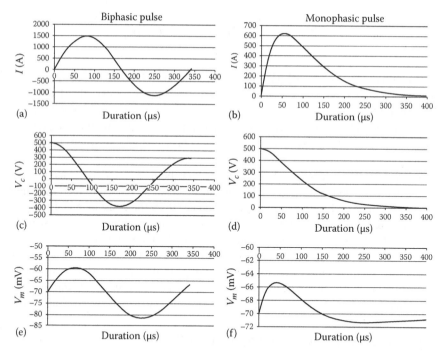

FIGURE 1.10   Typical pulse waveforms. (a) Biphasic coil current. (b) Monophasic coil current. (c) Capacitor voltage V_C for a biphasic pulse. The voltage changes polarity and is recharged to about 60% of its initial value at the end of the pulse. (d) The capacitor voltage V_C for a monophasic pulse. The voltage decays to close to zero at the end of the pulse. (e) The transmembrane potential V_m during a biphasic current pulse. There are phases of depolarization and hyperpolarization, with the second phase leading to larger swing in V_m. (f) The transmembrane potential V_m during a monophasic current pulse. Only the first phase leads to a significant modulation of V_m.

A broad variety of TMS pulse schemes are used in clinical and investigational protocols. We will classify the main modes of operation in the following text:

1. *Single pulses*: Single TMS pulses are used mainly for diagnostic, follow-up, and research purposes. In addition, single pulses to the motor cortex are used to determine the stimulation intensity required to reach the threshold for a motor response. This motor threshold (MT) may vary significantly in the population, hence the TMS treatment is usually calibrated based on the individual MT.

2. *Paired pulses*: In this method, two pulses are induced with short inter-pulse intervals (IPIs) on the order of 1–100 ms. The relative amplitudes of the two pulses may vary in different protocols. The modulation of the neuronal activity may be very sensitive to the IPIs, with different IPIs leading to different and opposite effects such as short-interval cortical inhibition and intra-cortical facilitation. Paired-pulse schemes are usually used for research, although clinical applications have recently been suggested and studied.

3. *Repetitive TMS (rTMS)*: Most clinical and investigational protocols apply rTMS treatment sessions. In such a sequence, trains of several pulses are delivered at a predefined frequency.

   An rTMS sequence is characterized by several operation parameters, which are listed in Table 1.2 with typical ranges. All or most of these parameters are user-controllable in most current TMS stimulators.

   A typical rTMS scheme is shown in Figure 1.11.

   Sequences with frequencies above 5 Hz are considered high-frequency protocols. Frequencies of 1 Hz and below are considered low. The distinction between these sequence types translates to a different modulation of the neurophysiological parameters, as will be discussed in the next chapters.

   A 1 Hz sequence typically incorporates trains of 5–15 min with a total of 300–900 pulses per train.

4. *Theta bursts*: This sequence is based on physiological theta-frequency pattern of neuronal firing at a 50 Hz frequency. In a typical sequence, a three-pulse burst at 50 Hz is repeated every 200 ms (5 Hz, which is at a theta frequency). In general, there are two basic patterns: intermittent theta burst stimulation (iTBS), composed of interleaved trains of bursts, and continuous TBS (cTBS). Interestingly, these two sequences often induce opposite

TABLE 1.2

Typical Ranges of rTMS

Sequence Parameters

Parameter	Typical Range
Frequency	1–25 Hz
Train duration	1–10 s
Inter-train interval (ITI)	10–40 s
Number of trains	10–60
Total number of pulses	400–3000

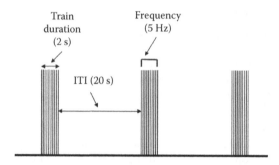

FIGURE 1.11 Typical rTMS scheme. In this example, the frequency is 5 Hz, the train duration is 2 s, and ITI is 20 s. The total number of trains in a session is typically 40–60, leading to 400–600 pulses delivered in a session time of 15–22 min.

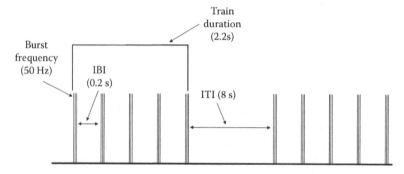

FIGURE 1.12 Typical iTBS scheme. Each burst includes three pulses at a 50-Hz frequency. The inter-burst interval (IBI) is 0.2 s, the train duration is 2.2 s (i.e., 10 bursts per train), and ITI is 8 s. Total number of trains in a session is typically 20, leading to total number of 600 pulses.

effects on the neuronal excitability; iTBS tends to increase the excitability, while cTBS decreases it. In iTBS, the inter-train interval and the number of trains determine the total session time and the number of pulses, which is typically 600. An example of an iTBS sequence is shown in Figure 1.12.

1.5 COIL DESIGN AND CONSTRUCTION

1.5.1 TYPES OF TMS COILS

The most commonly used coil in TMS studies is the figure-8 coil, sometimes referred as double-D or a butterfly coil. This shape allows relatively focal stimulation but the direct fields are limited to superficial layers of the cortex beneath the central portion of the coil. Neuronal fibers that are oriented parallel to the central segment of the coil are the most probable to be affected by the stimulation (Basser and Roth, 1991; Chen et al., 2003; Pell et al., 2011).

Coil elements that are not tangential to the scalp induce accumulation of charge on the surface and reduce coil effectiveness (Tofts, 1990; Branston and Tofts, 1991;

Eaton, 1992). Thus, coil designs that use less than 180° between the wings are more tangential to the scalp and hence more efficient (Thielscher and Kammer, 2004), but still, a one-plane design is the most popular because it is well suited for fine localization above most of scalp portions.

Many studies are still performed with circular coils of various sizes. Larger coils allow direct stimulation of deeper brain regions, but are less focal. The typical double cone coil is formed of two large adjacent circular wings at an angle of 95° so that the wiring of both wings is tangential to the head. This large coil induces a stronger and less focal electric field relative to a figure-8 coil (Lontis et al., 2006) and allows direct stimulation of deeper brain regions, but induce some discomfort especially when the higher intensities are required for reaching neural threshold of deeper brain areas.

Effectiveness and safety of the stimulation procedure are reduced by overheating of the coil during rTMS. Water, oil, and air cooling methods have been implanted to overcome this serious challenge. A figure-8 coil with a reduced resistance was designed recently (Weyh et al., 2005) achieving improved thermal characteristics. Ferromagnetic cores can serve as heat sinks, and coils with ferromagnetic cores were developed, producing significant reduction in heat generation and power consumption (Epstein and Davey, 2002). The safety of such iron-core coils, using a relatively high intensity (120% of MT) and frequency (10 Hz, 4 s trains), was recently demonstrated in a large multicenter study evaluating its antidepressant effects (O'Reardon et al., 2007).

Several coil designs for stimulation of deeper brain areas have been proposed and evaluated, termed H-coils (Roth et al., 2002; Zangen et al., 2005; Roth et al., 2007). Other theoretical designs for deep-brain TMS have been evaluated with computer simulations, such as stretched C-core coil (Davey and Riehl, 2006; Deng et al., 2008) and circular crown coil (Deng et al., 2008). All these coils are based on common design principles essential for effective deep-brain TMS (Roth et al., 2002; Zangen et al., 2005; Roth et al., 2007) and provide a significantly slower decay rate of the electric field with distance, though the stimulated area is wider. Thus, these coils are appropriate for relatively non-focal stimulation of deeper brain structures. However, it is important to remember that as in all TMS coils, the stimulation intensity of a single pulse is always maximal at the surface of the brain (Heller and van Hulsteyn, 1992) and that, because of the secondary electric field due to accumulation of charges at the boundaries, the effective electric field is always zero in the middle of a spherical conductor. The safety of H-coils and some of their cognitive effects, using a relatively high intensity (120% of MT) and frequency (20 Hz), have been carefully assessed (Levkovitz et al., 2007). In addition, several clinical studies showed promising effects of these coils in psychiatric disorders (Levkovitz et al., 2009; Harel et al., 2011).

1.5.2 TARGETING DEEP NEURONAL STRUCTURES

Until several years ago, the capacity of TMS to elicit neuronal responses has been limited to superficial structures. The coils used for TMS (such as round or figure-8 coils) induce stimulation in cortical regions mainly just superficially under the windings of the coil; the intensity of the electric field drops dramatically deeper in the brain as a function of the distance from the coil (Maccabee et al., 1990; Tofts, 1990; Tofts and Branston, 1991; Eaton, 1992). Therefore, to stimulate deep brain regions with such coils, a very high intensity would be needed, which is not feasible with standard magnetic stimulators. Moreover, the intensity needed to stimulate deeper brain regions effectively would stimulate cortical regions and facial nerves over the level that might lead to facial pain and facial and cervical muscle contractions and may cause epileptic seizures and other undesirable side effects (Roth et al., 2007).

The difficulty of efficiently activating deep neuronal structures using TMS emerges from physical properties of the brain and from physical and physiological aspects of the interaction of a TMS system with the human brain. As shown by Heller and Van Hulsteyn (1992), the three-dimensional maximum of the electric field intensity will always be located at the brain surface, for any configuration or superposition of TMS coils. However, both the TMS coils and the stimulator may be optimized for effective stimulation of deeper brain regions.

The construction of deep TMS coils should meet the following goals:

1. High-enough electric field intensity in the desired deep brain region that will surpass the threshold for neuronal activation
2. High percentage of electric field in the desired deep brain region relative to the maximal intensity in the cortex
3. Minimal adverse effects such as pain, motor activation, and activation of facial muscles

The design principles essential for effective stimulation of deeper brain regions include the following (Roth et al., 2002; Zangen et al., 2005; Roth et al., 2007):

1. Summation of electric impulses. The induced electric field in the desired deep brain regions is obtained by optimal summation of electric fields, induced by several coil elements with common direction, in different locations around the skull. The principle of summation may be applied in several ways (Roth et al., 2007).
2. Minimization of non-tangential components. Coil elements that are non-tangential to the surface induce accumulation of surface charge, which leads to the cancellation of the perpendicular component of the directly induced field at all points within the tissue, and usually to the reduction of the electric field in all other directions. In order to reduce accumulation of electrostatic charge, non-tangential elements in the coils are minimized, especially around the stimulation target. Therefore, these coils always include a flexible base complementary to the human head. The part of the coil close to the head (the base) must be optimally complementary to the human skull at the desired region.
3. Proper orientation of stimulating coil elements. Coils must be oriented such that they will produce a considerable field in a desired direction tangential to the surface, which should also be the preferable direction to activate the neuronal structures under consideration.

4. Remote location of return paths. The wires leading currents in a direction opposite to the preferred direction (the return paths) should be located far from the base and the desired brain region. This enables a higher absolute electric field in the desired brain region.

A comparison of the electric field profile along a line going from the coil surface into the center of a realistic phantom head model is shown in Figure 1.13 for five different TMS coils: a commercial figure-8 coil (70 mm diameter of each wing), commercial double cone coil (120 mm diameter of each wing, with an opening angle of 95°), large and small custom circular coils (with diameters of 160 and 55 mm, respectively), and a version of the H-coil that was used in a previous study (Zangen et al., 2005). The electric field distribution was measured in a realistic model of the human head ($x * y * z = 15 \times 13 \times 18$ cm, where x, y, and z are postero-anterior, right–left and inferior–superior axes, respectively), filled with physiological saline solution.

A sketch of the H-coil version is shown in Figure 1.13a.

The coil has 10 strips carrying a current in a common direction (postero-anterior direction) and located around the desired target site (segments A–B and G–H in Figure 1.13a). The average length of the strips is 110 mm. The only coil elements having radial current components are those connected to the return paths of five of the strips (segments C–I and J–F in Figure 1.13a). The length of these wires is 80 mm. The return paths of the other five strips are placed on the head at the contralateral hemisphere (segment D–E in Figure 1.13a). The wires connecting between the strips and the return paths (segments B–C and F–A) are on average 90 mm long.

The electric field amplitudes for all the coils were calculated along a line going downward (z axis in Figure 1.13a) with coordinates of $(x, y) = (0, 3)$, i.e., 3 cm laterally to the midline. For the figure-8 coil and the double cone coil, the line started at the coil center. For the two circular coils, the line started at the coil edge. For the H-coil, the line started at the center of elements A–B (Figure 1.13a).

From the plot in Figure 1.13b, it can be seen that the H-coil has the most favorable field profile (i.e., field attenuation to 66% at a distance of 4 cm). Among the rest of the coils, the large circular coil has the slowest rate of field decay with distance (i.e., field attenuation to 52% at a distance of 4 cm). Yet the circular coil induces a nonspecific effect over a complete cortical ring underneath the coil windings.

The other three coils present a much weaker depth penetration with a strong attenuation of the electric field with depth (i.e., field attenuation to 29%–37% at a distance of 4 cm). Note that the double cone coil is much larger than the figure-8 coil and produces a significantly higher absolute electric field amplitude at any distance. Yet the rate of decay of the field with distance from coil is similar between the two coils. This demonstrates that the coil size is not the only factor affecting the efficiency in activating deeper brain regions, and the principles detailed earlier must be accounted for.

1.5.3 ELECTRICAL SAFETY

During a TMS pulse, peak currents of the order of kiloamperes are delivered through the stimulating coil, and the capacitor voltage may be 2–3 kV. The coil windings and leads must have an electrical insulation rated far above the maximal voltage at 100%

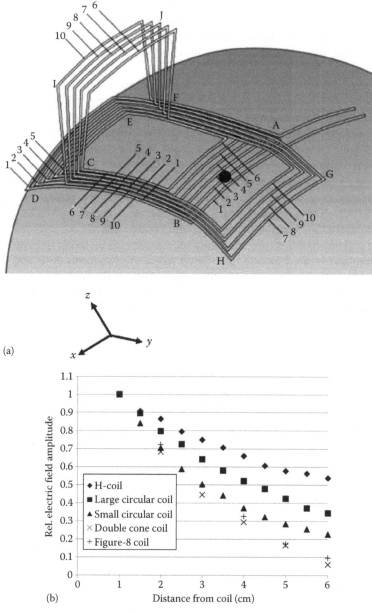

FIGURE 1.13 (a) A sketch of the H-coil version used in this comparison. (b) Plots of the electric field amplitude induced by several TMS coils, as a function of distance from the coil, normalized to the amplitude at 1 cm distance. The electric field was calculated in a phantom head model filled with a saline solution with physiological concentration. The coils are a H-coil version (diamonds), a large circular coil with 160 mm average diameter (squares), a small circular coil with 55 mm average diameter (triangles), a commercial figure-8 coil having 70 mm diameter of each wing (+'s), and a commercial double cone coil with 120 mm diameter of each wing (X's).

of the stimulator power output. The coil winding insulation must prevent any risk of electric shock to the patient or the operator, as well as any short circuit between the coil windings or any electrical arcing between the coil and any of the surrounding facilities such as temperature sensor leads or the positioning and attachment device.

The entire circuit of the stimulator and the coil must have a "floating" voltage and insulated from earth ground.

All TMS devices must comply with international standards such as IEC 60601. Safety regulations, guidelines, and recommendations for TMS studies are summarized in consensus papers (Wassermann et al., 1998; Rossi et al., 2009).

1.5.4 ENERGY CONSUMPTION

In a typical TMS stimulator, a charger circuit transforms the network AC voltage to high DC voltage on the capacitor. Hence the maximal energy consumption occurs during charging. Typical charging times are from tens to hundreds of milliseconds. The capacitor energy W is given by Equation 1.20.

The momentary power consumption P_{ch} during charging is given by

$$P_{ch} = \frac{W}{t_{ch}} \tag{1.31}$$

where t_{ch} is the charging time.

The RMS power during a train depends on the operation frequency f

$$P_{RMS} = fW \tag{1.32}$$

Consider a circuit with a capacitor of $C = 160\ \mu F$, which is charged to a voltage V_C of 1.5 kV. From Equation 1.20, $W = 180$ J.

Suppose the charging time in these conditions from 0 to 1.5 kV is 60 ms. Substitution in Equation 1.35 yields $P_{ch} = 3$ kW.

For operation frequency of 10 Hz, the average power consumption would be only $P_{RMS} = 1.8$ kW. From this example, it can be seen that the energy consumption of TMS stimulators is often not uniform with peaks of high power and current consumption. This may have an impact on the network and nearby electric devices such as flickering. These considerations must be accounted for in the TMS stimulator design.

The actual power consumption during a train of pulses may be significantly reduced in a biphasic pulse device thanks to capacitor recharging at the end of each pulse. Suppose that V_C at the pulse end is 60% of its initial value (i.e., 900 V in our example). Then the energy required for recharging is 115 instead of 180 J. Suppose the maximal P_{RMS} of the device is 3.2 kW. Looking at Equation 1.32, we deduce that the maximal attainable operation frequency during a train of 1.5 kV is 27 instead of 17 Hz.

The TMS coil configuration has a tremendous effect on the efficacy of neuronal activation. Thus, to obtain a certain neurological effect, an optimized coil would enable to achieve the goal with significantly lower energy consumption.

1.5.5 COIL HEATING

During rTMS operation, a large amount of heat may be produced in the stimulating coil. Hence a cooling system is often required in order to maintain an acceptable coil temperature during operation. The most widely used cooling systems are based on streaming cooled air. Water is an effective coolant because of its high specific heat. Yet water-based systems must cope with the need to prevent any accidental contact between the water and the high-voltage circuitry of the TMS coil and stimulator.

The heat dissipated in the coil during a pulse is given by

$$Q = R \int I(t)^2 dt \tag{1.33}$$

where
  R is the resistance in the coil
  $I(t)$ is the coil current
  the integration is over the pulse duration

Hence, a reduction in coil heating and in energy consumption may be achieved by one or more of the following methods:

1. Reducing coil (and lead) resistance. The resistance is inversely proportional to the wire cross section. On the other hand, too large cross section may be less favorable for achieving the accurate desired coil configuration and may make the manufacturing process more difficult. The coil resistance also increases with the number of windings N. Yet N also affects the coil inductance and the induced electric field. Hence each of these factors has to be optimized, accounting for these conflicting considerations.
2. Shortening the pulse duration, by reducing either the coil inductance L or the capacitance C (see Equation 1.22) or by reducing the peak current. This way has often a limited applicability, since the magnitude of the change in the neuron transmembrane potential V_m is proportional to the coil current.

1.5.6 MECHANICAL STRENGTH

The high currents flowing in the TMS coil induce significant mechanical forces between the coil elements. Each current element is affected by a Lorentz force F, which is proportional to the product of the current and the magnetic field B from all the other coil elements:

$$d\vec{F} = I d\vec{l} \times \vec{B} \tag{1.34}$$

where
  I is the current
  dl is the element length
  $\times$ indicates vector product

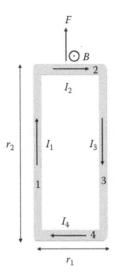

FIGURE 1.14 A simple loop of coil elements. The current directions are indicated by arrows. The total magnetic field B induced on element 2 by all the other elements is directed into the sheet, and the induced force F on element 2 tends to push it upward. In general, the forces tend to tear the coil apart. On the other hand, adjacent elements with currents in the same direction attract each other.

In order to understand the basic mechanism, let us consider the simple rectangular loop of Figure 1.14.

For the example shown in Figure 1.14, the magnetic field B can be expressed as

$$B = Ig(r_1, r_2) \tag{1.35}$$

where $g(r_1, r_2)$ is a geometric function of the loop dimensions.

Combining the last two equations, we obtain for the force per unit length

$$\frac{dF}{dl} = I^2 g(r_1, r_2) \tag{1.36}$$

Hence the mechanical force is proportional to I^2. Regarding the dependence on the physical dimensions of the problem (i.e., r_1 and r_2), we can state the following general assertions:

1. The forces would in general be stronger for coils with smaller dimensions.
2. Shorter elements would be exposed to stronger forces. Thus in Figure 1.14 where $r_1 < r_2$, elements 2 and 4 would "feel" higher mechanical forces than elements 1 and 3.

The design of any TMS coil should account for the mechanical forces and strains and include casings and mechanical elements that should guarantee the coil's mechanical stability under the most extreme conditions.

For TMS coils operating inside an MR scanner, there are additional Lorentz forces due to the interactions between the currents in the coil elements and the MR scanner's high steady magnetic field. Hence, such coils have to be designed to sustain even stronger mechanical forces.

1.5.7 ACOUSTIC ARTIFACT

The mechanical forces produced during operation lead to rapid vibrations of the coil elements. This in turn produces a broadband acoustic artifact, which may exceed 140 dB of sound pressure level (Counter and Borg, 1992). The strength and quality of the mechanical packing and casing of the coil elements may reduce significantly the audible artifact.

In any case, the use of hearing protection is recommended for all individuals receiving TMS stimulation or those being in the vicinity.

1.5.8 SHAM TMS COILS

In placebo-controlled clinical studies, it is required that the patients—and often also the operators—will be blinded with regard to whether they receive an active or a placebo TMS session. Hence various methods and devices of sham TMS stimulation have been developed.

Ideally, a sham device should

1. Have the same visual appearance as the active coil.
2. Produce a similar acoustic artifact.
3. Have the same sense of physical contact with the scalp.
4. Mimic the scalp sensations such as tingling.
5. Produce similar side effects whenever they exist as the active stimulation, including facial muscles activation. At the same time, the cerebral activation elicited by the sham stimulation should be minimal.

In practice, it is difficult to achieve simultaneously all the aforementioned goals.

The "classical" sham method used with figure-8 TMS coils was tilting the coil about 45° relative to the most favorable orientation for eliciting motor activation, when the coil is placed over the motor strip (Lisanby et al., 2001). This way, the acoustic artifact is approximately the same as in the active mode. Yet the contact sensation is not identical, the operator is clearly not blinded, and there is a significant electric field induced in the brain. Hence, even though the motor response is reduced, there may be neuronal effects of the sham session, which may limit its utility as a real control treatment.

In a recent study, a figure-8 coil with two separate wings was used, and an electronic switch controlled the relative current directions of the two wings. Hence, augmentation or cancellation of the electric field under the central segment of the coil was produced, depending on whether the two wings had currents in parallel or opposite directions (Hoeft et al., 2008). This way the operator may also be blinded. Yet under the sides of the wings, the induced electric field is close to 50% of the field produced in the active mode under the central segment.

A similar concept was used in a study with deep TMS H-coils (Levkovitz et al., 2007). In this system, the H2-coil was electrically divided into two identical coils, winded from two parallel wires. When the system was in the active mode, the two coils were connected in parallel. When the system was in the sham mode, the two coils were connected in parallel with opposite current directions. A shortcoming of this method was that the induced field was extremely low (about 3% of the active mode) and the scalp sensations and auditory artifact were weaker.

Another sham method (Amiaz et al., 2009) used several sheets of mu metal between the TMS coil and the scalp. These sheets are made of high permeability material and produce high attenuation of the induced field in the brain tissue. In this method, the acoustic artifact and the contact sensation are similar, yet the operator is not blinded.

A recent study (Rossi et al., 2007) demonstrated a placebo device added to a standard figure-8 TMS coil, based on local electrical stimulation with surface electrodes and attenuation of the field produced by the TMS coil. The electric field produced with this sham device was found to be 20%–25% of the field produced by a real TMS coil. This method enabled to mimic the auditory artifact and the scalp sensations evoked by a figure-8 TMS coil.

In a recently developed sham method, a sham coil is placed inside the same helmet encasing the active TMS coil (Enticott et al., 2011). An electronic system controls which of the two coils is connected to the stimulator in a certain session. The sham coil can mimic even the facial muscle activation induced by the active coil. The operation may be done by a magnetic card specific to each patient. This way both the patient and the operator may be blinded to the operation mode. This method is currently being used in several ongoing clinical trials with deep TMS H-coils.

1.6   TRADITIONAL NAVIGATION METHODS

Most TMS protocols establish the coil position and intensity individually for each subject. The intensity is usually based on the individual MT of the subject, which is determined by placing the TMS coil over the motor cortex and applying single probe pulses to search for contralateral hand or leg muscle movement either by visual inspection or by means of EMG. The MT is defined as the minimal stimulator output required to evoke a visible muscular response or a minimal motor-evoked potential of 50 μV as measured by the EMG device. In this procedure, the motor cortex region responsible for the specific hand muscle (such as the abductor policis brevis) is located. In some cases, the motor cortex of leg muscles (such as the tibialis) or the visual cortex (by measuring the threshold to evoke a phosphene) is used for localization.

After establishment of MT, coil position and orientation have to be reset to the treatment location. The classical method is to use Talairach coordinated in order to pre-evaluate the average distance of the target area from MT location and to simply measure the distance using a measuring tape. For instance, major depression is frequently treated with rTMS applied to either the left or right dorsolateral prefrontal cortex (DLPFC). The accepted navigation protocol (George et al., 1995; Pascual-Leone et al., 1996) is to reposition the coil 5 cm rostrally to the motor "hot spot" found earlier, thus presumably targeting the DLPFC.

Needless to say, because of individual differences in brain topography, application of the described navigation method is quite easy but rather inaccurate. For some individuals, coil should be repositioned 6 and not 5 cm away from the motor area to target the DLPFC and not the premotor cortex (Herwig et al., 2001; Johnson et al., 2006). A refined method is to use the nasion (the frontal and nasal bone intersection over the bridge of the nose), the inion (the tip of the occipital bone at the lower rear part of the skull), and the mastoid (a prominent bone just behind the ear) as topographic anchors to establish the desired position. But still, location errors of order of centimeters are a possibility because skull and brain topographies do not relate perfectly one with another. In some studies using deep TMS H-coils (Levkovitz et al., 2009; Enticott et al., 2011; Harel et al., 2011), the nasion and the inion were used to define a coordinate net of the subject head, as an aid for navigation. A special flexible cap is stretched on the subject head, and two rulers are attached to it. A medial ruler is attached along the head midline, with the zero mark at the nasion. A second lateral ruler is attached perpendicular to the first ruler, along a coronal slice. The H-coils are usually larger than standard TMS coils and are encased in a helmet. The lower circumference of the helmet is covered by coordinates with a resolution of millimeters. This way, the intersection points between the helmet coordinates and the two cap rulers give a unique definition of the coil location and orientation. By this method, the optimal position and orientation to evoke a motor response are recorded, and the coil is then navigated to the desired treatment location.

REFERENCES

Alberti, A. 1884. Contribucion al estudio de las Localizaciones cerebrales y a la Patogénesis de la epilepsia. July 31, 1884. Public Concourse, Circulo Médico Argentino, Buenos Aires, Argentina.

Alberti, A. 1886. Contribucion al estudio de las Localizaciones cerebrales y a la Patogénesis de la epilepsia. Biedma, Buenos Aires, Argentina.

Aldini, J. 1804. *Essai théorique et expérimental sur le galvanisme, avec une série d'expériences faites devant des commissaires de l'Institut national de France, et en divers amphithéâtres anatomiques de Londres.* Fournier Fils, Paris, France.

Amassian, V.E., Eberle, L., Maccabee, P.J., and Cracco, R.Q. 1992. Modelling magnetic coil excitation of human cerebral cortex with a peripheral nerve immersed in a brain shaped volume conductor: The significance of fiber bending in excitation. *Electroencephalogr Clin Neurophysiol* 85: 291–301.

Amiaz, R., Levy, D., Vainiger, D. et al. 2009. Repeated high-frequency transcranial magnetic stimulation over the dorsolateral prefrontal cortex reduces cigarette craving and consumption. *Addiction* 104: 653–660.

d'Arsonval, A. 1896. Dispositifs pour la mesure des courants alternatifs detoutes frequences. *CR Soc Biol (Paris)* 3: 450–451.

Barker, A.T., Freeston, I.L., Jalinous, R., and Jarratt, J.A. 1985a. Motor responses to non-invasive brain stimulation in clinical practice. *Electroencephalogr Clin Neurophysiol* 61: S70.

Barker, A.T., Jalinous, R., and Freeston, I.L. 1985b. Non-invasive magnetic stimulation of human motor cortex. *Lancet* 1: 1106–1107.

Bartholow, R. 1874. Experimental investigations into the functions of the human brain. *Am J Med Sci* 134: 305–313.

Basser, P.J. and Roth, B.J. 1991. Stimulation of a myelinated nerve axon by electromagnetic induction. *Med Biol Eng Comput* 29: 261–268.

Beer, B. 1902. Ueber das Auftraten einer objective Lichtempfindung in magnetischen Felde. *Klin Wochenschr* 15: 108–109.

Bickford, R.G. and Freming, B.D. 1965. Neural stimulation by pulsed magnetic fields in animals and man. In *Digest of the 6th International Conference on Medical Electronics and Biological Engineering*, Japan, Tokyo, Paper 7-6.

Bourland, J.D., Nyenhuis, J.A., Noe, W.A. et al. 1996. Motor and sensory strength-duration curves for MRI gradient fields. In *Proceedings of International Society Magnetic Resonance Medicine 4th Scientific Meeting and Exhibit*, New York, p. 1724.

Branston, N.M. and Tofts, P.S. 1991. Analysis of the distribution of currents induced by a changing magnetic field in a volume conductor. *Phys Med Biol* 36: 161–168.

Chen, R., Yung, D., and Li, J.Y. 2003. Organization of ipsilateral excitatory and inhibitory pathways in the human motor cortex. *J Neurophysiol* 89: 1256–1264.

Chen, R. et al. 1997. Depression of motor cortex excitability by low-frequency transcranial magnetic stimulation. *Neurology* 48: 1398–1403.

Counter, S.A. and Borg, E. 1992. Analysis of the coil generated impulse noise in extracranial magnetic stimulation. *Electroencephal Clin Neurophysiol* 85: 280–288.

Davey, K.R. and Riehl, M.E. 2006. Suppressing the surface field during transcranial magnetic stimulation. *IEEE Trans Biomed Eng* 53: 190–194.

Deng, Z.D., Peterchev, A., and Lisanby, S.H. 2008. Coil design considerations for deep-brain transcranial magnetic stimulation (dTMS). *Proc IEEE Eng Med Biol Soc* 2008: 5675–5679.

Dhuna, A., Gates, J., and Pascual-Leone, A. 1991. Transcranial magnetic stimulation in patients with epilepsy. *Neurology* 41: 1067–1071.

Du Bois-Reymond, E. Vol. I, 1848, Vol. II, 1849. *Untersuchungen fber Thiersche Ekktricitat*. Reimer, Berlin, Germany.

Dunlap, K. 1911. Visual sensations from the alternating magnetic field. *Science* 33: 68–71.

Eaton, H. 1992. Electric field induced in a spherical volume conductor from arbitrary coils: Application to magnetic stimulation and MEG. *Med Biol Eng Comput* 30: 433–440.

Enticott, P.G., Kennedy, H.A., Zangen, A. et al. 2011. Deep repetitive transcranial magnetic stimulation associated with improved social functioning in a young woman with an autism spectrum disorder. *J ECT* 27: 41–43.

Epstein, C.M. and Davey, K.R. 2002. Iron-core coils for transcranial magnetic stimulation. *J Clin Neurophysiol* 19: 376–381.

Esser, S.K., Hill, S.L., and Tononi, G. 2005. Modeling the effects of transcranial magnetic stimulation on cortical circuits. *J Neurophysiol* 94: 622–639.

Faraday, M. 1831. Effects on the production of electricity from magnetism. In *Michael Faraday*, Williams, L.P. (Ed.). Basic Books (Chapman Hall), New York, 1965, p. 531.

Faraday, M. 1839, 1844. *Experimental Researches in Electricity*, Vols. i. and ii. Richard and John Edward Taylor, London, U.K.

Ferrier, D. 1873. Experimental researches in cerebral physiology and pathology. *West Riding Lunatic Asylum Med Rep* 3: 30–96

Ferrier, D. 1873–1874. The localization of function in the brain. *Proc Roy Soc* 22: 229.

Ferrier, D. 1876. *The Function of the Brain*. Smith & Elder, London, U.K.

Fitzhugh, R. 1969. Mathematical models of excitation and propagation in nerve. In *Biological Engineering*, Schwan, H. (Ed.). McGraw-Hill, New York, pp. 1–83.

Fritsch, G. and Hitzig, E. 1870. Über die elecktrische Erregbarkeit des Grosshirns. *Arch Anat Physiol (Wiss Med)* 37: 300–333.

Galvani, L. 1791. De viribus electricitatis in motu musculari commentarius. *De Bononiensi Scientiarum et Artium Instituto atque Academia Commentarii* 7: 363–418.

George, M.S., Wassermann, E.M., Williams, W.A. et al. 1995. Daily repetitive transcranial magnetic stimulation (rTMS) improves mood in depression. *Neuroreport* 6: 1853–1856.

Gualtierotti, T.P. and Paterson, A.S. 1954. Electric stimulation of the unexpected cerebral cortex. *J Physiol* 125: 278–291.

Harel, E.V., Zangen, A., Roth, Y., Reti, I., Braw, Y., and Levkovitz, Y. 2011. H-coil repetitive transcranial magnetic stimulation for the treatment of bipolar depression: An add-on, safety and feasibility study. *World J Biol Psychiatry* 12: 119–126.

Heller, L. and van Hulsteyn, D.B. 1992. Brain stimulation using electromagnetic sources: Theoretical aspects. *Biophys J* 63: 129–138.

Herwig, U., Padberg, F., Unger, J., Spitzer, M., and Schönfeldt-Lecuona, C. 2001. Transcranial magnetic stimulation in therapy studies: Examination of the reliability of "standard" coil positioning by neuronavigation. *Biol Psychiatry* 50: 58–61.

Hodgkin, A.L. and Rushton, W.A.H. 1946. The electrical constants of a crustacean nerve fiber. *Proc Roy Soc Ser E* 133: 444–479.

Hoeft, F., Wu, D.A., Hernandez, A. et al. 2008. Electronically switchable sham transcranial magnetic stimulation (TMS) system. *PLoSONE* 3: e1923.

Iles, J.F. 2005. Simple models of stimulation of neurones in the brain by electric fields. *Prog Biophys Mol Biol* 87: 17–31.

Johnson, K.A., Ramsey, D., Kozel, F.A. et al. 2006. Using imaging to target the prefrontal cortex for transcranial magnetic stimulation studies in treatment-resistant depression. *Dialogues Clin Neurosci* 8: 266–268.

Katz, B. 1939. Nerve excitation by high-frequency alternating current. *J Physiol (Lond)* 96: 202–224.

Keck, M.E. 2007. Repetitive transcranial magnetic stimulation effects in vitro and in animal models. In *Transcranial Brain Stimulation for Treatment of Psychiatric Disorders*, Marcolin, M.A. and Padberg, F. (Eds.). Advances in Biological Psychiatry, Vol. 23, Karger, Basel, Switzerland, pp. 18–34.

Kobayashi, M., Ueno, S., and Kurokawa, T. 1997. Importance of soft tissue inhomogeneity in magnetic peripheral nerve stimulation. *Electroencephalogr Clin Neurophysiol* 105: 406–413.

Kolin, A., Brill, N.Q., and Broberg, P.J. 1959. Stimulation of irritable tissues by means of an alternating magnetic field. *Proc Soc Exp Biol Med* 102: 251–253.

Levkovitz, Y., Harel, E.V., Roth, Y. et al. 2009. Deep transcranial magnetic stimulation of the prefrontal cortex—Effectiveness in major depression. *Brain Stimul* 2: 188–200.

Levkovitz, Y. et al. 2007. A randomized controlled feasibility and safety study of deep transcranial magnetic stimulation. *Clin Neurophysiol* 118: 2730–2744.

Lisanby, S.H., Luber, D., Schroeder, B. et al. 2001. Sham TMS: Intracerebral measurement of the induced electrical field and the induction of motor-evoked potentials. *Biol Psychiatry* 49: 460–463.

Liu, R. and Ueno, S. 2000. Calculating the activation function of nerve excitation in inhomogeneous volume conductor during magnetic stimulation using the finite element method. *IEEE Trans Magn* 36: 1796–1799.

Lontis, E.R., Voigt, M., and Struijk, J.J. 2006. Focality assessment in transcranial magnetic stimulation with double and cone coils. *J Clin Neurophysiol* 23: 462–471.

Lorente de No, R. 1947. A study of nerve physiology. *Stud. Rockefeller Inst Med Res* 131–132: 1–548.

Maccabee, P.J., Amassian, V.E., Eberle, L.P., and Cracco, R.Q. 1993. Magnetic coil stimulation of straight and bent amphibian and mammalian peripheral nerve in vitro: Locus of excitation. *J Physiol* 460: 201–219.

Maccabee, P.J., Amassian, V.E., Eberle, L.P. et al. 1991. Measurement of the electric field induced into inhomogeneous volume conductors by magnetic coils: Application to human spinal neurogeometry. *Electroencephalogr Clin Neurophysiol* 81: 224–237.

Maccabee, P.J. et al. 1990. Spatial distribution of the electric field induced in volume by round and figure '8' magnetic coils: Relevance to activation of sensory nerve fibers. *Electroencephalogr Clin Neurophysiol* 76: 131–141.

Magnuson, C.E. and Stevens, H.C. 1911. Visual sensations caused by the changes in the strength of a magnetic field. *Am J Physiol* 29: 124–136.

Magnuson, C.E. and Stevens, H.C. 1914. Visual sensations created by a magnetic field. *Philos Mag* 28: 188–207.

Merton, P.A. and Morton, H.B. 1980. Stimulation of the cerebral cortex in the intact human subject. *Nature* 285: 227.

Miranda, P.C., Correia, L., Salvador, R., and Basser, P.J. 2007. Tissue heterogeneity as a mechanism for localized neural stimulation by applied electric fields. *Phys Med Biol* 52: 5603–5617.

Miranda, P.C., Hallett, M., and Basser, P.J. 2003. The electric field induced in the brain by magnetic stimulation: A 3-D finite-element analysis of the effect of tissue heterogeneity and anisotropy. *IEEE Trans Biomed Eng* 50: 1074–1085.

Nagarajan, S.S. and Durand, D.M. 1996. A generalized cable equation for magnetic stimulation of axons. *IEEE Trans Biomed Eng* 43: 304–312.

Nagarajan, S.S., Durand, D.M., and Hsuing-Hsu, K. 1997. Mapping location of excitation during magnetic stimulation: Effects of coil position. *Ann Biomed Eng* 25: 112–125.

Nagarajan, S.S., Durand, D.M., and Warman, E.N. 1993. Effects of induced electric fields on finite neuronal structures: A simulation study. *IEEE Trans Biomed Eng* 40: 1175–1188.

Opitz, A., Windhoff, M., Heidemann, R.M., Turner, R., and Thielscher, A. 2011. How the brain tissue shapes the electric field induced by transcranial magnetic stimulation. *NeuroImage* 58: 849–859.

O'Reardon, J., Solvason, B., Janicak, P. et al. 2007. Efficacy and safety of transcranial magnetic stimulation in the acute treatment of major depression: A multi-site randomized controlled trial. *Biol Psychiatry* 62: 1208–1216.

Padberg, F., Grossheinrich, N., Pogarell, O., Moller, H.-J., and Fregni, F. 2007. Efficacy and safety of prefrontal repetitive transcranial magnetic stimulation in affective disorders. In *Transcranial Brain Simulation for Treatment of Psychiatric Disorders*, Marcolin, M.A. and Padberg, F. (Eds.). Advances in Biological Psychiatry, Vol. 23, Karger, Basel, Switzerland, pp. 53–83.

Parent, A. 2004. Giovanni Aldini: From animal electricity to human brain stimulation. *Can J Neurol Sci* 31: 576–584.

Pascual-Leone, A., Gates, J.R., and Dhuna, A. 1991. Induction of speech arrest and counting errors with rapid-rate transcranial magnetic stimulation. *Neurology* 41: 697–702.

Pascual-Leone, A., Rubio, B., Pallardo, F. et al. 1996. Rapid-rate transcranial magnetic stimulation of left dorsolateral prefrontal cortex in drug-resistant depression. *Lancet* 348: 233–237.

Pascual-Leone, A. et al. 1994. Responses to rapid-rate transcranial magnetic stimulation of the human motor cortex. *Brain* 117: 847–858.

Pell, G.S., Roth, Y., and Zangen, A. 2011. Modulation of cortical excitability induced by repetitive transcranial magnetic stimulation: Influence of timing and geometrical parameters and underlying mechanisms. *Prog Neurobiol* 93: 59–98.

Penfield, W. and Boldrey, E. 1937. Somatic motor and sensory representation in the cerebral cortex of man as studied by electrical stimulation. *Brain* 60: 389–443.

Penfield, W.J. and Jasper, H. 1954. *Epilepsy and the Functional Anatomy of the Human Brain*. Little Brown, Boston, MA.

Peterchev, A.V., Jalinous, R., and Lisanby, S.H. 2008. A transcranial magnetic stimulator inducing near-rectangular pulses with controllable pulse width (cTMS). *IEEE Trans Biomed Eng* 55: 257–266.

Plonsey, R. and Heppner, D. 1967. Considerations of quasi-stationarity in electrophysiological systems. *Bull Math Biophys* 29: 657–664.

Rattay, F. 1986. Analysis of models for external stimulation of axons. *IEEE Trans Biomed Eng* 33: 974–977.

Reilly, J.P. 1989. Peripheral nerve stimulation by induced electric currents: Exposure to time-varying magnetic fields. *Med Biol Eng Comput* 27: 101–110.

Ridding, M.C. and Rothwell, J.C. 2007. Is there a future for therapeutic use of transcranial magnetic stimulation? *Nat Rev Neurosci* 8: 559–567.

Rossi, S., Ferro, M., Cincotta, M. et al. 2007. A real electro-magnetic placebo (REMP) device for sham transcranial magnetic stimulation (TMS). *Clin Neurophysiol* 118: 709–716.

Rossi, S., Hallett, M., Rossini, P.M. et al. 2009. Safety, ethical considerations, and application guidelines for the use of transcranial magnetic stimulation in clinical practice and research. *Clin Neurophysiol* 120: 2008–2039.

Rossini, P.M. and Rossi, S. 2007. Transcranial magnetic stimulation: Diagnostic, therapeutic and research potential. *Neurology* 68: 484–488.

Roth, B.J. 1994. Mechanisms for electrical stimulation of excitable tissue. *Crit Rev Biomed Eng* 22: 253–305.

Roth, B.J. and Basser, P.J. 1990. A model of the stimulation of a nerve fiber by electromagnetic induction. *IEEE Trans Biomed Eng* 37: 588–597.

Roth, B.J., Cohen, L.G., and Hallett, M. 1991a. The electric field induced during magnetic stimulation. *Electroencephalogr Clin Neurophysiol (Suppl)* 43: 268–278.

Roth, Y., Padberg, F., and Zangen, A. 2007. Transcranial magnetic stimulation of deep brain regions: Principles and methods. In *Transcranial Brain Stimulation for Treatment of Psychiatric Disorders*, Marcolin, M.A. and Padberg, F. (Eds.). Advances in Biological Psychiatry, Vol. 23. Karger, Basel, Switzerland, pp. 204–224.

Roth, B.J., Saypol, J.M., Hallett, M., and Cohen, L.G. 1991b. A theoretical calculation of the electric field induced in the cortex during magnetic stimulation. *Electroencephalogr Clin Neurophysiol* 81: 47–56.

Roth, Y., Zangen, A., and Hallett, M. 2002. A coil design for transcranial magnetic stimulation of deep brain regions. *J Clin Neurophysiol* 19: 361–370.

Rothwell, J.C. et al. 1991. Stimulation of the human motor cortex through the scalp. *Exp Physiol* 76: 159–200.

Ruohonen, J. 2003. Background physics for magnetic stimulation. *Suppl Clin Neurophysio* 15: 3–12.

Salvador, R., Silva, S., Basser, P.J., and Miranda, P.C. 2011. Determining which mechanisms lead to activation in the motor cortex: A modeling study of transcranial magnetic stimulation using realistic stimulus waveforms and sulcal geometry. *Clin Neurophysiol* 122: 748–758.

Sciamanna, E. 1882. Fenomeni prodotti dall'applicazione della corrente elettrica sulla dura madre e modificazione del polso cerebrale. Ricerche sperimentali sull'uomo. Atti della R. Accademia dei Lincei. *Memorie della Classe di scienze Fisiche, Matematiche e Naturali* 13: 25–42.

Silva, S., Basser, P.J., and Miranda, P.C. 2008. Elucidating the mechanisms and loci of neuronal excitation by transcranial magnetic stimulation using a finite element model of a cortical sulcus. *Clin Neurophysiol* 119: 2405–2413.

Terao, Y. and Ugawa, Y. 2002. Basic mechanisms of TMS. *J Clin Neurophysiol* 19: 322–343.

Thielscher, A. and Kammer, T. 2002. Linking physics with physiology in TMS: A sphere field model to determine the cortical stimulation site in TMS. *NeuroImage* 17: 1117–1130.

Thielscher, A. and Kammer, T. 2004. Electric field properties of two commercial figure-8 coils in TMS: Calculation of focality and efficiency. *Clin Neurophysiol* 115: 1697–1708.

Thielscher, A., Opitz, A., and Windhoff, M. 2011. Impact of the gyral geometry on the electric field induced by transcranial magnetic stimulation. *NeuroImage* 54: 234–243.

Thompson, S.P. 1910. A physiological effect of an alternating magnetic field. *Proc R Soc Lond [Biol]* B82: 396–399.

Tofts, P.S. 1990. The distribution of induced currents in magnetic stimulation of the nervous system. *Phys Med Biol* 35: 1119–1128.

Tofts, P.S. and Branston, N.M. 1991. The measurement of electric field, and the influence of surface charge, in magnetic stimulation. *Electroencephalogr Clin Neurophysiol* 81: 238–239.

Tranchina, D. and Nicholson, C. 1986. A model for the polarization of neurons by extrinsically applied electric fields. *Biophys J* 50: 1139–1156.

Wassermann, E.M. 1998. Risk and safety of repetitive transcranial magnetic stimulation: Report and suggested guidelines from the International Workshop on the Safety of Repetitive Transcranial Magnetic Stimulation, June 5–7, 1996. *Electroencephalogr Clin Neurophysiol* 108: 1–16.

Wassermann, E.M. and Lisanby, S.H. 2001. Therapeutic application of repetitive transcranial magnetic stimulation: A review. *Clin Neurophysiol* 112: 1367–1377.

Wassermann, E.M. et al. 1996a. Seizures in healthy people with repeated "safe" trains of transcranial magnetic stimuli. *Lancet* 347: 825–826.

Wassermann, E.M. et al. 1996b. Use and safety of a new repetitive transcranial magnetic stimulator. *Electroencephalogr Clin Neurophysiol* 101: 412–417.

Weyh, T. et al. 2005. Marked differences in the thermal characteristics of figure-of-eight shaped coils used for repetitive transcranial magnetic stimulation. *Clin Neurophysiol* 116: 1477–1486.

Zangen, A. et al. 2005. Transcranial magnetic stimulation of deep brain regions: Evidence for efficacy of the H-coil. *Clin Neurophysiol* 116: 775–779.

Ziemann, U. and Rothwell, J.C. 2000. I-waves in motor cortex. *J Clin Neurophysiol* 17: 397–405.

2 Physiological Basis of Transcranial Magnetic Stimulation

John C. Rothwell, Martin Sommer, and Vincenzo Di Lazzaro

CONTENTS

2.1 INTRODUCTION

Transcranial magnetic stimulation (TMS) is now accepted as a standard tool for stimulating the human brain through the intact scalp in fields ranging from preoperative neurosurgical mapping of cortex to cognitive neuroscience or therapy for psychiatric disease. Yet although the basic principles of the method are well understood, we still have relatively little information about which neurons are likely to be activated and how this interacts with ongoing activity in cortical circuits. Much of the uncertainty comes from the fact that the vast majority of experiments have been performed in humans using indirect methods; little detailed work is available in animal preparations. However, since activation by TMS depends on the anatomical arrangement of neural elements in the brain with respect to the position of the coil on the surface of the scalp, it seems likely that even if animal data were available, it could only provide part of the answer to questions about stimulation in the human brain.

There is another limitation to our understanding of the effects of TMS, and that is that most of our knowledge about its action comes from experiments on the primary motor cortex (M1). Here, the muscle twitch induced in peripheral muscles as a response to cortical TMS over the M1 allows for noninvasive assessment of the integrity of the corticospinal motor tract in vivo. A less frequently studied area is the occipital cortex, where TMS induces visual phenomena such as phosphenes or scotoma. Other areas do not have a direct measurable outcome and are therefore rarely used to explore mechanisms of action of TMS. In what follows, we make the tacit assumption that mechanisms relevant in M1 are likely to operate in other areas. However, given the differences in neural anatomy of the cortical layers between different areas, it may well be that there are subtle differences in response to a standard TMS pulse throughout the cortex. For example, M1 responds to a single pulse with a series of repetitive discharges known as I-waves followed by a longer period of inhibition. It could well be that in other cortical areas, I-waves occur less often or at a different frequency, or that the inhibition is longer or deeper than that in M1.

2.2 GETTING ELECTRIC CURRENTS TO FLOW IN THE BRAIN

It is difficult to use externally applied electric currents to stimulate the brain because the scalp and skull have a high electrical resistance. This means that most of the current flows between stimulating electrodes on the scalp, causing local muscle contraction and activation of sensory nerves in the skin. The result is that it can require a pulse of several hundred milliamps and 100–200 μs duration to activate M1 (transcranial electric stimulation, TES), and this is often perceived by volunteers as being uncomfortable (Merton and Morton, 1980). Note that this form of stimulation with a high-intensity, short, single pulse of electricity is quite different from the more commonly used transcranial direct current method, which uses very much smaller currents, usually applied for several minutes continuously.

TMS works by using a rapidly changing magnetic field, which can readily penetrate the scalp, to induce electric current in the brain (Barker et al., 1985; Barker, 1999). A typical stimulator consists of a large capacitor charged to a high voltage. Stimulation occurs by short-circuiting the capacitor across a coil of wire held over the scalp. This induces a magnetic field perpendicular to the coil, which lasts for the duration of the current, which is usually of the order of 1 ms. Peak coil current is reached at around 200 μs. The time course of the remainder of the pulse varies according to stimulator design. "Monophasic" current pulses slowly return to zero over the next 800 μs; biphasic currents have a large overshoot that can have an amplitude similar in size to the initial peak, before returning to zero. A "half sine" pulse is a biphasic pulse damped after the first half cycle and has been less frequently studied than the other types (Sommer et al., 2006).

Electrical current induced in the brain follows the time differential of the current in the coil, which means that in a monophasic stimulus, current in the brain peaks at around 100 μs, then reverses by around 10% maximum, and declines to zero with a slow tail over the remaining 1 ms. It is important to note that there is no net current flow in the brain: the initial large current in one direction is balanced by longer-duration smaller current in the opposite direction. It is usually assumed that this slow tail does not activate directly any neural elements. In a biphasic stimulus, the return current is of the same magnitude and duration as the initial current. In contrast to the monophasic pulse, neurons can potentially be activated during either phase of the current flow. In fact, both modeling and experimental studies show that stimulation occurs preferentially in the reverse cycle of current flow because this is the time at which there is maximum charge transfer (see in the following text) (Maccabee et al., 1993; Sommer et al., 2006). Again there is no net flow of electric current, and no net deposition of charge in the brain.

Induced currents in the brain are maximum under the coil windings and flow in the opposite direction to the current in the coil. There is no induced current in the center of the coil. In a figure-8 coil, which effectively consists of two overlapping circular coils, current is maximal under the junction region where the coils overlap and about half that value under the remainder of the windings (Barker, 1999). This type of coil is referred to as a "focal" coil because of the concentration of current at the junction. More complex designs are possible that may try, for example, to increase focality, such as a "clover leaf" design, where four coils overlap at one point to concentrate the stimulation still further.

TMS can readily activate neurons in superficial cortex, but stimulation of deeper structures is more difficult since the induced magnetic field falls off rapidly with distance from the coil. For a standard circular 9 cm diameter coil, the field is approximately halved some 4 cm from the coil surface (Mills et al., 1987). This can be compared with the distance from the coil to the surface of the brain, which is roughly 1 cm, and the depth of the central sulcus (2 cm).

The currents induced in the brain by TMS differ in one crucial feature from those produced by direct electrical stimulation of the surface: in a simple spherical model of the brain, they tend to run parallel to the surface with minimal radial currents that flow perpendicular to the surface (Tofts, 1990). The reason for this is that any radial current causes a buildup of electrical charge on the surface of the brain, which opposes the current that caused it, leaving only the current tangential to the surface. These surface charges also reduce the depth of stimulation within the brain. TMS coils that are specially designed to stimulate deep structures, such as the "H-coil," are constructed to minimize the accumulation of surface charge and to allow maximum penetration in depth (Roth, 1994).

Of course the brain is not a simple homogeneous sphere, so the actual distribution of currents is more complex than this. Surface charges build up at all regions where there is a discontinuity in resistances, such as at gray matter–white matter boundaries as well as at the brain surface itself. This, coupled with the complex geometric folding of the brain, leads to a very complex distribution of currents. One recent model (Thielscher et al., 2011) suggests that currents in gray matter will be maximal in gyral lips and crowns that are perpendicular to the orientation of the coil windings. This may explain why stimulation of the hand motor area is more efficient if the induced current flows in a direction perpendicular to the direction of the sulcus (see the following section). More unexpectedly, modeling suggests that there may also be hot spots of induced currents within the white matter, again dependent on the orientation of the coil.

2.3   WHAT IS STIMULATED?

Axons have the lowest threshold for conventional electrical stimulation because it is easier to depolarize their membranes than the cell body. Cable theory predicts that transmembrane current is proportional to the second spatial derivative of the external voltage (Roth, 1994). Since electric current is proportional to the first derivative (i.e., the voltage difference), then stimulation should occur where there is the maximal change of external current with length along an axon. In fact, since stimulation occurs with inward transmembrane current, activation happens at the negative spatial derivative of the longitudinal current.

If we imagine a circular coil with a circular axon underneath, then the current along the length of the axon would be the same everywhere: the spatial derivative of current flow along the axon length would be zero and the axon would not be stimulated. By contrast, if the axon bent out of the circle, then there would be a (negative) change in current along its length, which would be maximal at that point. Thus stimulation is likely to occur at the bend. Conversely, imagine a figure-8 coil with a straight nerve aligned along the junction region. There would be two points of maximal spatial derivative of current along the axon, at the two points of divergence of the circular windings. One of them would produce inward transmembrane current whereas the other would produce outward current. Maccabee et al. (1993) verified that this is the case by using a figure-8 coil to stimulate a straight length of phrenic nerve. They found that reversing the (monophasic) TMS current in the coil caused the point of stimulation to jump from one point of divergence to the other.

Maccabee et al. (1993) also explored the effect of stimulating with a biphasic current pulse, which is the typical waveform produced by many repetitive high-frequency TMS machines. In this case, they found that stimulation occurred preferentially during the reversing phase of the current pulse rather than on the rising edge as happens in a monophasic pulse. The direction of the stimulating current was thus opposite to what would have been expected (see the following section on motor-evoked potential (MEP) recruitment with different coil orientations). The stronger influence of the reverse phase of the biphasic pulse results primarily from its long duration, encompassing two quarter cycles, and to a lesser extent from an initial hyperpolarization induced by the initial quarter cycle (Sommer et al., 2006).

The conclusion is that with TMS, stimulation of axons in the brain will depend on their orientation with respect to the current induced in the brain. This explains why stimulation with a conventional round or figure-8 coils is so sensitive to the direction of current and orientation of the coil. A particularly important consequence is that in the hand area of M1, low intensities of TMS fail to activate the axons of corticospinal neurons directly. This is easy to understand in a simplified model where we imagine the corticospinal neurons to be perpendicular to the scalp. TMS-induced current will be parallel to the surface and hence will not cause any current flow parallel to the corticospinal axons. Instead it will activate axons oriented parallel to the surface, which may include presynaptic inputs to the corticospinal cells. It will therefore activate corticospinal neurons trans-synaptically and not directly.

However, as noted earlier, the complex geometry of the brain means that this simple model is a gross underestimation of the real situation. A more realistic approach was taken by Fox et al. (2004), who compared the site of positron emission tomography (PET)-evoked activity in M1 during volitional finger movement and evoked movements following TMS. In both cases, the maximal activation was in the gyral wall of the sulcus. Fox et al. took this to mean that TMS was likely to be activating directly at these points rather than at the gyral crown even though that is nearer the surface. They therefore postulated that TMS induced a posterior–anterior (PA) voltage gradient along cortical columns in the gyral wall. This would favor activation of pyramidal neurons perpendicular to the surface of the cortex. The advantage of this explanation is that it accounts for the directional sensitivity of TMS activation of M1, which is maximal when induced currents flow perpendicular to the sulcus. However, if some of the activation produced by TMS-evoked movement was due to sensory feedback rather than motor activation, the arguments could be incorrect. A second problem is that the hypothesis also implies that direct activation of corticospinal pyramidal neurons should occur relatively easily, giving rise to D-wave activation. However, this is only seen rarely with low intensities of TMS (see the following).

In other models, currents are induced in the gyral crowns and tend to activate local interneurons, many of which might be GABAergic inhibitory neurons, whereas current "hot spots" in the white matter stimulate inputs and outputs from the cortex. These could give rise to excitatory inputs to corticospinal output neurons and generate I-waves and even stimulate corticospinal neurons directly to produce D-waves. Importantly, because these thalamocortical, corticocortical, and corticospinal outputs are directionally oriented according to their source or target, the effects that are recruited would be sensitive to the direction of induced current in the brain. More detailed modeling studies of the distribution of current flow using finite element calculations and a geometrically accurate model of an individual head combined with high-resolution diffusion-weighted imaging for conductivity mapping may in future give a more informed picture of which elements are stimulated (Opitz et al., 2011; Thielscher et al., 2011).

Finally it should be noted that, as with conventional stimulation, large-diameter axons are more likely to be stimulated than smaller-diameter axons, which in turn are more sensitive to stimulation than the axon hillock region or cell body of a neuron. One piece of evidence that this actually occurs in practice comes from measurements made of the strength–duration characteristics for stimulation of the hand area of M1. Barker et al. (1991) constructed a special stimulator that allowed them to vary the duration of the induced current as well as its amplitude. They could therefore plot a strength–duration curve for recruitment of MEPs. This suggested that the time constant of the neural membrane being stimulated was of the order of 150 µs. Since this was the same as the estimated time constant for stimulation of peripheral nerve using the same device, they concluded that large-diameter myelinated axons are the most likely target of TMS in the brain.

There are two important conclusions from this complexity. The first is that most of the activity in the corticospinal output neurons is evoked indirectly, via synaptic connections from excitatory and inhibitory inputs in the cortex. TMS therefore gives an overall indication of the excitability of synaptic circuits in the cortex. The second is that given the directionality of induced currents and the spatial arrangement and size of cells and axons, it may be possible to bias activation to particular populations of cortical neurons, either by changing the direction of the stimulating coil and/or by adjusting the intensity of stimulation.

2.4   CORTICOSPINAL ACTIVITY EVOKED BY TRANSCRANIAL STIMULATION OF THE MOTOR CORTEX

Although TMS and TES activate a number of different neuronal types in the cortex, the final output of corticospinal neurons is surprisingly stereotyped. Direct recording from the pyramidal tract in cats and primates showed that in response to a single electrical stimulus to M1, an electrode placed in the medullary pyramid or on the dorsolateral surface of the cervical spinal cord records a series of high-frequency waves (Adrian and Moruzzi, 1939; Amassian et al., 1987; Kernell and Chien-Ping, 1967; Patton and Amassian, 1954). The earliest wave that persisted after cortical depression and after cortical ablation was thought to originate from the direct activation of the axons of fast pyramidal tract neurons (PTNs) and was therefore termed "D"-wave. The latter waves evoked by cortical stimulation required the integrity of the cortical gray matter and were thought to originate from indirect, trans-synaptic activation of PTNs and were therefore termed "I"-waves (Patton and Amassian, 1954). Recordings from individual PTN axons showed that a given axon may produce both a D- and a subsequent I-wave discharge (Patton and Amassian, 1954). More recently, direct recording of the activity evoked by transcranial stimulation was performed in humans. The initial studies were performed in anesthetized patients during surgery (Berardelli et al., 1990; Boyd et al., 1986; Burke et al., 1993; Thompson et al., 1991). These studies showed that both TES and TMS could evoke a series of waves traveling down the spinal cord. However, the level of anesthesia had pronounced effects on recruitment of descending waves making it difficult to characterize the output of M1 produced by transcranial stimulation. A few years later Kaneko et al. (1996) and Nakamura et al. (1996) recorded in conscious human subjects the descending volleys evoked by transcranial stimulation from epidural electrodes implanted chronically in the spinal cord for the relief of pain. From 1998 onward, Di Lazzaro and coworkers have also made a series of studies using the same approach (see Di Lazzaro et al., 2011 for a review). These studies have shown that the threshold for the activation of the different components of the descending volley is substantially different (Di Lazzaro et al., 2011). At around threshold intensity, TES evokes a short-latency wave that is not modified by changes in cortical excitability and is believed to originate from activation of corticospinal axons in the subcortical white matter at some distance from the cell body (Di Lazzaro et al., 2011). At low intensity, TMS with a focal coil and a posterior–anterior (PA) induced current in the brain evokes a single descending wave with a latency of about 1 ms longer than the D-wave, which is thought to originate from the activation of monosynaptic cortico-cortical connections projecting onto corticospinal neurons (Di Lazzaro et al., 2011). This descending wave produced by indirect activation of cortico-spinal cells is termed the I1 wave. At higher stimulus intensities, latter volleys appear: these are termed late I-waves and are thought to originate from a more complex circuit whose activation produces a repetitive discharge of PTNs (Di Lazzaro et al., 2011). A further increase in the stimulus intensity leads to a direct activation of the axons of the corticospinal cells evoking the D-wave (Di Lazzaro et al., 2011). When the orientation of the figure-8 coil is changed, so that monophasic currents in the brain are induced in a lateral to medial (LM) direction, TMS recruits a D-wave even at threshold intensities. This suggests that TMS with LM induced current in the brain excites PTNs directly at their axons, and, in addition, trans-synaptically.

Reversing the direction of stimulating current, from PA to an anterior–posterior (AP) direction, the descending volleys have slightly different peak latencies and/or longer duration than those seen after PA stimulation, and the order of recruitment of descending activity may change with later waves evoked at around threshold intensity (Di Lazzaro et al., 2011). These findings suggest that PA, AP, and LM TMS might activate different populations of cortical neurons/axons or the same population but at different sites.

The circuit generating the I1 wave and that generating the late I-waves have a different sensitivity to GABAA activity, have different behavior in several TMS protocols testing intracortical inhibition, and also have a different pattern of modulation after repetitive TMS (rTMS) protocols inducing LTP- and LTD-like phenomena in the human brain (Di Lazzaro et al., 2011). Benzodiazepines, positive modulators of the GABAA receptor, produce a selective suppression of late I-waves with no effect on the I1 wave (Di Lazzaro et al., 2011). Similarly, all the TMS inhibitory protocols, such as the short-interval and the long-interval intracortical inhibition produced by paired pulse stimulation of the motor cortex, the short-latency afferent inhibition produced by conditioning electrical stimuli of peripheral nerves, and the interhemispheric inhibition (produced by conditioning stimuli over the contralateral hemisphere), suppress selectively the late I-waves (Di Lazzaro et al., 2011).

Several models have been proposed to explain the origin and the nature of the descending waves evoked by TMS in humans: (1) periodic bombardment of corticospinal cells through chains of interneurons with fixed temporal characteristics (Amassian et al., 1987); (2) stimulus intensity-dependent activation of independent chains of interneurons, each responsible for generating a different I-wave (Day et al., 1989; Sakai et al., 1997); and (3) strong and synchronized depolarization of many corticospinal cells and/or interneurons, which leads to oscillatory activity and repetitive discharge of these cells as a consequence of their intrinsic membrane properties (Creutzfeldt et al., 1964; Phillips, 1987; Ziemann and Rothwell, 2000). None of these models can fully explain the properties and modulation of I-waves; moreover, all these models were built on the empirical observations provided by MEP and epidural activity recordings and not on an established model of cortical circuit organization. A model incorporating the detailed anatomy of M1 was proposed by Esser et al. (2005), who investigated the effects of TMS on cortical circuits by constructing a large-scale model of a portion of the thalamocortical system and simulating the delivery of TMS pulses. The authors suggested that the actual course of events underlying the generation of I-waves is best approximated as a combination of intrinsic neuronal properties (model of neural oscillator) and interactions among circuits of inhibitory and excitatory interneurons (Esser et al., 2005).

Recently, Di Lazzaro et al. (2011) evaluated whether the properties of the descending waves recorded in conscious humans after transcranial stimulation can be explained by an interaction of transcranial stimuli and the cortical circuits as characterized by an accepted anatomical model. To this end, they used the basic "canonical" circuit of cerebral cortex proposed by Douglas and Martin (1989), which can be applied to all neocortex. This model includes the superficial population of excitatory pyramidal neurons of layers II and III (P2–P3), the large PTN in layer V (P5), and the inhibitory GABA cells. It has been proposed that this represents the minimum architecture necessary for capturing the most essential cortical input–output operations (Douglas and Martin, 1989; Sheperd, 2004).

2.4.1 Transcranial Electrical Stimulation

The lowest intensities of anodal TES recruits an early descending volley, which has a latency of 2–2.6 ms when recorded from the high cervical cord (Di Lazzaro et al., 2008). In humans, this latency is compatible with direct activation of the pyramidal tract axons just below the gray matter, and hence this wave is referred to as a D-wave. This wave might originate from direct activation of P5 axons.

2.4.2 Monophasic Magnetic Stimulation

Using a focal coil and a monophasic waveform stimulus inducing a PA current across the central sulcus in the brain, the lowest threshold volley occurs at a latency of 1.0–1.4 ms longer than the volley recruited by electrical anodal stimulation (Di Lazzaro et al., 2011). The axons of the more superficial pyramidal neurons (P2, P3) are conceivably the most excitable elements; they also represent the main source of input to corticospinal neurons (Anderson et al., 2010). According to the canonical circuit of cerebral cortex, these neurons have a monosynaptic excitatory connection with the large PTNs of layer V (P5) (Figure 2.1). Thus, the I1 wave might originate from monosynaptic activation of the P5 cells by axons of P2 and P3 cells (Figure 2.1). The difference in latency between D-wave and I1 wave of about 1 ms is appropriate for a monosynaptic activation of P5 cells originating from presynaptic axons. The EPSP in P5 cells is followed by an IPSP after a further synaptic delay because the inhibitory connection between superficial excitatory neurons and P5 cells is disynaptic. Thus, the monosynaptic I1 activity is not influenced by TMS-related GABAergic discharge.

The I1 wave increases in size and is followed by later volleys as the intensity of stimulation increases. The interpeak interval between I-waves is about 1.5 ms, which indicates a discharge frequency of about 670 Hz. According to the canonical circuit, the P5 cells have excitatory monosynaptic reciprocal connections with layer II and III pyramidal neurons and interneurons, thus these cells are activated after P5 discharge and they will in turn reactivate the P5 cells themselves. The mean transmission delay between layer II–III neurons and layer V pyramidal neurons is about 1.5 ms (Thomson et al., 2002). With a synaptic delay of 1.5 ms, computational models of networks of highly connected excitatory and inhibitory neurons predict a peak of activity of 667 Hz (Brunel, 2000), which corresponds to the I-wave frequency. This high-frequency activity might be produced by the recruitment of fully synchronized clusters of excitatory and inhibitory neurons (Brunel, 2000). The frequency of I-waves is not compatible with the refractory period of excitatory and inhibitory neurons; however, it should be considered that single-cell discharge rates are typically much lower than those of neuronal networks. Computational models of spiking neurons have shown that single pyramidal cells may fire only once in every 15–20

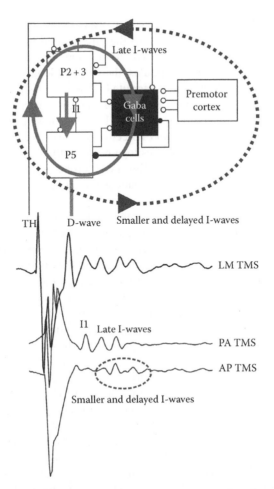

FIGURE 2.1   **(See color insert.)** A schematic view of the model of corticospinal volley generation based on canonical cortical circuit proposed by Douglas et al. (1989). This model includes
the superficial population of excitatory pyramidal neurons of layers II and III (P2–P3), the
large PTNs in layer V (P5), and the inhibitory GABA cells and the thalamocortical projections.
(Modified from Sheperd, G. *The Synaptic Organization of the Brain*, Cambridge University
Press, New York, Figure 1.14, 2004.) Magnetic stimulation with a PA induced current in the
brain produces monosynaptic activation of P5 cells by the axons of superficial pyramidal neurons (green arrow) evoking the I1 wave; at high intensities, it also produces recurrent activity
in a circuit composed of the layer II and III and layer V pyramidal neurons together with
their connections with local GABAergic interneurons (red ellipse and arrows) evoking later
I-waves, which include the same cortical elements. Magnetic stimulation with an LM induced
current in the brain produces a direct activation of the axons of corticospinal cells (blue line)
evoking the D-wave followed, at high intensities, by I-waves. Magnetic stimulation with an
AP induced current in the brain recruits smaller and delayed descending volleys with slightly
different peak latencies and longer duration than those seen after PA magnetic stimulation. It
is proposed that this more dispersed descending activity is produced by a more complex circuit
(violet-dotted ellipse and arrows) that might include cortico-cortical fibers originating from the
premotor cortex and projecting upon the M1 circuits generating the I-waves.

cycles of the population activity (Brunel and Wang, 2003). Polysynchronization of strongly interconnected groups of neurons that fire with millisecond precision may emerge in neuronal networks (Izhikevich, 2006). Now, it can be assumed that the stronger the intensity of the TMS stimulus, the more prolonged the activity will be with an increasing number of I-waves recruited at increasing stimulus intensities.

The model of I-wave generation based on canonical cortical circuit proposes a circuit for the I1 wave generation represented by the monosynaptic excitatory connections between P2 and P3 cells and P5 cells and a network for the late I-waves that includes the same cortical elements together with their connections with local GABAergic interneurons (Figure 2.1). This model might explain the fixed periodicity of the latter I-waves and the strict relationship between stimulus intensity and I-wave number.

The changes observed in I-waves after manipulation of cortical excitability might also be explained by this model. According to the proposed model, the GABAergic neurons are involved only in the production of the later I-waves while the I1 wave is not related to the activity of inhibitory neurons. This is in agreement with the changes observed in I-waves after pharmacological enhancement of inhibitory GABAergic activity through benzodiazepine administration. After lorazepam, there is a pronounced suppression of late waves while there is no change in I1 wave (Di Lazzaro et al., 2011). An increase in the excitability of P5 cells also results in a change in I-waves, which can be explained by the model. It has been shown that maximum voluntary contraction can substantially increase the amplitude of all I-waves including the I1 wave (Di Lazzaro et al., 1998). The increase in size of the I1 wave suggests an increase in excitability of PTNs. The large effect on the size of the descending corticospinal waves is not paralleled by a comparable effect on the threshold for evoking recognizable descending activity after TMS (Di Lazzaro et al., 1998). This suggests that the elements activated by PA TMS have a relatively constant threshold. The likely explanation is that magnetic stimulation activates axons projecting upon PTNs at some distance from the cell body so that threshold is unaffected by synaptic activity. According to the present model, the activated axons belong to the superficial P2 and P3 cells. At higher magnetic stimulus intensities, an earlier wave of small amplitude appears. This wave has the same latency as the D-wave evoked by TES at threshold, and as for the TES-evoked D-wave, it is proposed to be generated by direct activation of P5 axons. The D-wave can be evoked even at low stimulus intensity by changing the direction of the induced current in the brain (Di Lazzaro et al., 2011; Figure 2.1). At threshold stimulus intensities, TMS with LM induced current in the brain usually recruits a short-latency wave with the same latency of the D-wave evoked by electrical anodal stimulation (Di Lazzaro et al., 2011). This suggests that LM TMS excites P5 cells directly at their axons and, in addition, trans-synaptically.

Changing the parameters of the stimulus, M1 output might become more complex. When reversing the direction of the induced current in the brain from the usual PA direction to AP, smaller descending volleys are evoked with slightly different peak latencies and/or longer duration than those seen after PA stimulation (Figure 2.1; Di Lazzaro et al., 2011). When the peak latencies of the I-waves evoked by PA stimulation are compared with the peak latencies of the corresponding I-waves evoked by

AP stimulation, it appears that corresponding volleys differ in latency by 0.2–0.7 ms. These smaller and delayed late I-waves are similar to those recorded in monkeys by Shimazu et al. (2004) after premotor cortex stimulation. These authors showed that stimulation of the premotor cortex evokes I-wave responses smaller and later than those evoked by M1 stimulation and suggested that this activity may be mediated by corticocortical inputs to M1 impinging on interneurons generating late I-waves (Shimazu et al., 2004). A similar mechanism might be at the origin of the more temporally dispersed and delayed I-wave activity evoked by AP stimulation in humans. It can be hypothesized that AP magnetic stimulation activates populations of axons originating from the premotor cortex projecting upon the M1 circuits generating the I-waves.

2.4.3 BIPHASIC MAGNETIC STIMULATION

When using biphasic magnetic stimulation, the effects on cortical output are less consistent compared with monophasic stimulation (Di Lazzaro et al., 2011). At active threshold intensity, PA–AP stimulation evokes either an I1 wave, which has the same latency as the I1 wave after monophasic PA, or a delayed I1 wave together with a delayed D-wave, or a delayed I3 wave, which has the same latency as the I3 wave evoked by AP stimulation. At intermediate stimulus intensities, the recruitment of additional waves is observed: some of the waves are similar to those observed with monophasic AP stimulation and some have the same latency as the waves evoked by PA stimulation. At high stimulus intensities, the I-waves evoked by PA–AP stimulation are similar to those evoked by monophasic PA stimulation. At all intensities, the latency of the D-wave is longer than the latency of the D-wave from anodal stimulation. By analogy with non-focal stimulation of the brain, it is suggested that this proximal D-wave originates at the level of the axon hillock. Using AP–PA stimulation, active threshold is lower than the one to PA–AP stimulation, and the pattern of recruitment of D- and I-waves at increasing stimulus intensities resembles the one with monophasic PA stimulation more than that with AP stimulation. These findings suggest that both phases of the stimulus pulse can activate descending pathways, but the precise combination of neural elements activated by AP and PA directions depends on their relative threshold and amplitude.

2.4.4 NON-FOCAL STIMULATION OF THE BRAIN (CIRCULAR COIL)

In anesthetized human subjects, Burke et al. (1993) showed that, using a circular coil centered over the vertex and recording from the spinal cord, the D-wave is the component with lowest threshold.

The recording of epidural volleys in conscious humans showed different results in different subjects (Di Lazzaro et al., 2011). The earliest volley evoked by a circular coil centered over the vertex is a descending wave with a latency of about 0.2 ms longer than the LM or anodal D-wave. Either this wave or a wave corresponding to the I1 wave evoked by PA magnetic stimulation can be the lowest threshold volley evoked by a circular coil. At suprathreshold intensities, later waves can be recruited: these waves, in some cases, may have a latency that is outside the periodicity of

I-waves evoked by PA magnetic stimulation. A maximum voluntary contraction increases the amplitude of all descending volleys including the earliest (D) wave. These data show that there are major differences between non-focal stimulation of the hand area with a circular coil and focal stimulation with a figure-8 coil. Non-focal stimulation is more likely to evoke a D-wave than PA focal stimulation. Moreover, the early (D) volley recruited by non-focal stimulation has a slightly longer latency than the LM or anodal D-wave, and it is also facilitated by voluntary contraction. These features suggest that it is initiated closer to the cell body of the PTNs than the conventional D-wave evoked by anodal or LM magnetic stimulation, perhaps at axon hillock rather than at some distance down the axon.

2.5   CONCLUSION

By incorporating the known physiology of the epidural activity evoked by single and paired pulse and rTMS with the anatomical and computational characteristics of the canonical model of cerebral cortex circuit, composed of layer II and III and layer V excitatory pyramidal cells, inhibitory interneurons, cortico-cortical and thal-amocortical inputs, the characteristics and nature of the I-wave activity evoked by TMS including its regular and rhythmic nature, the dose dependence, and pharma-cological modulation of the discharge might be explained elegantly. A TMS-induced strong depolarization of the superficial excitatory cells of the circuit may lead to the recruitment of fully synchronized clusters of excitatory neurons, including layer V PTNs, and inhibitory neurons producing a high-frequency (~670 Hz) repetitive discharge of the corticospinal axons. The role of the inhibitory circuits is crucial to entrain the firing of the excitatory networks to produce a high-frequency discharge (Douglas and Martin, 1989). The integrative properties of the circuit might also pro-vide a good framework for the interpretation of the changes in the I-wave discharge produced by paired TMS.

REFERENCES

Adrian ED and G Moruzzi. Impulses in the pyramidal tract. *J Physiol* 1939; 97: 153–199.
Amassian VE, M Stewart, GJ Quirk, and JL Rosenthal. Physiological basis of motor effects of a transient stimulus to cerebral cortex. *Neurosurgery* 1987; 20: 74–93.
Anderson CT, PL Sheets, T Kiritani, and GM Shepherd. Sublayer-specific microcircuits of cor-ticospinal and corticostriatal neurons in motor cortex. *Nat Neurosci* 2010; 13: 739–744.
Barker AT. The history and basic principles of magnetic nerve stimulation. *Electroencephalogr Clin Neurophysiol Suppl* 1999; 51: 3–21.
Barker AT, CW Garnham, and IL Freeston. Magnetic nerve stimulation: The effect of waveform on efficiency, determination of neural membrane time constants and the measurement of stimulator output. *Electroencephalogr Clin Neurophysiol Suppl* 1991; 43: 227–237.
Barker AT, R Jalinous, and IL Freeston. Non-invasive magnetic stimulation of human motor cortex. *Lancet* 1985; 1: 1106–1107.
Berardelli A, M Inghilleri, G Cruccu, and M Manfredi. Descending volley after electrical and magnetic transcranial stimulation in man. *Neurosci Lett* 1990; 112: 54–58.
Boyd SG, JC Rothwell, JM Cowan, PJ Webb, T Morley, P Asselman et al. A method of moni-toring function in corticospinal pathways during scoliosis surgery with a note on motor conduction velocities. *J Neurol Neurosurg Psychiatry* 1986; 49: 251–257.

Brunel N. Dynamics of sparsely connected networks of excitatory and inhibitory spiking neurons. *J Comput Neurosci* 2000; 8: 183–208.

Brunel N and XJ Wang. What determines the frequency of fast network oscillations with irregular neural discharges? I. Synaptic dynamics and excitation-inhibition balance. *J Neurophysiol* 2003; 90: 415–430.

Burke D, R Hicks, SC Gandevia, J Stephen, I Woodforth, and M Crawford. Direct comparison of corticospinal volleys in human subjects to transcranial magnetic and electrical stimulation. *J Physiol (Lond)* 1993; 470: 383–393.

Creutzfeldt OD, HD Lux, and AC Nacimiento. Intracelluläre reizung corticaler nervenzellen. *Pflügers Arch* 1964; 281: 129–151.

Day BL, D Dressler, A Maertens de Noordhout, CD Marsden, K Nakashima, JC Rothwell et al. Electric and magnetic stimulation of human motor cortex: Surface emg and single motor unit responses. *J Physiol (Lond)* 1989; 412: 449–473.

Di Lazzaro V, P Profice, F Ranieri, F Capone, M Dileone, A Oliviero et al. I-wave origin and modulation. *Brain Stimul* 2011, in press.

Di Lazzaro V, D Restuccia, A Oliviero, P Profice, L Ferrara, A Insola et al. Effects of voluntary contraction on descending volleys evoked by transcranial stimulation in conscious humans. *J Physiol* 1998; 508 (Pt 2): 625–633.

Di Lazzaro V, U Ziemann, and RN Lemon. State of the art: Physiology of transcranial motor cortex stimulation. *Brain Stimul* 2008; 1: 345–362.

Douglas RJ and Martin KA. A canonical microcircuit of neocortex. *Neural Comput* 1989; 1: 480–488.

Esser SK, SL Hill, and G Tononi. Modeling the effects of transcranial magnetic stimulation on cortical circuits. *J Neurophysiol* 2005; 94: 622–639.

Fox PT, Narayana S, Tandon N, Fox SP, Sandoval H, Kochunov P et al. Intensity modulation of TMS-induced cortical excitation: primary motor cortex. *Hum Brain Mapp* 2006; 27: 478–487.

Izhikevich EM. Polychronization: Computation with spikes. *Neural Comput* 2006; 18: 245–282.

Kaneko K, S Kawai, Y Fuchigami, H Morita, and A Ofuji. The effect of current direction induced by transcranial magnetic stimulation on the corticospinal excitability in human brain. *Electroencephalogr Clin Neurophysiol* 1996; 101: 478–482.

Kernell D and WU Chien-Ping. Responses of the pyramidal tract to stimulation of the baboon's motor cortex. *J Physiol (Lond)* 1967; 191: 653–672.

Maccabee PJ, VE Amassian, LP Eberle, and RQ Cracco. Magnetic coil stimulation of straight and bent amphibian and mammalian peripheral nerve in vitro: Locus of excitation. *J Physiol* 1993; 460: 201–219.

Merton PA and HB Morton. Stimulation of the cerebral cortex in the intact human subject. *Nature* 1980; 285: 227.

Mills KR, NM Murray, and CW Hess. Magnetic and electrical transcranial brain stimulation: Physiological mechanisms and clinical applications. *Neurosurgery* 1987; 20: 164–168.

Nakamura H, H Kitagawa, Y Kawaguchi, and H Tsuji. Direct and indirect activation of human corticospinal neurons by transcranial magnetic and electrical stimulation. *Neurosci Lett* 1996; 210: 45–48.

Opitz A, M Windhoff, RM Heidemann, R Turner, and A Thielscher. How the brain tissue shapes the electric field induced by transcranial magnetic stimulation. *Neuroimage* 2011; 58: 849–859.

Patton HD and VE Amassian. Single- and multiple-unit analysis of cortical stage of pyramidal tract activation. *J Neurophysiol* 1954; 17: 345–363.

Roth BJ. Mechanisms for electrical stimulation of excitable tissue. *Crit Rev Biomed Eng* 1994; 22: 253–305.

Sakai K, Y Ugawa, Y Terao, R Hanajima, T Furabayashi, and I Kanazawa. Preferential activation of different i waves by transcranial magnetic stimulation with a figure-of-eight shaped coil. *Exp Brain Res* 1997; 113: 24–32.

Shepherd G. *The Synaptic Organization of the Brain*, Cambridge University Press, New York, 2004.

Shimazu H, MA Maier, G Cerri, PA Kirkwood, and RN Lemon. Macaque ventral premotor cortex exerts powerful facilitation of motor cortex outputs to upper limb motoneurons. *J Neurosci* 2004; 24: 1200–1211.

Sommer M, A Alfaro, M Rummel, S Speck, N Lang, T Tings et al. Half sine, monophasic and biphasic transcranial magnetic stimulation of the human motor cortex. *Clin Neurophysiol* 2006; 117: 838–844.

Thielscher A, A Opitz, and M Windhoff. Impact of the gyral geometry on the electric field induced by transcranial magnetic stimulation. *Neuroimage* 2011; 54: 234–243.

Thompson PD, BL Day, JC Rothwell, D Dressler, A Maertens de Noordhout, and CD Marsden. Further observations on the facilitation of muscle responses to cortical stimulation by voluntary contraction. *Electroencephalogr Clin Neurophysiol* 1991; 81: 397–402.

Thomson AM, DC West, Y Wang, and AP Bannister. Synaptic connections and small circuits involving excitatory and inhibitory neurons in layers 2–5 of adult rat and cat neocortex: Triple intracellular recordings and biocytin labelling in vitro. *Cereb Cortex* 2002; 12: 936–953.

Tofts PS. The distribution of induced currents in magnetic stimulation of the nervous system. *Phys Med Biol* 1990; 35: 1119–1128.

Ziemann U and JC Rothwell. I-waves in motor cortex. *J Clin Neurophysiol* 2000; 17: 397–405.

3 Cellular and Network Effects of Transcranial Direct Current Stimulation

Insights from Animal Models and Brain Slice

Marom Bikson, Davide Reato, and Asif Rahman

CONTENTS

This chapter addresses the contribution of animal research on direct current (DC) stimulation to current understanding of transcranial direct current stimulation (tDCS) mechanisms and prospects and pitfalls for ongoing translational research. Though we attempt to put in perspective key experiments in animals from the 1960s to the present, our goal is not an exhaustive cataloging of relevant animal studies, but rather to put them in the context of ongoing effort to improve tDCS. Similarly, though we point out essential features of meaningful animal studies, we refer readers to original work for methodological details. Though tDCS produces specific clinical neurophysiological changes and is therapeutically promising, fundamental questions remain about the mechanisms of tDCS and on the optimization of dose. As a result, a majority of clinical studies using tDCS employ a simplistic dose strategy where "excitability" is increased or decreased under the anode and cathode respectively. We discuss how this strategy, itself based on classic animal studies, may not account for the complexity of normal and pathological brain function, and how recent studies have already indicated more sophisticated approaches.

3.1   MEANINGFUL ANIMAL STUDIES OF tDCS AND THE QUASI-UNIFORM ASSUMPTION

The motivation for animal research of tDCS is evident and similar to other translational medical research efforts: to allow rapid and risk-free screening of stimulation protocols and to address the mechanisms of tDCS with the ultimate goal of informing clinical tDCS efficacy and safety. To have a meaningful relevance to clinical tDCS, animal studies must be designed with consideration of (1) conducting animal studies by correctly emulating the delivery of DC stimulation to the brain and (2) measuring responses that can be used to draw clinically relevant inferences. Before reviewing the main insights drawn from animal studies, we outline the basis and pitfalls of translational animal research on tDCS.

3.1.1   CLASSIFICATION OF ANIMAL STUDIES

The scope of this review includes any animal study exploring the behavioral, neurophysiological, or molecular response of the brain to DC currents; with a focus on macro-electrodes, relatively low intensity stimulation, and sustained (seconds to minutes) rather than pulsed (millisecond or less) waveforms. Animal studies can be broadly classified by the preparation and related method of stimulation, namely where the electrodes are placed, as (1) transcranial stimulation in animals, (2) intracranial stimulation in vivo including with one electrode on the cortex, and (3) stimulation of tissue in vitro, including brain slices.

1. Modern animal studies on tDCS use transcranial stimulation with a skull screw or skull mounted cup (Liebetanz et al. 2006a; Cambiaghi et al. 2010; Yoon et al. 2012)—advantages of transcranial stimulation include preventing electrochemical products from the electrodes from reaching the brain (which would confound any results). If the screw penetrates completely through the skull, stimulation is no longer in the transcranial category (see next). Rodents are typically used. A return electrode on the body, mounted in a "jacket" is typically used for "unipolar stimulation" (which is broadly analogous to a human tDCS extracephalic electrode). In a rabbit study four silver ball electrodes formed a single virtual electrode over the target (Marquez-Ruiz et al. 2012). Alternatively, two cranial electrodes produce bipolar stimulation (Ozen et al. 2010). Since the cranium is not penetrated, the effects of DC stimulation are probed through behavior, noninvasive recording (electroencephalogram, EEG), noninvasive electrical interrogation (e.g., transcranial magnetic stimulation, TMS; transcranial electrical stimulation, TES), or histology after sacrifice. In some older studies, transcranial DC stimulation was also applied in larger animals (monkeys) (Toleikis et al. 1974) but should be interpreted with caution when the skull was penetrated for recording electrodes (which distorts current flow) (Datta et al. 2010) or when recording between electrodes without control for cortical folding (leading to variation in current flow direction through cortical gyrations and inconsistent effects). Replacement of removed skull with insulating filler (e.g., dental cement) may correct shunting through the hole (Marquez-Ruiz et al. 2012).

2. Classic animal studies typically used an electrode on the brain (Creutzfeldt et al. 1962; Bindman et al. 1964), where the intracranial electrode was covered in something like a cotton wick (Redfearn et al. 1964) to buffer electrochemical changes. Cats and monkeys were typically used. Note that the protection of the skin from electrochemical product at the electrode is why saline-soaked sponges (or gel) are used in tDCS (Minhas et al. 2010)—and though improper set-up can result in skin irritation, these products can not reach the brain and so are not part of tDCS mechanisms. When an electrode is placed inside the cranium (on the animal brain), the potential interference from electrochemical changes at the electrodes diffusing into the brain cannot be automatically ignored. These electrochemical products can even be polarity specific (Merrill et al. 2005) and produce reversible changes, but still have no relevance to tDCS. Steps to reduce interference include using suitable electrode (e.g., Ag/AgCl) and wrapping the electrode in cotton to buffer chemical changes, protocols that where rationally used in many studies. Passage of prolonged DC current through a poorly selected electrode material (e.g., screw) is expected to produce significant electrochemical changes near the metal. It is generally assumed with cortical electrodes that current flow through nearby cortex will be

unidirectional (inward for anode, outward for cathode; see conventions in Section 3.1.3); however, the presence of CSF in convoluted gyri (especially in larger animals) will distort current flow patterns and can produce local direction inversions (Creutzfeldt et al. 1962). Despite these concerns, the rationale for invasive stimulation in classical animal studies may simply be that the cranium must be exposed regardless to facilitate insertion of recording electrodes, and a majority of these studies were interested in the general effects of DC current on brain function and not necessarily clinical *transcranial* DCS.

3. The use of brain slices to study the effects of weak DC stimulation dates to work by John G.R. Jefferys in 1981 (Jefferys 1981; Nitsche and Paulus 2000; Ardolino et al. 2005), with experimental techniques used to the present day established by Bruce Gluckman and Steven Schiff (Gluckman et al. 1996) and adapted by Dominique Durand and our group (Durand and Bikson 2001). The rationale for using a brain slice (usually rodents and ferrets) is the ability to probe brain function in detail using a range of electrophysiological, pharmacological, molecular, and imaging techniques. In isolated tissue, the direction of current flow is also known and precisely controlled. Lopez-Quintero et al. (2010) described techniques for stimulating cultured monolayers. In a seminal series of papers Chan and Nicholson used isolated turtle cerebellum (Chan and Nicholson 1986; Chan et al. 1988). For in vitro DC stimulation studies, electrodes are placed in the bath at some distance from the tissue to buffer electrochemical changes. As emphasized in the following, tDCS delivers electrical current not chemicals to the brain.

3.1.2   tDCS DOSE IN HUMAN AND ANIMALS, AND THE QUASI-UNIFORM ASSUMPTION

The clinical "dose" of tDCS has been defined as those aspects of stimulation that are externally controlled by the operator (Bikson et al. 2008; Peterchev et al. 2011), namely electrode montage (shape, location, etc.) and the specifics of the DC waveform (duration, intensity in mA applied, ramp, etc.). As explained next, it would be fundamentally misguided to simply replicate these dose parameters in animal studies. tDCS produces a complex pattern of current flow across the brain, which results in dose-specific electric field (current density) that varies significantly across brain regions. This brain electric-field distribution represents and determines the electrical actions of tDCS. The brain electric field is not a simple function of any dose parameter, for example the current density at the electrodes (total current/area) does not map simply to peak brain electric field (Miranda et al. 2007). There are fortunately well-established methods to predict the electric field generated in the brain using computational models (Miranda et al. 2006; Datta et al. 2009); though methodological approaches across groups vary those modeling studies using realistic anatomy have converged that the peak electrical field generated during tDCS is 0.2–0.5 V/m ($0.05–0.14$ A/m^2 current density) for a 1 mA intensity (Miranda et al. 2006; Datta et al. 2009; Sadleir et al. 2010). The electric field would scale linearly with current intensity such that 2 mA would produce up to 0.4–1 V/m ($0.1–0.28$ A/m^2 current density).

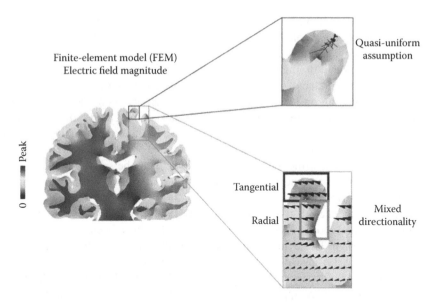

FIGURE 3.1 The quasi-uniform assumption is implicit in the majority of modeling and animal studies of tDCS. The first aspect of the quasi-uniform assumption is based on the electric field generated in the brain to not significantly change (be uniform) on the scale of a single cortical column or neuronal dendritic tree. Only in this way it is meaningful to represent, for a first approximation, neuromodulation by regional electric field. This assumption underpins the rational basis for replicating an electric field of interest in an animal model as described in the text. Shown is a high-resolution finite element model (FEM) computational model of current flow through the head with overlaid neuronal morphology.

These peaks represent specific hot-spots. Using conventional tDCS montages weaker electric fields are generated across much of the brain. In addition, due to subject-specific idiosyncratic cortical folding, the electric field is "clustered" (Datta et al. 2009), with many local maxima (Figure 3.1). There is thus no single electric field generated in the brain during tDCS but rather a range of distributed electric fields across the brain. The question therefore is: given this complexity of electric field distribution across brain structures, what should (and can) be mimicked in animal models? It is important to emphasize that simply mimicking tDCS clinical dose in animal models, or adjusting dose guidelines by an arbitrary rule of thumb (e.g., by head volume), may not be prudent.

One solution (which we term an aspect of the "quasi-uniform" assumption) is to consider only the peak electric field generated in the brain, or only the electric field in one brain region of interest, and then to replicate the electric field across an area of the animal brain or the entire animal brain/tissue. (It is impractical to replicate the electric field induced in each brain region during tDCS in all corresponding brain region in an animal model.) As it turns out, because of practical consideration, the quasi-uniform assumption is already adopted implicitly in most animal research of tDCS. This approach is partly supported by electric fields generated during tDCS being largely uniform across any specific cortical column (neuronal dendritic tree) of interest (Figure 3.1, inset)—hence one can speak of a single electric field in reference to a region of interest.

However, it is worthwhile to point out that considering the *peak* of the electric field (either across the whole brain or in a subregion) as basis for the field amplitude may be misleading. The field amplitude can change by orders of magnitudes in different brain areas and dramatically even across local gyri (Datta et al. 2009). The average (or median) value of the electric field can be up to 10 times smaller than the field peak (depending on local geometry and conductivity properties). Also, since usually the electric fields used in animal experiments are based on estimations from humans (using FEM), it is also necessary to consider how the coupling constant between neuronal polarization and electric field applied can vary across species (see next). For that reason, while the average electric field can be smaller than the peak value, the polarization of neurons in humans could be higher, assuming a higher coupling constant in humans (see next).

Generally, once a clinical (quasi-uniform) tDCS electric field that is to be replicated in an animal model is decided, three approaches have been taken in rational experimental design. These three approaches typically relate to (1) transcranial stimulation in animal, (2) invasive stimulation of animals with intracranial electrodes, and (3) stimulation of tissue/brain slices with bath electrodes. In each case, the quasi-uniform assumption is re-applied in the generation and control of a (quasi)uniform electric field in a targeted region of the animal brain or across isolated tissue.

1. In the first case of transcranial stimulation of animal, the same modeling approaches that predict electric fields during clinical tDCS can be used to model and guide stimulation design (Gasca et al. 2010). As the case in clinical tDCS, in DC transcranial stimulation in animals it is important to consider how the position of the "return/reference" electrode influences current flow even under the "active" electrode (Bikson et al. 2010; Brunoni et al. 2011). As anatomically precise animal models are under development, concentric sphere models (simply scaled to size) can be used to determine electric field intensity generated in the animal brain (Marquez-Ruiz et al. 2012); free tool available through neuralengr.com/BONSAI). In the absence of an specific modeling of current flow in animal, and in cases where the electrode is placed directly on the skull, one can, to a first approximation, assume a maximum potential brain current density equal to the average electrode current density (total current/electrode area; (Bikson et al. 2009). However, it is important to recognize that the direction (inward or outward) of the electric fields generated across the brain, including in deep brain structures (particularly in higher animals with increasing convoluted cortex) may also vary (as it does in human tDCS). The electric field in a region of interest may also be measured with invasive electrodes (Ozen et al. 2010), recognizing it is not uniform throughout the animal brain, and the insertion and presence of electrodes may itself distort current flow.

2. In the second case, for animal studies with an electrode placed on the brain surface, one might again assume that the (quasi-uniform) current density in the brain directly under the electrodes equals that *average* current density at the electrode (total current/electrode area). As with scalp electrodes in tDCS, when a sponge of cotton wrapper is used, its contact areas should be

used in calculations. But depending on the electrode design, current density may in fact be (orders of magnitude) higher at electrode edges (Miranda et al. 2006; Minhas et al. 2011)—an issue aggravated for small electrodes where electric field near a monopolar source can be very high leading to further potential complications (see discussion in Bindman et al. 1964). As with transcranial stimulation, current spread throughout the brain should be assumed with any return outside the head (Islam et al. 1995).

3. In the third approach, including in vitro brain slice studies, the task is simplified because using long parallel wires (or plates) placed in a bath across the entire tissue—with proper care, this generates a uniform electric field across the entire tissue (a truly uniform electric field) that can be readily calibrated to match tDCS levels (Gluckman et al. 1996; Francis et al. 2003; Bikson et al. 2004). Typically, the placement of the electrodes in the bath, away from the tissue of interest, protects from electrochemical products. The simplicity and versatility of this techniques, makes control of DC parameters in slice straightforward and allows analysis of function in detail not possible with other techniques (see next). It is interesting that the generation of uniform fields across an entire brain region can make the most invasive in vitro approaches analogous to regional electric field induced by tDCS.

In each of the aforementioned cases the quasi-uniform assumption applies by (1) assuming a uniform electric field in a region of interest during tDCS that is a function of scalp electrode position and applied current (Figure 3.1) though recognizing that the electric field varies across the brain; and (2) positioning electrodes and selecting current in translational research that replicate this electric field in a specific region of the animal brain (recognizing that electric field will vary across the entire animal brain) or across a brain slice in vitro (recognizing that the entire brain slice will be exposed to a single electric field; Figure 3.2).

3.1.3 STIMULATOR AND ELECTRODE TECHNIQUES, AND NOMENCLATURE

On a technical note, our opinion is that for reproducibility and precision current-controlled stimulation should be used in animal studies. Indeed, for the same reason tDCS with current-controlled stimulation is used in almost all translational studies of DC stimulation. The electrode-solution interface represents an unknown and changing impedance in series with the stimulator (Merrill et al. 2005). It is well established that current-control guarantees consistent stimulation of tissue through this interface; the electric field in the brain tracks the applied current and can be simply scaled to match clinical electric field values. Using voltage control, especially during DC stimulation, may result in an unknown variable, and indeed changing (not DC) electric field. If voltage control is used, the electric field generated in the tissue during the entire course of stimulation should be monitored. Simply using current control does not cancel the importance of considering (1) electrode size and position, which determines brain current flow pattern; and (2) electrode material and use of any buffer which determine electrochemical changes. Moreover, the two

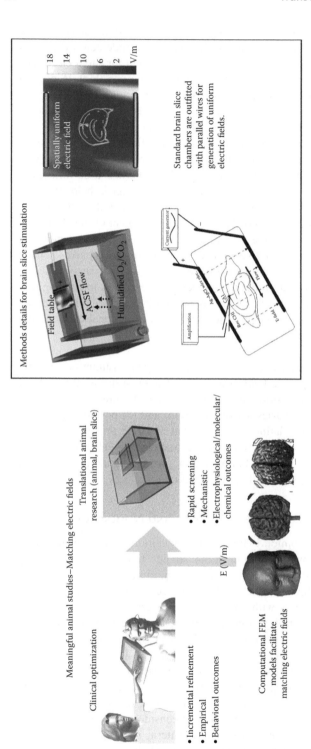

FIGURE 3.2 Animal studies on tDCS mechanism allow rapid screening of stimulation parameters and analysis of neurophysiological and molecular changes in ways not possible clinically. Meaningful translation research in animals required replication of electric fields generated clinically in animal brain/tissue. The electric field generated in the brain during tDCS is dependent on the stimulation dose (current intensity, electrode montage) and head anatomy. It is not trivial to relate externally controlled dose with internally generated electric fields (i.e., the current density in the brain is not the same as at the electrodes), but FEM computational models provide a method to do so. In the experimental design of animal studies, the electric field generated should correspond in intensity to that generated clinically; otherwise results should be applied to the clinical case with caution. In the case of in vitro brain slice studies, the replication of clinical electrical fields is experimentally straightforward with the use of two long parallel wires placed across the bath, generating uniform electric field. The single uniform electric field in the chamber can be simply calibrated, using a field-recording electrode, to the current applied to the wires. The position of the brain slice in the uniform field is not important to control; moreover, multiple slices can be screened at once.

can be interrelated as electrode degradation will make a portion of the electrode inactive causing current redistribution, while increasing electrode size (in particular electrode contact area) reduces electrochemical burden. We caution: as more diverse research groups apply increasingly sophisticated techniques to analyze the effects of DC stimulation to understand the mechanisms of tDCS, it is simultaneously necessary to apply rigor in the methods used to delivery stimulation for each animal model. At a minimum, the "dose" of stimulation needs to be reported in a manner that allows reproduction consistent with clinical rules of dose reporting and control (Peterchev et al. 2011).

Some comments on conventions used to indicate the polarity of stimulation may be useful. Firstly, in brain electrical stimulation *anode* and *cathode* terminology should always be used consistently for indicating the electrode where positive current is entering the body (anode) and the electrode where positive current is exiting the body (cathode)—there is no basis to confuse these terms in electrical stimulation literature (Merrill et al. 2005). For conventional tDCS with two electrodes, there is simply one anode and one cathode, with the anode at a positive voltage relative to the cathode. In clinical and animal studies, *anodal stimulation* or *cathodal stimulation* would indicate that a cortical region of interest (target) was nearer the anode or the cathode, respectively. In earlier animal literature the terms *surface positive* and *surface negative*, correspond to an anode or cathode, respectively, electrode placed the surface of the cortex, with the other electrode often placed on the neck or body. Considering the cortical surface, *inward current* and *outward current* are typically expected under the anode and cathode respectively (though cortical anatomy may produce deviations). When discussing *electric field*, the direction needs to be specified. In our FEM studies we use the convention that an inward current will produce a positive electric field measured from outside pointed in, while an outward current will produce a negative electric field measured from outside pointed in (Datta et al. 2008) (unless otherwise stated, it is implied that the current and electric fields are normal/orthogonal to the cortical surface, rather than tangential/parallel). *Current density* will always be in the same polarity/direction as the electric field, for example current density flow is positive inward (into the cortex) under the anode. In tissue/brain slices, though the terms anode and cathode remain unambiguous in regards to the electrodes, the electric field reference direction is arbitrary and needs to be defined. In our studies where uniform DC stimulation is applied to cortical slices, we always define the electric field as positive when the anode is on the pia surface side of the slice and the cathode on the midbrain side—while a negative electric field indicates the cathode on the pia side (Radman et al. 2009). In this way a positive electric field indicates stimulation polarity (direction) associated with clinical anodal tDCS, while negative electric field indicates a polarity associated with cathodal tDCS. Typically, in hippocampal slice studies using parallel wires, a positive electric field indicates the anode on the alveus side of CA1 (Gluckman et al. 1996; Ghai et al. 2000; Bikson et al. 2004; Ranieri et al. 2012). Finally, the term *polarizing*, or polarizing current, is used in classic animal literature and modern tDCS, and appears to refer to the use of prolonged (not pulsed) DC stimulation applied with macro-electrodes, with the polarization related to the electrodes, brain, and/or neurons.

3.1.4 DOSE CONTROL AND MEANINGFUL ANIMAL STUDIES

We emphasize caution when drawing conclusions from studies using any DC currents in animals that do not produces electric field magnitudes comparable to those generated during tDCS. These studies are valuable in suggesting mechanisms for tDCS but, just as with drugs, increasing dose beyond clinical levels (by orders of magnitude) can induce physiological changes not relevant clinically. For example, some animal studies have shown that application of DC can control neuronal process orientation and growth direction (Alexander et al. 2006; Li et al. 2008); however, both the intensity and duration of electric fields were orders of magnitude greater than tDCS. Similarly, electroporation and joule heating can be caused by electricity in general, but do not seem relevant for clinical tDCS electric fields (Bikson et al. 2009; Datta et al. 2009; Liebetanz et al. 2009). Thus, these mechanisms and related animal studies are not considered further here. Additional theories have been ventured regarding the role of concentration changes induced by DC current (e.g., iontophoresis of charged molecules/ions) (Gardner-Medwin 1983), and though intriguing, to our knowledge no quantitative analysis of plausibility, and much less experimental evidence, exists for tDCS relevant electric fields. Speculative direct electrochemical changes in the brain should not be confused with (1) established electrochemical reactions that occur at the electrode interface, which would not reach the brain using scalp electrodes (Merrill et al. 2005; Minhas et al. 2010) and (2) indirect chemical and molecular changes secondary to neuronal activation (Stagg and Nitsche 2011). We also caution against any theories that suggest violation of electroneutrality during DC stimulation (e.g., "accumulation" of positive charge near the cathode). Rather, as explained in the next section, our mechanistic considerations start with the well-established principle of membrane polarization induced by extracellular DC current flow, with all other changes secondary to this polarization. Interestingly, in this context, the coupling sensitivity for human neurons may be higher than animals.

Though important to our understanding of tDCS mechanism, most animal work on DC stimulation in the 1960s used current densities with invasive electrodes higher than used in tDCS at the scalp (most of these studies did not intend to mimic tDCS). Recent animal studies often used transcranial DC stimulation with current density at the skull higher than used in tDCS at electrodes (Fregni et al. 2007; Brunoni et al. 2011). Perhaps also motivated by magnifying effect size (and not necessarily motivated only by tDCS) many recent in vitro studies, including those by our own group, used electric fields higher than those generated clinically (Andreasen and Nedergaard 1996; Bikson et al. 2004). Because of the complexity (nonlinearity) of the nervous system function one cannot automatically assume a monotonic (more field = more response) relationship between intensity and outcome; however, in vitro studies that explore field strength-response curves indicate a surprisingly linear response curve over low intensities (Bikson et al. 2004; Reato et al. 2010), and membrane coupling constant certainly appears linear with field strength (see next). Those in vitro studies that have explicitly explored the lower electric field limit of sensitivity to fields (see Section 3.4; Francis et al. 2003; Jefferys et al. 2003; Reato et al. 2010) report statistically significant responses at <0.2 V/m, within tDCS ranges.

One example of how the use of high DC current intensities can produce effects opposite than expected at DC relevant intensities is noted. As discussed later, in a polarity-specific manner, DC fields can increase excitability and evoked responses (synaptic efficacy). But if the intensity of DC current is increased significantly, it may increase excitability to the point that the neuron generates (high frequency) discharges—the responsiveness of the very active neuron to an evoked response may then *decrease* because it is often in a refractory state. This was shown in brain slice (Bikson et al. 2004) and may explain results in animal (Purpura and McMurtry 1965a,b) using high DC current intensities.

3.1.5   OUTCOME MEASURES

The second issue to consider in the design of translational studies is the appropriateness of the outcome measure in animal models, which is not specific to tDCS and will not be discussed in detail here. In considering the use of tDCS in clinical treatment, animal models of disease can be used, not simply to validate outcomes, but to characterize mechanisms and optimize stimulation protocols (Sunderam et al. 2010; Yoon et al. 2012). One factor facilitating quantitative translational research is the noteworthy emphasis by tDCS clinical researcher to determine neurophysiological markers of tDCS including spontaneous EEG (Marshall et al. 2004; Marshall et al. 2006) and TMS motor evoked responses (Nitsche and Paulus 2000), including while screening different dosage and time-course. These generic clinical measures of "excitability" have rough animal analog in spontaneous firing rate, oscillations, and evoked responses—though "evoked responses" or oscillations of a given frequency may not have the same origin in animals and humans. Animal research in tDCS has only started to access the breadth of behavior and disease models that are available. As summarized by Brunoni et al. (2011)

> Although pre-clinical studies, including experiments with animals, are critical in developing novel human therapies, translational research also has several challenging aspects, as animal and human studies can differ in characteristics of disease (i.e., 'human disease' vs. 'experimental animal model'), definition of outcomes (especially for neurological research that often rely heavily on behavioral outcomes....

Having outlined potential pitfalls in translational tDCS studies, the need and value of well-designed animal research remains evident. Contributions of animal studies to our current understanding of tDCS and their importance as tDCS becomes more sophisticated are discussed in the next sections.

3.2   SOMATIC DOCTRINE AND NEED FOR AMPLIFICATION

Since 2000 (Nitsche and Paulus 2000; Ardolino et al. 2005; Fregni et al. 2005, 2007), there has been rapid acceleration in the use of tDCS in both clinical and cognitive-neuroscience research, encouraged by the simplicity of the technique (two electrodes and a battery powered stimulator) and the perception that tDCS protocols can be simply designed by placing the anode over the cortex to "excite," and the cathode over cortex to "inhibit." Starting with the consideration of single neurons and acute effects, in this section we (1) define this simplistic and ubiquitous "somatic doctrine"; (2) consider its origin in classic animal studies; and (3) describe modern efforts to quantify somatic polarization using brain slices; this in turn leads to an appreciation of the need for amplification mechanisms. Discussions focused specifically on mechanism of lasting changes following tDCS (e.g., plasticity) and consideration of network activity are left for the next sections.

3.2.1 NEURONAL POLARIZATION

DC stimulation with electrodes on the scalp leads to current flow across the brain, which in turn, results in polarization of cell membranes when some of this current crosses the membrane (Ohms' law). Flow into a specific compartment of membrane will result in local membrane hyperpolarization, and flow out of another compartment of membrane will result in local membrane depolarization (Andreasen and Nedergaard 1996; Bikson et al. 2004). It is fundamental to emphasize (especially as this concept is overlooked in clinical literature) that there is no such thing a purely depolarizing or purely hyperpolarizing weak DC stimulation—the physics of electrical stimulation dictate that any neuron exposed to extracellular DC stimulation will have some compartments that are depolarized and some that are hyperpolarized (Chan et al. 1988; Bikson et al. 2004). Which compartments are polarized in which direction depends on the neuronal morphology relative to the DC electric field. Simplistically, for a pyramidal type neuron, with a large apical dendrite pointed toward the cortical surface, a surface anode (positive electrode, generating a cortical inward current flow) will result in somatic (and basal dendrite) depolarization and apical dendrite hyperpolarization (Radman et al. 2009). For this same neuron, a surface cathode (negative electrode, generating cortical outward current flow) will result in somatic (and basal dendrite) hyperpolarization and apical dendrite depolarization. tDCS protocols based on the "somatic doctrine" simply assume that somatic polarization determines all relevant functional/clinical outcomes. This consensus of a generic excitation/inhibition by anodic/cathodic stimulation underpins a majority of clinical tDCS study design (Figure 3.2)—combined with the concept that brain (dis)function is a sliding scale of excitability that can be controlled in this fashion.

3.2.2 MODULATION OF EXCITABILITY, POLARITY-SPECIFIC EFFECTS

The application of DC stimulation (often as short pulses) to the neuromuscular system dates to the origin of batteries (indeed, as electrical energy sources must predate any electrical devices, human and animals made natural targets). The review of the history of DC stimulation is well beyond the scope of this chapter, but some highlights help position the origin of the "somatic doctrine." In 1870, Fritsch and Hitzig may have been the first to show that application of a positive current to the cortex had stimulating effects, while a negative current inhibits (a finding that itself contributed to early understanding that the cortex is electrically excitable) (Fritsch 1870; Carlson and Devinsky 2009). Terzuolo and Bullock (1956) and Creutzfeldt et al. (1962) helped establish that ongoing discharge frequency is enhanced by surface-positive current and decreased by surface-negative currents (it is curious how the debate over the role of endogenous electric fields is reflected in these early works in which Creutzfeldt et al. suggested they are epiphenomena while Terzuolo and Bullock suggested a physiological role—indeed modern work with weak transcranial stimulation has provided the strongest clinical evidence for a plausible role). The concept that threshold for electric field sensitivity would be "lower for modulation of the frequency of an already active neuron than for excitation of a silent one" was thus well established, with early observation of changes in discharge rate with fields as low as 0.8 V/m (Terzuolo and Bullock 1956).

Note, the electric fields induced by tDCS are considered far too weak to trigger action potentials in quiescent neurons (compare >100 V/m induced by TMS to <1 V/m by tDCS). It is thus not surprising that early animal studies on lower-intensity DC stimulation addressed modulation of ongoing normal or pathological neuronal firing rate, as well as evoked response. In the early 1960s, animal studies by Bindman and colleagues (Bindman et al. 1962, 1964) confirmed polarity-specific changes in discharge rate and further showed excitability changes that are both cumulative with time and out-last stimulation (discussed in next section)—this group went on to explore long-duration stimulation in early psychiatric treatment. It was also recognized that the direction of changes in discharge rate were consistent with presumed somatic polarization (and dependent on the orientation of the apical dendrites). Furthermore, animal studies in the 1950s and 1960s examining control of epileptic discharges (Purpura et al. 1966), evoked responses (Creutzfeldt et al. 1962; Bindman et al. 1964; Purpura and McMurtry 1965a,b), lasting effects and related molecular changes (see also next; Gartside 1968a,b), also reinforced the concept that the direction of somatic polarization determined the net effect on excitability/functional outcomes (Figure 3.3).

3.2.3 QUANTIFYING POLARIZATION WITH COUPLING CONSTANTS

A specific and predictive understanding of tDCS requires quantitative model, beginning with quantification of somatic (and dendritic) polarization during tDCS. In the 1980s, Chan and colleagues (Chan and Nicholson 1986; Chan et al. 1988) used electrophysiological recordings from turtle cerebellum and analytical modeling to quantify polarization under quasi-static (low-frequency sinusoid electric fields)—these seminal studies identified morphological determinants of neuron sensitivity to applied DC fields. We extended this work to rat hippocampal CA1 neurons and then to cortical neurons with the approach of quantifying cell-specific polarization by weak DC fields using a single number—the "coupling constant" (also called the "coupling strength" or "polarization length"). We assumed that for weak electric fields (stimulation intensities too weak to significantly activate voltage gated membrane channels, and well below action potential threshold) that the resulting membrane polarization at any given compartment, including the soma, is linear with stimulation intensity. For uniform electric fields, the membrane potential polarization can be expressed as $V_{tm} = G * E$ where V_{tm} is the polarization of the compartment of interest (in: V), G is the coupling constant (in: V per V/m, or simply: m) and E is the electric field (in: V/m) along the primary dendritic axis. For rat hippocampus and cortical neurons the

The principle of "somatic doctrine" in basic tDCS montage design

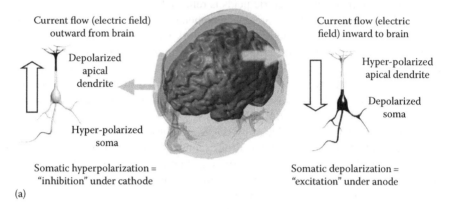

Current flow (electric field) outward from brain

Depolarized apical dendrite

Hyper-polarized soma

Current flow (electric field) inward to brain

Hyper-polarized apical dendrite

Depolarized soma

Somatic hyperpolarization = "inhibition" under cathode

Somatic depolarization = "excitation" under anode

(a)

Quantification of somatic (and dendrite) polarization under DC fields

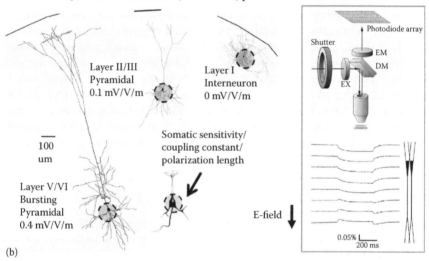

Layer II/III Pyramidal 0.1 mV/V/m

Layer I Interneuron 0 mV/V/m

100 um

Somatic sensitivity/ coupling constant/ polarization length

Layer V/VI Bursting Pyramidal 0.4 mV/V/m

E-field

Photodiode array

Shutter

EM

DM

EX

0.05%

200 ms

(b)

FIGURE 3.3 The principle and quantification of the somatic doctrine. (a) The somatic doctrine simplifies tDCS design by assuming inward current flow under the anode, leading to somatic depolarization, and a generic increase in excitability and function. Under the cathode, an outward current leads to somatic hyperpolarization and a generic decrease in excitability and function. (b) Modern efforts to quantify somatic polarization in animal models have confirmed some aspects of the somatic doctrine, at least under specific controlled and tested conditions, but indicated that the polarization produced by tDCS would be small.

somatic coupling constant is in the range of 0.1–0.3 mV polarization per V/m electric field (Figure 3.2c; Bikson et al. 2004; Deans et al. 2007; Radman et al. 2009). For ferret cortical neurons the coupling is similarly ~0.25 mV per V/m (Frohlich and McCormick 2010). For humans, assuming scaling of sensitivity with total neuronal length (Joucla and Yvert 2009) somatic depolarization per V/m might be higher than in animals.

The maximal depolarization occurs when the electric field is parallel with the somatic-dendritic axis which corresponds to radial to the cortical surface, while electric field orthogonal to the somatic-dendritic axis do not produce significant somatic polarization (Chan et al. 1988; Bikson et al. 2004; but see axon terminal polarization in Section 3.3.4). The somatic coupling strength is roughly related to the size of the cell and the dendritic asymmetry around the soma (Svirskis et al. 1997; Radman et al. 2009) making pyramidal neurons relatively sensitive. For cortical pyramidal neurons, the typical polarity of somatic polarization is consistent with the "somatic doctrine" (e.g., positive somatic depolarization for positive electric field). The polarity of the coupling constant is inverted (using our field direction convention) for CA1 pyramidal neurons due to their inverted morphology. Using experimental and modeling techniques the coupling constant of dendritic compartments can also be investigated; generally the maximal polarization is expected at dendritic tufts (Bikson et al. 2004), but should not exceed, in animals, ~1 mV polarization per V/m electric field (Chan et al. 1988; Radman et al. 2007b, 2009).

If tDCS produces a peak electric field of 0.3 V/m at 1 mA (with the majority of cortex at reduced values) then the maximal somatic polarization for the most sensitive cells is ~0.1 mV. Similarly, for 2 mA tDCS stimulation, the most sensitive cells in the brain region with the highest electric field would have somatic polarization of ~0.3 mV. Far from "closing the book" on tDCS mechanism, work by our group and others quantifying the sensitivity of neuron to weak DC fields, has raised questions about how such minimal polarization could result in functional/clinical changes especially considering that endogenous "background" synaptic noise can exceed these levels. In recent years, motivated by increased evidence that transcranial stimulation with weak currents has functional effects, as well as ongoing questions about the role of endogenous electric fields which can have comparable electric fields, the mechanisms of amplification have been explored in animal studies; we organize these efforts by nonlinear single cell properties (discussed next) as well as synaptic processing and network processing (addressed in the next sections).

3.2.4 Amplification through Rate and Timing

At the single cell level the most obvious nonlinear response that could provide a substrate for acute amplification is the action potential. As the electric fields induced by tDCS are far too weak to trigger action potentials (AP) in neurons at rest (i.e., ~15 mV depolarization from rest to AP threshold, one can consider instead modulation of ongoing AP activity. At the single cell level we (1) consider acute implication through the rate of action potential generation (rate effects) and (2) develop the concept of amplification through change in the timing of action potential (timing effects). As already discussed classic animal studies on weak DC stimulation addressed the rate of change of spontaneous action potential discharge rate in many systems changes roughly linearly with membrane polarization. The amplification (gain) would relate to sensitivity of discharge rate to membrane polarization. Terzuolo and Bullock (1956) reported a detectable change in discharge rate down to 0.8 V/m, and this detection threshold would likely decrease with longer experiments. Assuming a 0.6 V/m peak electric field during 2 mA tDCS leading to ~0.2 mV somatic polarization, and that across animal studies changes in firing rates of 7 Hz per mV membrane polarization are reported (Carandini and Ferster 2000), a change in firing rate of ~1.5 Hz is plausible. The aforementioned consideration is for isolated neurons, for neurons in an active network (see Section 3.4).

In 2007, we proposed that changes in action potential timing (rather than discharge rate) can act to amplify the effects of weak polarization (Radman et al. 2007b). Specifically we showed in brain slice recording and in a simple neuron model that the resulting change in timing was simply the induced membrane polarization times the inverse of the ramp slope (Figure 3.4a and b); the inverse of the ramp slope is thus a "gain/amplification" term because the more shallow a ramp, the larger the timing change for given small polarization (Figure 3.4c and d). For example, assuming as aforementioned a ~0.2 mV somatic polarization during 2 mA tDCS, then in response to a 1 mV/ms electric field, timing would change by 0.3 ms. We also extended these findings to AC fields (Radman et al. 2007a). Both the coupling sensitivity and the

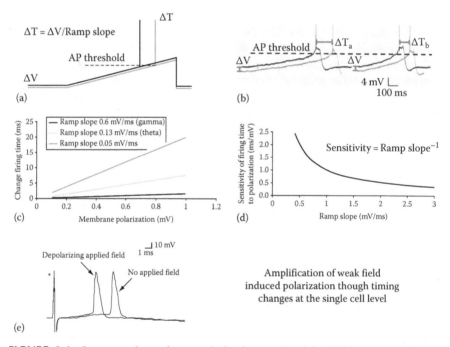

FIGURE 3.4 Incremental membrane polarization produced by tDCS may significantly affect the timing of AP in response to a ramp (synaptic, oscillation) input. Moreover, the amplification of effect (change in timing per change in membrane polarization) increased for more gradual input. (a) Schematic illustrates the principle of timing amplification. (From Radman, T. et al., *J. Neurosci.*, 27, 3030, 2007b.) (b) The timing amplification was validated in hippocampal CA1 neurons using intracellular injected current ramp of various slopes. (c and d) The timing change increased with membrane polarization with a sensitivity (amplification) that is the inverse of the input ramp slope. The amplification would function during processing of incoming synaptic input including oscillations. (e) Demonstration of timing change in response to an incoming EPSP. nA incremental depolarization produced by direct current led to significant change in action potential timing in response to a synaptic input.

timing changes were confirmed by Anastassiou et al. (2010) using a more complex model. Though the principle of timing amplification generalizes to other cell types and to synaptic input (Figure 3.4e; Bikson et al. 2004), the simple amplification equation (Figure 3.4a) makes specific assumptions about membrane dynamics (Radman et al. 2007a) that may not extend to all cell types (Radman et al. 2009).

Additional mechanisms of amplification at the single cell level remain an area of active investigation, especially when considering how exposure to long-duration fields (e.g., minutes as used in tDCS) may produce cumulative effects not observed during short term application (e.g., molecular changes; Gartside 1968b; Fritsch et al. 2010; Ranieri et al. 2012). It remains an open and key question how prolonged (minutes) polarization of both the soma and dendrites can then trigger specific chemical and molecular cascades, thereby leading to the induction of plasticity (see next section).

3.2.5 SEIZURE THRESHOLD AND MODULATION

The coupling constant also provides insight into the safety of tDCS in regards to triggering of seizures during stimulation. Whereas TMS produces 100 V/m pulsed electric fields that are suprathreshold, tDCS results in a static electric field <1 V/m at 2 mA producing <1 mV of polarization. Animal studies indicate that only application of DC fields >20 V/m (corresponding to >60 mA tDCS) would trigger action potential in the most sensitive quiescent cortical cells (Radman et al. 2009) while electric fields of ~100 V/m (corresponding to >500 mA tDCS) in the somatic depolarizing direction can trigger epileptiform activity in hippocampal slices (Bikson et al. 2004). This threshold would decrease for already active neurons. In brain slices, weak DC stimulation (on the order of 1 V/m) can modulate ongoing epileptiform activity (Gluckman et al. 1996; Ghai et al. 2000; Durand and Bikson 2001; Su et al. 2008; Sunderam et al. 2010), such that the cathodal tDCS may control ongoing seizures while anodal tDCS may aggravate seizure activity. In a polarity-specific fashion (consistent with somatic polarization) DC stimulation can also modulate the propagation of epileptiform activity in slices (Gluckman et al. 1996), spreading depression in vivo (Liebetanz et al. 2006a), and perhaps clinical epileptiform activity (Varga et al. 2011). Though the acute (during stimulation) effects of weak DC currents on epileptiform activity are well established in animal models, it remains an open question if and how prolonged DC stimulation modulates seizure propensity. Animal studies suggest that prolonged cathodal DC stimulation can be anticonvulsant, while reports of the effect of anodal tDCS are mixed (Hayashi et al. 1988; Liebetanz et al. 2006b), and pilot human studies suggest an antiepileptic effect.

3.2.6 LIMITATIONS OF THE SOMATIC DOCTRINE

The most evident limitation of the somatic doctrine is precisely what cell compartments it ignores: the dendrites and axons. While the basal dendrite will be polarized similarly as the soma, the apical dendrite will be polarized in the opposite direction (Figure 3.3; Andreasen and Nedergaard 1996; Bikson et al. 2004). The dendrites are electrically excitable. Animal studies with high-intensity applied DC fields (~100 V/m) have shown that with sufficiently strong stimulation, active processes (spikes) can be triggered in the dendrites (Chan et al. 1988; Wong and Stewart 1992; Andreasen and Nedergaard 1996; Delgado-Lezama et al. 1999). Even if the electric fields induced during tDCS are not sufficient in themselves to trigger dendritic spikes, the role of dendritic polarization during tDCS remains an open question especially when considering processing of synaptic input (next section).

It is well established that axons are sensitive to applied electric fields; the magnitude and direction of polarization is a function of neuronal and axonal morphology (Bullock and Hagiwara 1957; Takeuchi and Takeuchi 1962; Salvador et al. 2010). While the axon initial segment would likely be polarized in the same direction as the soma (Chan et al. 1988), for long axons this is not necessarily the case. Thus it is useful to separately consider the axon initial segments (within a membrane space constant of the soma) and more distal axonal processes, which can be further divided into "axons-of-passage" and afferent axons with terminations (discussed in the next section). Notably, for long straight axons-of-passage (e.g., Peripheral Nervous System, PNS) cathodal stimulation will be more effective than anodal stimulation in inducing depolarization (opposite to the somatic doctrine; Bishop and Erlanger 1926). It has been shown that lasting changes can be induced in PNS axons in humans (so by implication in CNS axons independent of somatic actions) and also, in brain slices, that weak DC fields can produce acute changes in CNS axon excitability (presynaptic/antidromic volley) (Jefferys 1981; Bikson et al. 2004; Kabakov et al. 2012). An important role for axon terminal polarization is introduced in the next section.

A presumption of the somatic doctrine is that under the anode currents are radial and inward through the cortex, while under the cathode current is radial and outward (Figure 3.3). However, high-resolution modeling suggests that in convoluted human cortex, current is neither unidirectional nor dominantly radial. Though the "somatic doctrine" is based only on radially directed electrical current flow (normal to the cortical surface), during tDCS significant tangential current flow is also generated (along the cortical surface). Indeed, recent work by our group suggests tangential currents may be more prevalent between and even under electrodes (Figure 3.1). As discussed next, tangential currents cannot be ignored in considering the effects of tDCS. Moreover, due to cortical folding the direction or radial current flow under tDCS electrodes is not consistent, meaning there are cluster of both inward (depolarizing) and outward (hyperpolarizing) cortical current flow under either the anode or the cathode! Due to the cortical convolutions, current is not unidirectional under electrodes thus under the cathode there may be isolated regions of inward cortical flow, and in those regions neuronal excitability may increase (Creutzfeldt et al. 1962). The relative uniformity of direction across a given patch of cortex depends on the electrode montage, with electrode across the head producing the most consistent polarization under each electrode (Turkeltaub et al. 2011) and closer electrodes, such as the classic M1-SO (anode on motor strip, cathode on contralateral supraorbital area) montage, producing bidirectional current flow with a slight directionality preference *on average* in some regions under the electrodes (Figure 3.1). This seems puzzling in light of the dependence on the somatic doctrine in tDCS montage design and study interpretation. The role of tangential and bidirectional current flow is addressed in the next two sections.

3.3 PLASTICITY, SYNAPTIC PROCESSING, AND A "TERMINAL DOCTRINE"

The clinical need for lasting changes by tDCS relates to the impracticality of constant stimulation with noninvasive technology (i.e., wearing a stimulation cap all the time). The desire for lasting change means tDCS should influence plasticity during or after stimulation in cognitive/therapeutic relevant way (Yoon et al. 2012). This section addresses the contribution of animal studies to understanding plasticity generated by weak DC electric fields. The contribution of early translational studies to tDCS protocols is notable as Bindman and colleagues (Bindman et al. 1962) recognized the importance or prolonged DC stimulation to produce lasting effects (>5 min), which informed their early work in tDCS of psychiatric disorders (Costain et al. 1964; Redfearn et al. 1964) and the multi-minute stimulation required in the Nitsche and Paulus (2000) report, that in turn established the use of prolonged stimulation across modern tDCS studies. The need for prolonged (minutes) stimulation to induce plasticity (mechanism) and protocols for optimizing long-lasting changes (clinical utility) remains a central question in tDCS research (Monte-Silva et al. 2010). Marquez-Ruiz recently summarized (Marquez-Ruiz et al. 2012)

> When tDCS is of sufficient length, synaptically driven after-effects are induced. The mechanisms underlying these after-effects are largely unknown, and there is a compelling need for animal models to test the immediate effects and after-effects induced by tDCS in different cortical areas and evaluate the implications in complex cerebral processes.

Animal studies in the 1960s also helped established that weak DC current produces plastic changes (a lasting physical change in the brain rather than a "reverberating circuit" of activation) (Gartside 1968a). As noted earlier, early animal studies also contributed to establishing the "somatic doctrine" in tDCS but modern clinical studies on tDCS have suggested that changes in excitability are not necessarily polarity specific (Marquez-Ruiz et al. 2012), or monotonic with intensity such that cathodal or anodal stimulation can produce variable effects depending on intensity, duration, or underlying activity. In the past decade, animal and computational studies are beginning to address these issues. Both in humans and animal studies changes in evoked (synaptically mediated) neurophysiological responses are considered reliable hallmarks of plastic changes that could support behavioral or clinical lasting changes (and are thus a focus of this section), though one recent report in rabbits indicated this was not simply the case (Marquez-Ruiz et al. 2012).

3.3.1 Paradigms for DC Modulation of Synaptic Efficacy

In the previous section it was discussed that as tDCS electric fields are subthreshold (too weak to trigger action potential in quiescent neurons); their acute role is thus truly neuromodulation through either rate or timing effects. Weak DC stimulation may generate plasticity through different paradigms, which are not necessarily exclusive:

1. Membrane polarization may trigger plastic changes in a manner independent of any ongoing synaptic input or action potential generation (i.e., simply holding the membrane at an offset polarization initiates changes). Though in a cortical brain slice model (with no background activity), weak polarization was not sufficient to induce plastic changes (Fritsch et al. 2010).
2. Changes in action potential discharge rate or timing, secondary to neuronal polarization, where the firing change (which is dependent on many factors including the direct polarization) actually determines plasticity. "There is some evidence that a determining factor in producing long-lasting after-effects is the change in the firing rate of neurones rather than the … current flow that produces the changes" (Bindman et al. 1964). Though classic animal studies indicated weak DC stimulation is sufficient to induce plastic changes (Gartside 1968b), it is important to note that polarization would affect a network of neurons such that increased firing in afferents would in fact increase synaptic input (see next point).
3. Polarization of the membrane in combination with ongoing synaptic input. Though it is established that weak DC stimulation can lead to acute and lasting changes in synaptic efficacy (see next), the specific hypothesis here is that the generation of plasticity requires synaptic coactivation during DC stimulation. Evidence from brain slices (Fritsch et al. 2010) shows potentiation under anodal stimulation only during specificity matched patterns (frequencies) of synaptic input. In a rabbit study, DC was combined with repeated somatosensory stimulation, leading to acute polarity-specific changes, and lasting changes for the cathodal case (Marquez-Ruiz et al. 2012). If dependent on combined polarization and synaptic input, then synapse specific changes are plausible. If one assumes DC exerts a postsynaptic priming effect (polarization of soma/dendrite) than coactivation of afferent synaptic input could be conceived as Hebbian reinforcement (except postsynaptic AP may or may not be required). Clinically this plasticity paradigm is broadly analogous to combining tDCS with a cognitive task or specific behavior that coactivates a targeted network or combining tDCS with TMS. Indeed, work showing the importance of coactivation in cortical slice (Rioult-Pedotti et al. 1998; Hess and Donoghue 1999), influenced Nitsche and Paulus (2000) in developing tDCS. However, though TMS is used to probe the effects of tDCS, it turned out not to be necessary to apply it during tDCS. We would note that, unlike in brain slice and anesthetized animal models, the human cortex is constantly active such that tDCS is always applied on conjunction with ongoing synaptic input.
4. Polarization of the membrane in combination with ongoing activity that itself is independently leading to potentiation (i.e., modulation of ongoing plasticity). For example, in the aforementioned rabbit study, DC stimulation modulated ongoing synaptic habituation, a model of associative learning (Marquez-Ruiz et al. 2012). Clinically this fourth paradigm is analogous to combing tDCS with learning/training (Bolognini et al. 2010). Evidence from brain slices (Ranieri

et al. 2012) shows DC modulation of LTP induced by tetanic stimulation, in a polarity-specific manner apparently opposite to the somatic doctrine (when one considers that CA1 neuron morphology results in "anodal" stimulation producing somatic hyperpolarization). Using intracellular current injection, Artola et al. (1990) showed that depending on the level of polarization of the postsynaptic neuron, the same tetanic stimulation can induce LTD or LTP.

A separate classification includes lasting changes that are (1) "synaptic": occur at synaptic processes (e.g., increased vesicle release, receptor density) and are blocked by synaptic antagonists; versus (2) "nonsynaptic": occur independent of synaptic processes (e.g., membrane polarization). Though the synapse is typically considered the locus of plastic changes, so-called nonsynaptic changes have been noted after DC stimulation in peripheral axons away from any synapse; importantly, in this case, cathodal stimulation inducing potentiation (Ardolino et al. 2005) which is consistent with cathodal stimulation induced preferential depolarization in long axons. Clinically, the question of "synaptic" versus "nonsynaptic" origin of tDCS modulation in the CNS has been explored and debated by distinguishing modulation of TMS versus TES evoked potentials—both of which, in our opinion, have variable origin (depending on methods). Moreover, using either TES or TMS, the presence of and role of background synaptic activity in priming excitability during tDCS, regardless of "synaptic" or "nonsynaptic" locus, muddles any effort to disambiguate the mechanism along these lines. However, in brain slice models, where background synaptic activity is absent, synaptic (orthodromic) and nonsynaptic (axon, antidromic) can be precisely isolated—see discussion on axon effects given earlier. It is important to note that as far as outcome, in the CNS changes of "nonsynaptic" origin would be expected to affect synaptic processing (Mozzachiodi and Byrne 2010).

When considering the complexity of (multiple forms of) tDCS plasticity, the need for animal models is evident. Animal models allow for synaptic efficacy to be quantitatively probed with pathway specify—the importance of which we discuss next. The mechanisms of plasticity can be analyzed using specific pharmacology not practical in people (for toxicity), not to mention the ability to resect tissue for detailed cellular and molecular analysis (Islam et al. 1995; Yoon et al. 2012). Though limited to timescales of hours, the use of brain slice further facilitates imaging, precise drug concentration control, control of the background level and nature of ongoing activity (from quiescent, to transient activation at specific frequencies, to oscillations, to epileptiform) and, especially relevant for tDCS, the control of electric field orientation relative to slice (Figure 3.5). It may not be prudent to revert to a one-dimensional "sliding scale of excitability" explanation where anodal/cathodal tDCS increases/decreases "function" leading to lasting increases/deceases in generic synaptic plasticity, which are then related to cognitive/behavior changes—this approach seems simplistic and unlikely to ultimately advance tDCS sophistication and efficacy.

3.3.2 RELATION WITH TETANIC STIMULATION INDUCED LTP/LTD

Animal studies using tetanic stimulation to induce long-term potentiation/depression (LTP/LTD) have suggested multiple forms of plasticity, involving distinct pre- and

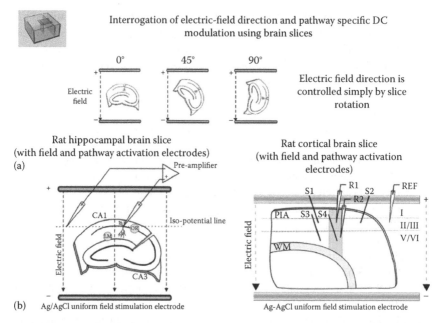

FIGURE 3.5 Further advantages of the brain slice preparation in studying mechanisms of weak DC stimulation. (a) A discussion in the text, the direction of the applied electric field relative to the somato-dendritic axis can be precisely controlled. (Adapted from Bikson, M. et al., *J. Physiol.*, 557(Pt 1), 175, 2004.) The effects of DC current on brain function may vary with orientation. (b) Synaptic function/efficacy is not "one thing," rather there are multiple distinct synaptic afferent to any brain region which can be evaluated in isolation in brain slices. The effects of DC current on synaptic function may be highly pathway specific.

post-synaptic mechanisms, on distinct time scales. When comparing the thousands of studies on tetanic LTP compared with the <50 animal studies on DC induced plasticity, one may speculate that, despite recent progress, there is much to investigate about (multiple potential forms) of weak DC stimulation induced plasticity (Bindman and colleague's conclusion in 1964 that "at the moment we are not in a position to discuss the way in which polarizing currents acts on neurons to alter long-term excitability" comes to mind [Bindman et al. 1964]). LTP/LTD induced by tetanic stimulation and by DC current may, not surprisingly, share some common molecular substrates (Gartside 1968b; Islam et al. 1995; Ranieri et al. 2012). It is remarkable that a decade before the lauded discovery of Long Term Potentiation by trains of suprathreshold pulses by Bliss and Lomo (1973), animal studies had shown lasting changes in excitability following DC stimulation lasting up to hours (Bindman et al. 1962) and moreover had begun established plastic changes and started to address the underlying molecular mechanisms (Gartside 1968b) and translating results to humans! It was recently suggested that tDCS shares some of the classical molecular mechanisms associated with tetanic stimulation LTP/LTD (Marquez-Ruiz et al. 2012) including adenosine-elicited accumulation of cAMP (Hattori et al. 1990) inducing increased protein kinase C and calcium levels (Islam et al. 1994, 1995). The wealth of techniques and tools developed by the "cottage industry" of tetanic stimulation LTP have yet to be fully leveraged to dissect the mechanisms of tDCS—but this seems a matter of time. In the context of translational importance, it is also interesting that protocols using tetanic stimulation in animals have influenced the design of TMS protocols (LTP/LTD, theta-burst, etc.).

3.3.3 WHICH COMPARTMENTS ARE INFLUENCED BY DC STIMULATION: SOMA AND DENDRITES

As discussed earlier, we assume that the actions of DC stimulation initiate with membrane polarization, with all other (complex) changes secondary to this polarization. We noted that DC polarization will influence all neurons in areas of the brain with current flow, with equal portions membrane depolarization and hyperpolarization. Our research is thus focused on which membrane compartments (soma, dendrites, axons process, axon terminals) when polarized by weak DC stimulation are relevant from both the perspective of locus of change and mechanism. Can the somatic doctrine be used to predict plasticity changes? Or is plasticity related to polarization of specific dendritic or axonal compartments?

If one considers the lasting effects of tDCS to be generally analogous to long-term potentiation then tDCS effects how information is processed by neurons though altered synaptic efficacy. Though the soma is important for integration of synaptic input, the dendrites evidently play a central role in synaptic processing and in the induction of plasticity. Several animal and clinical studies have implicated processes linked to the dendrites in tDCS (e.g., glutamatergic receptors like n-methyl-D-aspartic receptor, NMDAR) (Liebetanz et al. 2002; Nitsche et al. 2003; Ranieri et al. 2012; Yoon et al. 2012). A key question is thus: as half the dendrite will be polarized in the same direction at the soma and half of the dendrite will be polarized in the opposite direction (Figure 3.3), how do polarity-specific changes arise? Are changes in synaptic processing/plasticity always consistent with the somatic doctrine? As summarized next, the answer seems to be: "it depends."

Early work probing evoked responses in animal models indicated modulation in excitability, with the direction of evoked response change consistent with the somatic doctrine (Creutzfeldt et al. 1962; Bindman et al. 1964) though Bishop and O'Leary (1950) already noted deviations. Recent studies aimed at developing and validated animal models of transcranial electrical stimulation have shown modulation of TMS evoked potential and visual evoked potentials consistent with the somatic doctrine (Cambiaghi et al. 2010, 2011). In a pioneering work using uniform electric fields in brain slices, Jefferys showed acute modulation of evoked responses in the dentate gyrus of hippocampal slices when electric fields were parallel to the primary target cell dendritic axis, with polarity-specific changes consistent with somatic polarization, and no modulation when the electric field was applied orthogonal to the primary dendritic axis (Jefferys 1981). The precise control of electric field angle is possible in brain slices and leveraged in future work.

In Bikson et al. (2004) we used the hippocampal slice preparation, which was initially conceived as a series of straightforward experiments to confirm the validity of the somatic doctrine in predicting acute changes in excitability—to our surprise we found several deviations. Optical imaging with voltage sensitive dyes provided direct evidence that DC electric fields always produces bimodal polarization across target neurons such that somatic depolarization is associated with apical dendrite hyperpolarization, and vice versa—yet over longer timescales interactions across compartments were observed. In addition, for synaptic inputs to the soma and basal dendrite, we reported modulation consistent with the somatic doctrine (considering the inversion relative). But for strong synaptic input on to the apical dendritic tuft both DC field polarities enhanced synaptic efficacy—such that dendrite depolarization with somatic hyperpolarization also enhanced synaptic efficacy. Also in hippocampal slices, both Kabakov et al. (2012) and Ranieri et al. (2012) reported modulation of synaptic efficacy in a direction opposite to that expected from the somatic doctrine (again noting inversion of dendrite morphology in CA1 pyramids relative to cortex). In these cases, one may speculate the apical dendrite depolarization (despite somatic hyperpolarization) determines the direction of modulation (Bikson 2004, p. 74); though Kabakov et al. (2012) provides evidence suggesting dendritic polarization effects the magnitude but not direction of modulation. As noted, in cortical slices by Fritsch et al. (2010), modulation of evoked responses is indeed consistent with the somatic doctrine—a finding we have confirmed for four distinct afferent cortical synaptic pathways. These variations across animal studies could be simply ascribed to different in region/preparation, timescale (acute, long-term), and different forms of plasticity (BDNF dependent/independent), but this is speculative and provides little insight into tDCS. Rather, in attempt to reconcile these findings in a single framework, we site evidence for and define the "terminal doctrine" to compliment the "somatic doctrine."

3.3.4 WHICH COMPARTMENTS ARE INFLUENCED BY DC STIMULATION: SYNAPTIC TERMINALS

In the 2004 study (Bikson et al. 2004) we also investigated the effects of tangential fields on synaptic efficacy—tangential fields are oriented perpendicular to the primary somato-dendritic axis, so are expected to produce little polarization (which we directly confirmed with intracellular recording). Electric fields applied tangentially were as effective at modulating synaptic efficacy as radially directed fields. The afferent axons run tangentially, so we speculated that they might be the targets of stimulation. Exploring different pathways we found that axon pathways with terminal pointed toward the anode were potentiated, while axon pathways with terminals pointed toward the cathode were inhibited. Kabakov et al. (2012) reported similar pathway specific dependence summarizing "the fEPSP is maximally suppressed when the AP travels toward the cathode, and either facilitated or remains unchanged when the excitatory signal [AP] propagates toward the anode." In addition, Kabakov et al. (2012) observed changes in paired-pulse facilitation potentially consistent with pre-synaptic vesicular glutamate release. We recently confirmed a similar directional sensitivity in cortical slices across four distinct pathways (Figure 3.5) where electric field applied tangentially to the surface (and so producing minimal somatic polarization) (Radman et al. 2009), modulated synaptic efficacy. Interestingly, an in vivo study suggested axonal regrowth (as well as dendritic growth) in tDCS mediated neuroplasticity after cerebral ischemia (Yoon et al. 2012).

A role for presynaptic modulation during DC stimulation is indeed not surprising and historically noted. Purpura and McMurtry (1965a) observed

> although the [somatic] membrane changes produced by transcortical polarization current satisfactorily explains alterations in spontaneous discharges and evoked synaptic activities in [pyramidal tract] cell, it must be emphasized that the effects of polarizing current on other elements constituting the 'pre-synaptic', interneuronal pathway to [pyramidal tract] cells also appear to be determinants of the overt changes observed in [pyramidal tract] cells activities.

Bishop and O'Leary (1950) not only quantified presynaptic effects during DC stimulation in animals, they noted that presynaptic effects would complicate the interpretation of postsynaptic changes as well as themselves induce long-lasting after-effects.

It is well established that cellular process terminals including axon terminals are especially sensitive to electric fields as a result of their morphology (DelCastillo and Katz 1954; Bullock and Hagiwara 1957; Hubbard and Willis 1962a,b; Takeuchi and Takeuchi 1962; Awatramani et al. 2005) and that terminal polarization can modulate synaptic efficacy (independent of target soma polarization) (DelCastillo and Katz 1954; Bullock and Hagiwara 1957; Hubbard and Willis 1962a,b; Takeuchi and Takeuchi 1962; Awatramani et al. 2005). Moreover, this modulation is cumulative in time and endures after stimulation if stopped (Hubbard and Willis 1962a,b); a temporal profile noted in classic DC experiments (Bindman et al. 1964) and suggesting the possibility for plasticity. The direction of modulation in brain slice studies consistently suggests that terminal hyperpolarization enhanced efficacy, while depolarization inhibited efficacy. Paired-pulse analysis in a rabbit model suggested tDCS influences presynaptic sites (Marquez-Ruiz et al. 2012). Our proposed "terminal doctrine" postulates that afferent synaptic processes oriented toward the cathode (or more specifically parallel with the direction of electric field) will be potentiated (due to synaptic terminal hyperpolarization), while processes oriented toward the anode (or specifically antiparallel with the electric field) will be inhibited. As tDCS induced significant tangential fields (Figure 3.1), the role of terminal polarization (independent of the "somatic doctrine") remains a compelling and open question especially when taken together with the need for amplification and the role of synapses in plasticity.

This proposal of a "somatic doctrine" versus "terminal doctrine" can be conceptualized as generically analogous to the pre-/post-synaptic debate in tetanic stimulation induced LTD/LTP (Artola et al. 1990); and as with tetanic stimulation induced LTD/LTP, both mechanisms are likely to play a role. It is important to note that current crossing the grey matter is rarely purely radial or tangential, such that simultaneous somatic and terminal polarization is broadly expected. Even in the brain slice afferent axons and the target neurons are not perfectly orthogonal, which may explain some of the divergent findings in hippocampal brain slices noted earlier. During tDCS because of the complexity of current flow across the gray matter

(Figure 3.1) the situation is still more complex, especially considering that whether the terminal doctrine predicts excitation or inhibition depends on the direction of incoming axons. Perhaps careful considering on brain current flow patterns combing with extending thinking beyond the simple somatic doctrine to include both the role of (oppositely polarized) dendrite, axons, and axons terminals can reconcile divergent clinical findings showing inversion of classical direction effects (Nitsche and Paulus 2000; Ardolino et al. 2005) or direction-neutral effects (Nitsche et al. 2003). We emphasize that given the complexity of plasticity paradigms and stimulation targets, leading to potentially multiple forms of tDCS plasticity, translational animal studies are critical, alongside clinical neurophysiology, to understand tDCS and ultimately inform rational electrotherapy. Moreover, meaningful clinical outcome rely on specific and increasingly long-lasting changes; the basis of which can be studied in animals.

3.4 NETWORK EFFECTS

The consideration of how weak DC electric fields interact with active networks (e.g., oscillations) is a very compelling area of ongoing research because, just as network of coupled active neurons exhibit "emergent" network activity not apparent in isolated neurons, so does application of electrical stimulation to active networks often produces responses not expected by single neurons. These responses are specific to the architecture and activity of the network. Networks also provide a substrate for amplification beyond the cell/synapse level. Reports that DC current can alter "spontaneous rhythm" in animals span decades (Dubner and Gerard 1939; Antal et al. 2004; Marshall et al. 2011), while recent clinical work on tDCS has addressed modulation of EEG oscillations. New animal studies on DC stimulation, which addressed mechanism of this coupling, are reviewed in the section—with a focus on acute effects as the role of ongoing activity in plasticity is discussed earlier.

3.4.1 FURTHER AMPLIFICATION THROUGH ACTIVE NETWORKS

In principle, the initial action of DC stimulation remains to polarize all neurons sensing the electric field. As discussed earlier, pyramidal somas are more sensitive by virtue of their morphology, but axonal and dendritic polarization should not be ignored. Note that tDCS generates electric field across large areas of cortex. In networks, a key concept is that the entire population of coupled neurons is polarized—this *coherent* polarization of the population provides a substrate for signal detection and for amplification. Interestingly, the effective coupling constant for a neuron immersed in an active network may be enhanced compared to that neuron in isolated (Reato et al. 2010)—meaning that by virtue of being in a network a given compartment (soma) may be polarized directly by the field and indirectly by field actions on a collective of afferent neurons.

As noted earlier, the concept that threshold for electric field sensitivity would be "lower for modulation of the frequency of an already active neuron than for excitation of a silent one" (Terzuolo and Bullock 1956) is well established, but

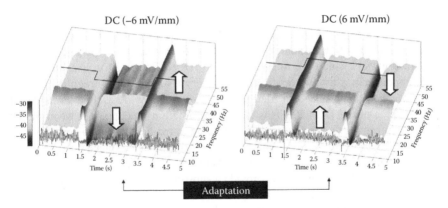

FIGURE 3.6 Modulation of gamma oscillations in brain slice by weak DC fields. Gamma oscillations were induced in the CA3 region by perfusion with carbachol. Negative fields which produce hyperpolarization of CA1 pyramidal neuron soma, attenuated oscillation, but interestingly the attenuation was most pronounced when the fields where turned on, after which oscillation activity partly rebounded even though the field was still on. This suggests homeostatic "adaptation" (arrows) to the DC field by neuronal network system. After the field is turn off, there is an excitatory rebound response consistent with this adaption. An opposite effects is observed for positive fields that would be depolarizing the soma of CA3 pyramidal neurons. This adaption at the network level is not expected from single neurons, so reflects an emergent response of an active network to DC fields.

network activity add another dimension to this. During many network activities, notably oscillations, neurons are constantly near threshold (e.g., primed for firing). If a neuron is near threshold by virtue of network drive, then a small polarization may be influential in modulating the likelihood of firing. For example, a relatively small depolarization may be sufficient to trigger an action potential. Moreover, because the network is interconnected, activated neurons could synaptically trigger AP in other neurons. The whole process can be feed-forward such that a small DC electric field can induce a robust action potential discharge in a population. This has been shown in the brain slice (Reato et al. 2010). This concept is interesting because it clouds the entire distinction between "suprathreshold" stimulation, such as TMS, and "subthreshold" stimulation, as tDCS is commonly considered. It remains the cases that the electric fields produced by tDCS are insufficient to trigger action potential in quiescent neurons (Figure 3.6).

3.4.2 OSCILLATIONS

A majority of work on weak DC electric fields and network activity in slice addressed epileptiform activity (in investigation of methods for seizure control). These reports generally observed a change in the rate of epileptiform discharge generation (the likelihood an event would initiate) rather than a change in event waveform once initiated. This finding is consistent with the concept that weak field polarize neurons (Bikson et al. 1999, 2004) and that weak stimulation is more likely to influence stochastic initial recruitment of neurons in the robust regenerative epileptiform event. DC electric fields may also influence the propagation rate (Francis et al. 2003; Varga et al. 2011). Reato et al. (2010) considered the effects of DC fields on gamma oscillations in brain slice and noted both transient effects when the field was turned on, and secondary sustained effects which are more relevant to tDCS (Figure 3.6). Sustained effects were characterized by a dramatic compensatory ("homeostatic") regulation by the network, such that the system tried to normalize activity to baseline levels despite the presence of the DC-field. This network adaptation was apparent when the DC field was turned off as the network was delayed in readjusting to the absence of the field—in this way, excitatory (somatic depolarizing) fields produced poststimulation inhibition of oscillations, and vice versa. Network level mechanisms (as opposed to single neuron behavior) may thus provide a substrate for activity dependent homeo-static-like observations during tDCS (Cosentino et al. 2012).

Weak stimulation of "physiological" activity work with AC or pulsed stimulation is more common (Deans et al. 2007; Frohlich and McCormick 2010). Though Reato et al. (2010) proposed the effects of AC stimulation at different frequencies and DC could be explained in a single continuous framework, it is important to distinguish between studies exploring the limits of network sensitivity to weak AC or pulse fields, and prolonged DC current (tDCS). When a network is generating spontaneous oscillations of epileptiform activity (regenerative events), then it is well established that an electrical pulse can trigger a regenerative network event; moreover, repetitive weak pulses or AC stimulation can entrain activity by aligning the phase of these events with that of the repetitive stimulation. By definition, (except for at the start) during prolonged DC stimulation there basis for entrainment (there is no phase to the DC) such that tDCS can affect average discharge rate or waveform, but not phase. Thus, though entrainment is central in AC/pulsed stimulation studies in animals (as well as clinically) (Marshall et al. 2006), its relevance to tDCS is limited.

3.5 INTERNEURONS AND NON-NEURONAL EFFECTS

The role of interneurons and non-neuronal cells, such as glia and endothelial cells, in tDCS remains both a wide open and critical question. We distinguish between (1) direct stimulation effects, reflecting direction polarization and modulation of these cell types by DC fields; (2) indirect stimulation effects, reflect change in function secondary to direct excitatory neuronal activation that then influences these other cell types; and (3) modulatory effects, where the sensitivity of neurons to direct effects (e.g., their excitability) is influenced by other cell types. In fact, the function of interneurons and non-neuronal cell types are so intricately wound together with excitatory neurons that the second and third aspects are presumed (though complex), and we here focus mostly on the first possibility of direct effects.

3.5.1 INTERNEURONS

Because of their relatively symmetric dendritic morphology, interneuron somas are expected to polarize less than pyramidal neurons (Radman et al. 2009). Based on the "somatic doctrine" their importance might then be considered diminished.

However, we cannot exclude polarizing effects of fields on dendrites and axons. Moreover, interneurons represent a wide range of morphologies and size, including asymmetric morphologies (Freund and Buzsaki 1996). Interneurons exert a powerful regional effect, including playing role in plasticity and oscillations. An effect of paired-pulse facilitation in hippocampal slice may also suggest modulation of the activity of interneurons (Kabakov et al. 2012). The role of interneurons in the direct effect of tDCS remains then an open question.

3.5.2 GLIA

Glia cells represent the majority of cell in the CNS—the concept that they are just "passive" support cells is outdated (Haydon and Carmignoto 2006) and their complex role in neuronal functions such as plasticity are being elucidated (DiCastro et al. 2011; Panatier et al. 2011). Some glia have distributed processes which would influence their sensitivity to applied electric fields (Ruohonen and Karhu 2012), but even more interesting is the notion that the glial syncytium (an electrically coupled population of glial cells), might act to amplify field polarization. One possible mechanism for DC modulation through glia cells relates to the concept of potassium "spatial buffering." Glia cells are thought to regulate extracellular potassium concentration through a polarization imbalance across their membrane, which is precisely the type of polarization induced by DC fields. How to explore possible effects of electric fields on this mechanism remains unclear. Gardner-Medwin induced extracellular potassium transport by passing DC current and noted concentration changes in saline near the electrodes, which is mechanistically distinct than tissue changes(Gardner-Medwin 1983). Studies in brain slice show no changes in extracellular potassium concentration with DC fields (Lian et al. 2003), though the brain slice preparation has distorted extracellular concentration control mechanisms (An et al. 2008). Neurons and glia can be cultured separately, but morphology and biophysics are altered in culture. In general, there are no "magic bullet" drugs for the glia function, and regardless any changes in glia function would influence neurons and so direct responses to DC fields may be difficult to determine. Still, given the growing interest in the role of glia cells in CNS function and the increased sophistication of experimental techniques, their role in tDCS is a worthwhile area of investigation (Ruohonen and Karhu 2012).

3.5.3 ENDOTHELIAL CELLS

Endothelial cells help form the blood-brain barrier that tightly regulates transport between the brain extracellular space and blood. Any direct action of DC stimulation on endothelial cells could have profound effects on brain function. Endothelial cells do not have processes and their spherical shape indicates peak polarization will be related to cell diameter (Kotnik and Miklavcic 2000)—during tDCS membrane polarization is expected to be well below the threshold for electroporation. The direct effects of tDCS current on vascular response are an open and compelling question. There are abundant evidences that DC current affects vascular function in skin (Ledger 1992; Prausnitz 1996; Berliner 1997; Malty and Petrofsky 2007) and indeed skin redness is typical under tDCS electrodes (Minhas et al. 2010). Vascular and neuronal functions

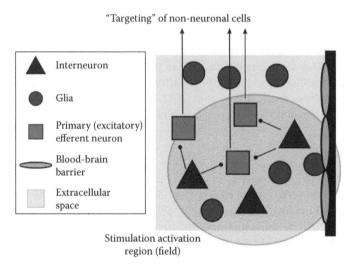

FIGURE 3.7 Could tDCS directly modulate the functional of interneurons or non-neuronal cells? The direct response of interneurons, glia, and endothelial cells (that form the blood-brain barrier) remains an open question—that is difficult to address even in animal models. It is expected that these cell types, by influencing excitatory neuronal cells can indirectly modulate the effects of tDCS on excitatory neurons, and also that through direct actions on excitatory neurons, these cells types will be indirectly affected. A direct effect on interneurons and non-neuronal cells is possible but at this time speculative. (Adapted from Bikson, M. et al., *Conf. Proc. IEEE Eng. Med. Biol. Soc.*, 1, 1616, 2006.)

in the brain are closely interrelated, as evidenced by functional Magnetic Resonance Imaging (fMRI). The relation is also complex, and it can be difficult to disentangle direct neuronal and potential direct vascular effects (a chicken-and-egg problem), including during tDCS. Wachter et al. (2011) found a polarity-specific change in blood perfusion during tDCS in rat, in a direction consistent with the somatic doctrine, and speculated the direction specificity was consistent with a primary neuronal action. The brain slice is compelling since the blood supply is not present such that findings in slice including acute (Bikson et al. 2004) and lasting synaptic efficacy (Fritsch et al. 2010) changes can exclude an endothelial contribution. Conversely, endothelial culture including models of the blood-brain-barrier can be electrically stimulated. We showed that high-intensity electrical stimulation could increase transport across such a model through a phenomena we called "electro-permeation" between cells, to distinguish it from electroporation of single cells (Lopez-Quintero et al. 2010). Investigation of DC stimulation in this model is ongoing (Figure 3.7).

3.6 SUMMARY AND A 3-TIER APPROACH

Clinical tDCS protocols continue to be largely designed and interpreted following the "somatic doctrine," namely that anode/cathode stimulation results in a generalized increase/decrease in neuronal excitability due to radial current flow and somatic polarization. Animal studies showed that current flow radial (normal) to the cortical surface can modulate spontaneous neuronal activity in a polarity-specific manner, with inward current (corresponding to somatic depolarization) increasing firing rate, and outward current (corresponding to somatic hyperpolarization) reducing firing rate (Creutzfeldt et al. 1962; Bindman et al. 1964; Purpura and McMurtry 1965a,b; Gartside 1968a); because of the dependence on the neuronal target, we refer to this as the "somatic doctrine." Indeed, modern tDCS was motivated by neurophysiologic studies showing that anodal/cathodal tDCS increase/decrease, respectively, responses to TMS evoked cortical and muscle potentials, which is consistent with the aforementioned "somatic doctrine" (Nitsche and Paulus 2000). Extensive clinical and cognitive-neuroscience studies have been largely rationalized based on the somatic doctrine. However, despite positive outcomes from many of these studies, emerging evidence suggests that neuromodulation by tDCS may be more complex; stemming largely from the recognition that brain function is evidently not a monolithic "sliding scale of excitability."

Modern animal research is beginning to explicate how modulation by tDCS cannot be explained as a monolithic "sliding-scale" of excitability (anode = up, cathode = down). Brain function/disease and so its influence by DC stimulation is complex. Neither polarization of dendrites of synaptic terminals can be ignored, which may result in differential modulation of specific synaptic inputs. This in turn, may lead to distinct forms of tDCS-induced plasticity. Moreover, which neuronal processes are affected and how, will depend on the tDCS montage used and the state of the underlying network. The rational advancement of tDCS requires departing from the sliding-scale approach (applied indiscriminately across cognitive applications and indications) and addressing these mechanistic and targeting issues. With increased recognition of complexity, the need for translational animal studies, that are properly designed, becomes increasingly clear. At the same times, these issues make the investigation daunting. Our approach, reflected generally in the organization of this review has been to consider changes on "3-tiers": neuronal compartment polarization, synaptic processing, and network effects. While the brain function is evidently understood to span these integrated tiers, this chapter introduces how a 3-tier approach can allow organization of concepts and framing of hypothesis.

REFERENCES

Alexander, J. K., B. Fuss, and R. J. Colello. 2006. Electric field-induced astrocyte alignment directs neurite outgrowth. *Neuron Glia Biol* 2 (2):93–103.

An, J. H., Y. Su, T. Radman, and M. Bikson. 2008. Effects of glucose and glutamine concentration in the formulation of the artificial cerebrospinal fluid (ACSF). *Brain Res* 1218:77–86.

Anastassiou, C. A., S. M. Montgomery, M. Barahona, G. Buzsaki, and C. Koch. 2010. The effect of spatially inhomogeneous extracellular electric fields on neurons. *J Neurosci* 30 (5):1925–1936.

Andreasen, M. and S. Nedergaard. 1996. Dendritic electrogenesis in rat hippocampal CA1 pyramidal neurons: Functional aspects of Na^+ and Ca^{2+} currents in apical dendrites. *Hippocampus* 6 (1):79–95.

Antal, A., E. T. Varga, T. Z. Kincses, M. A. Nitsche, and W. Paulus. 2004. Oscillatory brain activity and transcranial direct current stimulation in humans. *Neuroreport* 15 (8):1307–1310.

Ardolino, G., B. Bossi, S. Barbieri, and A. Priori. 2005. Non-synaptic mechanisms underlie the after-effects of cathodal transcutaneous direct current stimulation of the human brain. *J Physiol (Lond)* 568 (2):653–663.

Artola, A., S. Brocher, and W. Singer. 1990. Different voltage-dependent thresholds for inducing long-term depression and long-term potentiation in slices of rat visual cortex. *Nature* 347 (6288):69–72.

Awatramani, G. B., G. D. Price, and L. O. Trussell. 2005. Modulation of transmitter release by presynaptic resting potential and background calcium levels. *Neuron* 48 (1):109–121.

Berliner, M. N. 1997. Skin microcirculation during tapwater iontophoresis in humans: Cathode stimulates more than anode. *Microvasc Res* 54 (1):74–80.

Bikson, M., P. Bulow, J. W. Stiller, A. Datta, F. Battaglia, S. V. Karnup, and T. T. Postolache. 2008. Transcranial direct current stimulation for major depression: A general system for quantifying transcranial electrotherapy dosage. *Curr Treat Options Neurol* 10 (5):377–385.

Bikson, M., A. Datta, and M. Elwassif. 2009. Establishing safety limits for transcranial direct current stimulation. *Clin Neurophysiol* 120 (6):1033–1034.

Bikson, M., A. Datta, A. Rahman, and J. Scaturro. 2010. Electrode montages for tDCS and weak transcranial electrical stimulation: Role of "return" electrode's position and size. *Clin Neurophysiol* 121 (12):1976–1978.

Bikson, M., R. S. Ghai, S. C. Baraban, and D. M. Durand. 1999. Modulation of burst frequency, duration, and amplitude in the zero-Ca^{2+} model of epileptiform activity. *J Neurophysiol* 82 (5):2262–2270.

Bikson, M., M. Inoue, H. Akiyama, J. K. Deans, J. E. Fox, H. Miyakawa, and J. G. Jefferys. 2004. Effects of uniform extracellular DC electric fields on excitability in rat hippocampal slices in vitro. *J Physiol* 557 (Pt 1):175–190.

Bikson, M., T. Radman, and A. Datta. 2006. Rational modulation of neuronal processing with applied electric fields. *Conf Proc IEEE Eng Med Biol Soc* 1:1616–1629.

Bindman, L. J., O. C. Lippold, and J. W. Redfearn. 1962. Long-lasting changes in the level of the electrical activity of the cerebral cortex produced bypolarizing currents. *Nature* 196:584–585.

Bindman, L. J., O. C. Lippold, and J. W. Redfearn. 1964. The action of brief polarizing currents on the cerebral cortex of the rat (1) during current flow and (2) in the production of long-lasting after-effects. *J Physiol* 172:369–382.

Bishop, G. H. and J. Erlanger. 1926. The effects of polarization upon the activity of vertebrate nerve. *Am J Physiol* 78:630–657.

Bishop, G. H. and J. L. O'Leary. 1950. The effects of polarizing currents on cell potentials and their significance in the interpretation of central nervous system activity. *Electroencephalogr Clin Neurophysiol* 2 (4):401–416.

Bliss, T. V. and T. Lomo. 1973. Long-lasting potentiation of synaptic transmission in the dentate area of the anaesthetized rabbit following stimulation of the perforant path. *J Physiol* 232 (2):331–356.

Bolognini, N., F. Fregni, C. Casati, E. Olgiati, and G. Vallar. 2010. Brain polarization of parietal cortex augments training-induced improvement of visual exploratory and attentional skills. *Brain Res* 1349:76–89.

Brunoni, A. R., F. Fregni, and R. L. Pagano. 2011. Translational research in transcranial direct current stimulation (tDCS): A systematic review of studies in animals. *Rev Neurosci* 22 (4):471–481.

Bullock, T. H. and S. Hagiwara. 1957. Intracellular recording from the giant synapse of the squid. *J Gen Physiol* 40 (4):565–577.

Cambiaghi, M., L. Teneud, S. Velikova, J. J. Gonzalez-Rosa, M. Cursi, G. Comi, and L. Leocani. 2011. Flash visual evoked potentials in mice can be modulated by transcranial direct current stimulation. *Neuroscience* 185:161–165.

Cambiaghi, M., S. Velikova, J. J. Gonzalez-Rosa, M. Cursi, G. Comi, and L. Leocani. 2010. Brain transcranial direct current stimulation modulates motor excitability in mice. *Eur J Neurosci* 31 (4):704–709.

Carandini, M. and D. Ferster. 2000. Membrane potential and firing rate in cat primary visual cortex. *J Neurosci* 20 (1):470–484.

Carlson, C. and O. Devinsky. 2009. The excitable cerebral cortex Fritsch G, Hitzig E. Uber die elektrische Erregbarkeit des Grosshirns. *Arch Anat Physiol Wissen,* 1870 (37):300–332. *Epilepsy Behav* 15 (2):131–132.

Chan, C. Y., J. Hounsgaard, and C. Nicholson. 1988. Effects of electric fields on transmembrane potential and excitability of turtle cerebellar Purkinje cells in vitro. *J Physiol* 402:751–771.

Chan, C. Y. and C. Nicholson. 1986. Modulation by applied electric fields of Purkinje and stellate cell activity in the isolated turtle cerebellum. *J Physiol* 371:89–114.

Cosentino, G., B. Fierro, P. Paladino, S. Talamanca, S. Vigneri, A. Palermo, G. Giglia, and F. Brighina. 2012. Transcranial direct current stimulation preconditioning modulates the effect of high-frequency repetitive transcranial magnetic stimulation in the human motor cortex. *Eur J Neurosci* 35 (1):119–124.

Costain, R., J. W. Redfearn, and O. C. Lippold. 1964. A controlled trial of the therapeutic effect of polarization of the brain in depressive illness. *Br J Psychiatry* 110:786–799.

Creutzfeldt, O. D., G. H. Fromm, and H. Kapp. 1962. Influence of transcortical d-c currents on cortical neuronal activity. *Exp Neurol* 5:436–452.

Datta, A., V. Bansal, J. Diaz, J. Patel, D. Reato, and M. Bikson. 2009. Gyri-precise head model of transcranial direct current stimulation: Improved spatial focality using a ring electrode versus conventional rectangular pad. *Brain Stimul* 2 (4):201–207.

Datta, A., M. Bikson, and F. Fregni. 2010. Transcranial direct current stimulation in patients with skull defects and skull plates: High-resolution computational FEM study of factors altering cortical current flow. *Neuroimage* 52 (4):1268–1278.

Datta, A., M. Elwassif, F. Battaglia, and M. Bikson. 2008. Transcranial current stimulation focality using disc and ring electrode configurations: FEM analysis. *J Neural Eng* 5 (2):163–174.

Datta, A., M. Elwassif, and M. Bikson. 2009. Bio-heat transfer model of transcranial DC stimulation: Comparison of conventional pad versus ring electrode. *Conf Proc IEEE Eng Med Biol Soc* 2009:670–673.

Deans, J. K., A. D. Powell, and J. G. Jefferys. 2007. Sensitivity of coherent oscillations in rat hippocampus to AC electric fields. *J Physiol* 583 (Pt 2):555–565.

Del Castillo, J. and B. Katz. 1954. Changes in end-plate activity produced by presynaptic polarization. *J Physiol* 124 (3):586–604.

Delgado-Lezama, R., J. F. Perrier, and J. Hounsgaard. 1999. Local facilitation of plateau potentials in dendrites of turtle motoneurones by synaptic activation of metabotropic receptors. *J Physiol* 515 (Pt 1):203–207.

Di Castro, M. A., J. Chuquet, N. Liaudet, K. Bhaukaurally, M. Santello, D. Bouvier, P. Tiret, and A. Volterra. 2011. Local Ca^{2+} detection and modulation of synaptic release by astrocytes. *Nat Neurosci* 14 (10):1276–1284.

Dubner, H. H. and R. W. Gerard. 1939. Factors controlling brain potentials in the cat. *J Neurophysiol* 2 (2):142–152.

Durand, D. M. and M. Bikson. 2001. Suppression and control of epileptiform activity by electrical stimulation: A review. *Proc IEEE* 89 (7):1065–1082.

Francis, J. T., B. J. Gluckman, and S. J. Schiff. 2003. Sensitivity of neurons to weak electric fields. *J Neurosci* 23 (19):7255–7261.

Fregni, F., P. S. Boggio, M. Nitsche, F. Bermpohl, A. Antal, E. Feredoes, M. A. Marcolin et al. 2005. Anodal transcranial direct current stimulation of prefrontal cortex enhances working memory. *Exp Brain Res* 166 (1):23–30.

Fregni, F., D. Liebetanz, K. K. Monte-Silva, M. B. Oliveira, A. A. Santos, M. A. Nitsche, A. Pascual-Leone, and R. C. Guedes. 2007. Effects of transcranial direct current stimulation coupled with repetitive electrical stimulation on cortical spreading depression. *Exp Neurol* 204 (1):462–466.

Freund, T. F. and G. Buzsaki. 1996. Interneurons of the hippocampus. *Hippocampus* 6 (4):347–470.

Fritsch G. T. and Hitzig E. 1870. On the electrical excitability of the cerebrum. In: Von Bonin G. (1960) trans. *Some Papers on the Cerebral Cortex*, Charles C. Thomas, Springfield, IL.

Fritsch, B., J. Reis, K. Martinowich, H. M. Schambra, Y. Ji, L. G. Cohen, and B. Lu. 2010. Direct current stimulation promotes BDNF-dependent synaptic plasticity: Potential implications for motor learning. *Neuron* 66 (2):198–204.

Frohlich, F. and D. A. McCormick. 2010. Endogenous electric fields may guide neocortical network activity. *Neuron* 67 (1):129–143.

Gardner-Medwin, A. R. 1983. A study of the mechanisms by which potassium moves through brain tissue in the rat. *J Physiol* 335:353–374.

Gartside, I. B. 1968a. Mechanisms of sustained increases of firing rate of neurones in the rat cerebral cortex after polarization: Reverberating circuits or modification of synaptic conductance? *Nature* 220 (5165):382–383.

Gartside, I. B. 1968b. Mechanisms of sustained increases of firing rate of neurones in the rat cerebral cortex after polarization: Role of protein synthesis. *Nature* 220 (5165):383–384.

Gasca, F., L. Richter, and A. Schweikard. 2010. Simulation of a conductive shield plate for the focalization of transcranial magnetic stimulation in the rat. *Conf Proc IEEE Eng Med Biol Soc* 2010:1593–1596.

Ghai, R. S., M. Bikson, and D. M. Durand. 2000. Effects of applied electric fields on low-calcium epileptiform activity in the CA1 region of rat hippocampal slices. *J Neurophysiol* 84 (1):274–280.

Gluckman, B. J., E. J. Neel, T. I. Netoff, W. L. Ditto, M. L. Spano, and S. J. Schiff. 1996. Electric field suppression of epileptiform activity in hippocampal slices. *J Neurophysiol* 76 (6):4202–4205.

Hattori, Y., A. Moriwaki, and Y. Hori. 1990. Biphasic effects of polarizing current on adenosine-sensitive generation of cyclic AMP in rat cerebral cortex. *Neurosci Lett* 116 (3):320–324.

Hayashi, Y., Y. Hattori, A. Moriwaki, H. Asaki, and Y. Hori. 1988. Effects of prolonged weak anodal direct current on electrocorticogram in awake rabbit. *Acta Med Okayama* 42 (5):293–296.

Haydon, P. G. and G. Carmignoto. 2006. Astrocyte control of synaptic transmission and neurovascular coupling. *Physiol Rev* 86 (3):1009–1031.

Hess, G. and J. P. Donoghue. 1999. Facilitation of long-term potentiation in layer II/III horizontal connections of rat motor cortex following layer I stimulation: Route of effect and cholinergic contributions. *Exp Brain Res* 127 (3):279–290.

Hubbard, J. I. and W. D. Willis. 1962a. Hyperpolarization of mammalian motor nerve terminals. *J Physiol* 163:115–137.

Hubbard, J. I. and W. D. Willis. 1962b. Mobilization of transmitter by hyperpolarization. *Nature* 193:174–175.

Islam, N., M. Aftabuddin, A. Moriwaki, Y. Hattori, and Y. Hori. 1995. Increase in the calcium level following anodal polarization in the rat brain. *Brain Res* 684 (2):206–208.

Islam, N., A. Moriwaki, Y. Hattori, and Y. Hori. 1994. Anodal polarization induces protein kinase C gamma (PKC gamma)-like immunoreactivity in the rat cerebral cortex. *Neurosci Res* 21 (2):169–172.

Islam, N., A. Moriwaki, and Y. Hori. 1995. Co-localization of c-fos protein and protein kinase C gamma in the rat brain following anodal polarization. *Indian J Physiol Pharmacol* 39 (3):209–215.

Jefferys, J. G. 1981. Influence of electric fields on the excitability of granule cells in guinea-pig hippocampal slices. *J Physiol* 319:143–152.

Jefferys, J. G. R., J. Deans, M. Bikson, and J. Fox. 2003. Effects of weak electric fields on the activity of neurons and neuronal networks. *Radiat Prot Dosimetry* 106 (4):321–323.

Joucla, S. and B. Yvert. 2009. The "mirror" estimate: An intuitive predictor of membrane polarization during extracellular stimulation. *Biophys J* 96 (9):3495–3508.

Kabakov, A. Y., P. A. Muller, A. Pascual-Leone, F. E. Jensen, and A. Rotenberg. 2012. Contribution of axonal orientation to pathway-dependent modulation of excitatory transmission by direct current stimulation in isolated rat hippocampus. *J Neurophysiol* 107 (7):1881–1889.

Kotnik, T. and D. Miklavcic. 2000. Analytical description of transmembrane voltage induced by electric fields on spheroidal cells. *Biophys J* 79 (2):670–679.

Ledger, P. W. 1992. Skin biological issues in electrically enhanced transdermal delivery. *Adv Drug Deliv Rev* 9:289–307.

Li, L., Y. H. El-Hayek, B. Liu, Y. Chen, E. Gomez, X. Wu, K. Ning et al. 2008. Direct-current electrical field guides neuronal stem/progenitor cell migration. *Stem Cells* 26 (8):2193–2200.

Lian, J., M. Bikson, C. Sciortino, W. C. Stacey, and D. M. Durand. 2003. Local suppression of epileptiform activity by electrical stimulation in rat hippocampus in vitro. *J Physiol (Lond)* 547 (2):427–434.

Liebetanz, D., F. Fregni, K. K. Monte-Silva, M. B. Oliveira, A. Amancio-dos-Santos, M. A. Nitsche, and R. C. Guedes. 2006a. After-effects of transcranial direct current stimulation (tDCS) on cortical spreading depression. *Neurosci Lett* 398 (1–2):85–90.

Liebetanz, D., F. Klinker, D. Hering, R. Koch, M. A. Nitsche, H. Potschka, W. Loscher, W. Paulus, and F. Tergau. 2006b. Anticonvulsant effects of transcranial direct-current stimulation (tDCS) in the rat cortical ramp model of focal epilepsy. *Epilepsia* 47 (7):1216–1224.

Liebetanz, D., R. Koch, S. Mayenfels, F. Konig, W. Paulus, and M. A. Nitsche. 2009. Safety limits of cathodal transcranial direct current stimulation in rats. *Clin Neurophysiol* 120 (6):1161–1167.

Liebetanz, D., M. A. Nitsche, F. Tergau, and W. Paulus. 2002. Pharmacological approach to the mechanisms of transcranial DC-stimulation-induced after-effects of human motor cortex excitability. *Brain* 125 (Pt 10):2238–2247.

Lopez-Quintero, S. V., A. Datta, R. Amaya, M. Elwassif, M. Bikson, and J. M. Tarbell. 2010. DBS-relevant electric fields increase hydraulic conductivity of in vitro endothelial monolayers. *J Neural Eng* 7 (1):16005.

Malty, A. M. and J. Petrofsky. 2007. The effect of electrical stimulation on a normal skin blood flow in active young and older adults. *Med Sci Monit* 13 (4):CR147–CR155.

Marshall, L., H. Helgadottir, M. Molle, and J. Born. 2006. Boosting slow oscillations during sleep potentiates memory. *Nature* 444 (7119):610–613.

Marshall, L., R. Kirov, J. Brade, M. Molle, and J. Born. 2011. Transcranial electrical currents to probe EEG brain rhythms and memory consolidation during sleep in humans. *PLoS One* 6 (2):e16905.

Marshall, L., M. Molle, M. Hallschmid, and J. Born. 2004. Transcranial direct current stimulation during sleep improves declarative memory. *J Neurosci* 24 (44):9985–9992.

Márquez-Ruiz, J., R. Leal-Campanario, R. Sánchez-Campusano, B. Molaee-Ardekani, F. Wendling, P. C. Miranda, G. Ruffini, A. Gruart, and J. M. Delgado-Garcia. 2012. Transcranial direct-current stimulation modulates synaptic mechanisms involved in associative learning in behaving rabbits. *Proc Natl Acad Sci USA* 109 (17):6710–6715.

Merrill, D. R., M. Bikson, and J. G. R. Jefferys. 2005. Electrical stimulation of excitable tissue: design of efficacious and safe protocols. *J Neurosci Methods* 141 (2):171–198.

Minhas, P., V. Bansal, J. Patel, J. S. Ho, J. Diaz, A. Datta, and M. Bikson. 2010. Electrodes for high-definition transcutaneous DC stimulation for applications in drug delivery and electrotherapy, including tDCS. *J Neurosci Methods* 190 (2):188–197.

Minhas, P., A. Datta, and M. Bikson. 2011. Cutaneous perception during tDCS: Role of electrode shape and sponge salinity. *Clin Neurophysiol* 122 (4):637–638.

Miranda, P. C., L. Correia, R. Salvador, and P. J. Basser. 2007. The role of tissue heterogeneity in neural stimulation by applied electric fields. *Conf Proc IEEE Eng Med Biol Soc* 2007:1715–1718.

Miranda, P. C., M. Lomarev, and M. Hallett. 2006. Modeling the current distribution during transcranial direct current stimulation. *Clin Neurophysiol* 117 (7):1623–1629.

Monte-Silva, K., M. F. Kuo, D. Liebetanz, W. Paulus, and M. A. Nitsche. 2010. Shaping the optimal repetition interval for cathodal transcranial direct current stimulation (tDCS). *J Neurophysiol* 103 (4):1735–1740.

Mozzachiodi, R. and J. H. Byrne. 2010. More than synaptic plasticity: Role of nonsynaptic plasticity in learning and memory. *Trends Neurosci* 33 (1):17–26.

Nitsche, M. A., K. Fricke, U. Henschke, A. Schlitterlau, D. Liebetanz, N. Lang, S. Henning, F. Tergau, and W. Paulus. 2003. Pharmacological modulation of cortical excitability shifts induced by transcranial direct current stimulation in humans. *J Physiol* 553 (Pt 1):293–301.

Nitsche, M. A. and W. Paulus. 2000. Excitability changes induced in the human motor cortex by weak transcranial direct current stimulation. *J Physiol* 527 (Pt 3):633–639.

Nitsche, M. A., A. Schauenburg, N. Lang, D. Liebetanz, C. Exner, W. Paulus, and F. Tergau. 2003. Facilitation of implicit motor learning by weak transcranial direct current stimulation of the primary motor cortex in the human. *J Cogn Neurosci* 15 (4):619–626.

Ozen, S., A. Sirota, M. A. Belluscio, C. A. Anastassiou, E. Stark, C. Koch, and G. Buzsaki. 2010. Transcranial electric stimulation entrains cortical neuronal populations in rats. *J Neurosci* 30 (34):11476–11485.

Panatier, A., J. Vallee, M. Haber, K. K. Murai, J. C. Lacaille, and R. Robitaille. 2011. Astrocytes are endogenous regulators of basal transmission at central synapses. *Cell* 146 (5):785–798.

Peterchev, A. V., T. A. Wagner, P. C. Miranda, M. A. Nitsche, W. Paulus, S. H. Lisanby, A. Pascual-Leone, and M. Bikson. 2011. Fundamentals of transcranial electric and magnetic stimulation dose: Definition, selection, and reporting practices. *Brain Stimul.* Nov 1 [Epub. ahead of print].

Prausnitz, M. R. 1996. The effects of electric current applied to skin: A review for transdermal drug delivery. *Adv Drug Deliv Rev* 18:395–425.

Purpura, D. P. and J. G. McMurtry. 1965a. Intracellular activities and evoked potential changes during polarization of motor cortex. *J Neurophysiol* 28 (1):166–185.

Purpura, D. P. and J. G. McMurtry. 1965b. Intracellular potentials of cortical neurons during applied transcortical polarizing currents. *Electroen Clin Neuro* 18 (2):203.

Purpura, D. P., J. G. McMurtry, and C. F. Leonard. 1966. Intracellular spike potentials of dendritic origin during hippocampal seizures induced by subiculum stimulation. *Brain Res* 1 (1):109–113.

Radman, T., A. Datta, and A. V. Peterchev. 2007a. In vitro modulation of endogenous rhythms by AC electric fields: Syncing with clinical brain stimulation. *J Physiol* 584 (Pt 2):369–370.

Radman, T., R. L. Ramos, J. C. Brumberg, and M. Bikson. 2009. Role of cortical cell type and morphology in subthreshold and suprathreshold uniform electric field stimulation in vitro. *Brain Stimul* 2 (4):215–228.

Radman, T., Y. Z. Su, J. H. An, L. C. Parra, and M. Bikson. 2007b. Spike timing amplifies the effect of electric fields on neurons: Implications for endogenous field effects. *J Neurosci* 27:3030–3036.

Ranieri, F., M. V. Podda, E. Riccardi, G. Frisullo, M. Dileone, P. Profice, F. Pilato, V. Di Lazzaro, and C. Grassi. 2012. Modulation of LTP at rat hippocampal CA3-CA1 synapses by direct current stimulation. *J Neurophysiol* 107 (7):1868–1880.

Reato, D., A. Rahman, M. Bikson, and L. C. Parra. 2010. Low-intensity electrical stimulation affects network dynamics by modulating population rate and spike timing. *J Neurosci* 30 (45):15067–15079.

Redfearn, J. W., O. C. Lippold, and R. Costain. 1964. A preliminary account of the clinical effects of polarizing the brain in certain psychiatric disorders. *Br J Psychiatry* 110:773–785.

Rioult-Pedotti, M. S., D. Friedman, G. Hess, and J. P. Donoghue. 1998. Strengthening of horizontal cortical connections following skill learning. *Nat Neurosci* 1 (3):230–234.

Ruohonen, J. and J. Karhu. 2012. tDCS possibly stimulates glial cells. *Clin Neurophysiol.* Apr 3 [Epub. ahead of print].

Sadleir, R. J., T. D. Vannorsdall, D. J. Schretlen, and B. Gordon. 2010. Transcranial direct current stimulation (tDCS) in a realistic head model. *Neuroimage* 51 (4):1310–1318.

Salvador, R., A. Mekonnen, G. Ruffini, and P. C. Miranda. 2010. Modeling the electric field induced in a high resolution realistic head model during transcranial current stimulation. *Conf Proc IEEE Eng Med Biol Soc* 2010:2073–2076.

Stagg, C. J. and M. A. Nitsche. 2011. Physiological basis of transcranial direct current stimulation. *Neuroscientist* 17 (1):37–53.

Su, Y. Z., T. Radman, J. Vaynshteyn, L. C. Parra, and M. Bikson. 2008. Effects of high-frequency stimulation on epileptiform activity in vitro: ON/OFF control paradigm. *Epilepsia* 49 (9):1586–1593.

Sunderam, S., B. Gluckman, D. Reato, and M. Bikson. 2010. Toward rational design of electrical stimulation strategies for epilepsy control. *Epilepsy Behav* 17 (1):6–22.

Svirskis, G., A. Gutman, and J. Hounsgaard. 1997. Detection of a membrane shunt by DC field polarization during intracellular and whole cell recording. *J Neurophysiol* 77 (2):579–586.

Takeuchi, A. and N. Takeuchi. 1962. Electrical changes in pre- and postsynaptic axons of the giant synapse of Loligo. *J Gen Physiol* 45:1181–1193.

Terzuolo, C. A. and T. H. Bullock. 1956. Measurement of imposed voltage gradient adequate to modulate neuronal firing. *Proc Natl Acad Sci USA* 42 (9):687–694.

Toleikis, J. R., A. Sances, Jr. and S. J. Larson. 1974. Effects of diffuse transcerebral electrical currents on cortical unit potential activity. *Anesth Analg* 53 (1):48–55.

Turkeltaub, P. E., J. Benson, R. H. Hamilton, A. Datta, M. Bikson, and H. B. Coslett. 2011. Left lateralizing transcranial direct current stimulation improves reading efficiency. *Brain Stimul.*

Varga, E. T., D. Terney, M. D. Atkins, M. Nikanorova, D. S. Jeppesen, P. Uldall, H. Hjalgrim, and S. Beniczky. 2011. Transcranial direct current stimulation in refractory continuous spikes and waves during slow sleep: A controlled study. *Epilepsy Res* 97 (1–2):142–145.

Wachter, D., A. Wrede, W. Schulz-Schaeffer, A. Taghizadeh-Waghefi, M. A. Nitsche, A. Kutschenko, V. Rohde, and D. Liebetanz. 2011. Transcranial direct current stimulation induces polarity-specific changes of cortical blood perfusion in the rat. *Exp Neurol* 227 (2):322–327.

Wong, R. K. and M. Stewart. 1992. Different firing patterns generated in dendrites and somata of CA1 pyramidal neurones in guinea-pig hippocampus. *J Physiol* 457:675–687.

Yoon, K. J., B. M. Oh, and D. Y. Kim. 2012. Functional improvement and neuroplastic effects of anodal transcranial direct current stimulation (tDCS) delivered 1 day vs. 1 week after cerebral ischemia in rats. *Brain Res.* 1452:61–72.

4 Physiological Basis and Methodological Aspects of Transcranial Electric Stimulation (tDCS, tACS, and tRNS)

Walter Paulus, Andrea Antal,
and Michael A. Nitsche

CONTENTS

4.1 INTRODUCTION

In principle, electrical currents have been applied to modulate cerebral functions for several hundred years (for a review see Zaghi et al. 2010a). In these early times, the main interest of researchers was to induce observable effects, such as limb movements, changes in perception, or reduction of clinical symptoms in patients (Hellwag and Jacobi 1802) using electrical stimulation, which, secondarily, led to neuronal activation. The stimulation techniques discussed in this chapter refine some of these early attempts with much better evaluation methods. Because they aimed primarily at modulating spontaneous cortical activity and excitability, these approaches might be termed neuromodulation rather than stimulation techniques in a functional sense. Animal experiments with transcranial direct current stimulation (tDCS) in the 1960s showed that delivery of relatively weak currents over cortical areas could influence resting membrane potentials and consecutively spontaneous neuronal activity. A sufficiently long application furthermore results in longer-lasting after-effects (Bindman et al. 1964; Creutzfeldt et al. 1962; Purpura and McMurtry 1965). These after-effects aim to bring about neuroplastic alterations in the brain functions involved in learning and memory processes, as well as pathological alterations in neuropsychiatric diseases. Some 35 years later use of the "surrogate marker" transcranial magnetic stimulation (TMS) showed that non-invasive stimulation with direct currents in humans elicits reproducible excitability increases or decreases in the motor cortex (M1) during (Priori et al. 1998) and after stimulation (Nitsche and Paulus 2000). Even before this, anodal stimulation at 0.3 mA was shown to speed up human motor responses in a reaction paradigm (Elbert et al. 1981). On the basis of the increasing accumulation of TMS surrogate marker data in the last 10 years, experiments with tDCS could be more precisely designed. Since then it has been shown that tDCS, due to its neuroplasticity-inducing properties, alters motor and cognitive functions in humans as well as clinical symptoms in patients (Nitsche and Paulus 2011; Nitsche et al. 2008). Transcranial alternating current stimulation (tACS) and transcranial random noise stimulation (tRNS) were introduced by using comparable duration and intensities to tDCS; at particular frequencies they were also shown to induce plasticity-inducing results comparable to tDCS (Chaieb et al. 2011a,b; Moliadze et al. 2010a; Terney et al. 2008). However due to the alternating current flow, they are probably not able to accumulate membrane potential influence over time but instead modulate oscillating neural circuits or neural membranes. This chapter will be dedicated to the physiological modes of action of the aforementioned stimulation techniques and methodological aspects which are relevant for an optimized application.

4.2   tDCS

4.2.1   PHYSIOLOGY

The primary mechanism of DC stimulation is the alteration of neuronal resting membrane potentials. Depending on current flow direction relative to neuronal orientation, DC stimulation de- or hyperpolarises cortical neurons at a subthreshold level, thereby modulating spontaneous firing frequency (Bindman et al. 1964; Creutzfeldt et al. 1962; Radman et al. 2009). Interneurons seemed to be more susceptible to the stimulation than pyramidal neurons (Purpura and McMurtry 1965); however, in a recent study it was claimed that layer V/VI neuron absolute electric field action potential thresholds were lower than those of layer II/III pyramidal neurons and interneurons (Radman et al. 2009). Early studies already confirmed that DC stimulation, applied for a critical period of time, produces after-effects on cortical activity which last for hours after the end of stimulation (Bindman et al. 1964). For anodal stimulation, these after-effects are protein-synthesis-dependent, are accompanied by intracellular calcium accumulation, and enhance cAMP-levels (Gartside 1968;

Islam et al. 1995; Moriwaki et al. 1995). The physiological and cellular characteristics of the stimulation effects thus share some common features with long-term potentiation (LTP). Indeed, recent in vitro studies directly confirmed the synaptic effect of DCS in both inducing LTP (Fritsch et al. 2010) and modulating LTP induction (Ranieri et al. 2012).

Transcranial application of DC was already being used with humans in the 1960s and 1970s with the hope of modulating cognition, affect and the clinical symptoms of psychiatric patients—but with mixed results (for an overview see Lolas 1977; Nitsche et al. 2003b). Physiological effects have been explored more systematically in the last 12 years. Most of the physiological studies were conducted in the M1. Generally in accordance with the results of animal experiments, it could be shown that anodal tDCS enhances, while cathodal stimulation reduces, cortical excitability (Figure 4.1). Short stimulation lasting a few seconds generates these effects only during stimulation, while longer-lasting tDCS (about 10–15 min) induces excitability alterations which are stable for about 1 h or longer (Nitsche and Paulus 2000, 2001; Nitsche et al. 2003c; Figure 4.2). The direction of the current flow determines the kind of effects and after-effects produced, i.e., excitation or inhibition (Accornero et al. 2007; Antal et al. 2004a; Nitsche and Paulus 2000). This is, however, only valid for the resting state: as soon as the muscle is activated or attentional processes are initiated, anodal excitation (Antal et al. 2007; Figure 4.3) or tACS in the ripple range (Moliadze et al. 2010b) turns into inhibition.

The modulation of resting membrane potentials was indirectly confirmed by abolishing excitatory anodal effects when combining tDCS with the sodium channel blocker carbamazepine (Nitsche et al. 2003a). With regard to the after-effects of stimulation, anodal tDCS has been shown to reduce intracortical inhibition, while it enhances cortical facilitation, whereas cathodal tDCS has reverse effects (Nitsche et al. 2005). In addition, the after-effects seem to be synaptically driven and to depend

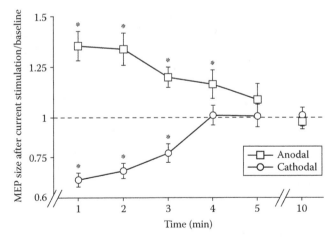

FIGURE 4.1 Time course of polarity-specific motor cortex excitability changes outlasting stimulation duration, shown after 5 min DC stimulation at 1 mA. Asterisks indicate significant differences between MEP amplitudes after stimulation and at baseline. (Reproduced from Nitsche, M.A. and Paulus W., *J. Physiol.*, 527(Pt 3), 633, 2000. With permission.)

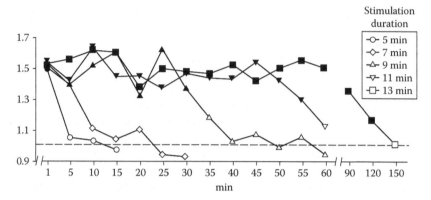

FIGURE 4.2 tDCS of the human motor cortex modulates the MEP-amplitudes after stimulation duration-dependently for up to an hour after tDCS. 5–7 min anodal stimulation induces short-lasting after-effects, while prolonged anodal tDCS increases the duration of the after-effects over-proportionally. (Reproduced from Nitsche M.A. and Paulus W., *Neurology*, 57, 1899, 2001. With permission). Filled symbols represent significant changes compared to baseline.

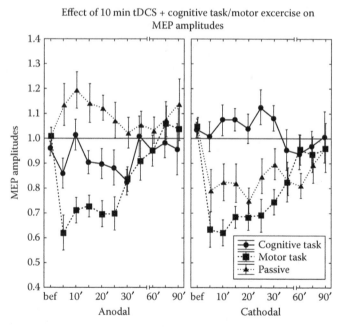

FIGURE 4.3 Effect of 10 min anodal and cathodal stimulation on MEP amplitudes. During the stimulation, the subjects were in a passive state (sitting), and were required to complete a cognitive test presented on a computer monitor or had to push a ball with their right hand. The figure shows mean MEP amplitudes and their SEMs. (Reproduced from Antal, A. et al., *Eur. J. Neurosci.*, 26, 2687, 2007. With permission.)

on the glutamatergic system. This is confirmed by the abolition of plasticity induced by the NMDA receptor blocker dextromethorphan (Nitsche et al. 2003a). In contrast, enhancement of NMDA receptor activity by D-Cycloserine prolongs the excitability enhancement achieved by anodal tDCS by a factor of 20 in the 24 h range (Nitsche et al. 2004). An MRI spectroscopy study showed that free GABA is reduced after both anodal and cathodal tDCS (Stagg et al. 2009). In particular, reduction of GABAergic activity might have a gating function to increase (glutamatergic) plasticity, which was shown after ischemic blockade of the forearm muscles (Ziemann et al. 1998).

tDCS can also alter oscillatory cortical activity at cortical areas beneath the electrodes. Beta and gamma oscillatory activity was enhanced in the visual cortex after anodal tDCS, while cathodal tDCS induced reverse effects (Antal et al. 2004a; Ardolino et al. 2005). For interregional effects on network activation, a positron emission tomography (PET) study with DC electrodes positioned at the M1 and the contralateral prefrontal cortex showed widespread cortical and subcortical activity alterations, whose origin—connectivity or current spread—could however not be clearly identified (Lang et al. 2005). Recently conducted studies using graph-theoretical approaches to track tDCS-induced alterations of functional connectivity are in accordance with tDCS-induced alterations of cortical network connectivity related to the stimulated area. In an electroencephalography (EEG) study, Polania and colleagues (Polania et al. 2011a) explored the effect of anodal tDCS over the M1 and found increased functional connectivity of motor-related areas in the gamma frequency band by combining anodal tDCS and simple finger movements, as opposed to stimulation or finger tapping alone (Figure 4.4a and b). Similarly, in a functional magnetic resonance imaging (fMRI) experiment, anodal tDCS of the primary M1 alone was shown to enhance functional coupling between premotor and superior parietal areas which are components of the motor network (Figure 4.4c), whereas connectivity of the primary M1 to more distant areas was reduced (Polania et al. 2011b). Moreover, tDCS-induced alterations of functional connectivity seem not to be limited to cortical networks, but also affect cortico-subcortical connections, as demonstrated in another fMRI study of the same group (Polania et al. 2011c).

Taken together, the aforementioned studies show that tDCS induces neuroplastic excitability alterations of the stimulated cortical regions. Furthermore, they have also demonstrate that tDCS alters oscillatory activity, that these activity alterations are present at the regional and interregional level, and that tDCS affects functional connectivity between remote areas.

4.2.2 METHODOLOGICAL ASPECTS

Transcranial electrical stimulation is not difficult to perform and the devices needed are technically simple and inexpensive if compared, e.g., to transcranial magnetic stimulators. Today's devices are increasingly complex as soon as they start to add supplementary features, such as automatized fading-in procedures to avoid retinal phosphenes from a frontal return electrode, or documentation of the time when current flow has been generated (for the Food and Drug Administration (FDA), or fulfilling requests to document parameter by print or devices that adjust to electric resistance and allow blinding. Nevertheless, current generation as such is simple

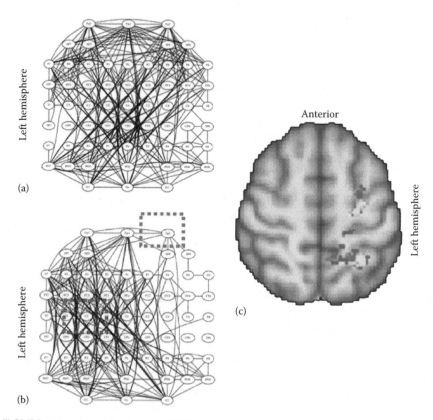

FIGURE 4.4 **(See color insert.)** EEG recordings were acquired during simple right-hand finger tapping before (a) and after (b) the application of 10 min anodal tDCS over the left primary motor cortex. After the end of stimulation, functional connectivity significantly increased in the left hemisphere in the high-gamma band (60–90 Hz), whereas the number of inter-hemispheric significantly decreased. Panel (c) shows regions that significantly increased the functional coupling with the left sensorimotor cortex following anodal stimulation of the left primary motor cortex as measured by BOLD-fMRI during resting state (axial image is displayed according to radiological convention, left is right).

and is usually battery driven. Most recently even a simple static magnet has been invented for inhibiting brain function (Oliviero et al. 2011). It is therefore not astonishing that early attempts to apply transcranial electric stimulation date back to the eighteenth century. First friction was used to generate the necessary electricity; with the discovery of the voltaic pile (first electric battery), transcranial direct current stimulation became a possibility. By around 1800, whole books appeared describing this technology and a series of treated patients (e.g., Hellwag and Jacobi 1802). Until the 1960s, the method did not make substantial progress, mainly due to the lack of a simple method of evaluating the biological effects. This situation changed markedly when TMS was combined with tDCS (Nitsche and Paulus 2000; Priori et al. 1998). Most of our methodological concepts available at present thus rely on TMS results, mainly in terms of motor evoked potential (MEP) changes.

4.2.2.1 Intensity and Electrode Size

There seems to be a narrow window of intensity for inducing tDCS-related after-effects. Current density is the critical parameter, usually varying between 0.029 and 0.08 mA/cm^2 in most published studies. Current density depends on electrode size; a standard size is 35 cm^2 as introduced by Nitsche and Paulus (2000). With this size, a commonly used intensity is 1 mA. If intensity is increased to 3 mA, tDCS starts to become painful (Furubayashi et al. 2008). Compared to tDCS, with tRNS and with 140 Hz tACS an interesting phenomenon was seen: whereas a peak-to-peak intensity of 0.2 mA is subthreshold, 0.4 mA leads to inhibition, 0.6 and 0.8 mA do not provide a significant effect, but with only 1 mA excitation can be seen (Moliadze et al. 2011).

The most commonly used montage targets the left M1 as identified by TMS, and the return electrode is placed at the contralateral forehead (Nitsche and Paulus 2000). According to (Nitsche and Paulus 2000), the lower threshold for inducing plastic after-effects of the M1 is 0.4 mA. Thus the range for inducing biological effects covers roughly a factor of 5. Electrode size can be decreased, e.g., to 3.5 cm^2 to produce a focal stimulation differentiating cortical representations of the thenar and hypothenar. It can also be increased in order to assure subthreshold stimulation at the return electrode (Nitsche et al. 2007).

4.2.2.2 Electrode Placement

When tDCS after-effects were observed, we placed the target electrode at the M1 and tried different return electrode positions (Nitsche and Paulus 2000). Only placing the return electrode at the contralateral forehead gave positive results. One of our explanations was that the direction of induced current flow also targets the new M1 in the central sulcus (overview in Salvador et al. 2011).

A recent modeling study using a high-resolution, finite-element model showed that the current density maxima are on localized hotspots at the bottom of the sulci and not on the cortical surface as would be expected from spherical models (Salvador et al. 2010). Furthermore, on the basis of modeling it was recommended that when working with conventional electrodes (25–35 cm^2), one of the electrodes should be placed just "behind" the target relative to the other electrode to give maximum current density on the target (Faria et al. 2011). The use of electrodes sizes in the range of 3.5–12 cm^2 was suggested to provide a better compromise between focality and current density. Finally, the use of multiple small return electrodes may be more efficient than the use of a single large return electrode (Faria et al. 2011). Individualized head models permit a targeted stimulation e.g., in stroke patients (Datta et al. 2011) and even in patients with skull defects and skull plates (Datta et al. 2010). Ring electrodes have been investigated for more precise location (Datta et al. 2009) although this concept was challenged as soon as tissue anisotropy was taken into account (Suh et al. 2009). The background here is that below about 200 Hz, electric current is conducted ten times better along white matter fiber tracts than perpendicular to them. Although most basic data have been acquired at the M1 because of the ease of quantification by TMS, nowadays virtually all cortical areas have been targeted, the visual cortex among the early studies (Antal et al. 2003a,b). Quantification of phosphene thresholds may be seen in analogy to the MEP at the M1 if tDCS or tACS induced modulations are explored.

Extracephalic return electrodes have been used with the aim to minimize stimulation effects at the return electrode (Priori et al. 2008). It appears that higher stimulation intensities are required in order to obtain comparable effects to cephalic electrodes (Moliadze et al. 2010b). Although tDCS keeps current constant with increasing electrode to electrode distance, a greater current flow was needed to guarantee after-effects of similar magnitude. Direct spinal stimulation has been accomplished as well (Cogiamanian et al. 2008; Roche et al. 2011; Winkler et al. 2010).

4.2.2.3 Sham Stimulation

Perception of prickling beneath the electrodes can occur. This sensation is greater at the beginning of stimulation, is more likely to occur with tDCS than with tACS or tRNS, and of course depends on stimulation intensity. Thus sham stimulation is usually accomplished by starting current flow in a way identical to the verum condition but fading it out after some 20 s. However, the perception threshold, at which 50% of the subjects detect the stimulation is 400 µA for tDCS and 1200 µA in the case of tRNS (Ambrus et al. 2010). Anodal and cathodal tDCS are indistinguishable regarding sites of perception (Ambrus et al. 2010). Experienced investigators show a significantly higher anodal stimulation detection rate than the naive group. Furthermore, investigators performed significantly better than naive subjects in nonstimulation discrimination (Ambrus et al. 2010). Further investigations in blinding methods e.g., using "placebo itching," may allow blinding at higher intensities.

4.2.2.4 Stimulation Duration

A minimal stimulation duration is necessary to induce measurable after-effects. This was 3 min with 1 mA in the "standard" condition (Nitsche and Paulus 2000), at least if MEPs are taken as marker. The next investigations showed that longer stimulation durations produced longer after-effects, both with anodal and cathodal stimulation (Nitsche and Paulus 2001; Nitsche et al. 2003c). It became clear only recently that this connection cannot be extended indefinitely. Prolonging stimulation protocols may not just prolong after-effects but may also reverse them. Doubling the duration of theta burst stimulation (TBS) ended in reversed after-effects, excitatory TBS became inhibitory and vice versa (Gamboa et al. 2010). When using cathodal tDCS it became clear that doubling stimulation time from 9 to 18 min only increased the duration of inhibitory after-effects from 60 to 90 min (Monte-Silva et al. 2010). In still unpublished work we were able to show that doubling 13 min anodal stimulation time to 26 min converted an MEP increase to inhibition. It is thus most likely that there is an optimal stimulation duration for both anodal and cathodal tDCS for maximizing the duration of after-effects. In order to get clinically relevant longer after-effects, one has to introduce intervals between two stimulation sessions. The after-effect duration could be increased to 120 min when two 9 min cathodal tACS sessions were applied at an interval of three or 20 min or 24 h. The boosting effect of repeated anodal tDCS with a 24 h stimulation interval had already been shown previously for a motor skill measure. Furthermore this skill measure remained greater with anodal tDCS at 3 months (Reis et al. 2009). Also, daily tDCS led to greater increases in cortical excitability than tDCS every second day (Alonzo et al. 2011).

4.2.2.5 tDCS, Neuroimaging, and Other Neurophysiological Techniques

For several years it was not possible to apply tDCS in the MRI scanner because of concerns about the remote possibility of sudden electrode heating. First studies thus compared pre-tDCS with post-tDCS scanning, tDCS being performed outside the scanner (Baudewig et al. 2001). Now that CE certified stimulators are available, fMRI data on tDCS effects during scanning are emerging (Antal et al. 2011; Jang et al. 2009; Kim et al. 2011; Kwon et al. 2008). In addition to fMRI, cerebral blood flow studies (Zheng et al. 2011) and MR spectroscopy have also been performed (Binkofski et al. 2011; Stagg et al. 2009). One study dealt with the effects of tDCS seen in the H_2O PET (Lang et al. 2005). Near-infrared spectroscopy is able to quantify prefrontal hemodynamic changes induced by tDCS (Merzagora et al. 2010). TDCS or tACS effects can be measured with the EEG (Ardolino et al. 2005; Keeser et al. 2011; Marshall et al. 2006; Matsumoto et al. 2010; Zaehle et al. 2010, 2011b) and by evoked potentials (Accornero et al. 2007; Antal et al. 2004a,b; Csifcsak et al. 2009; Dieckhofer et al. 2006; Matsunaga et al. 2004; Zaehle et al. 2011a).

4.3 OSCILLATORY STIMULATION

4.3.1 tACS AND tRNS

In the case of tDCS, *current polarity* is the main determinant of the direction of the after-effects, with anodal stimulation generally causing an increase and cathodal stimulation causing a decrease in cortical excitability. This is no longer an issue when considering tACS and tRNS, as long as they lack polarity in a superimposed DC component. In the case of tACS and tRNS, *frequency* or *frequency range* and *intensity* are the major parameters determining intervention outcome.

tACS tends to interfere with ongoing oscillations in the brain, at least in the EEG frequency range. It has been suggested that different, specialized brain areas have to be bound together by fluctuating oscillations, mainly in the gamma range, to provide a temporary functional network that is able to solve any required higher cognitive task (for a recent review see: Marceglia et al. 2011). External application of tACS might interact with these oscillations and thus allow an experimental validation of this "binding hypothesis" (Singer 2001). However, frequencies in the kHz range are much more likely to interfere with membrane excitability. In a recent study it was observed that stimulation in the low kHz range (1–5 kHz) over the human M1 allows for the generation of similar sustained cortical excitability increases, as previously seen with rTMS or tDCS paradigms (Chaieb et al. 2011a; Figure 4.5).

Inducing slow oscillation-like potential fields by transcranial application of oscillating currents with DC-offset at a frequency of 0.75 Hz during early nocturnal non-rapid-eye-movement sleep enhances the retention of hippocampus-dependent declarative memories in healthy humans (Marshall et al. 2006). The stimulation induced an immediate increase in endogenous cortical slow oscillations and slow spindle activity in the frontal cortex. In contrast, tACS plus DC overlay at 5 Hz, which normally predominates during rapid-eye-movement sleep, decreased slow oscillations and left declarative memory unchanged. A later study, in which combined tACS/tDCS

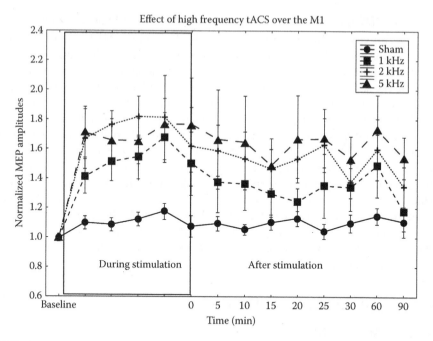

FIGURE 4.5 Effects of 1, 2, and 5 kHz tACS on cortical excitability of M1, as measured by mean MEP amplitude. MEP values are normalized to baseline. Timecourse over 90 min is shown. ± error bars indicate SEM. (Reproduced from Chaieb, L. et al., *Restor. Neurol. Neurosci.*, 29, 167, 2011a. With permission.)

was applied over the M1, argues in favor of a tDCS effect for the externally induced M1 plasticity at this case (Groppa et al. 2010). Kirov et al. (2009). Again, using combined tDCS and tACS similar effects were found by Marshall et al. (2006) during wakefulness, demonstrating an additional increase in theta (4–8 Hz) activity due to stimulation.

Earlier attempts to use complex patterns of alternating current stimulation such as cranial electrotherapy stimulation, Limoge's and Lebedev's current have recently been reviewed by (Zaghi et al. 2010a). The recent investigations however try to uncover mechanisms of tACS by confining physical parameters to as few as possible using pure sinusoidal stimulation (e.g., Moliadze et al. 2010a).

tACS performed on the M1 with 10 Hz improved implicit motor learning and modified motor cortical excitability, and this effect outlasted the duration of stimulation (Antal et al. 2008). Other tACS frequencies between 5 and 40 Hz failed to induce any measurable after-effect, probably also due to the low stimulation intensity (0.4 mA) that was used to avoid inducing retinal phosphenes. A marked decrease in MEP amplitude compared to sham stimulation was observed following 10 Hz AC stimulation, without a change of EEG power. The reason for the efficacy of 10 Hz is unclear though it fits quite well within the context of intrinsic M1 resonance frequencies (Castro-Alamancos et al. 2007).

Pogosyan et al. (2009) applied tACS at 5 and 20 Hz over the M1 during a concurrent visuomotor task and observed a retardation of voluntary movement only during the 20 Hz stimulation. Furthermore, the EEG recording after the stimulation showed that tACS induced a strong increase in the beta coherence in the contralateral M1. The authors suggested that the results can be considered as evidence of the causality between oscillatory brain activity and concurrent motor behavior. A recent study supported the efficacy of applying 20 Hz tACS over the M1. Although the stimulation was applied for a shorter duration (90 s) an increase in corticospinal excitability was observed (Feurra et al. 2011). Other tACS frequencies (5, 10, and 40 Hz) did not affect the size of the MEPs. Moreover, tACS applied to the parietal cortex and to the ulnar nerve also failed to modulate MEPs. In a further study it was also demonstrated that 15 Hz tACS of the M1 (electrodes placed at C3 and C4) can significantly *diminish* the amplitude of MEPs and decrease intracortical facilitation (ICF) compared to baseline and sham stimulation (Zaghi et al. 2010b). The authors proposed that tACS at 15 Hz may have a dampening effect on cortical networks and perhaps interfere with the temporal and spatial summation of weak subthreshold electric potentials.

With regard to other sensory areas, Zaehle et al. (2010) aimed to determine whether tACS at the alpha frequency of 8–12 Hz induced an entrainment of the applied oscillatory activity delivered over the visual system. Two electrodes were placed bilaterally over the parieto-occipital cortex (PO9 and PO10). Here tACS induced an increase in alpha power compared to the sham condition implying that it is possible to interact with ongoing visual cortical oscillatory activity by means of tACS. Using tACS over the occipital cortex with a reference electrode at Cz, phosphenes were elicited in a frequency-dependent manner, with a peak slightly lower in darkness than in brightness (Kanai et al. 2008). Subsequently, Schutter and Hortensius (2009; see also Schwiedrzik 2009) argued that even remote electrodes might stimulate the retina, with its clearly higher sensitivity, by far field potentials. Direct proof of the cortical phosphene origin is still lacking (Paulus 2010); but tACS to the visual cortex does influence cortical excitability as determined by TMS-induced phosphenes thresholds (Kanai et al. 2010), implying that tACS to the visual cortex at least has a modulating effect. In line with this, application of tACS over the visual cortex decreased subjects' contrast-discrimination thresholds significantly during 60 Hz tACS, while 40 and 80 Hz stimulation had no effect (Laczo et al. 2011).

The problem of inducing retinal phosphenes can be avoided using higher stimulation frequencies. Moliadze et al. (2010a) applied tACS over M1 using the so-called ripple frequency range. Ripples are short hippocampal oscillations in the frequency range between 100 and 250 Hz associated with memory encoding. After tACS applied for 10 min at 140 Hz and 1 mA intensity over the M1, an MEP increase lasting up to 1 h can be measured by single pulse TMS. Stimulation at 80 Hz remained without effect, while 250 Hz was clearly less efficient (Figure 4.6).

These data are concordant with the high frequency range above 100 Hz being able to elicit lasting MEP changes for 1.5 h in contrast to frequencies below 100 Hz (Terney et al. 2008). This study used 10min tRNS with the frequency varying as "white noise" in a Gaussian distribution between 0.1 and 640 Hz (Terney et al. 2008; Figure 4.7). After tRNS, a reduction of the blood-oxygen-level dependence (BOLD) response in the M1 could be seen in the fMRI during finger tapping (Chaieb et al. 2009). Using a probabilistic classification task and applying tRNS to

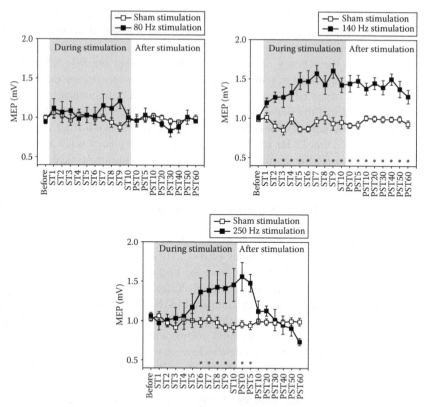

FIGURE 4.6 The figure shows mean amplitudes of MEPs and their SEMs during 10 min and after tACS up to 60 min. 140 Hz AC stimulation significantly increased MEPs at the ST2–ST10, and PST0–PST60 time points compared to the sham stimulation (*p < 0.05). Controls by sham and 80 Hz stimulation were without any effect. 250 Hz AC stimulation significantly increased MEPs at the ST6–ST10, PST0–PST5. (Reproduced from Moliadze, V. et al., *J. Physiol.*, 588, 4891, 2010a; Moliadze, V. et al., *Clin. Neurophysiol.*, 121, 2165, 2010b. With permission.)

the dorsolateral prefrontal cortex (DLPFC) it was found that tRNS can influence categorization performance (Ambrus et al. 2011). Interestingly, the results showed a disappearance of the prototype effect, meaning that during stimulation the subjects made more mistakes with regard to the categorization of the prototype pattern. These data are similarly to those with anodal tDCS to this area, further supporting the idea of similar effect mechanisms for tRNS and tDCS. On the other hand, it was recently demonstrated that tRNS improves neuroplasticity in a perceptual learning paradigm (Fertonani et al. 2011). Finally, a study investigating the effects of tDCS and tRNS on the n-back test showed that the significant improvement found in the anodal tDCS condition was not observable with tRNS (Mulquiney et al. 2011). Still, tRNS may be seen as an "early development Noninvasive Brain Stimulation (NIBS) device," suggesting that although promising initial results exist, clinical trials are yet to be conducted (Edelmuth et al. 2010).

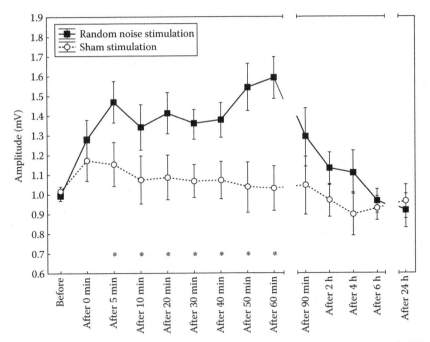

FIGURE 4.7 Effect of 10 min tRNS on MEPs. Time course of motor cortex excitability changes lasting for 60 min post-stimulation, shown after 10 min RN stimulation over M1 at 1 mA compared to sham stimulation. The figure shows mean amplitudes and their SEMs up to 60 min (including all subjects, n = 17) and between 90 min and 24 h (including eight subjects). Asterisks indicate significant differences between MEP amplitudes after 5, 10–60 min post-stimulation compared to baseline. (Reproduced from Terney, D. et al., *J. Neurosci.,* 28, 14147, 2008. With permission.)

The intensity of both tRNS and tACS can be used to direct excitatory or inhibitory after-effects. When performing tACS at 140 Hz or full-spectrum tRNS at five different intensities between 0.2 and 1.0 mA on separate study days, instead of finding a simple threshold for excitation, stimulation with 0.4 mA caused inhibition and only stimulation with 1 mA led to excitation (Moliadze et al. 2011). The intermediate currents of 0.6 and 0.8 mA had no effect at all on MEP intensity. The inhibition produced by 140 Hz tACS was somewhat stronger than that induced by tRNS. These findings open the potential to control enhancement or reduction of M1 excitability simply by altering stimulation intensity. The physiological basis for this must be explored. There is some evidence in the literature that lower intensities preferentially activate inhibitory cortical networks whereas higher intensities preferentially target excitatory neurons.

With regard to the frequency, it appears that transcranial techniques using even higher frequencies, outside of the physiological neuronal network oscillatory frequency range, have biological effects. Indeed, recent studies reported that the application of high frequency oscillating currents to cell cultures in vitro and animal studies in vivo are effective in entraining neuronal rhythmicity and propagating synchronicity in neuronal networks (Jinno 2007; Radman et al. 2007). Nevertheless, studies suggest that oscillations in a range of several hundred Hertz may still inter-fere with interneuronal rhythms (Ylinen et al. 1995) whereas frequencies in the kHz range are much more likely to interfere with membrane excitability. It was observed that stimulation in the low kHz range over the human M1 enables the gen-eration of sustained cortical excitability increases, as previously seen with rTMS or tDCS paradigms (Chaieb et al. 2011a; Figure 4.5). In this study, stimulation with 1 mA at 1, 2, and 5 kHz was applied over M1 for a duration of 10 min. All frequen-cies of tACS increased the amplitudes of MEPs up to 30–60 min post stimulation, compared to the baseline. Compared to sham stimulation, stimulation with 2 and 5 kHz was more effective than 1 kHz stimulation in inducing sustained changes in cortical excitability. This technique opens a new possibility to directly interfere with cortical excitability.

4.4   GENERAL CONCLUSION

Transcranial electric stimulation techniques now cover a broad range of stimula-tion frequencies, starting with direct current and now involving frequencies in the kHz range. Altogether an infinite number of possibilities exist, particularly when combining different frequencies. Strategies to explore the advantages and disad-vantages presently involve hypothesis-driven parameters. Direct current stimula-tion seems to be the most clearly defined. Alternating currents may be oriented at EEG frequency range or, in the kHz range, at physical membrane properties. The challenge in the future will be to refine and optimize techniques for distinct brain areas, diseases, genetic predispositions, cortical geometry and many more aspects.

REFERENCES

Accornero N, Li Voti P, La Riccia M, and Gregori B. 2007. Visual evoked potentials modula-tion during direct current cortical polarization. *Exp Brain Res* 178:261–266.
Alonzo A, Brassil J, Taylor JL, Martin D, and Loo CK. 2012. Daily transcranial direct current stimulation (tDCS) leads to greater increases in cortical excitability than second daily transcranial direct current stimulation. *Brain Stimul* 5(3):208–213.
Ambrus GG, Antal A, and Paulus W. 2010. Comparing cutaneous perception induced by elec-trical stimulation using rectangular and round shaped electrodes. *Clin Neurophysiol* 122(4):803–807.
Ambrus GG, Zimmer M, Kincses ZT, Harza I, Kovacs G et al. 2011. The enhancement of cortical excitability over the DLPFC before and during training impairs categorization in the prototype distortion task. *Neuropsychologia* 49:1974–1980.
Antal A, Boros K, Poreisz C, Chaieb L, Terney D, and Paulus W. 2008. Comparatively weak after-effects of transcranial alternating current stimulation (tACS) on cortical excitabil-ity in humans. *Brain Stimul* 1:97–105.
Antal A, Kincses TZ, Nitsche MA, Bartfai O, and Paulus W. 2004a. Excitability changes induced in the human primary visual cortex by transcranial direct current stimulation: Direct electrophysiological evidence. *Invest Ophthalmol Vis Sci* 45:702–707.
Antal A, Kincses TZ, Nitsche MA, and Paulus W. 2003a. Manipulation of phosphene thresh-olds by transcranial direct current stimulation in man. *Exp Brain Res* 150:375–378.
Antal A, Kincses TZ, Nitsche MA, and Paulus W. 2003b. Modulation of moving phosphene thresholds by transcranial direct current stimulation of V1 in human. *Neuropsychologia* 41:1802–1807.

Antal A, Polania R, Schmidt-Samoa C, Dechent P, and Paulus W. 2011. Transcranial direct current stimulation over the primary motor cortex during fMRI. *NeuroImage* 55:590–596.

Antal A, Terney D, Poreisz C, and Paulus W. 2007. Towards unravelling task-related modulations of neuroplastic changes induced in the human motor cortex. *Eur J Neurosci* 26:2687–2691.

Antal A, Varga ET, Kincses TZ, Nitsche MA, and Paulus W. 2004b. Oscillatory brain activity and transcranial direct current stimulation in humans. *Neuroreport* 15:1307–1310.

Ardolino G, Bossi B, Barbieri S, and Priori A. 2005. Non-synaptic mechanisms underlie the after-effects of cathodal transcutaneous direct current stimulation of the human brain. *J Physiol* 568:653–663.

Baudewig J, Nitsche MA, Paulus W, and Frahm J. 2001. Regional modulation of BOLD MRI responses to human sensorimotor activation by transcranial direct current stimulation. *Magn Reson Med* 45:196–201.

Bindman LJ, Lippold OC, and Redfearn JW. 1964. The action of brief polarizing currents on the cerebral cortex of the rat (1) during current flow and (2) in the production of long-lasting after-effects. *J Physiol* 172:369–382.

Binkofski F, Loebig M, Jauch-Chara K, Bergmann S, Melchert UH et al. 2011. Brain energy consumption induced by electrical stimulation promotes systemic glucose uptake. *Biol Psychiatry* 70(7):690–695.

Castro-Alamancos MA, Rigas P, and Tawara-Hirata Y. 2007. Resonance (approximately 10 Hz) of excitatory networks in motor cortex: Effects of voltage-dependent ion channel blockers. *J Physiol* 578:173–191.

Chaieb L, Antal A, and Paulus W. 2011a. Transcranial alternating current stimulation in the low kHz range increases motor cortex excitability. *Restor Neurol Neurosci* 29:167–175.

Chaieb L, Kovacs G, Cziraki C, Greenlee M, Paulus W, and Antal A. 2009. Short-duration transcranial random noise stimulation induces blood oxygenation level dependent response attenuation in the human motor cortex. *Exp Brain Res* 198:439–444.

Chaieb L, Paulus W, and Antal A. 2011b. Evaluating aftereffects of short-duration transcranial random noise stimulation on cortical excitability. *Neural Plast* 2011:105927.

Cogiamanian F, Vergari M, Pulecchi F, Marceglia S, and Priori A. 2008. Effect of spinal transcutaneous direct current stimulation on somatosensory evoked potentials in humans. *Clin Neurophysiol* 119:2636–2640.

Creutzfeldt OD, Fromm GH, and Kapp H. 1962. Influence of transcortical d-c currents on cortical neuronal activity. *Exp Neurol* 5:436–452.

Csifcsak G, Antal A, Hillers F, Levold M, Bachmann CG et al. 2009. Modulatory effects of transcranial direct current stimulation on laser-evoked potentials. *Pain Med* 10:122–132.

Datta A, Baker JM, Bikson M, and Fridriksson J. 2011. Individualized model predicts brain current flow during transcranial direct-current stimulation treatment in responsive stroke patient. *Brain Stimul* 4:169–174.

Datta A, Bansal V, Diaz J, Patel J, Reato D, and Bikson M. 2009. Gyri-precise head model of transcranial DC stimulation: Improved spatial focality using a ring electrode versus conventional rectangular pad. *Brain Stimul* 2:201–207.

Datta A, Bikson M, and Fregni F. 2010. Transcranial direct current stimulation in patients with skull defects and skull plates: High-resolution computational FEM study of factors altering cortical current flow. *Neuroimage* 52:1268–1278.

Dieckhofer A, Waberski TD, Nitsche M, Paulus W, Buchner H, and Gobbele R. 2006. Transcranial direct current stimulation applied over the somatosensory cortex—differential effect on low and high frequency SEPs. *Clin Neurophysiol* 117:2221–2227.

Edelmuth RC, Nitsche MA, Battistella L, and Fregni F. 2010. Why do some promising brain–stimulation devices fail the next steps of clinical development? *Expert Rev Med Devices* 7(1):67–97. Review.

Elbert T, Lutzenberger W, Rockstroh B, and Birbaumer N. 1981. The influence of low-level transcortical DC-currents on response speed in humans. *Intl J Neurosci* 14:101–114.

Faria P, Hallett M, and Miranda PC. 2011. A finite element analysis of the effect of electrode area and inter-electrode distance on the spatial distribution of the current density in tDCS. *J Neural Eng* 8:066017.

Fertonani A, Pirulli C, and Miniussi C. 2011. Random noise stimulation improves neuroplasticity in perceptual learning. *J Neurosci* 31:15416–15423.

Feurra M, Paulus W, Walsh V, and Kanai R. 2011. Frequency specific modulation of human somatosensory cortex. *Front Psychol* 2:13.

Fritsch B, Reis J, Martinowich K, Schambra HM, Ji Y, Cohen LG, and Lu B. 2010. Direct currentstimulationpromotesBDNF-dependentsynapticplasticity: Potential implications for motor learning. *Neuron* 66:198–204.

Furubayashi T, Terao Y, Arai N, Okabe S, Mochizuki H et al. 2008. Short and long duration transcranial direct current stimulation (tDCS) over the human hand motor area. *Exp Brain Res* 185:279–286.

Gamboa OL, Antal A, Moliadze V, and Paulus W. 2010. Simply longer is not better: Reversal of theta burst after-effect with prolonged stimulation. *Exp Brain Res* 204:181–187.

Gartside IB. 1968. Mechanisms of sustained increases of firing rate of neurones in the rat cerebral cortex after polarization: Role of protein synthesis. *Nature* 220:383–384.

Groppa S, Bergmann TO, Siems C, Molle M, Marshall L, and Siebner HR. 2010. Slow-oscillatory transcranial direct current stimulation can induce bidirectional shifts in motor cortical excitability in awake humans. *Neuroscience* 166:1219–1225.

Hellwag CF and Jacobi M. 1802. *Erfahrungen über die Heilkräfte des Galvanismus und Betrachtungen über desselben chemische und physiologische Wirkungen.* Hamburg, Germany: Friedrich Perthes.

Islam N, Aftabuddin M, Moriwaki A, Hattori Y, and Hori Y. 1995. Increase in the calcium level following anodal polarization in the rat brain. *Brain Res* 684:206–208.

Jang SH, Ahn SH, Byun WM, Kim CS, Lee MY, and Kwon YH. 2009. The effect of transcranial direct current stimulation on the cortical activation by motor task in the human brain: An fMRI study. *Neurosci Lett* 460:117–120.

Jinno S, Klausberger T, Marton LF, Dalezios Y, Roberts JD, Fuentealba P, Bushong EA, Henze D, Buzsáki G, and Somogyi P. 2007. Neuronal diversity in GABAergic long-range projections from the hippocampus. *J Neurosci* 27:8790–8804.

Kanai R, Chaieb L, Antal A, Walsh V, and Paulus W. 2008. Frequency-dependent electrical stimulation of the visual cortex. *Curr Biol* 18:1839–1843.

Kanai R, Paulus W, and Walsh V. 2010. Transcranial alternating current stimulation (tACS) modulates cortical excitability as assessed by TMS-induced phosphene thresholds. *Clin Neurophysiol* 121:1551–1554.

Keeser D, Padberg F, Reisinger E, Pogarell O, Kirsch V et al. 2011. Prefrontal direct current stimulation modulates resting EEG and event-related potentials in healthy subjects: A standardized low resolution tomography (sLORETA) study. *Neuroimage* 55:644–657.

Kim CR, Kim DY, Kim LS, Chun MH, Kim SJ, and Park CH. 2011. Modulation of cortical activity after anodal transcranial direct current stimulation of the lower limb motor cortex: A functional MRI study. *Brain Stimul.* [Epub. ahead of print].

Kirov R, Weiss C, Siebner HR, Born J, and Marshall L. 2009. Slow oscillation electrical brain stimulation during waking promotes EEG theta activity and memory encoding. *Proc Natl Acad Sci USA* 106:15460–15465.

Kwon YH, Ko MH, Ahn SH, Kim YH, Song JC et al. 2008. Primary motor cortex activation by transcranial direct current stimulation in the human brain. *Neurosci Lett* 435:56–59.

Laczo B, Antal A, Niebergall R, Treue S, and Paulus W. 2011. Transcranial alternating stimulation in a high gamma frequency range applied over V1 improves contrast perception but does not modulate spatial attention. *Brain Stimul.* [Epub. ahead of print].

Lang N, Siebner HR, Ward NS, Lee L, Nitsche MA et al. 2005. How does transcranial DC stimulation of the primary motor cortex alter regional neuronal activity in the human brain? *Eur J Neurosci* 22:495–504.

Lolas F. 1977. Brain polarization: Behavioral and therapeutic effects. *Biol Psychiatry* 12:37–47.

Marceglia S, Fumagalli M, and Priori A. 2011. What neurophysiological recordings tell us about cognitive and behavioral functions of the human subthalamic nucleus. *Expert Rev Neurother* 11:139–149.

Marshall L, Helgadottir H, Molle M, and Born J. 2006. Boosting slow oscillations during sleep potentiates memory. *Nature* 444:610–613.

Matsumoto J, Fujiwara T, Takahashi O, Liu M, Kimura A, and Ushiba J. 2010. Modulation of mu rhythm desynchronization during motor imagery by transcranial direct current stimulation. *J Neuroeng Rehabil* 7:27.

Matsunaga K, Nitsche MA, Tsuji S, and Rothwell JC. 2004. Effect of transcranial DC sensorimotor cortex stimulation on somatosensory evoked potentials in humans. *Clin Neurophysiol* 115:456–460.

Merzagora AC, Foffani G, Panyavin I, Mordillo-Mateos L, Aguilar J et al. 2010. Prefrontal hemodynamic changes produced by anodal direct current stimulation. *Neuroimage* 49:2304–2310.

Moliadze V, Antal A, and Paulus W. 2010a. Boosting brain excitability by transcranial high frequency stimulation in the ripple range. *J Physiol* 588:4891–4904.

Moliadze V, Antal A, and Paulus W. 2010b. Electrode-distance dependent after-effects of transcranial direct and random noise stimulation with extracephalic reference electrodes. *Clin Neurophysiol* 121:2165–2171.

Moliadze V, Atalay D, Antal A, and Paulus W. 2012. Close to threshold transcranial electrical stimulation preferentially activates inhibitory networks before switching to excitation with higher intensities. *Brain Stimul.* [Epub. ahead of print].

Monte-Silva K, Kuo MF, Liebetanz D, Paulus W, and Nitsche MA. 2010. Shaping the optimal repetition interval for cathodal transcranial direct current stimulation (tDCS). *J Neurophysiol* 103:1735–1740.

Moriwaki A, Islam N, Hattori Y, and Hori Y. 1995. Induction of Fos expression following anodal polarization in rat brain. *Psychiatry Clin Neurosci* 49:295–298.

Mulquiney PG, Hoy KE, Daskalakis ZJ, Fitzgerald PB. 2011. Improving working memory: exploring the effect of transcranial random noise stimulation and transcranial direct current stimulation on the dorsolateral prefrontal cortex. *Clin Neurophysiol* 122(12):2384–2389.

Nitsche MA, Cohen LG, Wassermann EM, Priori A, Lang N et al. 2008. Transcranial direct current stimulation: State of the art 2008. *Brain Stimul* 1:206–223.

Nitsche MA, Doemkes S, Karakose T, Antal A, Liebetanz D et al. 2007. Shaping the effects of transcranial direct current stimulation of the human motor cortex. *J Neurophysiol* 97:3109–3117.

Nitsche MA, Fricke K, Henschke U, Schlitterlau A, Liebetanz D et al. 2003a. Pharmacological modulation of cortical excitability shifts induced by transcranial direct current stimulation in humans. *J Physiol* 553:293–301.

Nitsche MA, Jaussi W, Liebetanz D, Lang N, Tergau F, and Paulus W. 2004. Consolidation of human motor cortical neuroplasticity by D-cycloserine. *Neuropsychopharmacology* 29:1573–1578.

Nitsche MA, Liebetanz D, Antal A, Lang N, Tergau F, and Paulus W. 2003b. Modulation of cortical excitability by weak direct current stimulation—Technical, safety and functional aspects. *Suppl Clin Neurophysiol* 56:255–276.

Nitsche MA, Nitsche MS, Klein CC, Tergau F, Rothwell JC, and Paulus W. 2003c. Level of action of cathodal DC polarisation induced inhibition of the human motor cortex. *Clin Neurophysiol* 114:600–604.

Nitsche MA and Paulus W. 2000. Excitability changes induced in the human motor cortex by weak transcranial direct current stimulation. *J Physiol* 527 (Pt 3):633–639.

Nitsche MA and Paulus W. 2001. Sustained excitability elevations induced by transcranial DC motor cortex stimulation in humans. *Neurology* 57:1899–1901.

Nitsche MA and Paulus W. 2011. Transcranial direct current stimulation—Update 2011. *Restor Neurol Neurosci* 29(6):463–492.

Nitsche MA, Seeber A, Frommann K, Klein CC, Rochford C et al. 2005. Modulating parameters of excitability during and after transcranial direct current stimulation of the human motor cortex. *J Physiol* 568:291–303.

Oliviero A, Mordillo-Mateos L, Arias P, Panyavin I, Foffani G, and Aguilar J. 2011. Transcranial static magnetic field stimulation (tSMS) of the human motor cortex. *J Physiol* 589 (Pt 20):4949–4958.

Paulus W. 2010. On the difficulties of separating retinal from cortical origins of phosphenes when using transcranial alternating current stimulation (tACS). *Clin Neurophysiol* 121:987–991.

Pogosyan A, Gaynor LD, Eusebio A, and Brown P. 2009. Boosting cortical activity at Beta-band frequencies slows movement in humans. *Curr Biol* 19:1637–1641.

Polania R, Nitsche MA, and Paulus W. 2011a. Modulating functional connectivity patterns and topological functional organization of the human brain with transcranial direct current stimulation. *Hum Brain Mapp* 32:1236–1249.

Polania R, Paulus W, Antal A, and Nitsche MA. 2011b. Introducing graph theory to track for neuroplastic alterations in the resting human brain: A transcranial direct current stimulation study. *Neuroimage* 54:2287–2296.

Polania R, Paulus W, and Nitsche MA. 2011c. Modulating cortico-striatal and thalamo-cortical functional connectivity with transcranial direct current stimulation. *Hum Brain Mapp.* doi: 10.1002/hbm.21380.

Priori A, Berardelli A, Rona S, Accornero N, and Manfredi M. 1998. Polarization of the human motor cortex through the scalp. *Neuroreport* 9:2257–2260.

Priori A, Mameli F, Cogiamanian F, Marceglia S, Tiriticco M et al. 2008. Lie-specific involvement of dorsolateral prefrontal cortex in deception. *Cereb Cortex* 18:451–455.

Purpura DP and McMurtry JG. 1965. Intracellular activities and evoked potential changes during polarization of motor cortex. *J Neurophysiol* 28:166–185.

Radman T, Datta A, and Peterchev AV. 2007. In vitro modulation of endogenous rhythms by AC electric fields: Syncing with clinical brain stimulation. *J Physiol* 584:369–370.

Radman T, Ramos RL, Brumberg JC, and Bikson M. 2009. Role of cortical cell type and morphology in subthreshold and suprathreshold uniform electric field stimulation in vitro. *Brain Stimul* 2:215–228.

Ranieri F, Podda MV, Riccardi E, Frisullo G, Dileone M, Profice P, Pilato F, Di Lazzaro V, and Grassi C. 2012. Modulation of LTP at rat hippocampal CA3-CA1 synapses by direct current stimulation. *J Neurophysiol* 107(7):1868–1880 [Epub. ahead of print].

Reis J, Schambra HM, Cohen LG, Buch ER, Fritsch B et al. 2009. Noninvasive cortical stimulation enhances motor skill acquisition over multiple days through an effect on consolidation. *PNAS* 106:1590–1595.

Roche N, Lackmy A, Achache V, Bussel B, and Katz R. 2011. Effects of anodal tDCS on lumbar propriospinal system in healthy subjects. *Clin Neurophysiol* 123(5):1027–1034.

Salvador R, Mekonnen A, Ruffini G, and Miranda PC. 2010. Modeling the electric field induced in a high resolution realistic head model during transcranial current stimulation. *Conf Proc IEEE Eng Med Biol Soc* 2010:2073–2076.

Salvador R, Silva S, Basser PJ, and Miranda PC. 2011. Determining which mechanisms lead to activation in the motor cortex: A modeling study of transcranial magnetic stimulation using realistic stimulus waveforms and sulcal geometry. *Clin Neurophysiol* 122:748–758.

Schutter DJLG and Hortensius R. 2010. Retinal origin of phosphenes to transcranial alternating current stimulation. *Clin Neurophysiol* 121(7):1080–1084.

Schwiedrzik CM. 2009. Retina or visual cortex? The site of phosphene induction by transcranial alternating current stimulation. *Front Integr Neurosci* 3:6.

Singer W. 2001. Consciousness and the binding problem. *Ann N Y Acad Sci* 929:123–146.

Stagg CJ, Best JG, Stephenson MC, O'Shea J, Wylezinska M et al. 2009. Polarity-sensitive modulation of cortical neurotransmitters by transcranial stimulation. *J Neurosci* 29:5202–5206.

Suh HS, Kim SH, Lee WH, and Kim TS. 2009. Realistic simulation of transcranial direct current stimulation via 3-d high-resolution finite element analysis: Effect of tissue anisotropy. *Conf Proc IEEE Eng Med Biol Soc* 2009:638–641.

Terney D, Chaieb L, Moliadze V, Antal A, and Paulus W. 2008. Increasing human brain excitability by transcranial high-frequency random noise stimulation. *J Neurosci* 28:14147–14155.

Winkler T, Hering P, and Straube A. 2010. Spinal DC stimulation in humans modulates post-activation depression of the H-reflex depending on current polarity. *Clin Neurophysiol* 121:957–961.

Ylinen A, Bragin A, Nadasdy Z, Jando G, Szabo I et al. 1995. Sharp wave-associated high-frequency oscillation (200 Hz) in the intact hippocampus: Network and intracellular mechanisms. *J Neurosci* 15:30–46.

Zaehle T, Beretta M, Jancke L, Herrmann CS, and Sandmann P. 2011a. Excitability changes induced in the human auditory cortex by transcranial direct current stimulation: Direct electrophysiological evidence. *Exp Brain Res* 215:135–140.

Zaehle T, Rach S, and Herrmann CS. 2010. Transcranial alternating current stimulation enhances individual alpha activity in human EEG. *PLoS One* 5:e13766.

Zaehle T, Sandmann P, Thorne JD, Jancke L, and Herrmann CS. 2011b. Transcranial direct current stimulation of the prefrontal cortex modulates working memory performance: Combined behavioural and electrophysiological evidence. *BMC Neurosci* 12:2.

Zaghi S, Acar M, Hultgren B, Boggio PS, and Fregni F. 2010a. Noninvasive brain stimulation with low-intensity electrical currents: Putative mechanisms of action for direct and alternating current stimulation. *Neuroscientist* 16:285–307.

Zaghi S, de Freitas Rezende L, de Oliveira LM, El-Nazer R, Menning S et al. 2010b. Inhibition of motor cortex excitability with 15Hz transcranial alternating current stimulation (tACS). *Neurosci Lett* 479:211–214.

Zheng X, Alsop DC, and Schlaug G. 2011. Effects of transcranial direct current stimulation (tDCS) on human regional cerebral blood flow. *Neuroimage* 58:26–33.

Ziemann U, Corwell B, and Cohen LG. 1998. Modulation of plasticity in human motor cortex after forearm ischemic nerve block. *J Neurosci* 18:1115–1123.

5 Biophysical Foundations of Transcranial Magnetic Stimulation and Transcranial Electric Stimulation
From Electromagnetic Fields to Neural Response

Timothy A. Wagner and Pedro Cavaleiro Miranda

CONTENTS

5.1 INTRODUCTION

Throughout history, doctors and physicists have been applying electromagnetic energy to the brain to modulate neural function and to treat a variety of neurological disorders. In 43 AD, the Greek physician, Scribonious Largus, began experimenting with the use of electrical currents to treat various physiological ailments such as gout and headaches by applying electric torpedo fish to the affected regions (Largus 1529). These treatments served as low-tech means of transcutaneous electrical nerve stimulation, nearly 1700 years before the discovery of the battery. As technology has evolved, so have the methods to electrically stimulate the nervous system. One can trace the development of the battery in parallel with early explorations into electrical neural stimulation from the works of Charles Le Roy (Pascual-Leone and Wagmer 2007), Duchenne de Boulogne (1855), Galvani (Galvani et al. 1793; Piccolino 1997), Volta (Piccolino 1997), and Aldini (1804). For example, in 1804, Giovanni Aldini employed voltaic cells to generate weak direct currents (DCs), which he applied transcranially to treat individuals suffering from depression (Aldini 1794, 1804). This is one of the first examples of transcranial DC stimulation (tDCS). The development of the laws of induction and inductive machines can be traced in parallel with the development of magnetic induction-based neural stimulation through the works of d'Arsonval (1896) and Thompson (1910). For example, d'Arsonval used an inductive device, with a coil that could encase a person's head, to generate pulsed magnetic fields that induced magneto-phosphenes in stimulated subjects (d'Arsonval 1896). This is one of the first examples of magnetic stimulation. Today, based on advancements in both electrophysiology and electromagnetic theory, numerous techniques and technologies have been developed to transcranially stimulate the nervous system. Transcranial magnetic stimulation (TMS) and transcranial electrical stimulation (tES) are currently the two most widely used forms of noninvasive brain stimulation.

Numerous TMS techniques have been proposed, such as single pulse TMS, repetitive TMS (rTMS), paired pulse stimulation, and theta burst stimulation (Barker et al. 1985; George et al. 1995; Huang et al. 2005; Paulus 2005; Fregni and Pascual-Leone 2007; Talelli et al. 2007; Valero-Cabre et al. 2008). These methods all work through electromagnetic induction, whereby pulsed magnetic fields are used to generate electric fields and currents in the brain. These currents can be of sufficient magnitude to depolarize neurons, and when applied repetitively (or in pulsed form, at various frequencies), they can modulate cortical excitability, decreasing or increasing it, depending on the parameters of stimulation (Wagner et al. 2007b). Similarly, many forms of tES have been proposed, such as tDCS, transcranial alternating current stimulation (tACS), and transcranial random noise stimulation (tRNS) (Nitsche et al. 2003; Paulus 2003; Priori 2003; Liebetanz et al. 2006; Fregni and Pascual-Leone 2007; Antal et al. 2008; Terney et al. 2008; Ambrus et al. 2010). During these types of tES, currents are applied via scalp electrodes and penetrate the skull to enter the brain. Most commonly, tES stimulation is applied in a passive modulatory way, i.e., stimulation is applied in such a manner as to alter the probability of a spontaneous neural firing, either inhibiting or facilitating the neural activity without actively initiating neural firing. However, it should be noted that tES as proposed by Merton and Morton in 1980 is an active technique, i.e., stimulation actively raises the neural membrane potential above the threshold and generates an action potential (Merton and Morton 1980). Both transcranial magnetic and electrical techniques can be provided in a manner that affects the targeted tissues past the duration of stimulation, which serves as the basis of most therapeutic applications.

In this chapter, we will review the biophysics and engineering of neural stimulation via electrical and magnetic means. We will first review the basic principles of an excitable membrane and the elements of a neuron that allow physicians to influence its function via electromagnetic fields. Second, we will review the basic electromagnetic principles of stimulation for magnetic and electrical sources and principles of stimulator device design. Next we will review modeling of transcranial electrical and magnetic stimulation.

5.2   EXCITABLE MEMBRANES

Neurons are excitable cells, i.e., they respond to appropriate stimuli by a change in their membrane potential. In the brain, neurons primarily receive and pass on stimuli at chemical synapses, where information is transmitted from a presynaptic to a postsynaptic neuron by means of neurotransmitters. Within neurons, the information is transmitted from the dendrites to the axon terminals in the form of electrical signals. Neuronal excitability is due to existence of different types of channels that cross the cell membrane and render the membrane selectively permeable to specific ions. Passive ionic channels and differences in intra- and extracellular ion concentrations establish a potential difference across the membrane, V, polarizing it. At rest, the membrane potential of typical mammalian neural cells is close to the Nernst potential for K^+, and so the potential in the intracellular side of the membrane, V_i, is typically 60 mV lower than the potential in its extracellular side, V_e, i.e., $V = V_i - V_e \approx -60$ mV. The membrane potential departs from its resting value when ligand-gated or voltage-gated channels open as a result of neurotransmitter release at synapses or of sufficiently large decreases in membrane polarization. When a neuron's membrane potential is raised above its threshold, it generates an action potential, an all-or-none electrical event characterized by the depolarizing opening of Na^+ channels and an inward Na^+ current followed by the opening of K^+ channels and an outflow of K^+ ions, which repolarizes the neuron.

The membrane potential can be altered by an external electric field that drives current to flow through the membrane. In transcranial stimulation, this external electric field is created either by a current pulse in a coil placed close to the scalp (electromagnetic induction) or by applying a potential difference between two or more electrodes on the scalp or proximal locations. Stimulation can either be "active" or be "passive." We define "active" stimulation as stimulation that can elicit an action potential in the neuron (i.e., stimulation that depolarizes the neuron above its membrane threshold during or immediately after the stimulus) and "passive" stimulation as stimulation that can alter the likelihood of a neuron spiking spontaneously. Traditionally, one would think of single-pulse TMS, such as that applied to the motor cortex to generate a motor-evoked potential (MEP), as being an "active" form of stimulation and tDCS, such as anodal tDCS applied to the motor cortex to increase its excitability, as being a "passive" form of stimulation. Both "active" and "passive" forms of stimulation can be applied in such a way that their effects last past the duration of stimulation. Examples are 1 Hz rTMS provided to inhibit the motor cortex or anodal tDCS to increase its excitability. For both "active" and "passive" stimulation, effects are brought about via electromagnetic field interactions with the neural tissue, and thus we review the neuron as described in typical electrical circuit form and the manner in which fields can interact with the neural circuit.

5.2.1 RESTING POTENTIAL

Under physiological conditions, a neural cell maintains a potential difference, V, across its membrane. This membrane potential is the result of unequal ionic concentrations between the inside and outside of the cell, which give rise to a flow of ions through the membrane channels along their electrochemical gradients. A steady state is reached when the net flow of ions through the membrane is zero, i.e., when the effect of the concentration difference is balanced by the effect of the potential difference. The membrane potential is then equal to its resting value, V_{rest}.

The resting membrane potential depends on the permeability of the membrane to the different ions and can be mathematically described by the Goldman equation. In the case of a mammalian neuron, the membrane potential is primarily determined by the flow of the Na^+, Cl^-, and K^+ ions:

$$V_{rest} = \frac{RT}{F} \ln \frac{P_K \left|K^+\right|_{out} + P_{Na} \left|Na^+\right|_{out} + P_{Cl} \left|Cl^-\right|_{in}}{P_K \left|K^+\right|_{in} + P_{Na} \left|Na^+\right|_{in} + P_{Cl} \left|Cl^-\right|_{out}} \tag{5.1}$$

where
  R is the universal gas constant
  T is the absolute temperature (K)
  F is the Faraday constant
  ln is the natural log
  in is the intracellular
  out is the extracellular
  $[ion]$ represents the ion's concentration (mM)
  P_{ion} indicates the permeability of the membrane to the ion (m/s)

In some cells, such as a squid giant axon, the Goldman equation can reduce to the Nernst equation, as unequal membrane ion permeabilities result in one ion having the dominant impact on the membrane potential.

5.2.2 PASSIVE AND EXCITABLE MEMBRANES

External stimuli, such as electromagnetic fields applied with transcranial stimulation technologies, can alter the membrane potential of neurons. A way to integrate the effects of the electromagnetic stimulation fields on the cells is to model the neuron as an electrical circuit with electrical elements that can respond to the applied fields.

Well below the threshold for action potential generation, the membrane's response to an applied electric field is determined by the passive properties of the membrane. The passive neural membrane is typically composed of passive ion channels, which act as resistors to the flow of ions. The membrane also acts as a capacitor, with equal

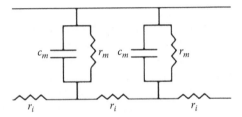

FIGURE 5.1 Passive cable model.

amounts of charge of opposite sign accumulating on either side of the membrane. As in a capacitor, the potential difference between both sides of the membrane, V, is proportional to the amount of charge, Q, on each side of the membrane, i.e., $Q = CV$, where C is the capacitance of the membrane. As an electrical circuit in its simplest form, a small patch of passive membrane can be represented by a resistor, R, and a capacitor, C, forming a parallel RC circuit (see Figure 5.1). Models of cellular elements with passive membranes can be built by interconnecting individual compartments, which represent the geometric and electrical properties of the modeled element. For example, models can be built to describe cellular elements whose length far exceeds the width. In such models, the spatial and temporal variations of the membrane potential, V, can be described by the typical cable equation used to describe transmission lines:

$$\lambda^2 \frac{\partial^2 V}{\partial x^2} - V = \tau \frac{\partial V}{\partial t} \tag{5.2}$$

where
$$\lambda = \sqrt{r_m / r_i}$$
$\tau = c_m^* r_m$, r_m is the membrane resistance in units of ohms* unit length, c_m is the membrane capacitance in units of Farads per unit length, and r_i is the axoplasm resistance in units of ohms per unit length (as in Figure 5.1)

With excitable membranes, one follows the same methods as earlier except that the modeling process accounts for the voltage-dependent characteristics of the individual ionic channels in the membrane (see Figure 5.2). Herein, the membrane is described by an electrical circuit composed of a capacitor, resistors, and batteries.

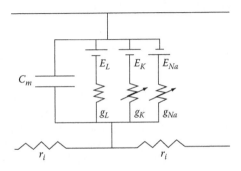

FIGURE 5.2 Excitable membrane model.

The capacitor is as described earlier. The Nernst potentials of individual ions can be used to determine the battery voltages for each ion. And the resistors are described for each ion, as time- and voltage-dependent functions. In this regime, changes in membrane potential can result in an action potential.

As initially described by Hodgkin and Huxley in 1952, an excitable membrane can be mathematically described by a set of nonlinear differential equations, which account for the time and voltage dependence of the individual ion channels (Hodgkin and Huxley 1952). The membrane current can be described by the following equation:

$$\lambda^2 \frac{\partial^2 V}{\partial x^2} + g_L(E_L - V) + g_{Na}(E_{Na} - V) + g_K(E_K - V) = C_m \frac{\partial V}{\partial t} \tag{5.3}$$

where

$\lambda = \sqrt{1/2\pi a^2 r_i}$, a is the axon radius, r_i is the axoplasm resistance per unit length
C_m is given as capacitance per unit area

This equation includes the voltage/time-dependent sodium and potassium channels represented by g_K and g_{Na} (conductances per unit area), the static leakage channels represented by g_L, and the Nernst potential for the sodium, potassium, and leakage ions represented by E_{Na}, E_K, and E_L. More details about the Hodgkin–Huxley model and its implementation can be found in Hodgkin and Huxley (1952) and Plonsey and Barr (2000).

5.2.3 STIMULATION BY APPLIED FIELDS

To determine the effects of transcranially applied stimulation fields, one must first determine the applied field distribution and then integrate those results with the earlier analysis. In essence, this amounts to adding an activation function to the preceding equations, which is representative of the applied stimulation fields. The importance of the field calculations is that they are used to determine the effect of the tissues on the applied fields and the relative orientation of the stimulation fields to the neural elements.

Mathematically, one could adapt the earlier equations to account for external electrical stimuli. For example, with a long peripheral nerve stimulated with a TMS field, one would adapt Equation 5.3 to now read thus:

$$\lambda^2 \frac{\partial^2 V}{\partial x^2} + g_L(E_L - V) + g_{Na}(E_{Na} - V) + g_K(E_K - V) = C_m \frac{\partial V}{\partial t} + \lambda^2 \frac{\partial^2 A}{\partial x \partial t} \tag{5.4}$$

where $-\partial A/\partial t$ represents the induced electric field. This model is a simplified one-dimensional model proposed originally by Roth (Roth and Basser 1990) based on the early Rattay model for electrical stimulation (Rattay 1989), which can be expanded in principle to other conditions. Depending on the system under study, the activating function can be adapted (such as for an electrical source) and boundary equations are developed for the neurons of finite length (or for boundaries between different cellular elements).

More details can be found in the literature. Most modern models are based on the work of McNeal (1976), which was later extended by Rattay (1989). Currently, analysis of stimulation fields is most fully explored for deep brain stimulation applications (McIntyre and Grill 1999, 2000; Durand and Bikson 2001; McIntyre and Grill 2001, 2002; Rattay et al. 2003; Durand and Bikson 2004; McIntyre et al. 2004a,b) but has been studied with TMS models (Roth and Basser 1990; Basser and Roth 1991; Roth et al. 1991, 1994; Nagarajan et al. 1993; Roth 1994; Nagarajan and Durand 1995; Butson and McIntyre 2005) and with subthreshold stimulation techniques (Tranchina and Nicholson 1986). While the impact of stimulation on cells and the computational aspects of this topic can serve as the basis of a complete textbook in and of itself, we direct the readers to the earlier articles, electrophysiology texts such as Johnston and Wu (1994) and Roth (1994), and deeper discussion included in other sections of this book.

5.3 BASIC ELECTROMAGNETIC PHYSICS CONCEPTS

The driving fields of stimulation are electric fields and currents, which can be generated with magnetic or electrical sources.

5.3.1 CURRENT INDUCTION (MAGNETIC SOURCE)

Magnetic stimulation is based on Faraday's law of electromagnetic induction (Faraday 1914, 1832; d'Arsonval 1896). A time-changing magnetic field induces an electric field, which drives currents in a conducting material exposed to it (see Figure 5.3).

In the case of TMS, the stimulation coil serves as an electromagnet, which generates a time-changing magnetic field due to a current in the coil. The magnetic field distribution is determined by both the current driving the electromagnet (magnitude and time course) and the physical properties of the stimulating coil (geometry and material properties). When the coil is placed on the scalp and the magnetic field is focused on the brain, an electric field is induced in the underlying neural and non-neural tissues. This electric field drives currents in the tissues, the characteristics

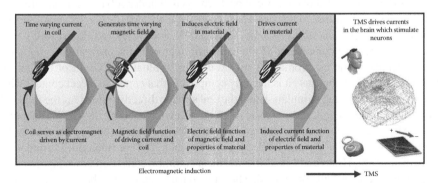

FIGURE 5.3   (See color insert.) Basics of TMS. (Partially adapted from Wagner, T. et al., *Cortex*, 45, 1025, 2009.)

of which are determined by their electrical impedance properties (i.e., electrical conductivity, electrical permittivity, and magnetic permeability, which is typically considered that of free space for the majority of TMS applications). Thus, the cortical current densities in TMS are determined by the stimulus waveform, the stimulating coil, and the tissue distribution that is unique to each subject being stimulated. Depending on the amplitude, dynamics, and duration of magnetic stimulation, the nervous system can be affected during and after stimulation.

5.3.1.1 Device Basics

Magnetic stimulation hardware is based upon a capacitive charge–discharge system and a stimulating coil. The charge–discharge system requires high-voltage and high-current capacity and is composed of a charging unit, a bank of storage capacitors, switching circuitry, and control electronics. Without the switching circuitry and control electronics, the circuit is essentially composed by a storage capacitor with capacitance C and a stimulating coil that is characterized by its inductance L. A higher inductance causes the current in the coil to rise more slowly and reduces the strength of the induced electric field. The resistance R of the coil and other components must also be taken into account. These three elements are connected in series to form an RLC circuit, where the resistance and the inductance are both set to the lowest practical values to minimize heating while generating the desired waveform. Stimulus duration is determined mainly by $\sqrt{LC}$. Reverse charging or ringing in the circuit is prevented by placing a diode thyristor between the capacitor and the inductor thereby increasing the current decay time and eliminating reverse currents.

For repetitive stimulators, the essential circuitry remains the same, except that modifications are made to the switching system to allow pulse rates of many times per second. Herein, thyristor switching schemes can be used to recover energy to the capacitive charging unit and increase charging rates. Recent generation devices allow upward of 100 Hz stimulation frequencies (Magstim 2012).

The difficulties in designing modern machines relate to overcoming the demands on the circuitry while optimizing the device components to generate the appropriate coil current waveforms for neural stimulation. Some of these demands are related to the high voltage (400 V to more than 3 kV), high current (4 kA to more than 10 kA), and high power (where over 500 J of energy can be discharged in under 100 μs, or approximately 5 MW) that are required to achieve TMS. For a detailed TMS device technology/electronics review, see Jalinous (2002), Hsu et al. (2003), and Davey and Riehl (2005).

A key hardware component of magnetic stimulators is the current-carrying coil, which serves as the electromagnet source during stimulation. Design of the coil is critically important because it is the only component that comes in direct contact with the subject undergoing stimulation and the coil's shape directly influences the induced current distribution and, thus, the site of stimulation (Cohen et al. 1990). While many researchers have explored unique coil designs for increased focality (Kraus et al. 1993; Carbunaru and Durand 2001; Hsu and Durand 2001) or functional stimulation (Lin et al. 2000; Hsiao and Lin 2001), the most common coils used nowadays are single circular loop or figure-8 shaped. They are constructed from tightly wound copper wire, which is adequately insulated and housed in plastic covers along with temperature sensors and safety switches. The choice of copper is primarily driven by its low electrical resistance, high heat capacity and tensile strength, availability, and relative low cost. Exploration of other materials seems desirable as it might lead to means to construct smaller, and thus more focal, stimulation coils. Commercially available coil diameters range from 4 to 9 cm, with anywhere from 10 to 20 turns, with typical coil inductances ranging from approximately 15 μH to approximately 150 μH.

Some have explored the design of coils to attain subcortical stimulation (Roth et al. 2002; Zangen et al. 2005). However, it has been shown analytically that TMS currents will always be maximum at the cortical surface (Heller and van Hulsteyn 1992). Nevertheless, Roth et al. (2002) developed a coil design where the rate of decay from the surface is attenuated, such that deeper structures can be stimulated (simultaneously with the overlaying cortical surface) without the need of excessive field strengths. More recently, researchers have been investigating the use of conducting shields, placed between the TMS coil and the patients' head, to alter and focus the stimulating fields (Davey and Riehl 2006; Kim et al. 2006). The use of nonlinear coil materials has also been explored, but has not been implemented commercially.

Numerous forms of magnetic stimulation exist, including single-pulse TMS, rTMS, paired-pulse TMS, and theta burst TMS. Single-pulse TMS provides single magnetic pulses. rTMS provides magnetic pulses continuously, usually at frequencies from 1 to 20 Hz, where the different frequencies can elicit different neural responses. Paired-pulse TMS provides two pulses in rapid succession (Rothwell 1999), from approximately 2 to 20 ms apart, where the first pulse conditions the effects of the second pulse. In theta burst TMS, three pulses of stimulation are given at 50 Hz, repeated every 200 ms, in different patterns to elicit different neural responses (Huang et al. 2005). All of these forms of stimulation work through inductive processes, but with different pulse patterns, and are detailed in part throughout different chapters in this book.

5.3.2 CURRENT INJECTION (ELECTRICAL SOURCE)

During tES, a low-amplitude current is applied to the cortex via surface-mounted scalp electrodes (see Figure 5.4). Neural activity can be affected during and after stimulation. Different applied current waveforms have been implemented, conventionally controlling current applied to the scalp. Stimulation pattern examples include DC (tDCS), balanced stepped signals (cranial electrical stimulation), low-frequency sinusoidal signals (tACS), and randomized signals (tRNS); these and other techniques are reviewed in the following text. Scalp currents used in tDCS and tACS generally range in magnitude from approximately 0.5 to 2 mA and are applied from seconds to minutes. Conventional therapies are usually applied in sessions lasting approximately 10–20 min. The electrodes can be simple saline-soaked cotton pads or specifically designed sponge patches covered with conductive gel and range in size from ~1 to 35 cm^2 depending on the application. With smaller electrodes, there is a greater current density in the scalp, which may prevent sham stimulation or increase the likelihood of complications such as scalp burns. One should note that dosing is just not a result of total current, but also influenced by current density, i.e., dosing is a result of relative electrode positions and size, in addition to the amount and duration of current.

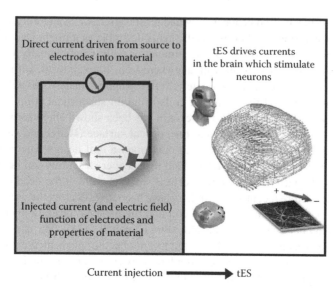

Current injection ⟶ tES

FIGURE 5.4   **(See color insert.)** Basics of tES.

5.3.2.1   Device Basics

There is no complex circuitry comprising these stimulators, and in their simplest form, a DC source is placed in series with the scalp electrodes and a potentiometer (variable resistor) to adjust for constant current. In practice, the constant current source is implemented with integrated circuits. Another circuit is used to generate a voltage of about 20–30 V from low-voltage batteries. This is required because of the skin's resistance and electrode polarization.

tDCS provides DC to the scalp where the polarity of the current determines the direction of neural effect. tACS provides fields of similar strength to those generated by tDCS, but that vary sinusoidally in time. And tRNS stimulation provides tDCS strength fields with randomized signal amplitude. One potential advantage of tACS and tRNS with respect to tDCS is that the charge transferred into the tissues is zero. This may make these two techniques safer than tDCS since tissue damage for tDCS is related to the total charge per unit electrode area injected into the tissue (Liebetanz et al. 2009). However, this needs further exploration as the manner in which the currents propagate through tissues is fundamentally different (ohmic for tDCS [static] vs. ohmic and possibly capacitive for tACS [time dependence]). All of these forms of stimulation work through current injection processes, but with different injection patterns, and are detailed in part throughout different chapters in this book.

5.4   FIELD MODELING

Modeling of the biophysics of stimulation can be divided into two key aspects: the cellular response, covered in the earlier part of the document, and modeling the electromagnetic fields, covered in this section. Early electromagnetic field models of transcranial brain stimulation were based on simplified geometries representing the human brain, such as infinite planes or spheres, which modeled the tissues as simple

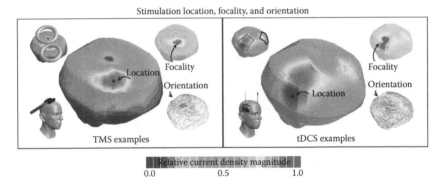

Stimulation location, focality, and orientation

TMS examples

tDCS examples

Relative current density magnitude
0.0   0.5   1.0

FIGURE 5.5   (See color insert.) TMS and tDCS field modeling examples of focality, orientation, and location. (Partially adapted from Wagner, T. et al., *Cortex*, 45, 1025, 2009.)

homogeneous conductors. With improved computational resources, models now include more realistic geometries (Nadeem et al. 2003; Lu et al. 2008; Toschi et al. 2008), tissue anisotropy (Miranda et al. 2003; De Lucia et al. 2007), and frequency-dependent conductivity and permittivity (Wagner et al. 2004). Tissue anisotropy can have a significant effect on the current-to-neural orientations, and the individual tissue anatomies and electromagnetic impedance parameters have proven necessary in predicting the maximum current density locations. These parameters are especially important in regions such as boundaries between sulci and gyri, as they can often lead to the perturbation of the predicted location (and magnitude) of current density maxima when compared with regions of continuous tissue (Miranda et al. 2003; Wagner et al. 2004). The electromagnetic field distributions that arise during stimulation are fundamental to understanding the resultant neural effects and have been studied to predict the cellular mechanism of activation (Roth and Basser 1990; Nagarajan et al. 1993), location (De Lucia et al. 2007; Wagner et al. 2007b), focality (Ueno et al. 1988; Cohen and Cuffin 1991; Toschi et al. 2008), depth of penetration (Heller and van Hulsteyn 1992; Zangen et al. 2005), and degree of stimulation (Bohning et al. 1997; Wagner et al. 2007b; see Figures 5.5 and 5.6). The field distributions can be also used for the quantitative analysis of the safety parameters (McCreery and Agnew 1990; McCreery et al. 1990; Wagner et al. 2007b) and the technological potential of brain stimulation (Wagner et al. 2007b).

5.4.1 GENERAL MODELING STRATEGIES

The electromagnetic equations that describe the fields that drive stimulation are based on Maxwell's equations. Computational modeling is needed, as it is not possible to analytically solve for the fields seen during stimulation in complex multi-tissue systems like the human brain, although early studies have analytically explored aspects of brain stimulation (Rush and Driscoll 1968; Tofts 1990; Eaton 1992; Heller and van Hulsteyn 1992). The process can be divided into modeling the source of the electromagnetic fields, modeling the head and brain tissues to be stimulated (defining tissue boundaries and electromagnetic properties of the tissues), and solving the set of equations describing the system.

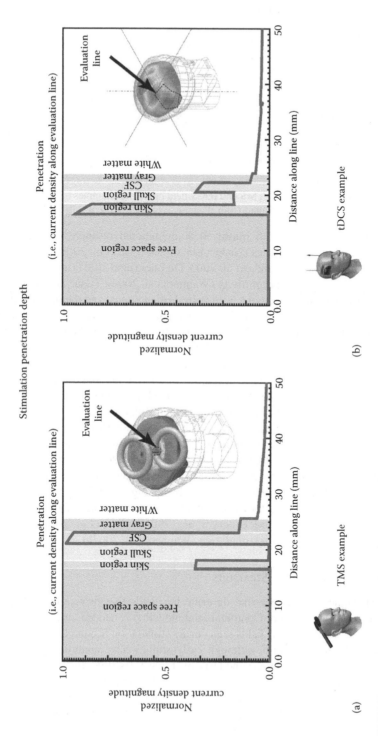

FIGURE 5.6 **(See color insert.)** (a) TMS and (b) tDCS field modeling example of depth of stimulation. (Adapted from Wagner, T. et al., *Annu. Rev. Biomed. Eng.*, 9, 527, 2007b.)

5.4.1.1 Source Modeling

In terms of modeling the driving source of the electromagnetic fields during magnetic stimulation, one must capture the coil location/position, geometry, material properties, and driving current parameters (amplitude and waveform). In the case of electrical stimulation, the same parameters must be captured with respect to the electrodes.

5.4.1.2 Anatomy and Tissue Modeling

Modeling the head/brain system in simulation space, which will serve as the basis for the field calculations, requires the adequate segmentation of the tissues to be included in the field solutions and assigning the appropriate tissue properties to the segmented tissues. The appropriate model choice depends on what question one plans on answering through the model solutions. Early models of electromagnetic brain stimulation, which were focused on simple field questions and the impact of different TMS coil shapes, ignored the head geometry/tissue distribution and represented the brain as free space (Grandori and Ravazzani 1991) or as infinite half planes (Branston and Tofts 1991; Esselle and Stuchly 1992). Recently, models focused on answering questions related to the effects of source parameters, tissue geometries, and/or tissue properties have also been pursued with spherical shell models (Rush and Driscoll 1968; Gandevia 1996; Miranda et al. 2003, 2006), basic head shapes as guided by MRI (Wagner et al. 2004, 2007a), or more resolved models based on direct segmentation of the MRI or other imaging methods (Datta et al. 2008, 2009; Chen and Mogul 2009; Sadleir et al. 2010; Salvador et al. 2010; Thielscher et al. 2011). A greater number of tissues and a higher geometric accuracy can lead to more accurate solutions to the fields of stimulation.

The next step is to assign the appropriate electromagnetic properties to the tissues. The power spectrum of typical stimulators is composed of components less than 10 kHz. Tissue conductivities, while somewhat attenuated below 1 kHz, have been considered essentially constant for the frequencies of brain stimulation and the permittivity values implemented during modeling generate quasi-static solutions with negligible displacement currents such that the permittivities can be disregarded (Cohen et al. 1991; Roth et al. 1991; Eaton 1992; Ueno and Liu 1997). However, with the permittivity values predicted via alpha dispersion, theorized to be caused by cellular organization and counterion diffusion effects, the charge relaxation times of the tissue can be of the same order of magnitude or greater than the time scale of the stimulating current source such that displacement currents need to be considered (Schwan 1957; Hasted 1973; Ludt and Herrmann 1973; Singh et al. 1979; Pethig 1984; Schwan 1985; Pethig and Kell 1987; Foster and Schwan 1989; Dissado 1990; Hart and Dunfree 1993; Foster and Schwan 1996; Gabriel et al. 1996a–c; Hart et al. 1996; Jonscher 1996; Martinsen et al. 2002). Furthermore, material science predicts the trend of increasing permittivity and decreasing conductivity with decreased frequency such that frequency dependence of the source and stimulation waveforms may need to be considered (see references in previous sentence).

5.4.1.3 Computational Aspects

In biologic systems at low frequencies, electromagnetic systems are usually analyzed via quasistatic forms of Maxwell's equations. Quasistatic approximations are often made when the time rates of change of the dynamic components of the system are slow compared with the processes under study, such that the wave nature of the fields can be neglected. In practice, the electromagnetics of low-frequency systems are generally addressed via either electroquasistatic or magnetoquasistatic methods where either the electric or the magnetic fields are the primary fields of importance. For a further review of quasistatics, see Melcher (1981). Ultimately, numerical techniques such as the boundary element method, finite element method or finite difference methods are used to solve the equations. See texts such as Sadiku (2000) for more details on computational methods.

Ultimately, one must implement and assess the validity of the models as appropriate for the systems/questions under study. While models have been validated in comparison with field measurements made in phantom, animal, and human recordings (Tay et al. 1989, 1991; Cohen et al. 1990; Maccabee et al. 1990; Yunokuchi and Cohen 1991; Tay 1992; Lisanby 1998; Yunokuchi et al. 1999; Wagner et al. 2004), many aspects of modeling cannot be easily validated with such methods due to limits in resolution and/or brain access of field modeling recording (e.g., recording probe size, surgical perturbation of recording site). Other means, such as magnetic resonance current density imaging (Joy et al. 1999) and poststimulation ^{14}C-2-deoxyglucose brain metabolism analysis (2DG) (Wagner et al. 2009) correlated with field modeling, have been suggested as methods to more fully depict the field distribution and/or impact of the stimulating field in the brain. Again, these methods also have limitations. But when appropriately applied, field models can answer many questions; thus, choosing the appropriate model for the situation under study is key, for instance, a less resolved model (Wagner et al. 2007a) would be inappropriate for calculating patient-specific field doses compared with a model with superior resolution (Salvador et al. 2010). However, the use of less resolved models could prove advantageous where computer resources are limited or the question under study does not depend on individual patient anatomical features).

5.4.2 Field Modeling Impact

Field modeling can serve to guide researchers beyond typical dosing/targeting concerns and impact considerations from safety limitations to supporting technologies.

5.4.2.1 Safety

The safety of TMS or tES is discussed elsewhere in this book. However, many topics of biophysical origin should be further considered as neurostimulation technologies evolve. For example, as TMS stimulation depths increase, it remains to be seen what effects will result from such overlying surface stimulation. Furthermore, as different waveforms are adapted for stimulation, the mechanisms by which charges are carried in tissues will need to be further explored in terms of neurohistotoxicity; while current metrics are typically based on the cofactors of charge per phase and total charge of stimulation pulses and electrode properties, further field parameters could prove important as further technologies and waveforms are explored (Yuen et al. 1981; Agnew and McCreery 1987; McCreery and Agnew 1990; McCreery et al. 1990; Wagner et al. 2004; Liebetanz et al. 2009). Another consideration that impacts safety is that, in the presence of pathologies, biophysical changes in the tissues and/or the replacement of tissues with new tissues with different electromagnetic properties can lead to altered stimulation current density distributions (Wagner et al. 2006a, 2007a,b, 2008). Depending on the pathology and the source of stimulation, currents could become unsafe to the tissue or directed/localized in regions not predicted by conventional targeting systems (Wagner et al. 2007b). Another safety-related area to consider during stimulation studies is the use of non-translatable metrics, such as machine output power (alone or as a relative measurement, e.g., MEP threshold), to quantify stimulation efficiency. As made clear from the preceding discussion, it is difficult to apply such metrics across different patients, devices, or brain regions in individual patients to describe variations in the biophysical or electrophysiological stimulatory effect. Until we have an objective measure that integrates the neural architecture, gauged cellular excitability, and induced current density distributions to be used with all studies, these metrics should be used cautiously. In the meantime while such predictive measures are unavailable, relative machine output values should at least be replaced with objective measures, such as pickup probe captured field dynamics measured at the coil interface, to account for device variability, and correlated with simplified field calculations based on relative coil to patient variables, to account for variability between individual patients and separate stimulation sessions. In tES, the question of how to adjust current intensity for electrodes of different sizes so as to produce equally effective fields in the brain remains open (Miranda et al. 2009).

5.4.2.2 Stereotactic Systems

MRI-guided neurotracking systems, which predict the location of stimulation by localizing the relative projection of the stimulating coil to the underlying brain anatomy, have been proposed. These systems can be further integrated with field-solver systems to make generalized predictions of the induced current magnitude, focality, and penetration depth. However, the majority of commercially available MRI-based neurotracking devices are based on simplified models, which ignore subject-specific electromagnetic field–tissue interactions or implement oversimplified approximations of the cortical current densities (Wagner et al. 2007b). Thus, these technologies provide little, if any, patient-specific information, no information about the predicted neural effect, and can in fact produce inaccurate predictions in regions of cortical inhomogeneity. This serves as an important area for future development, especially as technologies push the boundaries of what is possible with noninvasive stimulation. For instance, tracking systems have been explored that integrate other imaging modalities like diffusion tensor imaging (see Figure 5.7; Wagner et al. 2006b).

5.5 CONCLUDING COMMENTS

Ultimately, transcranial magnetic and electrical stimulation methodologies are limited in effect compared with invasive methods such as deep brain stimulation. These limitations will be improved upon with a greater knowledge and understanding of the fundamental biophysics of stimulation. With an understanding of these biophysical foundations, more advanced technologies will hopefully become available in the near future.

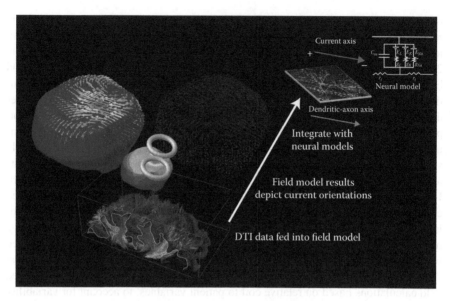

FIGURE 5.7 **(See color insert.)** From DTI to a neural model (a method for field localization and tracking). (From Wagner, T. et al., Transcranial magnetic stimulation: High resolution tracking of the induced current density in the individual human brain, *12th Annual Meeting of the Organization for Human Brain Mapping*, Florence, Italy, 2006b.)

REFERENCES

Agnew, W. F. and D. B. McCreery. 1987. Considerations for safety in the use of extracranial stimulation for motor evoked potentials. *Neurosurgery* 20:143–147.

Aldini, G. 1794. *De animali electricitate dissertationes duae*. Bologna, Italy: Lucheron.

Aldini, G. 1804. *Essai theorique et experimental sur le galvanisme*, Vol. 2. Paris, France: Fournier.

Ambrus, G. G., W. Paulus, and A. Antal. 2010. Cutaneous perception thresholds of electrical stimulation methods: Comparison of tDCS and tRNS. *Clin Neurophysiol* 121:1908–1914.

Antal, A., K. Boros, C. Poreisz, L. Chaieb, D. Terney, and W. Paulus. 2008. Comparatively weak after-effects of transcranial alternating current stimulation (tACS) on cortical excitability in humans. *Brain Stimul* 1:97–105.

d'Arsonval, A. 1896. Dispositifs pour la mesure des courants alternatifs de toutes fréquences. *C R Soc Biol (Paris)* 3:450–457.

Barker, A. T., I. L. Freeston, R. Jalinous, P. A. Merton, and H. B. Morton. 1985. Magnetic stimulation of the human brain. *J Physiol* 369:3P (abstract).

Basser, P. J. and B. J. Roth. 1991. Stimulation of a myelinated nerve axon by electromagnetic induction. *Med Biol Eng Comput* 29:261–268.

Bohning, D. E., A. P. Pecheny, C. M. Epstein et al. 1997. Mapping transcranial magnetic stimulation (TMS) fields in vivo with MRI. *Neuroreport* 8:2535–2538.

Branston, N. M. and P. S. Tofts. 1991. Analysis of the distribution of currents induced by magnetic field in a volume conductor. *Phys Med Biol* 36:161–168.

Butson, C. R. and C. C. McIntyre. 2005. Tissue and electrode capacitance reduce neural activation volumes during deep brain stimulation. *Clin Neurophysiol* 116:2490–2500.

Carbunaru, R. and D. M. Durand. 2001. Toroidal coil models for transcutaneous magnetic stimulation of nerves. *IEEE Trans Biomed Eng* 48:434–441.

Chen, M. and D. J. Mogul. 2009. A structurally detailed finite element human head model for simulation of transcranial magnetic stimulation. *J Neurosci Methods* 179:111–120.

Cohen, D. and B. N. Cuffin. 1991. Developing a more focal magnetic stimulator. Part 1: Some basic principles. *J Clin Neurophysiol* 8:102–111.

Cohen, L. G., B. J. Roth, J. Nilsson et al. 1990. Effects of coil design on delivery of focal magnetic stimulation. Technical considerations. *Electroencephalogr Clin Neurophysiol* 75:350–357.

Cohen, L. G., B. J. Roth, E. M. Wassermann et al. 1991. Magnetic stimulation of the human cerebral cortex as an indicator of reorganization in motor pathways in certain pathological conditions. *J Clin Neurophysiol* 8:56–65.

Datta, A., V. Bansal, J. Diaz, J. Patel, D. Reato, and M. Bikson. 2009. Gyri-precise head model of transcranial DC stimulation: Improved spatial focality using a ring electrode versus conventional rectangular pad. *Brain Stimul* 2:201–207.

Datta, A., M. Elwassif, F. Battaglia, and M. Bikson. 2008. Transcranial current stimulation focality using disc and ring electrode configurations: FEM analysis. *J Neural Eng* 5:163–174.

Davey, K. and M. Riehl. 2005. Designing transcranial magnetic stimulation systems. *IEEE Trans Magn* 41:1142–1148.

Davey, K. and M. Riehl. 2006. Suppressing the surface field during transcranial magnetic stimulation. *IEEE Trans Biomed Eng* 53:190–194.

De Lucia, M., G. J. Parker, K. Embleton, J. M. Newton, and V. Walsh. 2007. Diffusion tensor MRI-based estimation of the influence of brain tissue anisotropy on the effects of transcranial magnetic stimulation. *Neuroimage* 36:1159–1170.

Dissado, L. A. 1990. A fractal interpretation of the dielectric response of animal tissues. *Phys Med Biol* 35:1487–1503.

Duchenne (de Boulogne) GBA. *De l'électrisation localisée et de son application à la physiologie, à la pathologie et à la thérapeutique.* Paris: Baillière, 1855. [2nd edition, 1861; 3rd edition, 1872; English translation and edition by Poore GV. *Selections from the Clinical Works of Duchenne* (de Boulogne). London: The New Sydenham Society, 1883]; pp. 4–5.

Durand, D. M. and M. Bikson. 2001. Suppression and control of epileptiform activity by electrical stimulation: A review. *Proc IEEE* 89:1065–1082.

Durand, D. M. and M. Bikson. 2004. Control of neuronal activity by electrical fields: In vitro models of epilepsy. In *Deep Brain Stimulation and Epilepsy*, ed. H. Luders. London, U.K.: Taylor & Francis Group, pp. 67–86.

Eaton, H. 1992. Electric field induced in a spherical volume conductor from arbitrary coils: Applications to magnetic stimulation and MEG. *Med Biol Eng Comput* 30:433–440.

Esselle, K. P. and M. A. Stuchly. 1992. Neural stimulation with magnetic fields: Analysis of induced electrical fields. *IEEE Trans Biomed Eng* 39:693–700.

Faraday, M. 1914. *Experimental Researches in Electricity*, ed. J. Tyndall. London, U.K.: JM Dent & Son's.

Faraday, M. 1832. Experimental researches in electricity. *Phil Trans R Soc Lond* 122:125–162.

Foster, K. R. and H. P. Schwan. 1989. Dielectric properties of tissues and biological materials: A critical review. *Crit Rev Biomed Eng* 17:25–104.

Foster, K. R. and H. P. Schwan. 1996. Dielectric properties of tissues. In *CRC Handbook of Biological Effects of Electromagnetic Fields*, eds. C. Polk and E. Postow. New York: CRC Press, pp. 25–102.

Fregni, F. and A. Pascual-Leone. 2007. Technology insight: Noninvasive brain stimulation in neurology-perspectives on the therapeutic potential of rTMS and tDCS. *Nat Clin Pract Neurol* 3:383–393.

Gabriel, C., S. Gabriel, and E. Corthout. 1996a. The dielectric properties of biological tissues: I. Literature survey. *Phys Med Biol* 41:2231–2249.

Gabriel, S., R. W. Lau, and C. Gabriel. 1996b. The dielectric properties of biological tissues: II. Measurements in the frequency range 10 Hz to 20 GHz. *Phys Med Biol* 41:2251–2269.

Gabriel, S., R. W. Lau, and C. Gabriel. 1996c. The dielectric properties of biological tissues: III. Parametric models for the dielectric spectrum of tissues. *Phys Med Biol* 41:2271–2293.

Galvani, L., E. Valli, B. Carminati, A. Volta, and J. Mayer. 1793. *Aloysi Galvani abhandlung über die kräfte der thierischen elektrizität auf die bewegung der muskeln.* Prague, Czech Republic: J. G. Calve.

Gandevia, S. C. 1996. Insights into motor performance and muscle fatigue based on transcranial stimulation of the human motor cortex. *Clin Exp Pharmacol Physiol* 23:957–960.

George, M. S., E. M. Wassermann, W. A. Williams et al. 1995. Daily repetitive transcranial magnetic stimulation (rTMS) improves mood in depression. *Neuroreport* 6:1853–1856.

Grandori, F. and P. Ravazzani. 1991. Magnetic stimulation of the motor cortex – Theoretical considerations. *IEEE Trans Biomed Eng* 38:180–191.

Hart, F. X. and W. R. Dunfree. 1993. In vivo measurements of low frequency dielectric spectra of a frog skeletal muscle. *Phys Med Biol* 38:1099–1112.

Hart, F. X., R. B. Toll, N. J. Berner, and N. H. Bennett. 1996. The low frequency dielectric properties of octopus arm muscle measured in vivo. *Phys Med Biol* 41:2043–2252.

Hasted, J. B. 1973. Aqueous dielectrics. In *Studies in Chemical Physics*, 1 edn. New York: Halsted Press, pp. 318.

Heller, L. and D. B. van Hulsteyn. 1992. Brain stimulation using electromagnetic sources: Theoretical aspects. *Biophys J* 63:129–138.

Hodgkin, A. L. and A. F. Huxley. 1952. A quantitative description of membrane current and its application to conduction and excitation in nerve. *J Physiol* 117:500–544.

Hsiao, I. N. and V. W. Lin. 2001. Improved coil design for functional magnetic stimulation of expiratory muscles. *IEEE Trans Biomed Eng* 48:684–694.

Hsu, K. H. and D. M. Durand. 2001. A 3-D differential coil design for localized magnetic stimulation. *IEEE Trans Biomed Eng* 48:1162–1168.

Hsu, K. H., S. S. Nagarajan, and D. M. Durand. 2003. Analysis of efficiency of magnetic stimulation. *IEEE Trans Biomed Eng* 50:1276–1285.

Huang, Y. Z., M. J. Edwards, E. Rounis, K. P. Bhatia, and J. C. Rothwell. 2005. Theta burst stimulation of the human motor cortex. *Neuron* 45:201–206.

Jalinous, R. 2002. Principles of magnetic stimulator design. In *Handbook of Transcranial Magnetic Stimulation*, eds. A. Pascual-Leone, K. Davey, J. Rothwell, E. M. Wasserman, and B. K. Puri. London, U.K.: Arnold, pp. 30–38.

Johnston, D. and S. M.-S. Wu. 1994. *Foundations of Cellular Neurophysiology.* Cambridge, U.K.: The MIT Press.

Jonscher, A. K. 1996. *Universal Relaxation Law*, 1st edn. London, U.K.: Chelsea Dielectrics Press.

Joy, M. L., V. P. Lebedev, and J. S. Gati. 1999. Imaging of current density and current pathways in rabbit brain during transcranial electrostimulation. *IEEE Trans Biomed Eng* 46:1139–1149.

Kim, D. H., G. E. Georghiou, and C. Won. 2006. Improved field localization in transcranial magnetic stimulation of the brain with the utilization of a conductive shield plate in the stimulator. *IEEE Trans Biomed Eng* 53:720–725.

Kraus, K. H., L. D. Gugino, W. J. Levy, J. Cadwell, and B. J. Roth. 1993. The use of a cap-shaped coil for transcranial magnetic stimulation of the motor cortex. *J Clin Neurophysiol* 10:353–362.

Largus, S. 1529. De compositionibus medicamentorum. In *Scribonii Largi De Compositionibus Medicamentorum Liber Unus: Antehac Nusquam Excusus*, ed. J. Ruello. Paris, France: Excudebat Parisiis Simion Siluius, pp. 1–62.

Liebetanz, D., F. Klinker, D. Hering et al. 2006. Anticonvulsant effects of transcranial direct-current stimulation (tDCS) in the rat cortical ramp model of focal epilepsy. *Epilepsia* 47:1216–1224.

Liebetanz, D., R. Koch, S. Mayenfels, F. Konig, W. Paulus, and M. A. Nitsche. 2009. Safety limits of cathodal transcranial direct current stimulation in rats. *Clin Neurophysiol* 120:1161–1167.

Lin, V. W., I. N. Hsiao, and V. Dhaka. 2000. Magnetic coil design considerations for functional magnetic stimulation. *IEEE Trans Biomed Eng* 47:600–610.

Lisanby, S. H. 1998. Intercerebral measurements of rTMS and ECS induced voltage in vivo. *Biol Psychiatry* 43:100S.

Lu, M., S. Ueno, T. Thorlin, and M. Persson. 2008. Calculating the activating function in the human brain by transcranial magnetic stimulation. *IEEE Trans Magn* 44:1438–1441.

Ludt, H. and H. D. Herrmann. 1973. In vitro measurement of tissue impedance over a wide frequency range. *Biophysik* 10:337–345.

Maccabee, P. J., L. Eberle, V. E. Amassian, R. Q. Cracco, A. Rudell, and M. Jayachandra. 1990. Spatial distribution of the electric field induced in volume by round and figure '8' magnetic coils: Relevance to activation of sensory nerve fibers. *Electroenceph Clin Neurophysiol* 76:131–141.

Magstim. *Magstim Rapid 2*. http://www.magstim.com/transcranial-magnetic-stimulation/ magstim-rapid (accessed March 24, 2012).

Martinsen, O. G., S. Grimmes, and H. P. Schwan. 2002. Interface phenomena and dielectric properties of biological tissue. In *Encyclopedia of Surface and Colloid Science*. New York: Marcel Dekker, pp. 2643–2652.

McCreery, D. and W. Agnew. 1990. Neuronal and axonal injury during functional electrical stimulation; a review of the possible mechanisms. *Annual International Conference of the IEEE Engineering in Medicine and Biology Society*, Philadelphia, PA, pp.1488–1489.

McCreery, D., W. Agnew, T. G. Yuen, and L. Bullara. 1990. Charge density and charge per phase as cofactors in neural injury induced by electrical stimulation. *IEEE Trans Biomed Eng* 37:996–1001.

McIntyre, C. C. and W. M. Grill. 1999. Excitation of central nervous system neurons by non-uniform electric fields. *Biophys J* 76:878–888.

McIntyre, C. C. and W. M. Grill. 2000. Selective microstimulation of central nervous system neurons. *Ann Biomed Eng* 28:219–233.

McIntyre, C. C. and W. M. Grill. 2001. Finite element analysis of the current-density and electric field generated by metal microelectrodes. *Ann Biomed Eng* 29:227–235.

McIntyre, C. C. and W. M. Grill. 2002. Extracellular stimulation of central neurons: Influence of stimulus waveform and frequency on neuronal output. *J Neurophysiol* 88:1592–1604.

McIntyre, C. C., W. M. Grill, D. L. Sherman, and N. V. Thakor. 2004a. Cellular effects of deep brain stimulation: Model-based analysis of activation and inhibition. *J Neurophysiol* 91:1457–1669.

McIntyre, C. C., S. Mori, D. L. Sherman, N. V. Thakor, and J. L. Vitek. 2004b. Electric field and stimulating influence generated by deep brain stimulation of the subthalamic nucleus. *Clin Neurophysiol* 115:589–595.

McNeal, D. R. 1976. Analysis of a model for excitation of myelinated nerve. *IEEE Trans Biomed Eng* 23:329–337.

Melcher, J. R. 1981. *Continuum Electromechanics*. Cambridge, U.K.: MIT Press.

Merton, P. A. and H. B. Morton. 1980. Stimulation of the cerebral cortex in the intact human subject. *Nature* 285:227.

Miranda, P. C., P. Faria, and M. Hallett. 2009. What does the ratio of injected current to electrode area tell us about current density in the brain during tDCS? *Clin Neurophysiol* 120:1183–1187.

Miranda, P. C., M. Hallett, and P. J. Basser. 2003. The electric field induced in the brain by magnetic stimulation: A 3-D finite-element analysis of the effect of tissue heterogeneity and anisotropy. *IEEE Trans Biomed Eng* 50:1074–1085.

Miranda, P. C., M. Lomarev, and M. Hallett. 2006. Modeling the current distribution during transcranial direct current stimulation. *Clin Neurophysiol* 117:1623–1629.

Nadeem, M., T. Thorlin, O. P. Gandhi, and M. Persson. 2003. Computation of electric and magnetic stimulation in human head using the 3-D impedance method. *IEEE Trans Biomed Eng* 50:900–907.

Nagarajan, S. and D. M. Durand. 1995. Analysis of magnetic stimulation of a concentric axon in a nerve bundle. *IEEE Trans Biomed Eng* 42:926–933.

Nagarajan, S., D. M. Durand, and E. N. Warman. 1993. Effects of induced electric fields on finite neuronal structures: A simulation study. *IEEE Trans Biomed Eng* 40:1175–1188.

Nitsche, M. A., D. Liebetanz, N. Lang, A. Antal, F. Tergau, and W. Paulus. 2003. Safety criteria for transcranial direct current stimulation (tDCS) in humans. *Clin Neurophysiol* 114:2220–2222; author reply 2–3.

Pascual-Leone, A. and T. Wagner. 2007. A Brief Summary of the History of Noninvasive Brain Stimulation (supplement to Noninvasive Human Brain Stimulation). *Annu Rev Biomed Eng* 9:s1–s7.

Paulus, W. 2003. Transcranial direct current stimulation (tDCS). *Suppl Clin Neurophysiol* 56:249–254.

Paulus, W. 2005. Toward establishing a therapeutic window for rTMS by theta burst stimulation. *Neuron* 45:181–183.

Pethig, R. 1984. Dielectric properties of biological materials: Biophysical and medical applications. *IEEE Trans Electr Insul* 19:453–474.

Pethig, R. and D. B. Kell. 1987. The passive electrical properties of biological systems: Their significance in physiology, biophysics, and biotechnology. *Phys Med Biol* 32:933–970.

Piccolino, M. 1997. Luigi Galvani and animal electricity: Two centuries after the foundation of electrophysiology. *Trends Neurosci* 20:443–448.

Plonsey, R. and R. C. Barr. 2000. *Bioelectricity: A Quantitative Approach*, 2nd edn. New York: Kluwer Academic.

Priori, A. 2003. Brain polarization in humans: A reappraisal of an old tool for prolonged non-invasive modulation of brain excitability. *Clin Neurophysiol* 114:589–595.

Rattay, F. 1989. Analysis of models for extracellular fiber stimulation. *IEEE Trans Biomed Eng* 36:974–977.

Rattay, F., S. Resatz, P. Lutter, K. Minassian, B. Jilge, and M. R. Dimitrijevic. 2003. Mechanisms of electrical stimulation with neural prostheses. *Neuromodulation* 6:42–56.

Roth, B. J. 1994. Mechanisms for electrical stimulation of excitable tissue. *Crit Rev Biomed Eng* 22:253–305.

Roth, B. J. and P. J. Basser. 1990. A model of stimulation of a nerve fiber by electromagnetic induction. *IEEE Trans Biomed Eng* 37:588–597.

Roth, B. J., P. J. Maccabee, L. P. Eberle et al. 1994. in vitro evaluation of a 4-leaf coil design for magnetic stimulation of peripheral nerve. *Electroenceph Clin Neurophysiol* 93:68–74.

Roth, B. J., J. M. Saypol, M. Hallett, and L. G. Cohen. 1991. A theoretical calculation of the electric field induced in the cortex during magnetic stimulation. *Electroencephalogr Clin Neurophysiol* 81:47–56.

Roth, Y., A. Zangen, and M. Hallett. 2002. A coil design for transcranial magnetic stimulation of deep brain regions. *J Clin Neurophysiol* 19:361–370.

Rothwell, J. C. 1999. Paired-pulse investigations of short-latency intracortical facilitation using TMS in humans. *Electroencephalogr Clin Neurophysiol Suppl* 51:113–119.

Rush, S. and D. A. Driscoll. 1968. Current distribution in the brain from surface electrodes. *Anesth Analg* 47:717–723.

Sadiku, M. 2000. *Numerical Techniques in Electromagnetics.* Boca Raton, FL: CRC Press.

Sadleir, R. J., T. D. Vannorsdall, D. J. Schretlen, and B. Gordon. 2010. Transcranial direct current stimulation (tDCS) in a realistic head model. *Neuroimage* 51:1310–1318.

Salvador, R., A. Mekonnen, G. Ruffini, and P. C. Miranda. 2010. Modeling the electric field induced in a high resolution realistic head model during transcranial current stimulation. *Annual Conference of the IEEE Engineering in Medicine and Biology Society*, Buenos Aires, Argentina, pp. 2073–2076.

Schwan, H. P. 1957. Electrical properties of tissues and cell suspensions. *Adv Biol Med Phys* 5:147–209.

Schwan, H. P. 1985. Analysis of dielectric data:experience gained with biological materials. *IEEE Trans Electr Insul* 20:913–922.

Singh, C., C. W. Smith, and R. Hughes. 1979. In vivo dielectric spectrometer. *Med Biol Eng Comput* 17:45–60.

Talelli, P., R. J. Greenwood, and J. C. Rothwell. 2007. Exploring Theta Burst Stimulation as an intervention to improve motor recovery in chronic stroke. *Clin Neurophysiol* 118:333–342.

Tay, G. 1992. Measurement of current density distribution induced in vivo during magnetic stimulation. PhD thesis, Marquette University, Milwaukee, WI.

Tay, G., M. Chilbert, J. Battocletti, A. J. Sances, and T. Swiontek. 1991. Mapping of current densities induced in vivo during magnetic stimulation. *Annual International Conference of the IEEE Engineering in Medicine and Biology Society*, Orlando, FL, pp. 851–852.

Tay, G., M. Chilbert, J. Battocletti, A. Sances, Jr., T. Swiontek, and C. Kurakami. 1989. Measurement of magnetically induced current density in saline and in vivo. *Annual Conference of the IEEE Engineering in Medicine and Biology Society*, Seattle, WA, pp. 1167–1168.

Terney, D., L. Chaieb, V. Moliadze, A. Antal, and W. Paulus. 2008. Increasing human brain excitability by transcranial high-frequency random noise stimulation. *J Neurosci* 28:14147–14155.

Thielscher, A., A. Opitz, and M. Windhoff. 2011. Impact of the gyral geometry on the electric field induced by transcranial magnetic stimulation. *Neuroimage* 54:234–243.

Thompson, S. P. 1910. A physiological effect of an alternating magnetic field. *Proc R Soc Lond B* 82:396–398.

Tofts, P. S. 1990. The distribution of induced currents in magnetic stimulation of the nervous system. *Phys Med Biol* 35:1119–1128.

Toschi, N., T. Welt, M. Guerrisi, and M. E. Keck. 2008. Transcranial magnetic stimulation in heterogeneous brain tissue: Clinical impact on focality, reproducibility and true sham stimulation. *J Psychiatr Res* 43:255–264.

Tranchina, D. and C. Nicholson. 1986. A model for the polarization of neurons by extrinsically applied electric fields. *Biophys J* 50:1139–1156.

Ueno, S. and R. Liu. 1997. Determination of spatial distribution of induced electric fields and the gradients in inhomogeneous volume conductors exposed to time varying magnetic fields. *Annual Conference of the IEEE Engineering in Medicine and Biology Society*, Chicago, IL, October 30–November 2, pp. 2468–2469.

Ueno, S., T. Tashiro, and K. Harada. 1988. Localised stimulation of neural tissues in the brain by means of a paired configuration of time-varying magnetic fields. *J Appl Phys* 64:5862–5864.

Valero-Cabre, A., A. Pascual-Leone, and R. J. Rushmore. 2008. Cumulative sessions of repetitive transcranial magnetic stimulation (rTMS) build up facilitation to subsequent TMS-mediated behavioural disruptions. *Eur J Neurosci* 27:765–774.

Wagner, T., U. Eden, F. Fregni et al. 2008. Transcranial magnetic stimulation and brain atrophy: A computer-based human brain model study. *Exp Brain Res* 186:539–550.

Wagner, T., F. Fregni, U. Eden et al. 2006a. Transcranial magnetic stimulation and stroke: A computer-based human model study. *Neuroimage* 30:857–870.

Wagner, T., F. Fregni, S. Fecteau, A. Grodzinsky, M. Zahn, and A. Pascual-Leone. 2007a. Transcranial direct current stimulation: A computer-based human model study. *Neuroimage* 35:1113–1124.

Wagner, T., J. Rushmore, U. Eden, and A. Valero-Cabre. 2009. Biophysical foundations underlying TMS: Setting the stage for an effective use of neurostimulation in the cognitive neurosciences. *Cortex* 45:1025–1034.

Wagner, T., A. Valero-Cabre, and A. Pascual-Leone. 2007b. Noninvasive human brain stimulation. *Annu Rev Biomed Eng* 9:527–565.

Wagner, T., M. Zahn, A. J. Grodzinsky, and A. Pascual-Leone. 2004. Three-dimensional head model simulation of transcranial magnetic stimulation. *IEEE Trans Biomed Eng* 51:1586–1598.

Wagner, T., M. Zahn, V. J. Wedeen, A. Grodzinsky, and A. Pascual-Leone. 2006b. Transcranial magnetic stimulation: High resolution tracking of the induced current density in the individual human brain. *12th Annual Meeting of the Organization for Human Brain Mapping*, Florence, Italy, pp. S127.

Yuen, T. G., W. F. Agnew, L. A. Bullara, S. Jacques, and D. B. McCreery. 1981. Histological evaluation of neural damage from electrical stimulation: Considerations for the selection of parameters for clinical application. *Neurosurgery* 9:292–299.

Yunokuchi, K. and D. Cohen. 1991. Developing a more focal magnetic stimulator. Part 2: Fabricating coils and measuring induced current distributions. *J Clin Neurophysiol* 8:112–120.

Yunokuchi, K., R. Koyoshi, G. Wang et al. 1999. Estimation of focus of electric field in an inhomogenous medium exposed by pulsed magnetic field. *IEEE First Joint BMES/EMBS conference Serving Humanity and Advancing Technology*, Atlanta, GA, October 13–16, pp. 467.

Zangen, A., Y. Roth, B. Voller, and M. Hallett. 2005. Transcranial magnetic stimulation of deep brain regions: Evidence for efficacy of the H-coil. *Clin Neurophysiol* 116:775–779.

6 Patterned Protocols of Transcranial Magnetic Stimulation

Stefan Jun Groiss, Yoshikazu Ugawa,
Walter Paulus, and Ying-Zu Huang

CONTENTS

6.1 INTRODUCTION

For about two decades it has been well established that transcranial magnetic stimu-lation when applied repetitively (rTMS) is able to induce or modify cortical plasticity-like mechanisms during or beyond stimulation. As a general rule, higher frequencies distinctly above 1 Hz are expected to induce facilitation or poten-tiation, which is comparable with synaptic long-term potentiation (LTP) seen in experimental animals, and frequencies at or below 1 Hz inhibition or depression, which is comparable with long-term depression (LTD). For example, rTMS at 1 Hz given to the primary motor cortex (M1) for 25 min may suppress motor cortical excitability for about 25 min (Touge et al. 2001). By contrast, rTMS at 5 Hz given in separated short trains may facilitate motor cortical excitability for at least 30 min (Peinemann et al. 2004). The aftereffects of rTMS are believed to be plasticity like and share several common features of LTD and LTP (Ziemann 2004). These rules however may be violated or even reversed by a variety of factors such as duration of stimulation, homeostatic plasticity, or co-application of neuroactive drugs. Moreover, conventional forms of rTMS delivered at a regular fixed frequency usually require a lengthy stimulation and a high stimulus intensity to produce consistent aftereffects and commonly need to be interrupted for coil replacement because of overheating. Nevertheless, the aftereffects are still short-lived. In view of these difficulties, more efficient protocols, which are capable of producing consistent aftereffects with fewer stimuli or lower stimulus intensity, are demanded. Based on such demands, patterned rTMS protocols, e.g., theta burst stimulation (TBS), paired-pulse rTMS (pp rTMS), and more recently quadri-pulse stimulation (QPS), have been introduced.

One of the most interesting aspects in this context is the control of aftereffects by introducing patterned stimulation sequences. Since the possibilities for this are boundless, efforts in this direction are usually based on physiological assumptions in order to limit the approaches to the most promising. Talking about patterned stimulation is tantamount to cutting down hypothesis-driven protocols to meaningful intervals. The shortest intervals in this context are 1.5 ms determined by I-wave sequences and used by QPS or pp rTMS. Other intervals of interest are short-interval intracortical inhibition (SICI), intracortical facilitation (ICF), or long-interval intracortical inhibition (LICI), all of which originally refer to paired TMS protocols investigating intracortical excitability. ICF is closely related to the 20 ms interval, which carries the basic burst pattern in the theta burst paradigm. It is however also in the range at which QPS facilitation at shorter intervals switches to inhibition at 30 ms. Longer intervals of 200 ms carrying the theta in TBS play a role in LICI. Another interesting interval is the 8 s reversing TBS inhibition with continuous stimulation into excitation. With low-frequency single-pulse rTMS, these intervals are regarded to be inefficient, both in terms of induction of inhibition and in terms of time effort. However, in deafferentation-induced plasticity with assumed loss of GABAergic inhibition, a 10 s interstimulus interval (ISI) turned out to be very effective in inducing motor cortex facilitation (Ziemann et al. 1998). Further physiological anchors may be seen in the ranges in which short-term potentiation switches to LTP. There is also clear evidence that daily repetitive stimulation boosts clinically relevant effects and, from clinical studies, that at least 4 weeks of stimulation are necessary to obtain relevant effects with certain protocols, e.g., in depression or Parkinson's disease (O'Reardon et al. 2007; Hamada et al. 2008b).

Thus understanding the role of intervals in TMS may turn out to be one of the most relevant aspects in controlled guidance of aftereffects.

6.2 THETA BURST STIMULATION

In experimental animals, in addition to the stimulation at a regular frequency, short bursts of stimulation repeated at the theta range of frequencies in electroencephalography (EEG) terminology (4–7 Hz), evolved from observations of the natural firing pattern of pyramidal cells in the hippocampus of cats and rats (Kandel and Spencer 1961; Eichenbaum et al. 1987). They have proven to be an efficient protocol for LTP induction (Larson and Lynch 1986; Larson et al. 1986). Based on the success of the burst stimulation in animal studies, the first patterned rTMS protocol in the world, which has been called theta burst stimulation, was invented.

6.2.1 PROTOCOLS OF TBS

The basic element of TBS is a burst containing three TMS pulses at 50 Hz. Bursts are repeated at 5 Hz (e.g., every 200 ms). The most commonly used stimulus intensity of TBS is 80% of active motor threshold (AMT), which is the minimum stimulation intensity over the motor hot spot that could elicit a motor-evoked potential (MEP) of no less than 200 μV in 5 out of 10 trials during a voluntary contraction of the contralateral target muscle. Very short (20–192 s) TBS applied to M1 can produce both depression and potentiation effects on the motor system in conscious humans at an electrophysiological and behavioral level outlasting the period of stimulation by over an hour (Huang et al. 2005; Figure 6.1). When bursts are given intermittently (iTBS: intermittent TBS) with a 2 s TBS train every 10 s for 20 cycles, the cortical excitability is enhanced for 20 min or so. By contrast, when bursts are given continuously without interruption (cTBS: continuous TBS) for 20 or 40 s (cTBS300

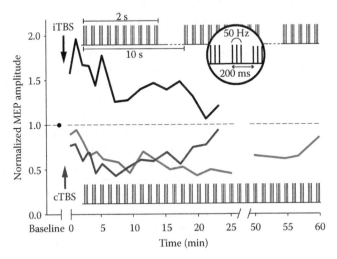

FIGURE 6.1 A typical form of TBS contains three-pulse 50 Hz bursts given at % Hz. If a 2 s train of TBS is given every 10 s (iTBS), an LTP-like effect is induced to enhance the cortical excitability. On the contrary, if bursts are given at 5 Hz continuously (cTBS), an LTD-like effect is induced to suppress the cortical excitability.

or cTBS600, respectively), a suppression effect for more than 20 or 60 min, respectively, is induced. Other than its efficiency and low stimulus intensity, as compared with other patterned rTMS protocols that usually need a combination of two or more TMS machines, TBS is convenient to use and only one high-output rTMS machine is required. As a variant, TBS is sometimes given as a 3-pulse burst at 30 Hz repeated at intervals of 100 ms for 44 s (Nyffeler et al. 2008). This variant has shown good inhibitory effect on the human frontal eye field and posterior parietal cortex (Nyffeler et al. 2006, 2008) and can be a compromise for people who do not have an rTMS machine for strong-enough 50 Hz output.

Although the original effects of TBS were measured when it was applied to M1, TBS has also been proven to have effects when applied to other brain areas. When cTBS was given to the premotor cortex, it modified the premotor area to slow down choice reaction time (Mochizuki et al. 2005) and suppressed M1 excitability remotely (Huang et al. 2009a). TBS to the cerebellum also showed the ability to modify the cerebellar efferent modulation on to the motor cortex (Koch et al. 2008). In addition, TBS has been successfully applied to visual cortex, frontal eye field, parietal cortex, and many other areas for behavioral studies (Nyffeler et al. 2006, 2008; Silvanto et al. 2007). Potential therapeutic benefits of TBS have also been demonstrated in patients with dystonia, tinnitus, pain, stroke, depression, and many other diseases (Talelli et al. 2007; Antal and Paulus 2010; Holzer and Padberg 2010; Huang et al. 2010b; Wu et al. 2010; Chung et al. 2011).

6.2.2 PATTERN AND LENGTH OF TBS

Interestingly, the pattern of delivery of TBS is crucial in determining the direction of change in cortical excitability. An LTD-like depression effect is induced when the bursts are delivered continuously as in cTBS, whereas an LTP-like potentiation effect is induced when the bursts are given in intermittent trains as in iTBS. When the length of the train of bursts and the pause between the trains are longer than those of iTBS and the train is shorter than that of cTBS, there may be no significant aftereffect (Huang et al. 2005, 2009a; Fang et al. 2010) (intermediate TBS [imTBS]). Such pattern dependency can be explained by the following: (1) a single burst of stimulation induces a mixture of excitatory and inhibitory effects; (2) those effects may cascade to produce long-lasting effects; and (3) the continuity and train length are crucial factors. Brief pauses between short trains seem to be critical for a protocol to produce potentiation effect, while a lengthy continuous protocol without interruptions favors a depression effect. The hypothesis is supported by a theoretical model based on a simplified description of the glutamatergic synapse in which postsynaptic Ca^{2+} entry initiates processes leading to different amounts of potentiation and depression of synaptic transmission (Huang et al. 2011). This mathematical model successfully simulates the pattern dependency and several other features of TBS.

To extend the viewpoint from the model considering the continuity and train length, if an excitatory paradigm (e.g., iTBS) is prolonged, potentiation may gradually decline because of a slower but then dominating buildup of inhibition. By contrast, if an inhibitory paradigm (e.g., cTBS) is prolonged, according to theory, then depression should be further enhanced because of a further buildup of inhibition. Indeed, compatible with this prediction, Gamboa et al. (2010) demonstrated that prolonged iTBS with 1200 pulses produced depression rather than potentiation effect on M1. However, against our prediction, the same study showed that prolonged cTBS potentiated instead of depressing the cortical excitability. By contrast, a more recent study showed that prolonged iTBS and cTBS did not reverse, but enhanced and prolonged their original potentiation and depression, respectively (Hsu et al. 2011). Therefore, further studies may be required to clarify the discrepancy of current results about prolonged TBS protocols.

6.2.3 Mechanism of TBS

Pharmacological studies have demonstrated that the aftereffects of TBS are dependent on N-methyl-D-aspartate (NMDA) receptors and Ca^{2+} channels. Memantine and dextromethorphan, NMDA receptor antagonists, prevent both cTBS and iTBS protocols from producing visible aftereffects in human subjects (Huang et al. 2007; Wankerl et al. 2010). D-Cycloserine, a partial NMDA receptor agonist, reverses the potentiation effect of iTBS to become suppressive (Teo et al. 2007). In addition, the polarity of the effect of cTBS can be reversed under the influence of nimodipine, an L-type voltage-gated Ca^{2+} channel antagonist (Wankerl et al. 2010). NMDA receptors are known to be one of the crucial components in the initiation of synaptic plasticity in animal studies (Larson and Lynch 1988; Lee et al. 1998; Liu et al. 2004). Dysfunction of NMDA receptors prevents synapses from undergoing LTP and LTD (Hirsch and Crepel 1991; Calabresi et al. 1992), and subtypes of NMDA receptors may govern the direction of plasticity induction (Liu et al. 2004). Calcium entry is also known to initiate both LTP and LTD. The rate and amount of calcium influx are critical factors governing the direction of plasticity that is induced in animal studies (Neveu and Zucker 1996; Malenka and Nicoll 1999; Yang et al. 1999; Kemp and Bashir 2001; Sheng and Kim 2002).

Cervical epidural recordings of descending motor volleys revealed that cTBS mainly suppressed the first indirect wave (I1-wave), while iTBS mainly enhanced the later I-waves but not the I1-wave (DiLazzaro et al. 2005, 2008). These findings suggest that TBS most likely works on synapses onto pyramidal cells within the motor cortex through plasticity-like mechanisms, although the mechanisms of cTBS and iTBS may be slightly different. Therefore, it is believed that the depression and potentiation effects of TBS are very close to LTD and LTP seen in animal models.

6.2.4 Regulation of TBS Aftereffects

The aftereffects induced by TBS are known to interact with physical activities performed before, during, and after the TBS conditioning. A slight (approximately 10% of maximum) voluntary tonic contraction of the target muscle during the whole period of TBS conditioning prevents cTBS and iTBS from producing visible effects on MEPs, although cTBS, but not iTBS, with the contraction still shows an effect on SICI tested by a paired-pulse technique (Huang et al. 2008). The difference in the effect of contraction on SICI and MEP may relate to different effect of contraction on MEP and SICI pathways, seeing that contraction of the target muscle increases excitability of MEP pathways but reduces excitability in the SICI pathway (Ridding et al. 1995). A similar modulation effect of the activity states of neural populations

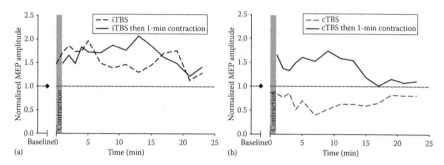

FIGURE 6.2 When voluntary contraction of the target muscle is performed immediately after TBS for 1 min, the potentiation effect of iTBS is enhanced (a) and the depression effect of cTBS300 is reversed to be potentiation (b).

under the coil has been also found in the behavioral effects induced by TBS given to the visual cortex. Modified cTBS given to V1/V2 region while subjects were looking toward one side only impaired detection of the incongruent direction (managed by less active neurons during TBS), but not congruent direction (managed by more active neurons during TBS), in a subsequent eye motion direction discrimination task (Silvanto et al. 2007).

When similar slight muscle contraction happens immediately after TBS for 1 min, the potentiation effect of iTBS is enhanced and the depression effect of cTBS300 is reversed to be potentiation (Figure 6.2; Huang et al. 2008). If the 1 min contraction happens at 10 min after cTBS300, the aftereffect of cTBS300 is only changed transiently. However, similar 1 min contraction immediately after cTBS600 for an EEG–EMG coherence study did not alter the suppression effect (Saglam et al. 2008). Moreover, tonic contraction of the target muscle before cTBS300 seems to be crucial for producing a suppression effect (Gentner et al. 2008). Without prior tonic contraction, cTBS300 may result in mild facilitation. However, cTBS600 is not affected by prior tonic contraction and always produces a reliable suppression effect. By contrast, phasic voluntary movements reverse the aftereffect of subsequent cTBS to be facilitation and that of subsequent iTBS to be inhibition (Iezzi et al. 2008).

Metaplasticity (Abraham and Bear 1996) is considered to be the underlying mechanism of the interactions between the muscle contraction and TBS. The voluntary muscle activity before and during TBS conditioning may modify the status of the neurons of synapses that are involved in the plasticity-like effects induced by TBS and alter the responses to TBS conditioning. However, according to the definition, metaplasticity may not be suitable for explaining the modulation effect of the contraction immediately after TBS. As an alternation, it is possible that part of the LTP-like effect of TBS may be caused by recruitment of such "silent" synapses, in which only LTP is possible and are resistant to modification by physiological activity (Montgomery and Madison 2004). By contrast, induction of LTD-like effects depends on other synapses, which are modifiable at all times. Another possibility is that activating the target muscle may cause a sudden large influx of calcium into postsynaptic cells, which is favorable for the development of LTP and the prevention of LTD (Bezprozvanny et al. 1991; Nishiyama et al. 2000).

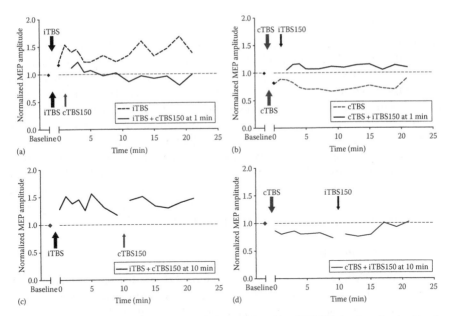

FIGURE 6.3 cTBS150 abolishes potentiation produced by iTBS if it is given 1 min after the end of iTBS (a); similarly, depression produced by cTBS300 can be abolished by iTBS150 at 1 min after the end of cTBS300 (b). By contrast, cTBS150 applied at 10 min after iTBS fails to modify the facilitation effect (c). iTBS150 applied at 10 min after cTBS300 takes longer than iTBS150 applied at 1 min after cTBS300 to reverse the depression effect (d).

TBS is modified not only by physical activity, but also by TBS itself (Huang et al. 2010a). A 10 s train of cTBS (cTBS150) abolishes potentiation produced by iTBS if it is given 1 min after the end of iTBS (Figure 6.3a); similarly, depression produced by cTBS300 can be abolished when a short protocol of iTBS containing 150 pulses only (iTBS150) is delivered 1 min after the end of cTBS300 (Figure 6.3b). Neither cTBS150 nor iTBS150 given alone has aftereffects on the size of MEPs or on the amount of SICI and ICF. By contrast, cTBS150 cannot modify the facilitation effect if it is applied 10 min after iTBS (Figure 6.3c). Although the reversal of depression is still seen when iTBS150 is applied at 10 min after cTBS, it now takes about 5 min (Figure 6.3d), in contrast to the almost immediate reversal when iTBS150 is given at 1 min, to build up. The later phenomenon suggests that the effects that reverse LTP-like and LTD-like effects of TBS depend on the time interval between the first and second protocols. These interactions between standard TBS and the second short form of TBS, including their time dependency, are compatible with the phenomena of depotentiation and de-depression that have been described in experiments on reduced and freely moving animals (Staubli and Scafidi 1999; Kulla and Manahan-Vaughan 2000; Huang et al. 2001; Picconi et al. 2003; Coesmans et al. 2004; Kumar et al. 2007). The reversal of plasticity, i.e., depotentiation and de-depression, is considered different from metaplasticity from the viewpoints of the protocols to induce them and the mechanism at the cellular level (Huang et al. 2010a).

6.2.5 Summary of TBS

As the first patterned rTMS in the world, protocols of TBS are effective and efficient to modulate cortical excitability bidirectionally through plasticity-like mechanisms. The effect of TBS is pattern dependent. Continuous bursts without interruption (i.e., cTBS) produces an LTD-like effect, while an LTP-like effect is induced when bursts are given in short trains intermittently (i.e., iTBS). The pattern dependency of TBS has been successfully simulated by a simple theoretical model based on the known knowledge on postsynaptic plasticity. Moreover, the aftereffects of TBS can be regulated by the physical activity before, during, and after TBS due to metaplasticity. A shorter form of TBS is able to reverse plasticity just induced by regular TBS to test depotentiation and de-depression in humans. In brief, TBS is an optimal protocol for studying the physiology and pathophysiology of human brain and has therapeutic potential for neurological and psychiatric disorders.

6.3 PAIRED-PULSE rTMS

Paired-pulse rTMS (pp rTMS) is another patterned rTMS that came out abound the same time as TBS was developed. Early studies on pp rTMS focused on effects during stimulation and then were extended to longer-lasting aftereffects.

6.3.1 Protocols of pp rTMS

The SICI paradigm with a subthreshold (e.g., 90% AMT) pulse followed by a suprathreshold (e.g., 120% RMT) pulse was applied repetitively with frequencies ranging from 0.17 to 5 Hz with altogether 80 single or 80 paired pulses (Sommer et al. 2001). The conditioning-test intervals were 2, 5, or 10 ms. pp rTMS at 5 Hz caused a strong enhancement of MEP size when compared with 5 Hz single pulse (Figure 6.4). In a subsequent study, MEPs showed less variability to pp rTMS than to single-pulse rTMS (Sommer et al. 2002c). With 900 inhibiting or facilitating paired or single pulses at 1 Hz over the motor cortex, MEPs were larger after facilitating paired pulses than after inhibiting paired pulses, both in controls and in Parkinson's disease patients (Sommer et al. 2002a).

Another protocol uses two pulses at equal stimulus intensity for pp rTMS. Khedr et al. (2004) applied 25 min pp rTMS at 0.6 Hz with 500 paired stimuli in total with an ISI of 3 ms at 80% of AMT. As expected with the inhibitory ISI of 3 ms, a significant increase in motor threshold with a concomitant decrease in MEP recruitment was shown. A special form of pp rTMS used ISIs of 1.5 ms in order to boost I-wave facilitation and has therefore been termed iTMS (Thickbroom et al. 2006). After 15 min of iTMS, an increase in MEP size compared with the control condition with single pulse was seen as well as an improvement in task performance (Benwell et al. 2006). When applying paired stimuli of equal strength at I-wave periodicity for 30 min at a rate of 0.2 Hz, paired-pulse-induced MEP amplitude increased fivefold by the end of the stimulation period whereas single-pulse MEP amplitude was increased fourfold for 10 min after stimulation (Thickbroom et al. 2006). In a similar experiment, MEPs were facilitated by about 90% for 10 min post iTMS intervention and returned to the baseline at 10–15 min post intervention (Hamada et al. 2007a).

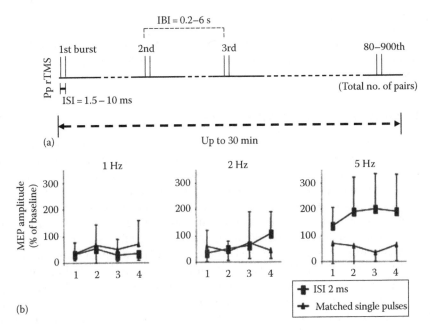

FIGURE 6.4 (a) Stimulation protocol of pp rTMS. Various protocols have been used with ISI from 1.5 to 10 ms and IBI from 0.2 to 6 s and applied for up to 30 min, resulting in a total number of stimuli of up to 900 paired pulses. (b) MEP amplitudes during the original paired-pulse paradigm: ISI 2 ms yielded smaller amplitudes than adjusted single stimuli at 1 Hz (left) and during early 2 Hz rTMS (center), but much larger amplitudes at 5 Hz (right). (Modified from Sommer, M et al., *Exp. Brain Res.,* 139, 465, 2001.)

Targeting the first three I-waves at ISIs of 1.3, 2.5, and 4.3 ms, the facilitation curves were increased in amplitude for all three. It was however concluded that an intervention with an ISI of 1.5 ms is able to target multiple I-waves (Cash et al. 2009). In a patient with an epidural electrode during iTMS, there was progressive increase in MEP amplitude by more than 300%, declining to 200% 3 min after the end of iTMS (DiLazzaro et al. 2007). Concerning the buildup of MEP increase, there is conflicting evidence: It was claimed that during the first 15 min, the main effect occurs and that there is no further change during the remainder of the intervention (Murray et al. 2011). By contrast, in a similar protocol using 15 min trains of pp rTMS with 180 pulse pairs at 0.2 Hz, 1.5 ms ISI, and supra-threshold intensity administered over 15 min, MEP size did not substantially increase post train (Fitzgerald et al. 2007). It is unclear if the facilitatory effect of 1.5 ms interval can be compensated by the ISI of 5 s, which is also neutral in TBS protocols.

6.3.2 MECHANISM OF PP rTMS

The underlying mechanism of pp rTMS has been less studied. PP rTMS effects have been confirmed by near infrared spectroscopy (NIRS). Concentration drops of oxyhemoglobin (HbO) and total hemoglobin (HbT) were observed, in particular a difference in the time taken to return to baseline (Thomson et al. 2011).

6.3.3 Summary of pp rTMS

The intra-stimulation effects and aftereffects of pp rTMS have been clearly demonstrated in several studies. However, data available so far have been less clear on pp rTMS when compared with TBS or QPS.

6.4 QUADRI-PULSE STIMULATION

The direction and the degree of induced neuroplasticity depend not only on parameters such as stimulation frequency, stimulation intensity, train duration, or length of inter-burst intervals (IBIs), but also on the direction of the applied current. Mostly sinusoidal biphasic pulses are used for rTMS due to technical reasons, but it has previously been shown that monophasic rTMS induces more distinct LTP-/LTD-like plasticity than biphasic rTMS (Sommer et al. 2002b; Tings et al. 2005; Arai et al. 2007). Whereas Sommer et al. (2002b) and Arai et al. (2007) compared the aftereffects of subthreshold monophasic and biphasic rTMS of the dominant motor cortex at 1 and 10 Hz, respectively, Tings et al. (2005) used 80 supra-threshold stimuli at 5 Hz. All three studies revealed that monophasic rTMS induces stronger and longer-lasting changes of the motor cortical excitability. The reasons for this are not yet clear. Sommer and colleagues argued that repetitive current flow in one direction only by monophasic stimuli may build up a kind of membrane polarization, whereas Arai and colleagues concluded that monophasic pulses might activate a more uniform population of cortical interneurons leading to a summation of short-lasting effects. QPS was therefore developed based on monophasic rTMS for the purpose of longer-lasting and more durable aftereffects.

6.4.1 Protocols of QPS

QPS is a patterned rTMS protocol to apply repeated bursts of monophasic stimuli. Four different stimulators are connected to one coil through a specially designed combining module, allowing various combinations of ISIs and IBIs of up to four single monophasic pulses to be selected.

In their original report, Hamada and colleagues compared the aftereffects of QPS with iTMS (Hamada et al. 2007b). They chose supra-threshold bursts at an ISI of 1.5 ms, which correspond to I-wave periodicity, and repeated them at 0.2 Hz (IBI = 5 s) for 30 min. The amplitudes of MEP to single-pulse TMS before and after intervention were compared. While they found a similar amount of facilitation of MEPs after both QPS and iTMS, the duration of facilitatory aftereffects was significantly longer for the QPS condition and lasted for more than 75 min. This implies that the number of pulses per burst is a crucial factor for the duration but not for the degree of aftereffects.

6.4.2 QPS and Cortical Plasticity

The long-lasting effect of QPS on the motor cortex has further been explored at different stimulation frequencies. Hamada et al. (2008a) compared seven different ISIs, which were 1.5, 5, 10, 30, 50, 100, and 1250 ms (Figure 6.5a). They delivered 1440

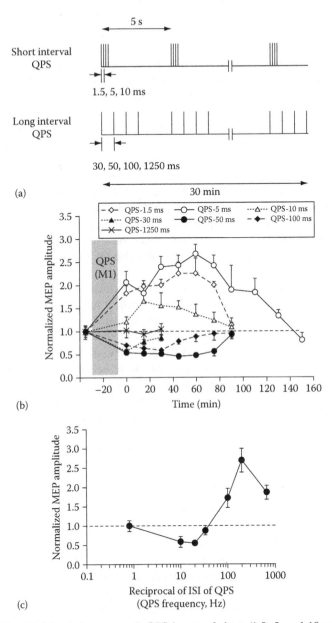

FIGURE 6.5 (a) Stimulation protocol. QPS bursts of short (1.5, 5, and 10 ms) and long (30, 50, 100, and 1250 ms) ISIs were applied at an IBI of 5 s for 30 min. (b) QPS at short ISI facilitated MEP for at least 75 min and QPS at long intervals suppressed MEPs. (c) Stimulus response curve illustrating the gradual shift from depression to potentiation by increasing the stimulation frequency. (From Hamada, M. and Ugawa, Y., *Restor. Neurol. Neurosci.*, 28, 419, 2010.)

pulses in 360 QPS bursts to M1 and evaluated the aftereffects on MEP amplitudes. For all conditions, 5 s IBI and 30 min total stimulation duration were kept constant. Short ISIs up to 10 ms significantly potentiated cortical excitability while higher ISIs from 30 to 100 ms resulted in a depression of MEP amplitudes (Figure 6.5b). Figure 6.5c shows the stimulus response curve illustrating the gradual shift of the aftereffects from depression to potentiation by increasing the stimulation frequency.

The mechanism of action for the QPS-induced neuroplasticity is not fully understood. However, there is increasing evidence supporting that it might be based on long-term modification of synaptic efficacy similar to LTP and LTD. The bidirectional induction of plasticity with the existence of a crossover point at approximately 40 Hz is consistent with mechanisms of synaptic plasticity describing a threshold for the induction of LTP in animal models (Bienenstock et al. 1982; Dudek and Bear 1992). It is believed that LTP and LTD can be subclassified into early LTP/LTD, which occurs when the postsynaptic calcium concentration reaches its threshold, and into late LTP/LTD, which requires protein synthesis and genetic transcription (Reymann and Frey 2007; Lu et al. 2008). Since the aftereffects of QPS seem to last longer than other rTMS protocols it is possible that QPS induces not only early but also late LTP/LTD.

Another important feature is that QPS solely modulates excitatory circuits within the motor cortex. ICF and short-interval intracortical facilitation (SICF), which are supposed to present glutamatergic facilitatory circuits, have been shown to be modulated by QPS significantly, while SICI and LICI, both of which measure GABAergic inhibitory function, were unchanged. Changes in membrane excitability, which can adequately be measured by motor thresholds, were not observed and are therefore supposed not to be involved in QPS-induced plasticity as well. Finally, input specificity, which is another basic property of synaptic plasticity (Bliss and Collingridge 1993), is also proved, since QPS-induced plasticity was topographically specific to the stimulation site.

The direction or degree of neuroplasticity induction may be determined by other stimulation parameters than ISI. In several experiments, Hamada and colleagues found that the total stimulation duration of 30 min was a crucial factor. When the stimulation duration was reduced to 10, 15, or 20 min, the post-interventional MEP sizes remained unchanged, indicating a threshold of total stimulation duration for plasticity induction (Hamada et al. 2008a). This is in line with early results of paired-pulse supra-threshold rTMS at frequencies between 0.17 and 5 Hz and ISIs of 2, 5, and 10 ms (Sommer et al. 2002a). Eighty paired stimuli irrespective of frequency and ISI were not able to induce aftereffects. On the other hand, longer QPS duration for 40 min failed to induce MEP amplitude changes as well. Animal studies supporting these data suggest an inverted U-shaped relation between the amount of stimulation and the degree of LTP. In hippocampal slices, a certain amount of stimulation is most effective for plasticity induction and overstimulation leads to reversal of LTP (Abraham and Huggett 1997; Christie et al. 1995). Depotentiation after tetanic stimulation might be responsible for this characteristic (Abraham and Huggett 1997). In analogy, it could be shown that doubling the stimulation duration and accordingly the number of stimuli of the classical theta burst paradigm led to a reversal of the direction of aftereffects; cTBS became excitatory and iTBS inhibitory (Gamboa et al. 2010).

The stimulation intensity seems to be less critical for QPS-induced neuroplasticity. A comparison of 360 trains of supra- and subthreshold QPS did not reveal any differences (Hamada et al. 2007b). However, at intensities close to threshold, it might be different. Excitatory cTBS300 turned out to switch to inhibition when the intensity was reduced from 70% to 65% resting motor threshold (Doeltgen and Ridding 2011).

6.4.3 Metaplasticity in QPS

Metaplasticity is a higher-order form of synaptic plasticity and describes the activity-dependent modification of synaptic plasticity. It is also called "plasticity of synaptic plasticity." It has been shown in animal models that the application of a priming stimulation, which itself does not induce plasticity, followed by an LTP/LTD inducing conditioning stimulation, leads to a shift of the LTP/LTD—inducing threshold. This is supposed to reflect a homeostatic mechanism of synaptic plasticity to prevent destabilization and to maintain modifiability of neuronal circuits and is well recognized as Bienenstock–Cooper–Munro (BCM) theory (Bienenstock et al. 1982; Abraham 2008). The shift of the crossover point of the frequency response function of synaptic plasticity strongly suggests the existence of a sliding modification threshold (Figure 6.6a).

There have been studies investigating metaplasticity in humans by using TMS revealing a dependence of TMS-induced plasticity on prior synaptic activity, which were all consistent with the BCM theory. However, all of these studies used priming protocols that themselves induced LTP or LTD and therefore already altered synaptic efficacy. This made it difficult to distinguish the effects from other homeostatic mechanisms such as ceiling/saturation or depotentiation. Therefore, Hamada and colleagues used their newly developed QPS protocol to investigate human cortical metaplasticity in detail. A 10 min priming stimulation at two different ISIs of 5 and 50 ms, which alone did not induce any plasticity, followed by the plasticity-inducing conditioning protocols at the aforementioned different ISIs (1.5–1250 ms) significantly shifted the stimulus response function along the abscissa (Figure 6.6b). Priming with QPS 5 ms shifted the curve to the right, favoring the induction of LTD-like plasticity through the conditioning stimulation, whereas priming with QPS 50 ms shifted the curve to the left, favoring the induction of LTP-like plasticity. Since priming stimulation has transiently influenced facilitatory circuits pertinent to SICF, it is assumed that it transiently modulates neuronal activity in respect to the BCM theory, determining the direction of the plasticity induced by the following conditioning stimulation (Hamada et al. 2008a). Both priming and conditioning protocols were applied at the same site with the same stimulation intensity favoring the modulation of the same synaptic connections, which can therefore be termed homotopic metaplasticity.

To study metaplastic interactions not only within the M1 but also within a wider motor network, the effect of priming stimulation of the supplementary motor area (SMA) on subsequent conditioning of M1 was investigated with QPS. Again priming stimulation with QPS 5 and 50 ms resulted in a shifting of the stimulus response

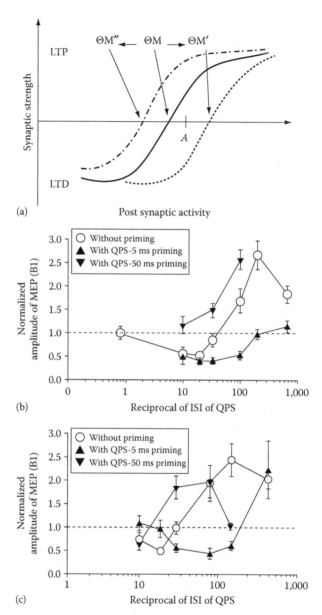

FIGURE 6.6 QPS and metaplasticity. (a) Model of metaplasticity based on the Bienenstock–Cooper–Monroe theory, illustrating a sliding of the modification threshold depending on the preceding postsynaptic neuronal activity. (b) Shift of the stimulus response function of QPS-induced plasticity after priming stimulation of M1. Priming with QPS 5 ms shifted the curve to the right and priming with QPS 50 ms shifted the curve to the left. (c) Shift of the stimulus response function of QPS-induced plasticity after priming stimulation of SMA. Priming with QPS 5 ms shifted the curve to the right and priming with QPS 50 ms shifted the curve to the left. (From Hamada, M. and Ugawa, Y., *Restor. Neurol. Neurosci.*, 28, 419, 2010.)

function to the right and left, respectively, by transient modulation of SICF-related facilitatory circuits (Figure 6.6c). In contrast to the metaplasticity within M1, it is suspected that heterotopic mechanisms are responsible for this kind of metaplastic interactions between SMA and M1 (Hamada et al. 2009). QPS has therefore been proven to be a valuable tool to study plasticity and metaplasticity mechanisms of the motor cortical networks in humans.

6.4.4 EFFECTS OF QPS ON CORTICAL HEMODYNAMICS

There are several studies investigating the influence of rTMS on cortical hemo-dynamics using functional imaging (Siebner et al. 2009). However, functional magnetic resonance imaging or positron emission tomography/single photon emission computed tomography may have some limitations: interference with the magnetic field, limited temporal or spatial resolution, or limitation of repeated investigations due to irradiation. Furthermore, the results between studies were incongruent and varied considerably depending on stimulation parameters such as intensity or frequency. NIRS is a promising new imaging method to measure cor-tical cerebral hemodynamics without any interference with electromagnetic fields and therefore suitable to investigate rTMS influences on cerebral hemodynamics. The effects of QPS 5 and 50 ms on cortical hemodynamic changes were studied. A new figure-of-eight stimulation coil and an NIRS recording probe, which were specially designed to fit each other, enabled online recordings of hemodynamic changes just beneath the stimulation coil during rTMS of the left M1 (Groiss et al. 2012). Using the standard stimulation intensity of 90% AMT, facilitatory QPS 5 ms compared with sham stimulation led to a deactivation pattern with HbO concentration decrement while inhibitory QPS 50 ms did not significantly influ-ence HbO concentration. Although these results support the frequency-dependent influence of QPS on the stimulated motor cortex, the changes were not bidirec-tional. While bidirectional long-term aftereffects of QPS reflect synaptic efficacy changes at least partly requiring protein synthesis, online effects during QPS and subsequent hemodynamic changes may reflect pure electrophysiological changes within the membrane or synapse without requiring protein syntheses. Since the neuronal inhibitory postsynaptic potential right after the excitatory postsynap-tic potential (postexcitatory inhibition) typically peak within the first 15–20 ms, pulses delivered at higher frequencies may lead to significant summation of the inhibitory effects. Hence, the decrease in HbO during QPS 5 ms may reflect a summation of postexcitatory inhibition at cortical pyramidal neurons (Groiss et al. 2012).

In another experiment, the online effects of QPS on the contralateral hemisphere were investigated using NIRS. When the stimulation intensity was increased to 110% of AMT, significant HbO concentration decrease was observed not only in the con-tralateral M1 but also in the surrounding primary sensory (S1), secondary sensory (S2), and ventral premotor cortex (PMv) during QPS 5 ms (Hirose et al. 2011). By contrast, QPS 50 ms decreased HbO concentration only in the contralateral M1 and not in the surrounding areas. These unidirectional remote effects on hemodynam-ics of the contralateral hemisphere again underline the difference between online effects during QPS probably reflecting pure electrophysiological property changes and long-term aftereffects reflecting synaptic efficacy changes (Hirose et al. 2011).

6.4.5 QPS AND BRAIN-DERIVED NEUROTROPIC FACTOR

Brain-derived neurotropic factor (BDNF) is a protein belonging to the neurotropins, and it is considered to be important for the development of the nervous system (Bath and Lee 2006). It and its precursor pro-BDNF play significant roles for plasticity such as LTP or LTD induction (Lu et al. 2005). Recently, it has been reported that a naturally occurring BDNF polymorphism with valine to methionine substitution at position 66 influences the BDNF secretion and leads to impairment of hippocampal plasticity, i.e., long-term memory (Egan et al. 2003; Bath and Lee 2006). In humans, BDNF polymorphism has been described to be an important modulation factor for the induction of LTP- or LTD-like plasticity by rTMS as well. Several plasticity-inducing protocols such as TBS, PAS, or tDCS were differently affected by BDNF polymorphism, non-Val–Val polymorphism mainly showing less stable plasticity induction (Cheeran et al. 2008; Antal et al. 2010). In contrast to these results, QPS has been shown to induce stable plasticity for both LTP and LTD regardless of BDNF polymorphism (Nakamura et al. 2011; Figure 6.7). One main difference between QPS and the other stimulation protocols is the stimulation duration. Given the fact that induction of late LTP/LTD (Lu et al. 2008) and morphological synaptic changes (Nägerl et al. 2004) need at least 30 min to occur, it may be of particular importance that 30 min of stimulation are necessary to induce long-term plasticity by QPS (Hamada et al. 2008a). Neither 10, 15, or 20 min of QPS was sufficient to induce plasticity. Since BDNF polymorphism has a strong influence on the speed and amount of BDNF secretion but not on the function itself (Egan et al. 2003), longer stimulation duration as 30 min may be crucial to induce the same magnitude of long-term plasticity in subjects with non-Val–Val BDNF polymorphism (Nakamura et al. 2011).

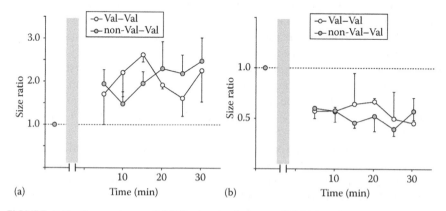

FIGURE 6.7 Comparisons of MEP changes between Val–Val and non-Val–Val BDNF polymorphisms after QPS 5 ms (a) and QPS 50 ms (b). No differences were found between Val–Val and non-Val–Val BDNF polymorphisms. (From Nakamura, K. et al., *Neurosci. Lett.,* 487, 264, 2011.)

6.4.6 SUMMARY AND PERSPECTIVES OF QPS

QPS is a novel powerful and safe rTMS method that induces both facilitatory and depressive cortical plasticity depending on ISI length. Its mechanism of action fits well with homeostatic plasticity proposed with the BCM theory. It has been shown that QPS is a meaningful method to study plasticity and metaplasticity mechanisms in humans. In addition, the long duration of aftereffects might be useful to modulate altered cortical activity in patients with neurological or psychiatric diseases.

6.5 SAFETY OF PATTERNED rTMS PROTOCOLS

Compared with conventional rTMS protocols, for which we already have well-established safety guidelines (Rossi et al. 2009, see also Chapter 16), we still have only few experiences with newer patterned rTMS protocols such as QPS or TBS. Besides transient feelings of dizziness or transient headache, the most important side effects induced by rTMS are epileptic seizures.

A recent review on the safety of TBS including 67 studies with a total of 1001 individuals (776 healthy controls and 225 patients with various neurological disorders) found transient headaches and neck pain to be the most common side effects similar to those after conventional rTMS protocols (Oberman et al. 2011). However, these occurred in less than 3% after TBS while it was reported in up to 40% after conventional rTMS (Rossi et al. 2009; Oberman et al. 2011). Until now, only a single case of seizure after TBS has been reported (Oberman and Pascual-Leone 2009). Although the subject did not have any risk factors for epilepsy, he had an overseas flight 2 days before the event, which might have altered his sleep pattern. Moreover, in this case, stimulation intensity was set at 100% RMT, which is considerably higher than the commonly used intensity of 80% AMT.

For pp rTMS, no detailed safety studies exist. In the first report, Sommer et al. observed a spread of excitation in one subject (Sommer et al. 2001). However, no epileptic seizures have been reported in any of the different pp rTMS paradigms until now.

In the original reports on QPS, all subjects tolerated the intervention without any side effects and the authors neither found changes in the occurrence rate of afterdischarges nor did they observe any spread of excitation. More recently, Nakatani-Enomoto et al. (2010) have investigated safety of QPS in more detail. The standard protocol of either QPS 5 or 50 ms, which has been proven to be the protocols inducing the most powerful changes in cortical excitability, applying a total of 1440 pulses at an intensity of 90% AMT, did not influence vital signs, neurological status, serum prolactin level, or EEG in healthy subjects immediately after the intervention (Nakatani-Enomoto et al. 2010). Even though there is no large safety study on patients with neurological disorders, personal observational data on patients with cortical myoclonus and basal ganglia disorders such as Parkinson's disease or Huntington's disease suggest a high tolerability of the protocol without any complications. We have also observational data on healthy subjects, who underwent unproblematic stimulation up to 2880 pulses in 360–720 bursts (up to 8 pulses per burst) during one session.

Further investigations on safety of patterned rTMS are warranted, but it is supposed that the low subthreshold stimulation intensity of 80%–90% AMT contributes to the safety of QPS or TBS (Nakatani-Enomoto et al. 2010).

6.6 NEGLECTED ROLE OF INTERVALS IN HIGH-FREQUENCY rTMS

Although usually high-frequency rTMS is assumed to be non-patterned, in most cases some kind of patterning essentially exists. Intervals are introduced in order to avoid coil heating and sometimes with a vague assumption that they may be protective against seizures. In a most important safety consensus on TMS (Rossi et al. 2009) in the parameter section of high-frequency rTMS, the role of intervals was described as follows: "... whereas protocols of fast rTMS (e.g., >5 Hz stimulation frequency) apply shorter periods of rTMS separated by periods of no stimulation (e.g., 1200 pulses at 20 Hz and subthreshold stimulation intensity might be delivered as 30 trains of 40 pulses (2 s duration) separated by 28 s intertrain intervals). There is only limited safety information on the effect of inserting pauses (intertrain intervals) into rTMS protocols (Chen et al. 1997). However, considering metaplasticity arguments (Abraham and Bear 1996; Bear 2003), it is likely that such pauses also have a significant impact on the effect of rTMS, both in terms of efficacy and safety. Therefore, further investigations are needed."

In table S3 of the supplement section (Rossi et al. 2009), an overview over a decade of high-frequency rTMS studies is provided, many studies not having reported on intervals and the others using time ranges between about 1 s to a few min. It seems to be most important to systematically clarify the role of these intervals; TBS may serve as a good example. The cTBS protocol induces inhibition when 8 s interval excitation is applied, but with 5 s interval, this is neutralized to the control condition (Huang et al. 2005). In a study addressing this point, a standard 5 Hz rTMS protocol using 1200 pulses in a block design with interstimulus train intervals of 60 s was compared with a continuous rTMS protocol using the same number of pulses (Rothkegel et al. 2010). Only the 5 Hz rTMS in the standard block design induces facilitatory aftereffects on corticospinal excitability, the continuous train instead induced inhibition, very much like the TBS results. Interestingly, when comparing a positive clinical trial in depression (O'Reardon et al. 2007) with a negative one (Herwig et al. 2007), the former used intertrial intervals of 26 s, the latter intervals of 8 s. The contribution of this parameter to outcome is however difficult to decide since these studies differed in many other aspects. Thus, for better understanding of the role of intervals in clinical studies, clearly more data are needed here.

6.7 CONCLUSION AND OUTLOOK

Although still at the beginning, patterned rTMS has opened fascinating perspectives. By introducing IBIs in any combination into the sequence of TMS pulses, several effects can be targeted and improvements achieved. Efficacy can be clearly increased when compared with regular rTMS. Along this, the burden of applied magnetic charge can be reduced for subjects or patients and concomitantly coil heating reduced. In general, the protocols seem to provide more robust aftereffects, in particular in separating into excitation or inhibition. Some questions remain to be solved here as well. Looking at TBS, four different intervals play a role, 20 ms, 200 ms, 8 s, and total stimulation duration. So far, it is unclear if each of these intervals is already adjusted to the individual optimum. For QPS, it seems to be necessary to apply four pulses, the number so far limited because of technical reasons. It may however turn out that five or six or more may be even better suited. Many other factors play a role as well, such as genetic variations in the BDNF gene, for which QPS seems to be more robust than TBS. It is also not clear to which extent the MEP as a biomarker is predictive for behavioral or clinical effects as well. Or, in other words, do M1 results predict effects in other brain areas? So far most of the progress was made on an empirical basis pursuing hypotheses predetermined by biological data such as I-waves or theta rhythms. Really most helpful for generating future protocols would be a theoretical framework providing testable hypotheses; a first step has been published recently (Huang et al. 2011).

REFERENCES

Abraham WC. 2008. Metaplasticity: Tuning synapses and networks for plasticity. *Nat Rev Neurosci* 9: 387.

Abraham WC and Bear MF. 1996. Metaplasticity: The plasticity of synaptic plasticity. *Trends Neurosci* 19: 126–130.

Abraham WC and Huggett A. 1997. Induction and reversal of long-term potentiation by repeated high-frequency stimulation in rat hippocampal slices. *Hippocampus* 7: 137–145.

Antal A, Chaieb L, Moliadze V et al. 2010. Brain-derived neurotrophic factor (BDNF) gene polymorphisms shape cortical plasticity in humans. *Brain Stimul* 3: 230–237.

Antal A and Paulus W. 2010. Effects of transcranial theta-burst stimulation on acute pain perception. *Restor Neurol Neurosci* 28: 477–484.

Arai N, Okabe S, Furubayashi T et al. 2007. Differences in after-effect between monophasic and biphasic high-frequency rTMS of the human motor cortex. *Clin Neurophysiol* 118: 2227–2233.

Bath KG and Lee FS. 2006. Variant BDNF (Val66Met) impact on brain structure and function. *Cogn Affect Behav Neurosci* 6: 79–85.

Bear MF. 2003. Bidirectional synaptic plasticity: From theory to reality. *Philos Trans R Soc Lond B Biol Sci* 358: 649–655.

Benwell NM, Mastaglia FL, and Thickbroom GW. 2006. Paired-pulse rTMS at trans-synaptic intervals increases corticomotor excitability and reduces the rate of force loss during a fatiguing exercise of the hand. *Exp Brain Res* 175: 626–632.

Bezprozvanny I, Watras J, and Ehrlich BE. 1991. Bell-shaped calcium-response curves of Ins(1,4,5)P3- and calcium-gated channels from endoplasmic reticulum of cerebellum. *Nature* 351: 751–754.

Bienenstock EL, Cooper LN, and Munro PW. 1982. Theory for the development of neuron selectivity: Orientation specificity and binocular interaction in visual cortex. *J Neurosci* 2: 32–48.

Bliss TVP and Collingridge GL. 1993. A synaptic model of memory: Long-term potentiation in the hippocampus. *Nature* 361: 31–39.

Calabresi P, Pisani A, Mercuri NB, and Bernardi G. 1992. Long-term potentiation in the striatum is unmasked by removing the voltage-dependent magnesium block of NMDA receptor channels. *Eur J Neurosci* 4: 929–935.

Cash RF, Benwell NM, Murray K, Mastaglia FL, and Thickbroom GW. 2009. Neuromodulation by paired-pulse TMS at an I-wave interval facilitates multiple I-waves. *Exp Brain Res* 193: 1–7.

Cheeran B, Talelli P, Mori F et al. 2008. A common polymorphism in the brain-derived neu-rotrophic factor gene (BDNF) modulates human cortical plasticity and the response to rTMS. *J Physiol* 586: 5717–5725.

Chen R, Gerloff C, Classen J, Wassermann EM, Hallett M, and Cohen LG. 1997. Safety of different inter-train intervals for repetitive transcranial magnetic stimulation and rec-ommendations for safe ranges of stimulation parameters. *Electroencephalogr Clin Neurophysiol* 105: 415–421.

Christie BR, Stellwagen D, and Abraham WC. 1995. Reduction of the threshold for long-term potentiation by prior theta-frequency synaptic activity. *Hippocampus* 5: 52–59.

Chung HK, Tsai CH, Lin YC et al. 2011. Effectiveness of Theta-Burst Repetitive Transcranial Magnetic Stimulation for Treating Chronic Tinnitus. *Audiol Neurootol* 17: 112–120.

Coesmans M, Weber JT, De Zeeuw CI, and Hansel C. 2004. Bidirectional parallel fiber plastic-ity in the cerebellum under climbing fiber control. *Neuron* 44: 691–700.

Di Lazzaro V, Pilato F, Dileone M et al. 2008. The physiological basis of the effects of inter-mittent theta burst stimulation of the human motor cortex. *J Physiol* 586: 3871–3879.

Di Lazzaro V, Pilato F, Saturno E et al. 2005. Theta-burst repetitive transcranial magnetic stimulation suppresses specific excitatory circuits in the human motor cortex. *J Physiol* 565: 945–950.

Di Lazzaro V, Thickbroom GW, Pilato F et al. 2007. Direct demonstration of the effects of repetitive paired-pulse transcranial magnetic stimulation at I-wave periodicity. *Clin Neurophysiol* 118: 1193–1197.

Doeltgen SH and Ridding MC. 2011. Low-intensity, short-interval theta burst stimulation mod-ulates excitatory but not inhibitory motor networks. *Clin Neurophysiol* 122: 1411–1416.

Dudek SM and Bear MF. 1992. Homosynaptic long-term depression in area CA1 of hippocampus and effects of N-methyl-D-aspartate receptor blockade. *PNAS* 89: 4363–4367.

Eichenbaum H, Kuperstein M, Fagan A, and Nagode J. 1987. Cue-sampling and goal-approach correlates of hippocampal unit activity in rats performing an odor-discrimination task. *J Neurosci* 7: 716–732.

Egan MF, Kojima M, Callicott JH et al. 2003. The BDNF val66met polymorphism affects activity-dependent secretion of BDNF and human memory and hippocampal function. *Cell* 112: 257–269

Fang JH, Chen JJ, Hwang IS, and Huang YZ. 2010. Repetitive transcranial magnetic stimu-lation over the human primary motor cortex for modulating motor control and motor learning. *J Med Biol Eng* 30: 193–201.

Fitzgerald PB, Fountain S, and Daskalakis ZJ. 2006. A comprehensive review of the effects of rTMS on motor cortical excitability and inhibition. *Clin Neurophysiol* 117: 2584–2596.

Fitzgerald PB, Fountain S, Hoy K et al. 2007. A comparative study of the effects of repeti-tive paired transcranial magnetic stimulation on motor cortical excitability. *J Neurosci Methods* 165: 265–269.

Gamboa OL, Antal A, Moliadze V, and Paulus W. 2010. Simply longer is not better: Reversal of theta burst after-effect with prolonged stimulation. *Exp Brain Res* 204: 181–187.

Gentner R, Wankerl K, Reinsberger C, Zeller D, and Classen J. 2008. Depression of human corticospinal excitability induced by magnetic theta-burst stimulation: Evidence of rapid polarity-reversing metaplasticity. *Cereb Cortex* 18: 2046–2053.

Groiss SJ, Mochizuki H, Furubayashi T et al. 2012. Quadri-pulse stimulation induces stimulation frequency dependent cortical hemoglobin concentration changes within the ipsilateral motor cortical network. *Brain Stimul* DOI: 10.1016/j.brs.2011.12.004 (Epub. ahead of print).

Hamada M, Hanajima R, Terao Y et al. 2007a. Origin of facilitation in repetitive, 1.5 ms inter-val, paired pulse transcranial magnetic stimulation (rPPS) of the human motor cortex. *Clin Neurophysiol* 118: 1596–1601.

Hamada M, Hanajima R, Terao Y et al. 2007b. Quadro-pulse stimulation is more effective than paired-pulse stimulation for plasticity induction of the human motor cortex. *Clin Neurophysiol* 118: 2672–2682.

Hamada M, Hanajima R, Terao Y et al. 2009. Primary motor cortical metaplasticity induced by priming over the supplementary motor area. *J Physiol* 587: 4845–4862.

Hamada M, Terao Y, Hanajima R et al. 2008a. Bidirectional long-term motor cortical plasticity and metaplasticity induced by quadripulse transcranial magnetic stimulation. *J Physiol* 586: 3927–3947.

Hamada M and Ugawa Y. 2010. Quadri-pulse stimulation—A new patterned rTMS. *Restor Neurol Neurosci* 28: 419–424.

Hamada M, Ugawa Y, Tsuji S, and the Effectiveness of rTMS on Parkinson's Disease Study Group Japan. 2008b. High-frequency rTMS over the supplementary motor area for treatment of Parkinson's disease. *Mov Disord* 23: 1524–1531.

Herwig U, Fallgatter AJ, Hoppner J et al. 2007. Antidepressant effects of augmentative transcranial magnetic stimulation: Randomised multicentre trial. *Br J Psychiatry* 191: 441–448.

Hirose M, Mochizuki H, Groiss SJ et al. 2011. On-line effects of quadripulse transcranial magnetic stimulation (QPS) on the contralateral hemisphere studied with somatosensory evoked potentials and near infrared spectroscopy. *Exp Brain Res* 214: 577–586.

Hirsch JC and Crepel F. 1991. Blockade of NMDA receptors unmasks a long-term depression in synaptic efficacy in rat prefrontal neurons in vitro. *Exp Brain Res* 85: 621–624.

Holzer M and Padberg F. 2010. Intermittent theta burst stimulation (iTBS) ameliorates therapy-resistant depression: A case series. *Brain Stimul* 3: 181–183.

Hsu YF, Liao KK, Lee PL et al. 2011. Intermittent theta burst stimulation over primary motor cortex enhances movement-related beta synchronisation. *Clin Nneurophysiol* 122: 2260–2267.

Huang YZ, Chen RS, Rothwell JC, and Wen HY. 2007. The after-effect of human theta burst stimulation is NMDA receptor dependent. *Clin Neurophysiol* 118: 1028–1032.

Huang YZ, Edwards MJ, Rounis E, Bhatia KP, and Rothwell JC. 2005. Theta burst stimulation of the human motor cortex. *Neuron* 45: 201–206.

Huang CC, Liang YC, and Hsu KS. 2001. Characterization of the mechanism underlying the reversal of long term potentiation by low frequency stimulation at hippocampal CA1 synapses. *J Biol Chem* 276: 48108–48117.

Huang YZ, Rothwell JC, Chen RS, Lu CS, and Chuang WL. 2011. The theoretical model of theta burst form of repetitive transcranial magnetic stimulation. *Clin Neurophysiol* 122: 1011–1018.

Huang YZ, Rothwell JC, Edwards MJ, and Chen RS. 2008. Effect of physiological activity on an NMDA-dependent form of cortical plasticity in human. *Cereb Cortex* 18: 563–570.

Huang YZ, Rothwell JC, Lu CS et al. 2009a. The effect of continuous theta burst stimulation over premotor cortex on circuits in primary motor cortex and spinal cord. *Clin Neurophysiol* 120: 796–801.

Huang YZ, Rothwell JC, Lu CS, Chuang WL, Lin WY, and Chen RS. 2010a. Reversal of plasticity-like effects in the human motor cortex. *J Physiol* 588: 3683–3693.

Huang YZ, Rothwell JC, Lu CS, Wang J, and Chen RS. 2010b. Restoration of motor inhibition through an abnormal premotor-motor connection in dystonia. *Mov Disord* 25: 689–696.

Huang YZ, Sommer M, Thickbroom GW et al. 2009b. Consensus: New methodologies for brain stimulation. *Brain Stimul* 2: 2–13.

Iezzi E, Conte A, Suppa A et al. 2008. Phasic voluntary movements reverse the aftereffects of subsequent theta-burst stimulation in humans. *J Neurophysiol* 100: 2070–2076.

Kandel ER and Spencer WA. 1961. Electrophysiology of hippocampal neurons. II. After-potentials and repetitive firing. *J Neurophysiol* 24: 243–259.

Kemp N and Bashir ZI. 2001. Long-term depression: A cascade of induction and expression mechanisms. *Prog Neurobiol* 65: 339–365.

Khedr EM, Gilio F, and Rothwell J. 2004. Effects of low frequency and low intensity repetitive paired pulse stimulation of the primary motor cortex. *Clin Neurophysiol* 115: 1259–1263.

Koch G, Mori F, Marconi B et al. 2008. Changes in intracortical circuits of the human motor cortex following theta burst stimulation of the lateral cerebellum. *Clin Neurophysiol* 131: 3147–3155.

Kulla A and Manahan-Vaughan D. 2000. Depotentiation in the dentate gyrus of freely moving rats is modulated by D1/D5 dopamine receptors. *Cereb Cortex* 10: 614–620.

Kumar A, Thinschmidt JS, Foster TC, and King MA. 2007. Aging effects on the limits and stability of long-term synaptic potentiation and depression in rat hippocampal area CA1. *J Neurophysiol* 98: 594–601.

Larson J and Lynch G. 1986. Induction of synaptic potentiation in hippocampus by patterned stimulation involves two events. *Science* 232: 985–988.

Larson J and Lynch G. 1988. Role of N-methyl-D-aspartate receptors in the induction of synaptic potentiation by burst stimulation patterned after the hippocampal theta-rhythm. *Brain Res* 441: 111–118.

Larson J, Wong D, and Lynch G. 1986. Patterned stimulation at the theta frequency is optimal for the induction of hippocampal long-term potentiation. *Brain Res* 368: 347–350.

Lee HK, Kameyama K, Huganir RL, and Bear MF. 1998. NMDA induces long-term synaptic depression and dephosphorylation of the GluR1 subunit of AMPA receptors in hippocampus. *Neuron* 21: 1151–1162.

Liu L, Wong TP, Pozza MF et al. 2004. Role of NMDA receptor subtypes in governing the direction of hippocampal synaptic plasticity. *Science* 304: 1021–1024.

Lu Y, Christian K, and Lu B. 2008. BDNF: A key regulator for protein synthesis-dependent LTP and long-term memory? *Neurobiol Learn Mem* 89: 312–323.

Lu B, Pang PT, and Woo NH. 2005. The yin and yang of neurotrophin action. *Nat Rev Neurosci* 6: 603–614.

Malenka RC and Nicoll RA. 1999. Long-term potentiation—A decade of progress? *Science* 285: 1870–1874.

Mochizuki H, Franca M, Huang YZ, and Rothwell JC. 2005. The role of dorsal premotor area in reaction task: Comparing the "virtual lesion" effect of paired pulse or theta burst transcranial magnetic stimulation. *Exp Brain Res* 167: 414–421.

Montgomery JM and Madison DV. 2004. Discrete synaptic states define a major mechanism of synapse plasticity. *Trends Neurosci* 27: 744–750.

Murray LM, Nosaka K, and Thickbroom GW. 2011. Interventional repetitive I-wave transcranial magnetic stimulation (TMS): The dimension of stimulation duration. *Brain Stimul* 4: 261–265.

Nakatani-Enomoto S, Hanajima R, Hamada M et al. 2010. Some evidence supporting the safety of quadripulse stimulation (QPS). *Brain Stimul* 4: 303–305.

Nakamura K, Enomoto H, Hanajima R et al. 2011. Quadri-pulse stimulation (QPS) induced LTP/LTD was not affected by Val66Met polymorphism in the brain-derived neurotrophic factor (BDNF) gene. *Neurosci Lett* 487: 264–267.

Nägerl UV, Eberhorn N, Cambridge SB, and Bonhoeffer T. 2004. Bidirectional activity-dependent morphological plasticity in hippocampal neurons. *Neuron* 44: 759–767.

Neveu D and Zucker RS. 1996. Postsynaptic levels of [Ca2+]i needed to trigger LTD and LTP. *Neuron* 16: 619–629.

Nishiyama M, Hong K, Mikoshiba K, Poo MM, and Kato K. 2000. Calcium stores regulate the polarity and input specificity of synaptic modification. *Nature* 408: 584–588.

Nyffeler T, Cazzoli D, Wurtz P et al. 2008. Neglect-like visual exploration behaviour after theta burst transcranial magnetic stimulation of the right posterior parietal cortex. *Eur J Neurosci* 27: 1809–1813.

Nyffeler T, Wurtz P, Luscher HR et al. 2006. Extending lifetime of plastic changes in the human brain. *Eur J Neurosci* 24: 2961–2966.

Obermann L, Edwards D, Eldaief M, and Pascual-Leone A. 2011. Safety of theta burst magnetic stimulation: A systematic review of the literature. *J Clin Neurophysiol* 28: 67–74.

Obermann L and Pascual-Leone A. 2009. Report on seizure induced by continuous theta burst stimulation. *Brain Stimul* 2: 246–247.

O'Reardon JP, Solvason HB, Janicak PG et al. 2007. Efficacy and safety of transcranial magnetic stimulation in the acute treatment of major depression: A multisite randomized controlled trial. *Biol Psychiatry* 62: 1208–1216.

Pascual-Leone A, Gates JR, and Dhuna A. 1991. Induction of speech arrest and counting errors with rapid-rate transcranial magnetic stimulation. *Neurology* 41: 697–702.

Peinemann A, Reimer B, Loer C et al. 2004. Long-lasting increase in corticospinal excitability after 1800 pulses of subthreshold 5 Hz repetitive TMS to the primary motor cortex. *Clin Neurophysiol* 115: 1519–1526.

Picconi B, Centonze D, Hakansson, K et al. 2003. Loss of bidirectional striatal synaptic plasticity in L-DOPA-induced dyskinesia. *Nat Neurosci* 6: 501–506.

Reymann KG and Frey JU. 2007. The late maintenance of hippocampal LTP: Requirements, phases, 'synaptic tagging', 'late-associativity' and implications. *Neuropharmacology* 52: 24–40.

Ridding MC, Taylor JL, and Rothwell JC. 1995. The effect of voluntary contraction on cortico-cortical inhibition in human motor cortex. *J Physiol* 487: 541–548.

Rossi S, Hallett M, Rossini PM, Pascual-Leone A, and The Safety of TMS Consensus Group. 2009. Safety, ethical considerations, and application guidelines for the use of transcranial magnetic stimulation in clinical practice and research. *Clin Neurophysiol* 120: 2008–2039.

Rothkegel H, Sommer M, and Paulus W. 2010. Breaks during 5 Hz rTMS are essential for facilitatory after effects. *Clin Neurophysiol* 121: 426–430.

Saglam M, Matsunaga K, Murayama N, Hayashida Y, Huang YZ, and Nakanishi R. 2008. Parallel inhibition of cortico-muscular synchronization and cortico-spinal excitability by theta burst TMS in humans. *Clin Neurophysiol* 119: 2829–2838.

Sheng M and Kim MJ. 2002. Postsynaptic signaling and plasticity mechanisms. *Science* 298: 776–780.

Siebner HR, Bergmann TO, Bestmann S et al. 2009. Consensus paper: Combining transcranial stimulation with neuroimaging. *Brain Stimul* 2: 58–80.

Silvanto J, Muggleton NG, Cowey A, and Walsh V. 2007. Neural activation state determines behavioral susceptibility to modified theta burst transcranial magnetic stimulation. *Eur J Neurosci* 26: 523–528.

Sommer M, Kamm T, Tergau F, Ulm G, and Paulus W. 2002a. Repetitive paired-pulse transcranial magnetic stimulation affects corticospinal excitability and finger tapping in Parkinson's disease. *Clin Neurophysiol* 113: 944–950.

Sommer M, Lang N, Tergau F, and Paulus W. 2002b. Neuronal tissue polarization induced by repetitive transcranial magnetic stimulation? *Neuroreport* 13: 809–811.

Sommer M, Tergau F, Wischer S, and Paulus W. 2001. Paired-pulse repetitive transcranial magnetic stimulation of the human motor cortex. *Exp Brain Res* 139: 465–472.

Sommer M, Wu T, Tergau F, and Paulus W. 2002c. Intra- and interindividual variability of motor responses to repetitive transcranial magnetic stimulation. *Clin Neurophysiol* 113: 265–269.

Staubli U and Scafidi J. 1999. Time-dependent reversal of long-term potentiation in area CA1 of the freely moving rat induced by theta pulse stimulation. *J Neurosci* 19: 8712–8719.

Talelli P, Greenwood RJ, and Rothwell JC. 2007. Exploring Theta Burst Stimulation as an intervention to improve motor recovery in chronic stroke. *Clin Neurophysiol* 118: 333–342.

Teo JT, Swayne OB, and Rothwell JC. 2007. Further evidence for NMDA-dependence of the after-effects of human theta burst stimulation. *Clin Neurophysiol* 118: 1649–1651.

Thickbroom GW, Byrnes ML, Edwards DJ, and Mastaglia FL. 2006. Repetitive paired-pulse TMS at I-wave periodicity markedly increases corticospinal excitability: A new technique for modulating synaptic plasticity. *Clin Neurophysiol* 117: 61–66.

Thomson RH, Daskalakis ZJ, and Fitzgerald PB. 2011. A near infra-red spectroscopy study of the effects of pre-frontal single and paired pulse transcranial magnetic stimulation. *Clin Neurophysiol* 122: 378–382.

Tings T, Lang N, Tergau F, Paulus W, and Sommer M. 2005. Orientation-specific fast rTMS maximizes corticospinal inhibition and facilitation. *Exp Brain Res* 164: 323–333.

Touge T, Gerschlager W, Brown P, and Rothwell JC. 2001. Are the after-effects of low-frequency rTMS on motor cortex excitability due to changes in the efficacy of cortical synapses? *Clin Neurophysiol* 112: 2138–2145.

Wankerl K, Weise D, Gentner R, Rumpf JJ, and Classen J. 2010. L-type voltage-gated Ca^{2+} channels: A single molecular switch for long-term potentiation/long-term depression-like plasticity and activity-dependent metaplasticity in humans. *J Neurosci* 30: 6197–6204.

Wu CC, Tsai CH, Lu MK, Chen CM, Shen WC, and Su KP. 2010. Theta-burst repetitive transcranial magnetic stimulation for treatment-resistant obsessive-compulsive disorder with concomitant depression. *J Clin Psychiatry* 71: 504–506.

Yang SN, Tang YG, and Zucker RS. 1999. Selective induction of LTP and LTD by postsynaptic $[Ca^{2+}]i$ elevation. *J Neurophysiol* 81: 781–787.

Ziemann, U. 2004. TMS induced plasticity in human cortex. *Rev Neurosci* 15: 253–266.

Ziemann U, Hallett M, and Cohen LG. 1998. Mechanisms of deafferentation-induced plasticity in human motor cortex. *J Neurosci* 18: 7000–7007.

7 Motor Cortical and Corticospinal Measures in Health and Disease

Massimo Cincotta, Angelo Quartarone, and Giovanni Abbruzzese

CONTENTS

7.1   INTRODUCTION

In 1980, Merton and Morton first reported muscle responses in an unanesthetized human following motor cortex stimulation with high-voltage electric shocks delivered through the intact skull. Although noninvasive, this high-intensity transcranial electrical stimulation (TES) is painful because of activation of pain fibers in the scalp. In 1985, Barker et al. showed that painless transcranial stimulation of the human brain could be obtained using the small electric field induced by a time-varying magnetic field, which in turn is produced by passing a large current through an insulated copper wire coil placed above the scalp. As magnetic field is not significantly attenuated by the scalp and skull, the induced electric current could activate superficial neural elements (Roth et al., 1991).

TMS has first been used to study corticospinal motor conduction. Basically, a single TMS pulse of sufficient intensity delivered to the primary motor cortex (M1) activates corticospinal (or corticobulbar) neurons and produces multiple descending volleys, which are referred to as indirect (I) waves and numbered according to their latencies. These descending volleys in turn depolarize spinal (or bulbar) motoneurons. The resulting peripheral volleys produce a relatively synchronous response in the target muscle, the motor-evoked potential (MEP), which can be recorded through conventional surface electromyographic (EMG) electrodes. As high-intensity TES mainly activates corticospinal neurons directly, TMS-induced activation of corticofugal pathways largely depends on synaptic inputs through interneuronal elements (at least with the standard magnetic coil placements and orientations). In conscious humans, direct demonstration was provided by recording corticospinal volleys through cervical, epidural electrodes implanted for pain relief (Di Lazzaro et al., 1998a) or for intraoperative neurophysiological monitoring (Nakamura et al., 1997). The most largely employed measure of conduction along the corticospinal pathways is the central motor conduction time (CMCT), which is calculated by subtracting the peripheral motor conduction time (PMCT) from the MEP latency. PMCT can be estimated by the F-wave method (Kimura, 1983) or by spinal root stimulation (Rossini et al., 1994). Additional information is provided by the MEP amplitude, which is usually expressed as percentage of the compound muscle action potential

(CMAP) elicited after supramaximal electrical stimulation of the peripheral nerve in an attempt to reduce the considerable variability of this measure among healthy individuals. More recently, Magistris and coworkers (Magistris et al., 1998; Rosler et al., 2002) have proposed a collision method known as triple stimulation technique (TST) in order to circumvent a number of problems encountered with MEP size assessment (see the "Advanced TMS Techniques" paragraph for details).

In the meantime that TMS was applied to study corticospinal conduction in health and disease, it became evident that TMS also provided a noninvasive evaluation of distinct excitatory and inhibitory functions of the human cerebral cortex (for review, see Hallett, 1996; Rossini and Rossi, 1998; Chen, 2000; Hallett, 2000; Abbruzzese and Trompetto, 2002; Chen et al., 2008; Rothwell, 2010). Therefore, TMS has been largely used to assess the physiology of human motor control and pathophysiological mechanisms in patients with neurological diseases. Besides the evaluation of the threshold stimulus intensity to elicit the MEP and of MEP size changes following single magnetic pulses at different stimulus intensities (input–output [I–O] curve), two main groups of TMS paradigms have been employed to this purpose. First, the modulatory effects of TMS on voluntary ongoing EMG activity (i.e., the contralateral and ipsilateral silent periods) have been evaluated. Second, a large number of different conditioning stimuli have been used to modulate the size of the MEP elicited by a test magnetic pulse delivered to the M1 at different interstimulus intervals (ISIs). In these conditioning-test protocols, the conditioning stimulus can be another magnetic pulse, which is delivered to the same scalp position of the test stimulus through the same coil (i.e., the technique used to assess short- [SICI] and long-interval intracortical inhibition [LICI]) or at different scalp positions by another focal coil (e.g., for interhemispheric inhibition of the MEP). Alternatively, the conditioning stimulus is a non-magnetic one, for instance, an electric pulse delivered to the peripheral nerve to assess short and long afferent inhibition. Furthermore, in some studies, the modulation of MEP (Berardelli et al., 1998) and of contralateral (Berardelli et al., 1999; Romeo et al., 2000; Inghilleri et al., 2004) and ipsilateral silent periods (Cincotta et al., 2006b) produced by short trains of repetitive TMS at different frequencies has also been investigated.

In the present chapter, we summarize the available data on the main application of single- and paired-pulse TMS measures of corticospinal conduction and motor cortical excitability in health and neurological disease. In Sections 7.2 and 7.3, the main TMS variables along with the current knowledge on their physiological significance are described. As studies tailored to investigate the effects of drugs with well-known modes of action on cortical excitability contributed to understand the mechanisms underlying TMS parameters (Ziemann, 2004), such data are resumed in these sections. These studies have similar protocols: in normal subjects, TMS variables were measured before a single-dose intake (baseline) and retested at delays compatible with the pharmacokinetics of a given drug. In addition, the relevant contribution in this field of recordings of TMS-induced corticospinal volleys in conscious humans (Nakamura et al., 1997; Di Lazzaro et al., 1998a) and studying interactions of different measures (Chen, 2004) is also underlined. In Section 7.4, the most relevant applications of these TMS measures in patients with neurological disorders are discussed.

7.2 ROUTINE TMS TECHNIQUES

7.2.1 MEASURES OF CORTICOSPINAL EXCITABILITY

7.2.1.1 Motor Threshold

The motor threshold (MT) is defined as the minimum stimulus strength that produces a small MEP (usually 50–100 μV) in the target muscle, in at least half of 10 (or more) consecutive trials (Rossini et al., 1994). MT can be determined at rest (RMT, Figure 7.1a) or during slight isometric activation of the target muscle (AMT). MT reflects the excitability and the local density of a central core of excitatory interneurons and corticospinal neurons in the muscle representation in the M1, as well as the excitability of the target spinal (or brainstem) motoneurons (Hallett et al., 2000; Tassinari et al., 2003). AMT is lower than RMT. This facilitatory effect of tonic muscle contraction mainly depends on enhanced excitability of the motoneuron pool, although corticospinal neuron excitability is also slightly increased (Mazzocchio et al., 1994; Di Lazzaro et al., 1998b).

In healthy subjects, acute intake of voltage-dependent sodium channel blocker drugs increases the MT (Mavroudakis et al., 1994; Ziemann et al., 1996b; Chen et al., 1997; Boroojerdi et al., 2001). The maximum MT enhancement has been reported at the plasma peak time (Ziemann et al., 1996b). In contrast, Schulze-Bonhage et al. (1996) did not find significant MT modifications after carbamazepine (CBZ) administration. One possible explanation of this discrepancy is that these authors used stimulus intensity steps of 5% of maximum stimulator output (MSO), which might have been too large to detect MT changes reported to be less than 10% of MSO in the other studies (Ziemann et al., 1998c). Acute intake of drugs enhancing γ-aminobutyric acid (GABA)-mediated inhibition does not produce MT

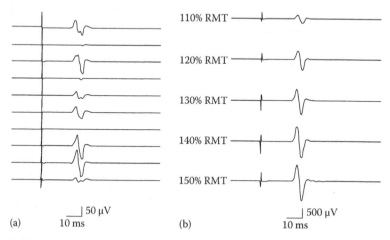

FIGURE 7.1 (a) A representative example of resting motor threshold (RMT) in a healthy volunteer. Recordings were made from the relaxed first dorsal interosseous muscle (FDI) after focal TMS of the contralateral primary motor cortex. Ten trials at a stimulus intensity of 36% of maximum stimulator output are shown. (b) Input–output (I–O) curve of the MEP in the FDI of a normal subject. Stimulus intensity is expressed as the percentage of the RMT. Traces show the average of five consecutive trials.

changes in normal subjects (Ziemann et al., 1995; Inghilleri et al., 1996; Ziemann et al., 1996b,c; Mavroudakis et al., 1997; Palmieri et al., 1999; Werhahn et al., 1999). Moreover, MT is not modified by dextromethorphan (DM), which reduces the glutamate-mediated excitation by blocking the N-methyl-D-aspartate (NMDA) receptor (Ziemann et al., 1998a). As membrane excitability of axons mainly depends on the permeability of voltage-gated sodium channels (Hodgkin and Huxley, 1952a,b), these findings strongly suggest that MT reflects membrane excitability in the aforementioned motor system neurons (Ziemann et al., 1996b). More recently, Di Lazzaro et al. (2003) reported a dose-dependent reduction of MT after subanesthetic infusion of the non-NMDA glutamatergic agent ketamine. This suggests that MT also reflects the effect of non-NMDA glutamatergic transmission on corticospinal excitability.

7.2.1.2 Input–Output Curve

I–O curve refers to the increase in MEP size as a function of TMS intensity (Figure 7.1b) and is also known as stimulus-response or recruitment curve. In the resting target muscle, MEP amplitude shows a sigmoid increment with increasing stimulus strength up to a plateau level (Hess et al., 1987; Devanne et al., 1997). The I–O curve slope is related to the number of corticospinal neurons activated at the different stimulus intensities (Ridding and Rothwell, 1997) and depends on the strength of the corticospinal projections relative to the target muscle (Abbruzzese et al., 1999). Compared to the neurons tested at MT, the neural elements assessed by the I–O curve are intrinsically less excitable by TMS or further away from the central core of the representation of the target muscle in the M1 (Chen, 2000; Hallett, 2000). In healthy volunteers, the slope of the I–O curve was reduced by acute intake of the ion channel blocker lamotrigine or the GABA$_A$ agonist lorazepam (LRZ) and increased by the dopaminergic–adrenergic agonist D-amphetamine (Boroojerdi et al., 2001). Although spinal mechanisms could not be definitely ruled out, the fact that these drugs did not affect the F waves suggested that changes in I–O curves may occur at the supraspinal level. Direct recordings of corticospinal volleys supported this view, showing that LRZ intake mainly suppressed the amplitude of the later I-waves and the size of the MEP produced by high-intensity TMS, whereas the early I-wave and the MT were unaffected (Di Lazzaro et al., 2000a).

In conclusion, the I–O curve represents an overall sensitive but relatively nonspecific measure of the excitability of the corticospinal system.

7.2.1.3 Central Motor Conduction Time

In clinical practice, TMS is widely used to study the CMCT (Hallett, 2000). CMCT aims to estimate conduction time between motor cortex and spinal (or bulbar) motoneurons and is calculated by subtracting the PMCT from the MEP latency.

The latency of the MEP following TMS includes the times for trans-synaptic depolarization of pyramidal cells in the M1, conduction via the fast-conducting corticospinal (or corticobulbar) pathways, excitation of the motoneurons sufficient to exceed their firing threshold, and conduction from the motoneurons to the target muscle (Chen et al., 2008). During slight contraction of the target muscle, the motoneuron firing threshold is lower and the opportunity for the TMS-induced corticospinal volleys to depolarize them is higher than that at rest. Consequently, "contracted" MEPs are shorter in latency and larger in amplitude than MEPs obtained during full muscle relaxation (Rossini and Rossi, 1998). Therefore, CMCT is usually measured with the target muscle active (Chen et al., 2008). Although MEP facilitation is maximum when the target muscle is the prime mover for the voluntary movement (Tomberg and Caramia, 1991) and is task dependent (Schieppati et al., 1996), this phenomenon can also be observed in other conditions, such as vibration of the target muscle and contraction of the homologous contralateral muscle (Rossini et al., 1987; Mariorenzi et al., 1991). These alternative facilitatory maneuvers can be used in neurological patients unable to activate the muscle under examination.

Two methods are available to estimate the PMCT (Rossini et al., 1994). First, the latency from motoneuron to muscle is calculated as $(F+M-1)/2$ (Kimura, 1983), where F is the latency (ms) of F wave and M is the latency (ms) of the CMAP (M wave) elicited by supramaximal electrical stimulation of the peripheral nerve. F latency reflects the antidromic motor conduction from the stimulation site to motoneurons, the estimated time (1 ms) taken by this afferent volley to depolarize a subset of α-motoneuronal cell bodies, and the orthodromic conduction from these motoneurons to the target muscle. Although experiments in macaques suggested that the longest F latency should be used to avoid CMTC overestimation (Oliver et al., 2002), in clinical evaluation, the latency of the shortest F wave is usually used as it is much easier to measure. The major limitation of this method is that usually F wave cannot be obtained in proximal muscles. The second method to estimate the PMCT is electrical or magnetic stimulation over the vertebral column. However, this procedure depolarizes motor roots at the level of the intervertebral foramen (Mills and Murray, 1986). Therefore, using this method, the CMCT includes the conduction time in proximal root segment between cord and exit foramen. This "peripheral" segment is about 3 cm for the cervical roots and even longer for the lumbosacral roots (Chokroverty et al., 1991).

A strong relationship between CMCT to lumbosacral segments and height has been reported (Rossini et al., 1987; Claus, 1990). In contrast, CMCT to upper limb muscles is scarcely related to height, likely because the distance from motor cortex to the cervical segments is much shorter.

7.2.1.4 TMS of Cranial Nerves

Although less commonly performed than TMS of the corticospinal projections to limb motoneurons, TMS has also been used for painless study of corticonuclear function and motor conduction along the cranial nerves since the second half of the 1980s (Murray et al., 1987; Schriefer et al., 1988; Rosler et al., 1989). Projections to facial and tongue muscles have been particularly investigated due to their relatively large representation in the M1. Detailed descriptions of stimulation and recording techniques are provided by a number of authors (Rimpilainen et al., 1992; Glocker et al., 1994; Rosler et al., 1995; Urban, 2003).

While conventional electrical stimulation of the facial nerve can be performed only at distal sites (i.e., the stylomastoid foramen), TMS allows stimulation of the facial nerve at the more proximal, extra-axial intracranial segment. To achieve this, TMS is delivered by applying a circular coil over the ipsilateral parieto-occipital region (see Figure 3 by Chen et al., 2008). Comparison with intraoperative direct electrical stimulation suggests that TMS-induced depolarization of the facial nerve occurs at the end of the labyrinthine segment (canalicular stimulation) where the nerve leaves the low-resistance cerebrospinal fluid and enters the high-resistance petrous bone (Schmid et al., 1991, 1992). CMAPs have been recorded from several sites, such as the nasalis, mentalis, orbicularis oculi, frontalis, triangularis, levator labii, and buccinators muscles.

The MEPs elicited by TMS of the facial representation in the M1 are reproducible, although they have a high threshold, usually need contraction of the target muscle, and are small compared to the CMAP obtained by electrical and magnetic stimulation of the facial nerve (Rosler et al., 1989). Possible technical pitfalls are contamination by activation of uncrossed, ipsilateral corticonuclear projections, peripheral stimulation of the ipsilateral facial nerve, responses of masticatory muscles, and blinking (Turk et al., 1994; Paradiso et al., 2005; Chen et al., 2008).

7.3 ADVANCED TMS TECHNIQUES

7.3.1 Advanced Measure of Corticospinal Conduction

7.3.1.1 Triple Stimulation Technique

The size of MEP is employed as a measure of corticospinal conduction in healthy subjects and patients with neurological diseases and usually expressed as percentage of the CMAP evoked by supramaximal peripheral nerve stimulation. However, several other factors besides the number of discharging corticospinal neurons and the amount of activated spinal (or brainstem) motoneurons influence the amplitude, area, and duration of the MEP (Magistris and Rösler, 2003). First, desynchronization of the TMS-induced motoneurons discharges induces phase cancellation (i.e., the negative phases of biphasic motor unit potentials are partly cancelled by positive phases of others; Kimura et al., 1986). Hence, the MEP size is reduced as compared to the maximal CMAP, even if the TMS pulse is delivered during contraction of the target muscle (Magistris et al., 1998). Second, motoneurons may unpredictably discharge more than once after a single TMS pulse. In addition, the size of the MEP varies considerably among subjects and from one stimulus to the next (Magistris et al., 1998; Rosler et al., 2002). In order to obviate part of these difficulties of conventional MEP assessment, the TST has been proposed. First developed to evaluate conduction blocks in peripheral nerves (Roth and Magistris, 1989) and subsequently adapted to be used with TMS, this double collision technique aims to synchronize the action potentials of the motoneurons driven to discharge by the transcranial magnetic pulse (see Figures 2 and 3 of Magistris et al., 1998, for detailed representation of the experimental setup and principles, respectively). Briefly, the TMS pulse is followed by two supramaximal stimuli given in sequence to the nerve supplying the target muscle with appropriate ISI. The first nerve stimulation is delivered distally (close to the target muscle) and the second peripheral stimulation is given as proximally as possible on the nerve. If a given spinal motoneuron is excited by the corticospinal volleys evoked by TMS, its descending action potential collides with the antidromic action potential elicited by the first distal nerve stimulus. Hence, the action potential produced by the second proximal nerve stimulus descends to the target muscle. In contrast, if a spinal motoneuron is not excited by TMS, the antidromic potential evoked by the distal nerve stimulus ascends and collides with the action potential elicited by the proximal nerve stimulus (Magistris et al., 1998). In conclusion, the larger is the number of spinal motoneurons that TMS-induced corticospinal volleys are capable to excite, the larger is the size of the CMAP evoked by the second proximal nerve stimulus. At difference from the desynchronized action potentials elicited by the magnetic pulse, the action potentials elicited by the single supramaximal proximal nerve stimulus are synchronized. The amplitude ratio between the CMAP obtained by TST and the CMAP obtained by a control TST procedure, where the TMS pulse is replaced by a supramaximal proximal nerve stimulus, is calculated. In healthy subjects, a TST amplitude ratio of nearly 100% can be obtained with TMS pulse of appropriate intensity (Magistris et al., 1998). Limitations of TST in routine clinical practice are the quite complex technical setup, the fact that it is not applicable to proximal muscle because they do not allow for sufficient ISI, and the uncomfortable proximal nerve stimulation, whose discomfort can be reduced using "monopolar" nerve stimulation.

7.3.2 Measures of Intracortical Excitability

7.3.2.1 Short-Interval Intracortical Inhibition

The most widely used paired-pulse TMS paradigm involves a subthreshold conditioning stimulus that does not elicit an MEP in the target muscle followed by a suprathreshold test pulse delivered by the same coil applied to the M1 (Kujirai et al., 1993). At rest, the effect of the conditioning stimulus on the test MEP is inhibitory at ISIs of 1–5 ms and facilitatory at ISIs of 8–30 ms (Figure 7.2). Direct evidence that no corticospinal volley is evoked by the conditioning pulse when the stimulus intensity is just below the AMT comes from epidural recordings (Di Lazzaro et al., 1998c). This strongly suggests that both the inhibitory and excitatory effects observed at different ISIs are of cortical origin. The two phenomena are referred to as SICI and intracortical facilitation (ICF), respectively. Epidural recordings provide further evidence that SICI depends on intracortical mechanisms showing that the I3 wave and subsequent I-waves evoked by the unconditioned test pulse are suppressed by the conditioning stimulus (Nakamura et al., 1997; Di Lazzaro et al., 1998a). This paired-pulse TMS paradigm is usually performed using the focal, figure-of-eight coil. However, it has been shown that reliable SICI and ICF measurements can be obtained also by the large circular coil (Abbruzzese et al., 1999; Badawy et al., 2011).

The threshold for eliciting ICF is slightly higher than that for SICI (Kujirai et al., 1993; Chen et al., 1998; Awiszus et al., 1999; Ilic et al., 2002; Orth et al., 2003). In addition, contraction of the target muscle suppresses both SICI and ICF compared with the resting condition (Ridding et al., 1995c; Abbruzzese et al., 1999). Finally, when the two pulses are delivered using two different coils on top of each other, SICI and ICF can be dissociated by different orientation of the "conditioning" coil (Ziemann et al., 1996d). These data suggest that SICI and ICF do not represent two aspects of the same physiological phenomenon and, conversely, are mediated by distinct neural substrates within the M1. Therefore, SICI and ICF will be treated separately.

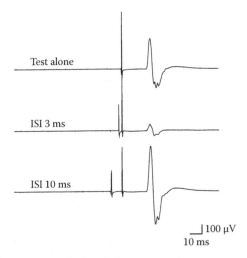

FIGURE 7.2   Short-interval intracortical inhibition (SICI) and intracortical facilitation (ICF) tested by a paired-pulse TMS paradigm with a conditioning pulse below the active motor threshold followed by a supra-threshold test stimulus. Motor-evoked potentials (MEPs) were recorded from the first dorsal interosseous muscle of a healthy volunteer. Traces are the averages of 10 trials and show the unconditioned test MEP, SICI of the test MEP at 3 ms interstimulus interval (ISI), and ICF of the test MEP at 10 ms ISI.

Although more extensively investigated in the intrinsic hand muscles, SICI has also been reported in proximal arm (Abbruzzese et al., 1999), truncal, lower limb (Chen et al., 1998), and facial muscles (Paradiso et al., 2005), as well as in diaphragm (Demoule et al., 2003) and anal sphincter (Lefaucheur, 2005). The amount of SICI is influenced by the unconditioned test MEP size (Roshan et al., 2003) and by the intensity of the conditioning pulse (Schafer et al., 1997; Chen et al., 1998; Butefisch et al., 2003). Namely, in the hand muscles, SICI increases with increasing intensity of the conditioning stimulus up to a certain level, then further increments of the conditioning pulse intensity result in lower inhibition and even facilitation.

Besides getting reduced during contraction of the target muscle, SICI is also downregulated just before movement onset (Ridding et al., 1995c; Reynolds and Ashby, 1999), whereas it is increased by active inhibition of the muscle under examination (Sohn et al., 2002). Taken together, these findings suggest that the neural circuits underlying SICI are involved in focusing the motor output to produce the intended movement and prevent unwanted muscle activation during selective hand movements (Floeter and Rothwell, 1999; Stinear and Byblow, 2003; Zoghi et al., 2003; Rosenkranz and Rothwell, 2004).

In healthy subjects, acute administration of most $GABA_A$ergic drugs enhanced SICI (Ziemann et al., 1995, 1996b,c). The facilitatory effect of LRZ on SICI has been also documented by epidural recording of TMS-induced corticospinal volleys (Di Lazzaro et al., 2000a). These findings, along with the time course of SICI, strongly support the view that SICI reflects $GABA_A$-mediated inhibition in the M1. Werhahn et al. (1999) found "paradoxical" SICI reduction after tiagabine (TGB) intake. As TGB increases the GABA$_B$-mediated long-lasting component of the inhibitory postsynaptic potential (IPSP) (Thompson and Gahwiler, 1992), it has been suggested that TGB effects on SICI may be due to inhibition of the GABA$_A$-mediated IPSPs by presynaptic GABA$_B$ receptors. However, this hypothesis cannot explain the finding that acute intake of the GABA$_B$ agonist baclofen (BAC) increased SICI (Ziemann et al., 1996b). It has been reported that also anti-glutamatergic drugs (Ziemann et al., 1998a; Schwenkreis et al., 1999, 2000), dopamine agonists (Ziemann et al., 1996a, 1997a), and the norepinephrine antagonist guanfacine (Korchounov et al., 2003) enhance SICI, whereas norepinephrine agonist drugs decrease it (Herwig et al., 2002; Ilic et al., 2003). Finally, a number of studies show that acute administration of ion channel blocker drugs has no effect on this parameter (Ziemann et al., 1996c; Chen et al., 1997; Boroojerdi et al., 2001).

While SICI provides relevant information on cortical excitability in health and disease, the wide interindividual variability reduces the value of SICI in clinical use. It has been suggested that this variability may be somewhat reduced if the intensity of the conditioning pulse is expressed as a percentage of the individual SICI threshold instead of a percentage of the AMT in the target muscle (Stinear and Byblow, 2004). However, several other factors should be taken into account to interpret and deal with the variability of SICI and other TMS measures. For instance, in women, SICI is modulated by physiological phenomena such as the menstrual cycle and is greater in the luteal phase than in the follicular phase (Smith et al., 1999). In addition, in a general population sample, it has been shown that the balance between SICI and ICF is correlated with neuroticism, which is considered a stable measure of trait-level anxiety (Wasserman et al., 2001).

Finally, the possible role of genetic polymorphisms in modulating SICI needs further investigation.

7.3.2.2 Long-Interval Intracortical Inhibition

At difference from the SICI paradigm, in this paired-pulse TMS protocol, the conditioning and test pulses have the same supra-threshold intensity and are delivered at longer ISI (Claus et al., 1992; Valls-Solè et al., 1992; Valzania et al., 1994). Test MEP inhibition is observed at ISIs of 100–200 ms (Figure 7.3). Recordings of the corticospinal volleys have demonstrated a cortical component of this long-ISI inhibition (Nakamura et al., 1997; Chen et al., 1999b). Therefore, this measure is known as LICI. However, concomitant influences of the supra-threshold conditioning pulse at the spinal level cannot be excluded. LICI as well as cortical silent period (CSP) duration are increased by TGB (Werhahn et al., 1999). This finding, as well as the similar time course, suggests that the two measures could share the same GABA$_B$-mediated cortical mechanisms. However, in patients with Parkinson's disease (PD), CSP shortening and increased LICI have been reported (Berardelli et al., 1996; Valzania et al., 1997). This LICI and CSP dissociation is in keeping with the view that fast-conducting corticospinal pathways underlying the TMS-induced MEP may partly differ from the pathways involved in maintaining tonic voluntary muscle contraction (Tergau et al., 1999).

7.3.2.3 Intracortical Facilitation

The term ICF refers to the increase in the test MEP size observed when the test magnetic pulse is preceded by a conditioning TMS pulse delivered through the same

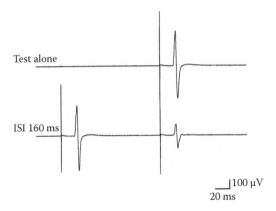

Test alone

ISI 160 ms

_|100 µV
20 ms

FIGURE 7.3 Long-interval intracortical inhibition (LICI) tested by a paired-pulse TMS paradigm with the conditioning and test pulses delivered at the same supra-threshold intensity. Motor-evoked potentials (MEPs) were recorded from the first dorsal interosseous muscle of a normal subject. Traces are the averages of 10 trials and show the unconditioned test MEP and LICI of the test MEP at 160 ms interstimulus interval.

coil at ISIs of 8–30 ms, using the paradigm described in Section 7.3.2.1 (Kujirai et al., 1993) (Figure 7.2). As the conditioning pulse at an intensity below the AMT does not evoke descending corticospinal volleys (Di Lazzaro et al., 1998c), ICF is thought to be due to cortical mechanisms. Nevertheless, epidural recordings have not shown a significant enhancement of descending volleys related to ICF (Di Lazzaro et al., 2006a). Several experimental data (see Section 7.3.2.1) indicate that the cortical circuitries underlying ICF and SICI can be dissociated (Kujirai et al., 1993; Ridding et al., 1995c; Ziemann et al., 1996d; Chen et al., 1998; Abbruzzese et al., 1999; Awiszus et al., 1999; Ilic et al., 2002; Orth et al., 2003).

In neurons of the rat sensorimotor cortex, the latency of the fast component of the excitatory postsynaptic potentials, which are mediated by the NMDA glutamatergic receptors, is in the order of 10 ms (Hwa and Avoli, 1992). This latency is consistent with the ICF time course. Hence, it has been hypothesized that ICF may reflect NMDA glutamatergic transmission in the M1 (Ziemann, 2004). This hypothesis is supported by experimental data showing that NMDA antagonists reduce ICF (Ziemann et al., 1998c; Schwenkreis et al., 1999). Acute administration of $GABA_A$ergic drugs also suppresses ICF (Ziemann et al., 1995, 1996b,c; Inghilleri et al., 1996), although not unanimously (Werhahn et al., 1999; Boroojerdi et al., 2001). These data suggest that also suppression of $GABA_A$ transmission may play a role in mediating ICF and are consistent with the fact that the tail of the $GABA_A$-mediated IPSP reaches approximately 20 ms (Connors et al., 1988). The possibility that the paradoxical effects of TGB on this paired-pulse TMS paradigm (Werhahn et al., 1999) may reflect inhibition of the $GABA_A$-mediated IPSPs by presynaptic $GABA_B$ receptors has been discussed in Section 7.3.2.1. Several authors have found no effect of acute intake of ion channel blocker drugs on ICF (Ziemann et al., 1996b; Chen et al., 1997; Boroojerdi et al., 2001). Chronic intake of paroxetine has been reported to enhance ICF without modification of SICI (Gerdelat-Mas et al., 2005). The effect of this serotonin reuptake inhibitor further supports the view that SICI and ICF reflect distinct neural substrates.

7.3.2.4   Facilitatory I-Wave Interaction

This measure is tested using a paired-pulse TMS paradigm in which, at difference from the classical protocol by Kujirai et al. (1993), both the magnetic stimuli are approximately of threshold intensity (Tokimura et al., 1996) or the conditioning pulse is supra-threshold and the test pulse is subthreshold (Ziemann et al., 1998d). MEP size facilitation is observed at ISIs of about 1.1–1.5, 2.3–3.0, and 4.1–4.5 ms. As the mean latency difference between these ISIs is similar to the interpeak latency between I-waves, it has been argued that this facilitatory effect (also known as short-interval intracortical facilitation) may depend on the cortical neurons responsible for the generation of these descending volleys (Tokimura et al., 1996; Ziemann et al., 1998d). Namely, the hypothesis is that the initial axonal segments of those neurons generating late I-waves, which are partly depolarized by the first pulse but do not fire an action potential, are excited by the second pulse (Hanajima et al., 2002; Ilic et al., 2002). Facilitatory I-wave interaction is reduced by $GABA_A$ agonists (Ziemann et al., 1998e; Ilic et al., 2002), whereas it is not modified by ion channel blocker (Ziemann et al., 1998e) or muscarinic receptor blocker (Di Lazzaro et al., 2000b) intake.

7.3.2.5   Cortical Silent Period

The term CSP indicates a transient silence in isometric voluntary EMG activity of the target muscle produced by single-pulse TMS of the contralateral M1 (Figure 7.4). At stimulus intensity above the AMT, the CSP is preceded by the MEP, but CSP threshold can be lower than AMT (Davey et al., 1994). CSP duration increases with the stimulus intensity. While spinal mechanisms such as after-hyperpolarization and recurrent inhibition may play a role in the first part of the CSP, the latter part is due to the activity of inhibitory circuits at the M1 level, resulting in a transient failure of the cortical motor output (Fuhr et al., 1991; Cantello et al., 1992; Inghilleri et al., 1993; Roick et al., 1993; Triggs et al., 1993; Wassermann et al., 1993; Chen et al., 1999b).

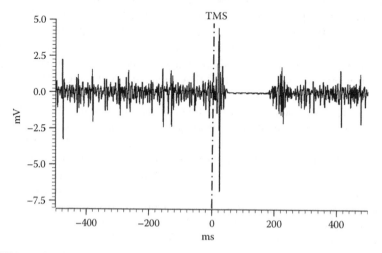

FIGURE 7.4   Cortical silent period (CSP) recorded from the first dorsal interosseous muscle of a representative healthy volunteer. Focal TMS of the contralateral primary motor cortex was delivered at 120% resting motor threshold during isometric contraction of the target muscle.

Various cortical and subcortical areas projecting to the M1 modulate these inhibitory phenomena (von Giesen et al., 1994; Classen et al., 1997; Cincotta et al., 2000a).

Stimulation of GABAergic interneurons in human neocortex slices evokes biphasic inhibitory IPSPs (for review, see McCormick, 1992). While GABA$_A$ receptors mediate the short-lasting IPSP (in the order of tens of milliseconds), the long-lasting component of the IPSP (in the order of hundreds of milliseconds) depends on GABA$_B$ receptors. Due to the similar time profile, it has been hypothesized that the latter part of the CSP could be mediated by GABA$_B$ inhibition at the M1 level (Roick et al., 1993; Classen et al., 1997). Experimental data support this view. Siebner et al. (1998) found a strong, dose-dependent CSP lengthening after continuous intrathecal infusions of the GABA$_B$ agonist BAC in a patient with generalized dystonia. This CSP prolongation was not seen with intravenous (i.v.) and oral BAC administration in normal subjects (Inghilleri et al., 1996; Ziemann et al., 1996b). However, the portion of BAC that penetrates the blood–brain barrier after enteral or i.v. application is low (Knutsson et al., 1974). Moreover, TGB, which is known to block the GABA uptake from the synaptic cleft (Brodie, 1995) and increase GABA$_B$-mediated IPSP (Thompson and Gahwiler, 1992), lengthens the CSP duration in healthy subjects (Werhahn et al., 1999). In contrast, the peripheral silent period, which depends on spinal inhibitory mechanisms (Shahani and Young, 1973), is not affected by TGB. As to GABA$_A$ agonists, conflicting data have been reported. Ethanol and LRZ lengthened the CSP (Ziemann et al., 1995, 1996c), whereas oral intake of diazepam (DZP) (Palmieri et al., 1999) and vigabatrin (Ziemann et al., 1996b; Mavroudakis et al., 1997) and i.v. administration of thiopental (Inghilleri et al., 1996) did not affect this measure. Moreover, a short-lasting CSP shortening (in the order of tens of minutes) after i.v. DZP injection has been reported (Inghilleri et al., 1996). It has been hypothesized that, just after the i.v. DZP bolus, GABA$_A$-mediated inhibition could transiently reduce facilitatory thalamo-cortical influences on inhibitory interneurons of the M1 (Inghilleri et al., 1996). The hypothesis of a subcortical origin of CSP modulation is supported by the CSP shortening also found in patients with PD (Cantello et al., 1991; Priori et al., 1994a; Valls-Solè et al., 1994). Most studies show that ion channel blockers leave the CSP unaffected (Mavroudakis et al., 1994; Ziemann et al., 1996b; Chen et al., 1997). However, it has been reported that CBZ can slightly prolong the CSP duration in normal subjects (Schulze-Bonhage et al., 1996; Ziemann et al., 1996b). As additional CBZ effects on adenosine (Marangos et al., 1983), glutamate (Hough et al., 1996), and serotonin (Dailey et al., 1997) systems have been reported, one possibility is that this minor CSP modification could be due to CBZ modes of action distinct from the main mechanism of voltage-dependent sodium channel blockade (Tassinari et al., 2003). Finally, a slight lengthening of CSP duration has been found after a single dose of the NMDA antagonist DM (Ziemann et al., 1998a). In conclusion, neuropharmacologic data support the notion that CSP mainly reflect GABA$_B$-mediated cortical inhibition, although it is also evident that the physiological mechanisms underlying this complex phenomenon have still to be fully understood (Hallett, 1995).

7.3.2.6 Short-Latency Afferent Inhibition

Peripheral afferent inputs produce a complex modulation of the size of MEP elicited by TMS of the contralateral M1 (Delwaide and Olivier, 1990; Mariorenzi et al., 1991;

Tokimura et al., 2000; Roy and Gorassini, 2008; Fischer and Orth, 2011). When a magnetic pulse of the hand area of one M1 is preceded by an electrical stimulus of the contralateral median (or ulnar) nerve at wrist at an ISI exceeding by 1–8 ms, the latency of the N20 component of the somatosensory evoked potential (EP), the MEP size in the hand muscles, is reduced compared to the baseline (Delwaide and Olivier, 1990). This effect is named short-latency afferent inhibition (SAI) and is replaced by MEP facilitation at slightly longer ISI (Mariorenzi et al., 1991; Fischer and Orth, 2011).

Afferent inputs have no effects on H-reflexes in forearm muscles (Delwaide and Olivier, 1990). Moreover, recordings of descending corticospinal volleys from patients with cervical epidural electrodes showed that later I-waves were strongly suppressed by afferent inputs while earlier descending waves were unaffected (Tokimura et al., 2000). These data support the view that SAI occurs at cortical level.

Cholinergic cerebral circuits have been proposed to be involved in regulating SAI as the inhibition is reduced or abolished by intravenous injection of the muscarinic antagonist scopolamine (Di Lazzaro et al., 2000b). In addition, $GABA_A$ activity enhancement by administration of LRZ and zolpidem reduces SAI (Di Lazzaro et al., 2005b, 2007). This suggests a concomitant $GABA_A$-mediated inhibition of the circuits responsible for SAI. As the effects of zolpidem, LRZ, and DZP on SAI and SICI are dissociated, it has been hypothesized that $GABA_A$-mediated modulation of SAI may rely on specific $GABA_A$ receptor subtypes, which are partly distinct from those mediating SICI (Di Lazzaro et al., 2007). This view is supported by the data on the interactions between SICI and SAI obtained by the co-application of the two protocols (Alle et al., 2009).

7.3.2.7 Long-Latency Afferent Inhibition

MEP size reduction following peripheral nerve stimulation has been reported also at longer ISI and is referred to as long-latency afferent inhibition (LAI) (Chen et al., 1999a; Abbruzzese et al., 2001). Suppression of the MEP amplitude is more consistent when the conditioning sensory stimulation precedes the TMS pulse of approximately 200 ms (Chen et al., 1999a; Sailer et al., 2002). As no spinal cord excitability modification is observed at this interval (Chen et al., 1999a), LAI is likely due to cortical mechanisms.

7.3.3 MEASURES OF CORTICO-CORTICAL EXCITABILITY

7.3.3.1 Interhemispheric Inhibition

The MEP elicited by focal TMS of one M1 can be suppressed if a conditioning magnetic pulse is delivered by another focal coil to the opposite M1 6–50 ms earlier (Ferbert et al., 1992). Recordings of the corticospinal volleys during the application of this paired-pulse TMS protocol at ISIs of 6–11 ms showed that the conditioning stimulus reduced the size of the later I-waves evoked by the test pulse (Di Lazzaro et al., 1999). This finding strongly supports the view that this interhemispheric inhibition (IHI) of the MEP occurs at the cortical level, likely through transcallosal fibers. A slight interhemispheric facilitation of the MEP has been also reported at shorter ISI (4–6 ms) (Hanajima et al., 2001).

IHI at short ISI of around 10 ms (S-IHI) and IHI at long ISI of 20–50 ms (L-IHI) have been dissociated using interaction between LAI and IHI (Kukaswadia et al., 2005), pharmacological modulation (Irlbacher et al., 2007), and different coil orientations and intensities of the conditioning pulse (Ni et al., 2009). These data indicate that S-IHI and L-IHI are physiologically distinct forms of interhemispheric inhibitory phenomena. L-IHI is thought to be mediated by $GABA_B$ receptors (Irlbacher et al., 2007).

7.3.3.2 Ipsilateral Silent Period

Besides IHI of the MEP, interhemispheric inhibitory influences between the M1 can also be investigated by short suppression of ongoing voluntary EMG activity in hand muscles induced by single-pulse focal TMS of the ipsilateral M1 (Wassermann et al., 1991; Ferbert et al., 1992; Meyer et al. 1995; Trompetto et al., 2004; Cincotta et al., 2006b). This ipsilateral silent period (iSP, Figure 7.5) starts 30–40 ms after the magnetic pulse and lasts, on average, 25 ms (Meyer et al., 1995).

Studies in patients with callosal lesions indicate that the iSP mainly reflects cortical inhibitory mechanisms mediated by fibers passing through the posterior half of the trunk of the corpus callosum (Meyer et al., 1995, 1998). However, a minor role of subcortical circuits is suggested by the fact that, in preactivated muscles, focal TMS of the ipsilateral M1 also reduces the size of MEP elicited by electrical stimulation at the level of the pyramidal decussation (Gerloff et al., 1998). In addition, using electrodes implanted on the subthalamic nucleus for deep brain stimulation (DBS) in PD patients, Compta et al. (2006) have shown that subcortical stimulation of the corticospinal tract can also produce an iSP.

By examining the effects of different stimulus intensities and current directions, Chen et al. (2003) demonstrated that the neural mechanisms underlying the iSP differ from those responsible for paired-pulse S-IHI of the MEP. It was concluded that the iSP is a measure that provides information on interhemispheric inhibition

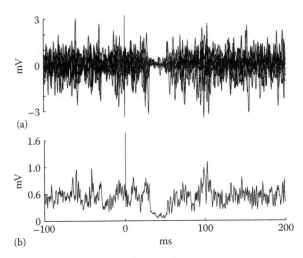

FIGURE 7.5 Ipsilateral silent period (iSP) recorded from the first dorsal interosseous muscle of a normal subject. Focal TMS of the ipsilateral primary motor cortex was delivered at 130% resting motor threshold during isometric contraction of the target muscle. Ten trials were recorded. Superimposition of row recordings (a) and the average of the rectified traces (b) are shown.

complementary but not identical to the information gained by S-IHI measurements (Chen et al., 2003). On the other hand, Chen et al. (2003) and Avanzino et al. (2007) found similarities between ISP and L-IHI tested at ISIs of 40 ms (L-IHI).

In conclusion, the iSP appears to be an original and appropriate tool to investigate interhemispheric control of voluntary cortical motor output.

7.3.3.3 Cerebellar Inhibition

Cerebellar stimulation can induce an inhibitory effect over the motor cortex. Such effect, known as cerebellar inhibition (CBI), has been studied evaluating the amplitude of the MEP elicited by TMS of the M1 after a conditioning stimulation of the cerebellum. This inhibitory influence was first demonstrated by Ugawa et al. (1991) using electrical conditioning stimulation. Subsequent studies showed that a conditioning magnetic pulse applied over the lateral cerebellum by a double-cone coil consistently reduces the amplitudes of MEP elicited by a test pulse to the contralateral M1 at ISIs of 5–8 ms (Ugawa et al., 1995; Werhahn et al., 1996; Pinto and Chen, 2001; Daskalakis et al., 2004; Iwata and Ugawa, 2005; Luft et al., 2005). CBI is thought to depend on inhibition of the dentato-thalamo-cortical pathway induced by activation of cerebellar Purkinje cells (Ugawa et al., 1991). This hypothesis is supported by data showing that the inhibitory effect is absent in patients with lesions of the cerebellum or its efferent pathway, but is preserved in patients with lesions limited to the cerebellar afferent pathway (Di Lazzaro et al., 1994a, 1995; Ugawa et al., 1994, 1997).

CBI is more pronounced when the TMS pulse applied over the M1 is at near threshold intensities than at higher intensities of the test pulse (Ugawa et al., 1995; Pinto and Chen, 2001). Moreover, CBI is reduced with voluntary contraction (Pinto and Chen, 2001). Daskalakis et al. (2004) reported a reduction of SICI and an increase of ICF in the contralateral M1 after cerebellar stimulation.

Long-lasting change of the M1 excitability has been reported after low-frequency rTMS (Oliveri et al., 2005; Fierro et al., 2007; Koch et al., 2008b) or theta burst stimulation (Popa et al., 2010) applied over the lateral cerebellum. Moreover, execution and observation of a visuomotor procedural learning task are related to modulation of cerebello-motor connectivity (Torriero et al., 2011). Namely, during specific phases of procedural learning, cerebellar conditioning stimulation induces a selective facilitation of contralateral M1 excitability (Torriero et al., 2011).

7.3.3.4 Premotor–Motor Interactions

Paired-pulse TMS using two coils over distinct scalp sites (bifocal or dual-site TMS approach) has been used to test cortico-cortical connections between premotor areas and M1 at rest and during movement preparation (for review, see Rothwell, 2010). Both facilitation and inhibition may be detected in the M1, ipsilaterally or contralaterally to the site of the conditioning pulse, and such effects critically depend on the intensity of the conditioning stimulus and on ISI. The first evidence has been obtained by Civardi et al. (2001) using small TMS stimulating eight-shaped coils (4 cm diameter internal loop). These authors found that a low-intensity (90% AMT) conditioning pulse to the dorsal premotor (PMd) cortex reduced the MEP amplitude elicited in the hand muscles by a test pulse delivered to the ipsilateral M1 when the ISI was 6 ms. Increasing the intensity of the conditioning pulse to 110%–120% AMT, a facilitatory effect was seen. Reduction of M1 excitability has also been reported delivering a conditioning TMS pulse over the contralateral PMd at 90% or 110% of the RMT when the ISI was 8–10 ms (Mochizuki et al., 2004). On the contrary, an enhancement of MEP amplitude was found by Baumer et al. (2006) delivering a conditioning stimulus of lower intensity (80% AMT) over the contralateral PMd at ISIs of 8 ms.

Similar facilitatory (conditioning stimulus at low intensities) and inhibitory (conditioning stimulus at higher intensities) effects on the M1 excitability at rest have been observed when the conditioning TMS pulses were applied over the ventral premotor cortex (PMv) (Davare et al., 2008, 2009; Baumer et al., 2009).

Some studies investigated cortico-cortical connections between premotor (PMd and PMv) cortex and M1 during movement preparation (Davare et al., 2006; Koch et al., 2006).

7.3.3.5 Parieto-Motor Interactions

Recently, bifocal paired-pulse TMS approach consented to demonstrate functional connections between specific portions of the posterior parietal cortex (PPC) and the ipsilateral, or even the contralateral M1, at rest and during the reaction time of a task requiring reaching movements (Koch et al., 2007, 2008a, 2009). Namely, a conditioning TMS pulse applied over the caudal part of the intraparietal sulcus (cIPS) at an intensity of 90% RMT produces a facilitation of the MEP elicited by a test pulse to the ipsilateral M1 when the ISI between the pulses is 4–6 ms (Koch et al., 2007). In contrast, conditioning pulses applied over the anterior part of the intraparietal sulcus (aIPS) was found to activate an inhibitory projection toward ipsilateral M1 at the same intensity of stimulation (90% RMT) and ISI (4 ms) (Koch et al., 2007). Similar facilitatory and inhibitory effects were observed toward the contralateral M1 at rest (Koch et al., 2009). A conditioning pulse over the right cIPS at 90% RMT enhanced the MEP amplitude evoked by left M1 stimulation while SICI was reduced. Facilitation was maximal when the ISI was 6 or 12 ms (Koch et al., 2009). In contrast, conditioning pulse over the aIPS at 90% RMT induced inhibition of the excitability of the contralateral M1 at ISIs of 10–12 ms.

Koch et al. (2008a) also tested the functional interplay between PPC and ipsilateral M1 during the reaction time of a task requiring reaching movements toward a left or right visual target. These authors found time-related and task-dependent modulations of the PPC–M1 connectivity. Namely, facilitation was induced in the M1 by a conditioning pulse to the PPC only when subjects planned a reaching movement toward the contralateral hemispace at specific time intervals after an auditory cue that indicated where to reach.

7.4 CLINICAL APPLICATIONS

7.4.1 Myelopathy

TMS has a very good sensitivity (100%) and specificity (85%) in detecting spinal cord abnormality when the MRI is chosen as gold standard (Lo et al., 2004). TMS may disclose incipient spinal cord compression before the development of any clinical or radiological sign (Travlos et al., 1992). In addition, upper and lower limb CMCT correlated with the severity of cord compression. Since myelopathy and radiculopathy may commonly coexist, TMS can also be helpful to detect a peripheral involvement (Abbruzzese et al., 1988). This subset of patients, with such a mixed damage, had prolonged latencies from both cortical and root stimulation, and the overall sensitivity of the technique is better than abnormalities detected by somatosensory EPs (Kameyama et al., 1995; Kaneko et al., 1997).

Finally, TMS can be used to define the most important segment of compression, especially in cervical myelopathy (Chan et al., 1998).

For instance, in C6 myelopathy, CMCT to first dorsal interosseous (C8, T1) and extensor digitorum communis (C7) muscles would be prolonged, but that to the deltoid (C5) muscle would be normal. TMS is not useful in clarifying the nature of the spinal cord lesion. Another limitation is that CMCT findings have no predictive value for the final clinical outcome (Jaskolski et al., 1990).

7.4.2 AMYOTROPHIC LATERAL SCLEROSIS

The clinical diagnosis of amyotrophic lateral sclerosis (ALS) is made on the basis of combined upper and lower motoneuron signs. TMS can be useful in detecting subclinical upper motoneuron dysfunctions that are sometimes elusive and obscured by the lower motoneuron involvement (Triggs and Eggar, 1995).

Prolonged CMCT and MEP latencies were the abnormalities reported more frequently (Eisen et al., 1990). The sensitivity of TMS for demonstrating upper motoneuron dysfunction in MND may be improved by the choice of the target muscle.

For instance, ALS patients have a prominent loss of dexterity of thumb and index finger. This is the reason why ALS patients demonstrated preferential involvement of the cortical control of the thenar eminence muscles compared to the hypothenar muscles (Weber et al., 2000).

A frequent differential diagnosis of ALS is cervical myelopathy especially when it is associated with radiculopathy. A clever strategy is to demonstrate an abnormal central motor conduction to muscles innervated by nerves emerging above the foramen magnum such as trapezius (Truffert et al., 2000), tongue (Urban et al., 1998), or masseter (Trompetto et al., 1998).

An abnormal CMCT recorded from these muscles can exclude cervical myelopathy and make the diagnosis of ALS more likely.

Finally, the sensitivity of TMS for detecting upper motoneuron abnormalities in ALS can be improved by using the TST (Magistris et al., 1998; Rosler et al., 2000, 2004).

Several studies have demonstrated a cortical hyperexcitability in the early phases of the disease, which could be related to glutamate-induced excitatory neurotoxicity in ALS. This hyperexcitability is indexed by a decrease in MT (Eisen et al., 1993; Mills and Nithi, 1997). In keeping with that, several groups have described, in the early stages of the disease, a reduction of the duration of CSP (Prout and Eisen, 1994; Mills, 2003) and reduction of SICI (Yokota et al., 1996; Enterzari-Taher et al., 1997; Ziemann et al., 1997b; Zanette et al., 2002).

However, MT tends to increase during the course of disease as a result of the progressive neuronal loss, and in the late stages of the disease, in some patients,

MEPs could not even been elicited (Triggs et al., 1999; Pouget et al., 2000; Mills, 2003; Attarian et al., 2005).

The assessment of multiple TMS parameters may increase the sensitivity of the technique even if the specificity is very low (Triggs et al., 1999).

7.4.3 CEREBELLAR DISEASES

In patients with unilateral cerebral lesions, MT of the contralateral motor cortex is increased (Di Lazzaro et al., 1994b, 1995; Cruz-Martinez and Arpa, 1997). Some studies reported normal (Ugawa et al., 1994) or increased SICI, and reduced ICF (Liepert et al., 1998; Restivo et al., 2002; Schwenkreis et al., 2002; Tamburin et al., 2004) in cerebellar ataxia. Interestingly, different genetic defects may result in different patterns of TMS abnormalities.

For example, in patients with inherited spinocerebellar ataxia (SCA), reduced ICF may be more specific for SCA2 and SCA3 subtypes (Schwenkreis et al., 2002). Similarly, CMCT was found to be prolonged in patients with Friedreich's ataxia (Cruz-Martinez and Palau, 1997) and SCA types 1, 2 (Restivo et al., 2000), and 6 (Lee et al., 2003). These peculiarities may orient the genetic counseling in the presence of ataxia.

7.4.4 ALZHEIMER'S DISEASE AND OTHER DEMENTIAS

Motor cortex is hyperexcitable in Alzheimer's disease (AD) patients. This has been reported extensively (de Carvalho et al., 1997; Pepin et al., 1999; Alagona et al., 2001a; Di Lazzaro et al., 2002, 2004; Pennisi et al., 2002; Ferreri et al., 2003). This is suggested by a decreased RMT and decreased SICI and SAI (Liepert et al., 2001; Di Lazzaro et al., 2004) even if SICI has been reported normal in another study (Pepin et al., 1999).

The reduction of SAI, which is reported in 70% of patients with a clinical diagnosis of AD, may be related to cholinergic deficit (Di Lazzaro et al., 2002, 2004, 2005a).

An abnormal SAI has also been reported in Lewy body dementia while SAI was found normal in fronto-temporal dementia (Di Lazzaro et al., 2006a), a non-cholinergic form of dementia.

Interestingly, SAI is quickly normalized by a single oral dose of rivastigmine. A good restoration of SAI after rivastigmine could predict a good response to treatment. Conversely, a lack of normalization of SAI after a single oral dose of rivastigmine is associated with a poor response to long-term treatment (Di Lazzaro et al., 2005a).

7.4.5 MULTIPLE SCLEROSIS

CMCT is a relatively sensitive measure in detecting demyelinating lesions along corticospinal tracts even though it may vary across studies (56%–93%, Barker et al., 1986; Ingram et al., 1988; Rossini et al., 1989; Eisen and Shtybel, 1990; Van Der Kamp et al., 1991; Jones et al., 1991; Mayr et al., 1991; Kandler et al., 1991; Ravnborg et al., 1992; Michels et al., 1993; Beer et al., 1995). Sensitivity has been reported to be increased if MEPs are recorded in the lower limb muscles (Jones et al., 1991; Kandler et al., 1991; Mayr et al., 1991). The type of multiple sclerosis (MS) also influences CMCT sensitivity. Namely, CMCT prolongation is more pronounced in progressive MS than in relapsing–remitting MS (Filippi et al., 1995; Facchetti et al., 1997; Kidd et al., 1998; Humm et al., 2003).

TST may reveal a frequent occurrence of central conduction failure due to focal central conduction block (Humm et al., 2003, 2004) or loss of corticospinal axons in the presence of normal CMCT and MEP measures (Magistris et al., 1999).

TMS can reveal some subclinical deficit, although to a lesser degree than visual EPs and MRI.

Interestingly, most studies indicated a significant correlation between CMCT or TST abnormalities and clinical motor signs or motor disability (Ingram et al., 1988; Britton et al., 1991; Jones et al., 1991; Van Der Kamp et al., 1991; Facchetti et al., 1997; Kidd et al., 1998; Magistris et al., 1999).

Finally, it has been reported that a multimodal EP score including CMCT measurements predicted the Expanded Disability Status Scale (EDSS) of clinically definite MS patients 6–24 months later (Fuhr et al., 2001; Kallmann et al., 2006; Leocani et al., 2006; Feuillet et al., 2007). If confirmed, this may be an important information because it may influence therapeutic decisions.

In summary, the standard TMS measures (CMCT, TST) have moderately high nosologic sensitivity and correlate better than MRI with motor impairment and motor disability both cross-sectionally and longitudinally. TMS can be used to study fatigue in patients with MS. Our group has recently investigated this issue (Morgante et al., 2011). In particular, in this study, TMS was used to evaluate the increase in cortical excitability prior to movement (pre-movement facilitation) in MS patients with and without fatigue. Interestingly, in MS patients with fatigue, we found a lack of pre-movement facilitation, which correlated with lesion burden within frontal lobe and significantly correlated to the FSS score, suggesting that central fatigue in MS is probably due to a dysfunction of cortical motor areas involved in movement preparation.

In addition, other studies reported reduced ICF (Ho et al., 1999), reduced SICI (Caramia et al., 2004), and prolonged transcallosal conduction time (Boroojerdi et al., 1998) and iSP duration (Boroojerdi et al., 1998; Hoppner et al., 1999; Schmierer et al., 2000). It is possible to elaborate a composite TMS index that takes into account CMCT and iSP that have been reported to correlate with EDSS at least in progressive forms of MS (Schmierer et al., 2002).

7.4.6 MOVEMENT DISORDERS

TMS studies have contributed to a large extent in the understanding of the pathophysiology of several movement disorders even if they have a limited role as diagnostic test.

7.4.6.1 Parkinson's Disease

RMT is significantly decreased in very rigid parkinsonian patients (Cantello et al., 1991) while AMT may be increased in very bradykinetic patients (Ellaway et al., 1995). CMCT abnormalities may be useful in identifying Parkinson-plus syndromes or genetic PD forms where pyramidal signs are equivocal.

An alteration of CMCT may help orient the diagnosis toward a genetic PD related to parkin mutation (De Rosa et al., 2006). A prolonged CMCT has been reported also in patients with multiple system atrophy (MSA) and progressive supranuclear palsy (PSP; Abbruzzese et al., 1991, 1997b; Eusebio et al., 2007).

SICI is reduced in PD patients at rest and is normalized by dopaminergic medications (Ridding et al., 1995a), DBS of the subthalamus nucleus (Cunic et al., 2002), or lower frequency motor cortex rTMS (Lefaucheur et al., 2004). The alterations of LICI are controversial, as it has been reported either an increase (Berardelli et al., 1996) or a decrease (Pierantozzi et al., 2001) even if they are normalized by dopaminergic medications. SICI is similar abnormal in other parkinsonian syndromes such as corticobasal degeneration (CBD), MSA, and PSP (Kuhn et al., 2004); therefore, it is not diagnostically valuable.

In contrast, patients with essential tremor (ET) (Romeo et al., 1998) and patients with primary writing tremor (Modugno et al., 2002) have normal SICI and LICI. SP is abnormally shortened in PD patients (Cantello et al., 1991; Valls-Sole et al., 1994; Siebner et al., 2000) but can be normalized either by dopaminergic medications (Priori et al., 1994a) or by surgical lesions of the internal globus pallidus (Strafella et al., 1997; Young et al., 1997) and rTMS of motor cortex (Siebner et al., 2000; Lefaucheur et al., 2004).

The SP is normal in ET and task-specific tremor, but shortened in cortical myoclonus (Brown et al., 1996; Inghilleri et al., 1998) and in Tourette syndrome (TS) (Greenberg et al., 2000).

Clinicopathologic evidence suggests differential involvement of cortex and corpus callosum in various disorders presenting with a parkinsonian syndrome. TMS, and in particular iSP, may be a useful clinical test to differentiate patients with different parkinsonian syndromes. CBD patients have increased iSP threshold and reduced iSP duration, and these findings correlate with the atrophy of the corpus callosum on MRI (Trompetto et al., 2003). In PD, IHI is reduced in patients without mirror movement (MM), especially at long ISIs of 20–50 ms (Li et al., 2007).

SAI is a parameter underlying the physiology of cholinergic circuits. SAI was found to be normal in patients off medications, but administration of dopaminergic medication led to SAI reduction (Sailer et al., 2003). In addition, it has been recently reported that SAI was significantly reduced in patients with visual hallucinations (VHs) compared with controls and patients without VHs (Manganelli et al., 2009).

PD patients with dementia have a reduced SAI (Manganelli et al., 2009) whereas in PSP patients SAI is normal (Nardone et al., 2005). In contrast, MSA patients with parkinsonism features have reduced SAI obtained after digit stimulation.

LAI is reduced in PD and is unaffected by dopaminergic medications (Sailer et al., 2003). Subthalamic nucleus DBS normalized both SAI and LAI in PD patients (Sailer et al., 2007).

7.4.6.2 Dystonia

RMT is normal in primary dystonia. SICI is altered in both the affected and unaffected sides in focal hand dystonia (Ridding et al., 1995b), in hand muscles in patients with focal hand dystonia (Kanovsky et al., 2003), or in blepharospasm (Sommer et al., 2002). The effect of botulinum toxin on SICI is debated as some authors have reported either a normalization (Gilio et al., 2003) or no effects (Boroojerdi et al., 2003).

SICI has been reported as abnormal in genetic forms of dystonia both in the DOPA-responsive dystonia (Huang et al., 2006) and in the asymptomatic carriers of the DYT1 gene mutation (Edwards et al., 2003).

Both SICI and LICI were found to be abnormal in psychogenic dystonia (Espay et al., 2006). Therefore, testing of cortical inhibition may not distinguish between organic and psychogenic disorders.

We have recently used paired associative stimulation to test sensorimotor plasticity. Sensorimotor plasticity is altered both in focal dystonia (Writer's cramp, cervical dystonia) and in generalized dystonia (Quartarone et al., 2006a,b; Weise et al., 2006; Ruge et al., 2011). On the other hand, sensorimotor plasticity is normal in patients with psychogenic dystonia (Quartarone et al., 2009).

SP is also shortened in patients with focal hand dystonia (Filipovic et al., 1997) or cranial dystonia (Currà et al., 2000) and is not normalized by botulinum toxin injection (Allam et al., 2005). SP duration is even shorter in the Meige syndrome. In patients with focal hand dystonia, iSP is prolonged indicating increased transcallosal inhibition (Niehaus et al., 2001).

LAI is diminished or absent in patients with focal hand dystonia while SAI is normal (Abbruzzese et al., 2001). Other studies have revealed an altered surround inhibition in focal hand dystonia (Sohn and Hallett, 2004).

7.4.6.3 Huntington's Disease

A normal RMT has been reported in HD. An alteration of SICI and LICI has been reported by some researchers in HD (Tegenthoff et al., 1996; Abbruzzese et al., 1997a) but not confirmed by others (Priori et al., 1994b; Hanajima et al., 1996).

SP shortens with functional decline in HD and could be considered as a marker of disease progression (Lefaucheur et al., 2006).

7.4.6.4 Tourette Syndrome

RMT is reduced in patients with obsessive-compulsive disorders associated with tics (Greenberg et al., 2000).

SICI is also reduced in patients with TS and correlated with motor hyperactivity whereas tic occurrence is related to SP shortening (Greenberg et al., 2000).

7.4.6.5 Mirror Movements

Over the last two decades, TMS provided a valuable insight into the neural substrate of intended unimanual movements in healthy humans and patients affected by neurological disorders (for review, see Cincotta and Ziemann, 2008).

In patients with congenital MMs abnormally persisting into adulthood, the neurophysiological hallmark is the presence of fast-conducting corticospinal pathways connecting the hand area of one M1 with both sides of the spinal cord. This was first demonstrated by TES (Farmer et al., 1990; Cohen et al., 1991) and consistently confirmed by more than 20 TMS studies, showing that focal stimulation of the M1 elicits bilateral MEP of normal and symmetrical latency in the resting hand muscles (Cincotta and Ziemann, 2008). Moreover, in patients with persistent congenital MM not associated with other relevant motor abnormalities, CSP recordings show an abnormal contribution of the ipsilateral M1 to the motor output during intended unilateral hand muscle contraction (Cincotta et al., 2002). Finally, during an intended unimanual task, dissociation of SICI downregulation between the "voluntary" and the "mirror" hand strongly suggests that the ipsilateral corticospinal fibers are separate from the crossed ones, providing a rationale for successful rehabilitation (Cincotta et al., 2003b).

In patients with acquired MM associated with PD, MEP study and investigation of CSP, SICI, and online interference by rTMS during intended unilateral tasks indicated that MM do not depend on unmasking of ipsilateral projections but are explained by motor output along the crossed corticospinal projection from the mirror M1 (Cincotta et al., 2006a; Li et al., 2007). It has been hypothesized that in PD, enhanced mirroring may depend on a failure of basal ganglia output to support the cortical network underlying lateralization of voluntary movements (Cincotta and Ziemann, 2008).

7.4.6.6   Clinical Usefulness of TMS in Movement Disorders

Despite TMS can be a useful tool in understanding the pathophysiology of many movement disorders, unfortunately the usefulness in clinical practice is very limited. This is due to the considerable overlap between values obtained in controls and patients and due to the nonspecificity of the findings.

For instance, the reduction of SICI and SP observed in some movement disorders has also been reported in other neurological and psychiatric disorders. In addition, in some conditions, the reduction in cortical inhibition could represent a compensatory response rather than a direct effect of the disease. On the other hand, there are also few examples where TMS can be useful in a clinical setting: in atypical Parkinsonism where it can demonstrate equivocal pyramidal signs and in distinguishing patients with psychogenic dystonia even if this observation needs to be confirmed in future studies.

7.4.7   STROKE

TMS studies may have a prognostic value. Indeed, several evidences suggest that the presence of MEPs in the paretic limb in the first week after stroke can predict a good motor recovery (Heald et al., 1993; Catano et al., 1995; D'Olhaberriague et al., 1997; Escudero et al., 1998; Trompetto et al., 2000; Delvaux et al., 2003; Hendricks et al., 2003). On the other hand, the absence of MEPs in the paretic limb, along with an increase in motor cortex excitability of the unaffected hemisphere, may predict a poor motor recovery.

Another factor in favor of a poor recovery is the presence of ipsilateral MEPs in the paretic limb after stimulation of the primary motor area of the unaffected hemisphere (Turton et al., 1996; Netz et al., 1997; Gerloff et al., 2006). Interestingly, the occurrence of ipsilateral MEPs obtained after stimulation of the premotor cortex is associated with good motor recovery (Alagona et al., 2001b). This is keeping with other studies suggesting that premotor cortex plays a role in functional recovery after stroke (Delvaux et al., 2003; Fridman et al., 2004). Cortical excitability parameters, such as MT, I–O curve, and cortico-cortical inhibition, seem to correlate with lesion location and motor performance (Liepert et al., 2005).

SICI is reduced in the affected hemisphere in the acute phase of a motor cortical stroke (Liepert et al., 2000, 2005; Manganotti et al., 2002; Niehaus et al., 2003) and remains unchanged regardless of motor recovery.

On the other hand, SICI is initially reduced in the unaffected hemisphere but then revert back to normal or even increase in patients with good motor recovery (Manganotti et al., 2002; Cicinelli et al., 2003). In keeping with these findings, several authors have found a loss of IHI from the affected side to the unaffected side, which could shift the balance toward an increased excitability of the unaffected hemisphere in acute stroke (Boroojerdi et al., 1996; Shimizu et al., 2002; Niehaus et al., 2003).

Finally, TMS can be used for rehabilitation purpose looking at spatial cortical reorganization after stroke (Thickbroom et al., 2004).

7.4.8 EPILEPSY

Epileptic conditions are characterized by heterogeneous and dynamic pathophysiological processes leading to imbalances between excitatory and inhibitory influences at the cortical level (Engel, 1995). Antiepileptic drugs (AEDs) counteract these altered balances by different mechanisms (Kwan et al., 2001). Hence, TMS has been regarded as a valuable tool to assess changes in cortical excitability in patients affected by epilepsy (for review, see Ziemann et al., 1998c; Tassinari et al., 2003; Richardson and Lopes da Silva, 2011). As AEDs can influence TMS measures, the optimal approach to investigate epilepsies is to evaluate untreated patients. However, this is not always possible, and TMS studies should be tailored to control for the AED treatment. As AEDs largely produce bilateral changes, one approach is to investigate interside differences. Another possibility is to compare distinct group of patients receiving similar antiepileptic treatments (Cincotta et al., 1998; Aguglia et al., 2000). Furthermore, if treated patients show a modification of TMS measures opposite to the known effect of a given drug, it more likely depends on the epileptic process. The main data in this field are summarized in the following text.

Reutens and coworkers (1992, 1993a) found reduced MT in untreated patients with idiopathic generalized epilepsy (IGE) and increased MT in IGE patients chronically treated with valproic acid. These findings suggested cortical hyperexcitability due to the epileptic process. However, this was not a unanimous finding. Increased MT has been reported in untreated IGE patients, who mainly showed absence seizures (Gianelli et al., 1994). Interestingly, Delvaux et al. (2001) found MT increase in 48 h after the first generalized seizure, which normalized 2–4 weeks later. This finding suggested a protective mechanism against seizure recurrence and raised the possibility that the discrepancy among different studies could be due to dynamic MT variations locked to the timing of seizures. Another explanation of this discrepancy is the heterogeneity of IGE. In keeping with this view, Reutens et al. (1993b) found lower MT in patients with myoclonic seizures compared to patients with absence seizures. Finally, the increased interhemispheric difference in MT observed in patients with versive or circling seizures (Aguglia et al., 2000) supports a role for interhemispheric imbalance of cortical excitability in lateralized ictal phenomena of IGE patients.

Increased CSP duration has been reported in untreated IGE patients (Macdonell et al., 2001). A bilateral CSP prolongation was also seen in patients with cryptogenic partial epilepsy (PE), whose ictal pattern strongly suggested that the epileptic focus involved the M1 (Cincotta et al., 1998). This was likely not due to AED therapy, because no CSP lengthening was observed in patients receiving similar treatment who suffered from cryptogenic PE not involving the motor cortex. Moreover, a marked CSP prolongation was observed in a patient with PE due to a lesion within the supplementary motor area (Classen et al., 1995). In epileptic patients, CSP prolongation could be due to hyperexcitability of inhibitory circuits (Macdonell et al., 2001) or, alternatively, may represent compensatory interictal phenomena acting to prevent seizure reoccurrence and to contrast spread of epileptogenic influences from the affected M1 to the contralateral one (Cincotta et al., 1998). The second hypothesis is in agreement with the data from animal experiments (Engel, 1995) and is supported by a long-lasting follow-up study in a patient with a rolandic region meningioma and preoperative seizure with complete postsurgical remission, in which the CSP prolongation was related to the risk of seizures (Cincotta et al., 2002). Unlike these findings, CSP shortening has been reported in a group of six patients with post-stroke focal motor seizures (Kessler et al., 2002). However, it should be considered that a given epileptic syndrome can depend on different processes and, consequently, the CSP (as well as other TMS measures) can be either enhanced or reduced, if its neural substrate is involved in the epileptogenesis or in the interictal compensatory mechanisms (Cincotta et al., 2003a).

In untreated patients tested after their first generalized seizure, ICF was reduced, whereas SICI was normal (Delvaux et al., 2001). This suggested a protective mechanism against the spread or recurrence of seizures. In contrast, SICI reduction (Caramia et al., 1996; Manganotti et al., 2000) with normal ICF (Manganotti et al., 2000) was found in both treated and untreated patients with juvenile myoclonic epilepsy (JME), suggesting impaired inhibitory function. In untreated patients with symptomatic PE, in most of whom the epileptogenic zone was located outside the M1, SICI was prominently reduced in the "unaffected" hemisphere, whereas ICF was reduced in the "affected" hemisphere (Werhahn et al., 2000). These SICI and ICF data could reflect spread of epileptogenic influences and interictal compensatory mechanisms, respectively, and suggested a remote effect of the epileptic processes onto the M1. In addition, these findings are in keeping with the separate origin of SICI and ICF (see also Section 7.3.2.1). Using cluster analysis in 18 treated patients with cryptogenic PE, Cantello et al. (2000) found SICI reduction and ICF increase in a subgroup of seven patients in which seizure frequency was higher than the subgroup of patients with no ICI and ICF changes. Finally, in the group of PE patients undergoing AED withdrawal for presurgical evaluation, SICI and ICF reduction was seen in the subgroup of patients that had seizures within 48 h of TMS study, but not in patients that did not have a seizure (Wright et al., 2006).

In a group of IGE patients, Brodtmann et al. (1999) found reduced LICI at ISIs of 100–200 ms. Interestingly, the same group reported increased CSP in IGE, supporting the view that CSP and LICI can be dissociated (Berardelli et al., 1996; Valzania et al., 1997). In contrast, Manganotti et al. (2000) found no significant LICI modification in a group of IGE patients.

The hypothesis that in epilepsies TMS measures may reflect dynamic changes in the balance between excitatory and inhibitory mechanisms at the cortical level (Tassinari et al., 2003) is further supported by a recent study, in which repeated evaluations of MT, SICI, ICF, and LICI have been systematically performed in the interictal, preictal, and postictal states in a group of 58 untreated patients with new-onset IGE or PE (Badawy et al., 2009). Increased M1 excitability (decreased MT, SICI, and LICI and increased ICF) was seen in the preictal phase, and the opposite occurred in the postictal state.

In cortical epileptic myoclonus, several TMS data suggest an altered balance between excitatory and inhibitory mechanisms, resulting in hyperexcitability of the M1. Namely, SICI reduction (Brown et al., 1996; Inghilleri et al., 1998; Manganotti et al., 2001) with normal ICF (Manganotti et al., 2001) was found. Reduced LICI has also been reported (Inghilleri et al., 1998; Valzania et al., 1999). Moreover, both SICI and LICI can be differently affected in distinct epileptic syndromes with myoclonus (Badawy et al., 2010; Canafoglia et al., 2010). CSP shortening was consistently seen in patients presenting with continuous/semi-continuous epileptic myoclonus (Guerrini et al., 1996, 1998, 2001; Inghilleri et al., 1998), which can be considered an ictal phenomenon. Again, this supports the notion that, in epilepsies, interictal CSP prolongation could reflect compensatory mechanisms. Finally, S-IHI was reduced in patients whose excitatory myoclonic activity spreads transcallosally to the contralateral hemisphere (Brown et al., 1996). Altered sensorimotor integration has been found in cortical reflex myoclonus and progressive myoclonic epilepsy, indicating a possible spread of hyperexcitability from somatosensory to motor networks at the cortical or at a subcortical level (Cantello et al., 1997; Manganotti et al., 2001). In contrast, normal sensorimotor integration was seen in JME (Manganotti et al., 2004).

Finally, EEG recordings of TMS-evoked cortical potentials appear a promising approach to investigate cortical excitability outside the M1 in patients with epilepsy (Del Felice et al., 2011).

7.4.9 MIGRAINE

In patients with migraine, TMS has been used to test both visual and motor cortex excitability. Overall, TMS reveals a cortical hyperexcitability.

TMS studies in migraineurs are controversial: some authors have reported increased MT (Afra et al., 1998) while others did not (Bohotin et al., 2003; Gunaydin et al., 2006).

SP has been reported shortened in hand muscles (Aurora et al., 1998) and normal in others (Afra et al., 1998; Werhahn et al., 2000; Ozturk et al., 2002; Gunaydin et al., 2006; Currà et al., 2007). Conversely, in patients with chronic migraine, a prolonged SP was reported (Ozturk et al., 2002). In migraineurs with aura, SICI was reduced with normal ICF (Brighina et al., 2005).

The threshold of phosphenes is reduced in occipital cortex in keeping with the notion of a visual cortex hyperexcitability in migraineurs (Aurora et al., 1998; Battelli et al., 2002; Young et al., 2004; Gerwig et al., 2005; Gunaydin et al., 2006). This threshold may increase in the interictal period.

Chronic pain may alter cortical excitability. A reduction of SICI and LICI in the painful side of patients with fibromyalgia (Salerno et al., 2000) and complex pain regional syndrome (Schwenkreis et al., 2003; Eisenberg et al., 2005) has been reported.

ACKNOWLEDGMENT

The authors thank Dr. Fabio Giovannelli for his assistance in preparing this chapter.

REFERENCES

Abbruzzese, G., Assini, A., Buccolieri, A., Schieppati, M., and Trompetto, C. 1999. Comparison of intracortical inhibition and facilitation in distal and proximal arm muscles in humans. *Journal of Physiology* 514:895–903.

Abbruzzese, G., Buccolieri, A., Marchese, R., Trompetto, C., Mandich, P., and Schieppati, M. 1997a. Intracortical inhibition and facilitation are abnormal in Huntington's disease: A paired magnetic stimulation study. *Neuroscience Letters* 228:87–90.

Abbruzzese, G., Dall'Agata, D., Morena, M. et al. 1988. Electrical stimulation of the motor tracts in cervical spondylosis. *Journal of Neurology, Neurosurgery and Psychiatry* 51:796–802.

Abbruzzese, G., Marchese, R., Buccolieri, A., Gasparetto, B., and Trompetto, C. 2001. Abnormalities of sensorimotor integration in focal dystonia: A transcranial magnetic stimulation study. *Brain* 124:537–545.

Abbruzzese, G., Marchese, R., and Trompetto, C. 1997b. Sensory and motor evoked potentials in multiple system atrophy: A comparative study with Parkinson's disease. *Movement Disorders* 12:315–321.

Abbruzzese, G., Tabaton, M., Morena, M., Dall'Agata, D., and Favale, E. 1991. Motor and sensory evoked potentials in progressive supranuclear palsy. *Movement Disorders* 6:49–54.

Abbruzzese, G. and Trompetto, C. 2002. Clinical and research methods for evaluating cortical excitability. *Journal of Clinical Neurophysiology* 19:307–321.

Afra, J., Mascia, A., Gerard, P., Maertens, d. N., and Schoenen, J. 1998. Interictal cortical excitability in migraine: A study using transcranial magnetic stimulation of motor and visual cortices. *Annals of Neurology* 44:209–215.

Aguglia, U., Gambardella, A., Quartarone, A. et al. 2000. Interhemispheric threshold differences in idiopathic generalized epilepsies with versive or circling seizures determined with focal magnetic transcranial stimulation. *Epilepsy Research* 40:1–6.

Alagona, G., Bella, R., Ferri, R. et al. 2001a. Transcranial magnetic stimulation in Alzheimer disease: Motor cortex excitability and cognitive severity. *Neuroscience Letters* 314:57–60.

Alagona, G., Delvaux, V., Gerard, P. et al. 2001b. Ipsilateral motor responses to focal transcranial magnetic stimulation in healthy subjects and acute-stroke patients. *Stroke* 32:1304–1309.

Allam, N., Fonte-Boa, P. M., Tomaz, C. A., and Brasil-Neto, J. P. 2005. Lack of effect of botulinum toxin on cortical excitability in patients with cranial dystonia. *Clinical Neurophysiology* 28:1–5.

Alle, H., Heidegger, T., Kriváneková, L., and Ziemann, U. 2009. Interactions between short-interval intracortical inhibition and short-latency afferent inhibition in human motor cortex. *Journal of Physiology* 587:5163–5176.

Attarian, S., Azulay, J. P., Lardillier, D., Verschueren, A., and Pouget, J. 2005. Transcranial magnetic stimulation in lower motor neuron diseases. *Clinical Neurophysiology* 116:35–42.

Aurora, S. K., Ahmad, B. K., Welch, K. M., Bhardhwaj, P., and Ramadan, N. M. 1998. Transcranial magnetic stimulation confirms hyperexcitability of occipital cortex in migraine. *Neurology* 50:1111–1114.

Avanzino, L., Teo, J. T. H., and Rothwell, J. C. 2007. Intracortical circuits modulate transcallosal inhibition in humans. *Journal of Physiology* 583:99–114.

Awiszus, F., Feistner, H., Urbach, D., and Bostock, H. 1999. Characterisation of paired pulse transcranial magnetic stimulation conditions yielding intracortical inhibition or I-wave facilitation using a threshold-hunting paradigm. *Experimental Brain Research* 129:317–324.

Badawy, R., Macdonell, R., Jackson, G., and Berkovic, S. 2009. The peri-ictal state: Cortical excitability changes within 24 h of a seizure. *Brain* 132:1013–1021.

Badawy, R. A., Macdonell, R. A., Jackson, G. D., and Berkovic, S. F. 2010. Can changes in cortical excitability distinguish progressive from juvenile myoclonic epilepsy? *Epilepsia* 51:2084–2088.

Badawy, R. A., Tarletti, R., Mula, M., Varrasi, C., and Cantello, R. 2011. The routine circular coil is reliable in paired-TMS studies. *Clinical Neurophysiology* 122:784–788.

Barker, A. T., Freeston, I. L., Jalinous, R., and Jarratt, J. A. 1986. Clinical evaluation of conduction time measurements in central motor pathways using magnetic stimulation of human brain. *Lancet* 1:1325–1326.

Barker, A. T., Jalinous, R., and Freeston, I. L. 1985. Non-invasive magnetic stimulation of human motor cortex. *Lancet* 1:1106–1107.

Battelli, L., Black, K. R., and Wray, S. H. 2002. Transcranial magnetic stimulation of visual area V5 in migraine. *Neurology* 58:1066–1069.

Baumer, T., Bock, F., Koch, G. et al. 2006. Magnetic stimulation of human premotor or motor cortex produces interhemispheric facilitation through distinct pathways. *Journal of Physiology* 572:857–868.

Baumer, T., Schippling, S., Kroeger, J. et al. 2009. Inhibitory and facilitatory connectivity from ventral premotor to primary motor cortex in healthy humans at rest—A bifocal TMS study. *Clinical Neurophysiology* 120:1724–1731.

Beer, S., Rosler, K. M., and Hess, C. W. 1995. Diagnostic value of paraclinical tests in multiple sclerosis: Relative sensitivities and specificities for reclassification according to the Poser committee criteria. *Journal of Neurology, Neurosurgery and Psychiatry* 59:152–159.

Berardelli, A., Inghilleri, M., Gilio, F. et al. 1999. Effects of repetitive cortical stimulation on the silent period evoked by magnetic stimulation. *Experimental Brain Research* 125:82–86.

Berardelli, A., Inghilleri, M., Rothwell, J. C. et al. 1998. Facilitation of muscle evoked responses after repetitive cortical stimulation in man. *Experimental Brain Research* 122:79–84.

Berardelli, A., Rona, S., Inghilleri, M., and Manfredi, M. 1996. Cortical inhibition in Parkinson's disease. A study with paired magnetic stimulation. *Brain* 119: 71–77.

Bohotin, V., Fumal, A., Vandenheede, M., Bohotin, C., and Schoenen, J. 2003. Excitability of visual V1–V2 and motor cortices to single transcranial magnetic stimuli in migraine: A reappraisal using a figure-of-eight coil. *Cephalalgia* 23:264–270.

Boroojerdi, B., Battaglia, F., Muellbacher, W., and Cohen, L. G. 2001. Mechanisms influencing stimulus-response properties of the human corticospinal system. *Clinical Neurophysiology* 112: 931–937.

Boroojerdi, B., Cohen, L. G., and Hallett, M. 2003. Effects of botulinum toxin on motor system excitability in patients with writer's cramp. *Neurology* 61:1546–1550.

Boroojerdi, B., Diefenbach, K., and Ferbert, A. 1996. Transcallosal inhibition in cortical and subcortical cerebral vascular lesions. *Journal of Neurological Science* 144:160–170.

Boroojerdi, B., Hungs, M., Mull, M., Topper, R., and Noth, J. 1998. Interhemispheric inhibition in patients with multiple sclerosis. *Electroencephalography and Clinical Neurophysiology* 109:230–237.

Brighina, F., Giglia, G., Scalia, S., Francolini, M., Palermo, A., and Fierro, B. 2005. Facilitatory effects of 1 Hz rTMS in motor cortex of patients affected by migraine with aura. *Experimental Brain Research* 161:34–38.

Britton, T. C., Meyer, B. U., and Benecke, R. 1991. Variability of cortically evoked motor responses in multiple sclerosis. *Electroencephalography and Clinical Neurophysiology* 81:186–194.

Brodie, M. J. 1995. Tiagabine pharmacology in profile. *Epilepsia* 36 (Suppl 6):S7–S9.

Brodtmann, A., Macdonell, R. A., Gilligan, A. K., Curatolo, J., and Berkovic, S. F. 1999. Cortical excitability and recovery curve analysis in generalized epilepsy. *Neurology* 53:1347–1349.

Brown, P., Ridding, M. C., Werhahn, K., Rothwell, J. C., and Marsden, C. D. 1996. Abnormalities in the balance between inhibition and excitation in the motor cortex of patients with cortical myoclonus. *Brain* 119:309–317.

Butefisch, C. M., Netz, J., Wessling, M., Seitz, R. J., and Homberg, V. 2003. Remote changes in cortical excitability after stroke. *Brain* 126:470–481.

Canafoglia, L., Ciano, C., Visani, E. et al. 2010. Short and long interval cortical inhibition in patients with Unverricht-Lundborg and Lafora body disease. *Epilepsy Research* 89:232–237.

Cantello, R., Civardi, C., Cavalli, A., Varrasi, C., Tarletti, R., Monaco, F., and Migliaretti, G. 2000. Cortical excitability in cryptogenic localization-related epilepsy: interictal transcranial magnetic stimulation studies. *Epilepsia* 41:694–704.

Cantello, R., Granelli, M., Bettucci, D., Civardi, C., De Angelis, M. S., and Mutani, R. 1991. Parkinson's disease rigidity: Magnetic motor evoked potentials in a small hand muscle. *Neurology* 41:1449–1456.

Cantello, R., Gianelli, M., Civardi, C., and Mutani, R. 1992. Magnetic brain stimulation: The silent period after the motor evoked potential. *Neurology* 42:1951–1959.

Cantello, R., Gianelli, M., Civardi, C., and Mutani, R. 1997. Focal subcortical reflex myoclonus. A clinical and neurophysiological study. *Archives of Neurology* 54:187–196.

Caramia, M. D., Gigli, G., Iani, C., Desiato, M. T., Diomedi, M., Palmieri, M. G., and Bernardi, G. 1996. Distinguishing forms of generalized epilepsy using magnetic brain stimulation. *Electroencephalography and Clinical Neurophysiology* 98:14–19.

Caramia, A. D., Palmieri, M. G., Desiato, M. T. et al. 2004. Brain excitability changes in the relapsing and remitting phases of multiple sclerosis: A study with transcranial magnetic stimulation. *Clinical Neurophysiology* 115:956–965.

de Carvalho, M., de Mendonca, A., Miranda, P. C., Garcia, C., and Luis, M. L. 1997. Magnetic stimulation in Alzheimer's disease. *Journal of Neurology* 244:304–307.

Catano, A., Houa, M., Caroyer, J. M., Ducarne, H., and Noel, P. 1995. Magnetic transcranial stimulation in non-haemorrhagic sylvian strokes: Interest of facilitation for early functional prognosis. *Electroencephalography and Clinical Neurophysiology* 97:349–354.

Chan, K. M., Nasathurai, S., Chavin, J. M., and Brown, W. F. 1998. The usefulness of central motor conduction studies in the localization of cord involvement in cervical spondylytic myelopathy. *Muscle & Nerve* 21:1220–1223.

Chen, R. 2000. Studies of human motor physiology with transcranial magnetic stimulation. *Muscle & Nerve* 9:S26–S32.

Chen, R. 2004. Interactions between inhibitory and excitatory circuits in the human motor cortex. *Experimental Brain Research* 154:1–10.

Chen, R., Corwell, B., and Hallett, M. 1999a. Modulation of motor cortex excitability by median nerve and digit stimulation. *Experimental Brain Research* 129:77–86.

Chen, R., Cros, D., Currà, A. et al. 2008. The clinical diagnostic utility of transcranial magnetic stimulation: Report of an IFCN committee. *Clinical Neurophysiology* 119:504–532.

Chen, R., Lozano, A. M., and Ashby, P. 1999b. Mechanism of the silent period following transcranial magnetic stimulation. Evidence from epidural recordings. *Experimental Brain Research* 128:539–542.

Chen, R., Samii, A., Canos, M., Wassermann, E. M., and Hallett, M. 1997. Effects of phenytoin on cortical excitability in humans. *Neurology* 49:881–883.

Chen, R., Tam, A., Butefisch, C. et al. 1998. Intracortical inhibition and facilitation in different representations of the human motor cortex. *Journal of Neurophysiology* 80:2870–2881.

Chen, R., Yung, D., and Jie-Yuan, L. 2003. Organization of ipsilateral excitatory and inhibitory pathways in the human motor cortex. *Journal of Neurophysiology* 89:1256–1264.

Chokroverty, S., Picone, M. A., and Chokroverty, M. 1991. Percutaneous magnetic coil stimulation of human cervical vertebral column: Site of stimulation and clinical application. *Electroencephalography and Clinical Neurophysiology* 81:359–365.

Cicinelli, P., Pasqualetti, P., Zaccagnini, M., Traversa, R., Oliveti, M., and Rossigni, P. M. 2003. Interhemispheric asymmetries of motor cortex excitability in the postacute stroke stage: A paired-pulse transcranial magnetic stimulation study. *Stroke* 34:2653–2658.

Cincotta, M., Borgheresi, A., Balestrieri, F. et al. 2006a. Mechanisms underlying mirror movements in Parkinson's disease: A transcranial magnetic stimulation study. *Movement Disorders* 21:1019–1025.

Cincotta, M., Borgheresi, A., Balestrieri, F., and Zaccara, G. 2003a. Reduced inhibition within primary motor cortex in patients with poststroke focal motor seizures (Letter). *Neurology* 60:527–528.

Cincotta, M., Borgheresi, A., Balzini, L. et al. 2003b. Separate ipsilateral and contralateral corticospinal projections in congenital mirror movements: Neurophysiological evidence and significance for motor rehabilitation. *Movement Disorders* 18:1294–1300.

Cincotta, M., Borgheresi, A., Boffi, P. et al. 2002. Bilateral motor cortex output with intended unimanual contraction in congenital mirror movements. *Neurology* 58:1290–1293.

Cincotta, M., Borgheresi, A., Guidi, L. et al. 2000. Remote effects of cortical dysgenesis on the primary motor cortex: Evidence from the silent period following transcranial magnetic stimulation. *Clinical Neurophysiology* 111:1340–1345.

Cincotta, M., Borgheresi, A., Lori, S., Fabbri, M., and Zaccara, G. 1998. Interictal inhibitory mechanisms in patients with cryptogenic motor cortex epilepsy: A study of the silent period following transcranial magnetic stimulation. *Electroencephalography and Clinical Neurophysiology* 107:1–7.

Cincotta, M., Giovannelli, F., Borgheresi, A. et al. 2006b. Modulatory effects of high-frequency repetitive transcranial magnetic stimulation on the ipsilateral silent period. *Experimental Brain Research* 171:490–496.

Cincotta, M. and Ziemann, U. 2008. Neurophysiology of unimanual motor control and mirror movements. *Clinical Neurophysiology* 119:744–762.

Civardi, C., Cantello, R., Asselman, P., and Rothwell, J. C. 2001. Transcranial magnetic stimulation can be used to test connections to primary motor areas from frontal and medial cortex in humans. *Neuroimage* 14:1444–1453.

Classen, J., Schnitzler, A., Binkofski, F. et al. 1997. The motor syndrome associated with exaggerated inhibition within the primary motor cortex of patients with hemiparetic stroke. *Brain* 120:605–619.

Classen, J., Witte, O. W., Schlaug, G., Seitz, R. J., Holthausen, H., and Benecke, R. 1995. Epileptic seizures triggered directly by focal transcranial magnetic stimulation. *Electroencephalography and Clinical Neurophysiology* 94:19–25.

Claus, D. 1990. Central motor conduction: method and normal results. *Muscle & Nerve* 13:1125–1132.

Claus, D., Weis, M., Jahnke, U., Plewe, A., and Brunholzl, C. 1992. Corticospinal conduction studied with magnetic double stimulation in the intact human. *Journal of Neurological Science* 111:180–188.

Cohen, L. G., Meer, J., Tarkka, I. et al. 1991. Congenital mirror movements. Abnormal organization of motor pathways in two patients. *Brain* 114:381–403.

Compta, Y., Valls-Sole, J., Valldeoriola, F., Kumru, H., and Rumia, J. 2006. The silent period of the thenar muscles to contralateral and ipsilateral deep brain stimulation. *Clinical Neurophysiology* 117:2512–2520.

Connors, B. W., Malenka, R. C., and Silva, L. R. 1988. Two inhibitory postsynaptic potentials, and GABA$_A$ and GABA$_B$ receptor- mediated responses in neocortex of rat and cat. *Journal of Physiology* 406:443–468.

Cruz-Martinez, A. and Arpa, J. 1997. Transcranial magnetic stimulation in patients with cerebellar stroke. *European Journal of Neurology* 38:82–87.

Cruz-Martinez, A. and Palau, F. 1997. Central motor conduction time by magnetic stimulation of the cortex and peripheral nerve conduction follow-up studies in Friedreich's ataxia. *Electroencephalography and Clinical Neurophysiology* 105:458–461.

Cunic, D., Roshan, L., Khan, F. I., Lozano, A. M., Lang, A. E., and Chen, R. 2002. Effects of subthalamic nucleus stimulation on motor cortex excitability in Parkinson's disease. *Neurology* 58:1665–1672.

Currà, A., Pierelli, F., Coppola, G. et al. 2007. Shortened cortical silent period in facial muscles of patients with migraine. *Pain* 132:124–131.

Currà, A., Romaniello, A., Berardelli, A., Cruccu, G., and Manfredi, M. 2000. Shortened cortical silent period in facial muscles of patients with cranial dystonia. *Neurology* 54:130–135.

D'Olhaberriague, L., Espadaler Gamissans, J. M., Marrugat, J., Valls, A., Oliveras, L. C., and Seoane, J. L. 1997. Transcranial magnetic stimulation as a prognostic tool in stroke. *Journal of Neurological Science* 147:73–80.

Dailey, J. W., Reith, M. E., Yan, Q. S., Li, M. Y., and Jobe, P. C. 1997. Anticonvulsant doses of carbamazepine increase hippocampal extracellular serotonin in genetically epilepsy-prone rats: Dose response relationships. *Neuroscience Letters* 227:13–16.

Daskalakis, Z. J., Paradiso, G. O., Christensen, B. K., Fitzgerald, P. B., Gunraj, C., and Chen, R. 2004. Exploring the connectivity between the cerebellum and motor cortex in humans. *Journal of Physiology* 1:689–700.

Davare, M., Andres, M., Cosnard, G., Thonnard, J. L., and Olivier, E. 2006. Dissociating the role of ventral and dorsal premotor cortex in precision grasping. *Journal of Neuroscience* 26:2260–2268.

Davare, M., Lemon, R., and Olivier, E. 2008. Selective modulation of interactions between ventral premotor cortex and primary motor cortex during precision grasping in humans. *Journal of Physiology* 586:2735–2742.

Davare, M., Montague, K., Olivier, E., Rothwell, J. C., and Lemon, R. N. 2009. Ventral premotor to primary motor cortical interactions during object-driven grasp in humans. *Cortex* 45:1050–1057.

Davey, N. J., Romaiguere, P., Maskill, D. W., and Ellaway, P. H. 1994. Suppression of voluntary motor activity revealed using transcranial magnetic stimulation of the motor cortex in man. *Journal of Physiology* 477:223–235.

De Rosa, A., Volpe, G., Marcantonio, L. et al. 2006. Neurophysiological evidence of corticospinal tract abnormality in patients with Parkin mutations. *Journal of Neurology* 253:275–279.

Del Felice, A., Fiaschi, A., Bongiovanni, G. L., Savazz,i S., and Manganotti, P. 2011. The sleep-deprived brain in normals and patients with juvenile myoclonic epilepsy: A perturbational approach to measuring cortical reactivity. *Epilepsy Research* 96:123–131.

Delvaux, V., Alagona, G., Gerard, P., De Pasqua, V., Delwaide, P. J., and Maertens de Noordhout, A. 2001. Reduced excitability of the motor cortex in untreated patients with de novo idiopathic "grand mal" seizures. *Journal of Neurology, Neurosurgery and Psychiatry* 71:772–776.

Delvaux, V., Alagona, G., Gerard, P., De Pasqua, V., Pennisi, G., and Maertens de Noordhout, A. 2003. Post-stroke reorganization of hand motor area: A 1-year prospective follow-up with focal transcranial magnetic stimulation. *Clinical Neurophysiology* 114:1217–1225.

Delwaide, P. J. and Olivier, E. 1990. Conditioning transcranial cortical stimulation (TCCS) by exteroceptive stimulation in parkinsonian patients. *Advances in Neurology* 53:175–181.

Demoule, A., Verin, E., and Ross, E. et al. 2003. Intracortical inhibition and facilitation of the response of the diaphragm to transcranial magnetic stimulation. *Journal of Clinical Neurophysiology* 20:59–64.

Devanne, H., Lavoie, B. A., and Capaday, C. 1997. Input-output properties and gain changes in the human corticospinal pathway. *Experimental Brain Research* 114:329–338.

Di Lazzaro, V., Molinari, M., Restuccia, D., Leggio, M. G., Tardone, R., and Fogli, D. 1994a. Cerebro-cerebellar interactions in man: Neurophysiological studies in patients with focal cerebellar lesions. *Electroencephalography and Clinical Neurophysiology* 93:27–34.

Di Lazzaro, V., Oliviero, A., Meglio, M. et al. 2000a. Direct demonstration of the effect of lorazepam on the excitability of the human motor cortex. *Clinical Neurophysiology* 111:794–799.

Di Lazzaro, V., Oliviero, A., Pilato, F. et al. 2004. Motor cortex hyperexcitability to transcranial magnetic stimulation in Alzheimer's disease. *Journal of Neurology, Neurosurgery and Psychiatry* 75:555–559.

Di Lazzaro, V., Oliviero, A., Pilato, F. et al. 2005a. Neurophysiological predictors of long term response to AchE inhibitors in AD patients. *Journal of Neurology, Neurosurgery and Psychiatry* 76:1064–1069.

Di Lazzaro, V., Oliviero, A., Profice, P. et al. 1998a. Comparison of descending volleys evoked by transcranial magnetic and electric stimulation in conscious humans. *Electroencephalography and Clinical Neurophysiology* 109:397–401.

Di Lazzaro, V., Oliviero, A., Profice, P. et al. 1999. Direct demonstration of interhemispheric inhibition of the human motor cortex produced by transcranial magnetic stimulation. *Experimental Brain Research* 124:520–524.

Di Lazzaro, V., Oliviero, A., Profice, P. et al. 2000b. Muscarinic receptor blockade has differential effects on the excitability of intracortical circuits in the human motor cortex. *Experimental Brain Research* 135:455–461.

Di Lazzaro, V., Oliviero, A., Profice, P. et al. 2003. Ketamine increases human motor cortex excitability to transcranial magnetic stimulation. *Journal of Physiology* 547:485–496.

Di Lazzaro, V., Oliviero, A., Saturno, E. et al. 2005b. Effects of lorazepam on short latency afferent inhibition and short latency intracortical inhibition in humans. *Journal of Physiology* 564:661–668.

Di Lazzaro, V., Oliviero, A., Tonali, P. A. et al. 2002. Noninvasive in vivo assessment of cholinergic cortical circuits in AD using transcranial magnetic stimulation. *Neurology* 59:392–397.

Di Lazzaro, V., Pilato, F., Dileone, M. et al. 2006a. in vivo cholinergic circuit evaluation in frontotemporal and Alzheimer dementias. *Neurology* 66:1111–1113.

Di Lazzaro, V., Pilato, F., Dileone, M. et al. 2007. Segregating two inhibitory circuits in human motor cortex at the level of GABAA receptor subtypes: A TMS study. *Clinical Neurophysiology* 118:2207–2214.

Di Lazzaro, V., Pilato, F., Oliviero, A. et al. 2006b. Origin of facilitation of motor-evoked potentials after paired magnetic stimulation: Direct recording of epidural activity in conscious humans. *Journal of Neurophysiology* 96:1765–1771.

Di Lazzaro, V., Restuccia, D., Molinari, M. et al. 1994b. Excitability of the motor cortex to magnetic stimulation in patients with cerebellar lesions. *Journal of Neurology, Neurosurgery and Psychiatry* 57:108–110.

Di Lazzaro, V., Restuccia, D., Nardone, R. et al. 1995. Motor cortex changes in a patient with hemicerebellectomy. *Electroencephalography and Clinical Neurophysiology* 97:259–263.

Di Lazzaro, V., Restuccia, D., Oliviero, A. et al. 1998b. Effects of voluntary contraction on descending volleys evoked by transcranial stimulation in conscious humans. *Journal of Physiology* 508:625–633.

Di Lazzaro, V., Restuccia, D., Oliviero, A. et al. 1998c. Magnetic transcranial stimulation at intensities below active motor threshold activates intracortical inhibitory circuits. *Experimental Brain Research* 119:265–268.

Edwards, M. J., Huang, Y. Z., Wood, N. W., Rothwell, J. C., and Bhatia, K. P. 2003. Different patterns of electrophysiological deficits in manifesting and nonmanifesting carriers of the DYT1 gene mutation. *Brain* 126:2074–2080.

Eisen, A., Pant, B., and Stewart, H. 1993. Cortical excitability in amyotrophic lateral sclerosis: A clue to pathogenesis. *Canadian Journal of Neurological Science* 20:11–16.

Eisen, A. A. and Shtybel, W. 1990. AAEMminimonograph #35: Clinical experience with transcranial magnetic stimulation. *Muscle & Nerve* 13:995–1011.

Eisen, A., Shytbel, W., Murphy, K., and Hoirch, M. 1990. Cortical magnetic stimulation in amyotrophic lateral sclerosis. *Muscle & Nerve* 13:146–151.

Eisenberg, E., Chistyakov, A. V., Yudashkin, M., Kaplan, B., Hafner, H., and Feinsod, M. 2005. Evidence for cortical hyperexcitability of the affected limb representation area in CRPS: A psychophysical and transcranial magnetic stimulation study. *Pain* 113:99–105.

Ellaway, P. H., Davey, N. J., Maskill, D. W., and Dick, J. P. 1995. The relation between bradykinesia and excitability of the motor cortex assessed using transcranial magnetic stimulation in normal and parkinsonian subjects. *Electroencephalography and Clinical Neurophysiology* 97:169–178.

Engel, J. Jr. 1995. Inhibitory mechanisms of epileptic seizure generation. In: *Advances in Neurology. Vol. 67. Negative Motor Phenomena*, eds. S. Fahn, M. Hallett, H. O. Lüders, and C. D. Marsden, pp. 157–171. Philadelphia, PA: Lippincott-Raven Publishers.

Enterzari-Taher, M., Eisen, A., Stewart, H., and Nakajima, M. 1997. Abnormalities of cortical inhibitory neurons in amyotrophic lateral sclerosis. *Muscle & Nerve* 20:65–71.

Escudero, J. V., Sancho, J., Bautista, D., Escudero, M., and Lopez-Trigo, J. 1998. Prognostic value of motor evoked potential obtained by transcranial magnetic brain stimulation in motor function recovery in patients with acute ischemic stroke. *Stroke* 29:1854–1859.

Espay, A. J., Morgante, F., Purzner, J., Gunraj, C. A., Lang, A. E., and Chen, R. 2006. Cortical and spinal abnormalities in psychogenic dystonia. *Annals of Neurology* 59:825–834.

Eusebio, A., Azulay, J. P., Witjas, T., Rico, A., and Attarian, S. 2007. Assessment of corticospinal tract impairment in multiple system atrophy using transcranial magnetic stimulation. *Clinical Neurophysiology* 118:815–823.

Facchetti, D., Mai, R., Micheli, A. et al. 1997. Motor evoked potentials and disability in secondary progressive multiple sclerosis. *Can Journal of Neurological Science* 24:332–337.

Farmer, S. F., Ingram, D. A., and Stephens, J. A. 1990. Mirror movements studied in a patient with Klippel-Feil syndrome. *Journal of Physiology* 428:467–484.

Ferbert, A., Priori, A., Rothwell, J. C., Day, B. L., Colebatch, J. G., and Marsden, C. D. 1992. Interhemispheric inhibition of the human motor cortex. *Journal of Physiology* 453:525–546.

Ferreri, F., Pauri, F., Pasqualetti, P., Fini, R., Dal Forno, G., and Rossini, P. M. 2003. Motor cortex excitability in Alzheimer's disease: A transcranial magnetic stimulation study. *Annals of Neurology* 53:102–108.

Feuillet, L., Pelletier, J., Suchet, L. et al. 2007. Prospective clinical and electrophysiological follow-up on a multiple sclerosis population treated with interferon beta-1 a: A pilot study. *Multiple Sclerosis Journal* 13:348–356.

Fierro, B., Giglia, G., Palermo, A., Pecoraro, C., Scalia, S., and Brighina, F. 2007. Modulatory effects of 1 Hz rTMS over the cerebellum on motor cortex excitability. *Experimental Brain Research* 176:440–447.

Filipovic, S. R., Ljubisavljevic, M., Svetel, M., Milanovic, S., Kacar, A., and Kostic, V. S. 1997. Impairment of cortical inhibition in writer's cramp as revealed by changes in electromyographic silent period after transcranial magnetic stimulation. *Neuroscience Letters* 222:167–170.

Filippi, M., Campi, A., Mammi, S. et al. 1995. Brain magnetic resonance imaging and multi-modal evoked potentials in benign and secondary progressive multiple sclerosis. *Journal of Neurology, Neurosurgery and Psychiatry* 58:31–37.

Fischer, M. and Orth, M. 2011. Short-latency sensory afferent inhibition: Conditioning stimulus intensity, recording site, and effects of 1 Hz repetitive TMS. *Brain Stimulation* 4:202–209.

Floeter, M. K. and Rothwell, J. C. 1999. Releasing the brakes before pressing the gas pedal. *Neurology* 53:664–665.

Fridman, E. A., Hanakawa, T., Chung, M., Hummel, F., Leiguarda, R. C., and Cohen, L. G. 2004. Reorganization of the human ipsilesional premotor cortex after stroke. *Brain* 127:747–758.

Fuhr, P., Agostino, R., and Hallett, M. 1991. Spinal motor neuron excitability during the silent period after cortical stimulation. *Electroencephalography and Clinical Neurophysiology* 81:257–262.

Fuhr, P., Borggrefe-Chappuis, A., Schindler, C., and Kappos, L. 2001. Visual and motor evoked potentials in the course of multiple sclerosis. *Brain* 124:2162–2168.

Gerdelat-Mas, A., Loubinoux, I., Tombari, D., Rascol, O., Chollet, F., and Simonetta-Moreau, M. 2005. Chronic administration of selective serotonin re-uptake inhibitor (SSRI) paroxetine modulates human motor cortex excitability in healthy subjects. *NeuroImage* 27:314–322.

Gerloff, C., Bushara, K., Sailer, A. et al. 2006. Multimodal imaging of brain reorganization in motor areas of the contralesional hemisphere of well recovered patients after capsular stroke. *Brain* 129:791–808.

Gerloff, C., Cohen, L. G., Floeter, M. K., Chen, R., Corwell, B., and Hallett, M. 1998. Inhibitory influence of the ipsilateral motor cortex on responses to stimulation of the human cortex and pyramidal tract. *Journal of Physiology* 510:249–259.

Gerwig, M., Niehaus, L., Kastrup, O., Stude, P., and Diener, H. C. 2005. Visual cortex excitability in migraine evaluated by single and paired magnetic stimuli. *Headache* 45:1394–1399.

von Giesen, H. J., Roick, H., and Benecke, R. 1994. Inhibitory actions of motor cortex following unilateral brain lesions as studied by magnetic brain stimulation. *Experimental Brain Research* 99:84–96.

Gianelli, M., Cantello, R., Civardi, C., Naldi, P., Bettucci, D., Schiavella, M.P., and Mutani, R. 1994. Idiopathic generalized epilepsy: magnetic stimulation of motor cortex time-locked and unlocked to 3-Hz spike-and-wave discharges. *Epilepsia* 35:53–60.

Gilio, F., Currà, A., Inghilleri, M. et al. 2003. Abnormalities of motor cortex excitability preceding movement in patients with dystonia. *Brain* 126:1745–1754.

Glocker, F. X., Magistris, M. R., Rosler, K. M., and Hess, C. W. 1994. Magnetic transcranial and electrical stylomastoidal stimulation of the facial motor pathways in Bell's palsy: Time course and relevance of electrophysiological parameters. *Electroencephalography and Clinical Neurophysiology* 93:113–120.

Greenberg, B. D., Ziemann, U., Cora-Locatelli, G. et al. 2000. Altered cortical excitability in obsessive–compulsive disorder. *Neurology* 54:142–147.

Guerrini, R., Bonanni, P., Parmeggiani, L., Santucci, M., Parmeggiani, A., and Sartucci, F. 1998. Cortical reflex myoclonus in Rett syndrome. *Annals of Neurology* 43:472–479.

Guerrini, R., Bonanni, P., Patrignani, A. et al. 2001. Autosomal dominant cortical myoclonus and epilepsy (ADCME) with complex partial and generalized seizures: A newly recognized epilepsy syndrome with linkage to chromosome 2p11.1–q12.2. *Brain* 124:2459–2475.

Guerrini, R., De Lorey, T. M., Bonanni, P. et al. 1996. Cortical myoclonus in Angelman syndrome. *Annals of Neurology* 40:39–48.

Gunaydin, S., Soysal, A., Atay, T., and Arpaci, B. 2006. Motor and occipital cortex excitability in migraine patients. *Can Journal of Neurological Science* 33:63–67.

Hallett, M. 1995. Transcranial magnetic stimulation. Negative effects. In: *Advances in Neurology. Vol. 67. Negative Motor Phenomena*, eds. S. Fahn, M. Hallett, H. O. Lüders, and C. D. Marsden, pp. 107–113. Philadelphia, PA: Lippincott-Raven Publishers.

Hallett, M. 1996. Transcranial magnetic stimulation: A useful tool for Clinical Neurophysiologyogy. *Annals of Neurology* 40:344–345.

Hallett, M. 2000. Transcranial magnetic stimulation and the human brain. *Nature* 406:147–150.

Hanajima, R., Ugawa, Y., Machii, K. et al. 2001. Interhemispheric facilitation of the hand motor area in humans. *Journal of Physiology* 531:849–859.

Hanajima, R., Ugawa, Y., Terao, Y. et al. 2002. Mechanisms of intracortical I-wave facilitation elicited with paired-pulse magnetic stimulation in humans. *Journal of Physiology* 538:253–261.

Hanajima, R., Ugawa, Y., Terao, Y., Ogata, K., and Kanazawa, I. 1996. Ipsilateral cortico-cortical inhibition of the motor cortex in various neurological disorders. *Journal of Neurological Science* 140:109–116.

Heald, A., Bates, D., Cartlidge, N. E., French, J. M., and Miller, S. 1993. Longitudinal study of central motor conduction time following stroke. 2. Central motor conduction measured within 72 h after stroke as a predictor of functional outcome at 12 months. *Brain* 116:1371–1385.

Hendricks, H. T., Pasman, J. W., Merx, J. L., van Limbeek, J., and Zwarts, M. J. 2003. Analysis of recovery processes after stroke by means of transcranial magnetic stimulation. *Journal of Clinical Neurophysiology* 20:188–195.

Herwig, U., Brauer, K., Connemann, B., Spitzer, M., and Schonfeldt-Lecuona, C. 2002. Intracortical excitability is modulated by a norepinephrine-reuptake inhibitor as measured with paired-pulse transcranial magnetic stimulation. *Psychopharmacology* 164:228–232.

Hess, C. W., Mills, K. R., and Murray, N. M. 1987. Responses in small hand muscles from magnetic stimulation of the human brain. *Journal of Physiology (London)* 388:397–419.

Ho, K. H., Lee, M., Nithi, K., Palace, J., and Mills, K. 1999. Changes in motor evoked potentials to short-interval paired transcranial magnetic stimuli in multiple sclerosis. *Clinical Neurophysiology* 110:712–719.

Hodgkin, A. L. and Huxley, A. F. 1952a. Current carried by sodium and potassium ions through the membrane of the giant axon of Loligo. *Journal of Physiology* 116:449–472.

Hodgkin, A. L. and Huxley, A. F. 1952b. A quantitative description of membrane current and its application to conduction and excitation in nerve. *Journal of Physiology* 116:500–544.

Hoppner, J., Kunesch, E., Buchmann, J., Hess, A., Grossmann, A., and Benecke, R. 1999. Demyelination and axonal degeneration in corpus callosum assessed by analysis of transcallosally mediated inhibition in multiple sclerosis. *Clinical Neurophysiology* 110:748–756.

Hough, C. J., Irwin, R. P., Gao, X. M., Rogawski, M. A., and Chuang, D. M. 1996. Carbamazepine inhibition of N-methyl-D-aspartate-evoked calcium influx in rat cerebellar granule cells. *Journal of Pharmacology and Experimental Therapeutics* 276:143–149.

Huang, Y. Z., Trender-Gerhard, I., Edwards, M. J., Mir, P., Rothwell, J. C., and Bhatia, K. P. 2006. Motor system inhibition in dopa-responsive dystonia and its modulation by treatment. *Neurology* 66:1088–1090.

Humm, A. M., Beer, S., Kool, J., Magistris, M. R., Kesselring, J., and Rosler, K. M. 2004. Quantification of Uhthoff's phenomenon in multiple sclerosis: A magnetic stimulation study. *Clinical Neurophysiology* 115:2493–2501.

Humm, A. M., Magistris, M. R., Truffert, A., Hess, C. W., and Rosler, K. M. 2003. Central motor conduction differs between acute relapsing–remitting and chronic progressive multiple sclerosis. *Clinical Neurophysiology* 114:2196–2203.

Hwa, G. G. and Avoli, M. 1992. Excitatory postsynaptic potentials recorded from regular-spiking cells in layers II/III of rat sensorimotor cortex. *Journal of Neurophysiology* 67:728–737.

Ilic, T. V., Korchounov, A., and Ziemann, U. 2003. Methylphenidate facilitates and disinhibits the motor cortex in intact humans. *Neuroreport* 14:773–776.

Ilic, T. V., Meintzschel, F., Cleff, U., Ruge, D., Kessler, K. R., and Ziemann, U. 2002. Short interval paired-pulse inhibition and facilitation of human motor cortex: The dimension of stimulus intensity. *Journal of Physiology* 545:153–167.

Inghilleri, M., Berardelli, A., Cruccu, G., and Manfredi, M. 1993. Silent period evoked by transcranial magnetic stimulation of the human cortex and cervicomedullary junction. *Journal of Physiology* 466:521–534.

Inghilleri, M., Berardelli, A., Marchetti, P., and Manfredi, M. 1996. Effects of diazepam, baclofen, and thiopental on the silent period evoked by transcranial magnetic stimulation in humans. *Experimental Brain Research* 109:467–472.

Inghilleri, M., Conte, A., Frasca, V. et al. 2004. Antiepileptic drugs and cortical excitability: A study with repetitive transcranial stimulation. *Experimental Brain Research* 154:488–493.

Inghilleri, M., Mattia, D., Berardelli, A., and Manfredi, M. 1998. Asymmetry of cortical excitability revealed by transcranial stimulation in a patient with focal motor epilepsy and cortical myoclonus. *Electroencephalography and Clinical Neurophysiology: Electromyography and Motor Control* 109:70–72.

Ingram, D. A., Thompson, A. J., and Swash, M. 1988. Central motor conduction in multiple-sclerosis—Evaluation of abnormalities revealed by transcutaneous magnetic stimulation of the brain. *Journal of Neurology, Neurosurgery and Psychiatry* 51:487–494.

Irlbacher, K., Brocke, J., Mechow, J. V., and Brandt, S. A. 2007. Effects of GABA(A) and GABA(B) agonists on interhemispheric inhibition in man. *Clinical Neurophysiology* 118:308–316.

Iwata, N. K. and Ugawa, Y. 2005. The effects of cerebellar stimulation on the motor cortical excitability in neurological disorders: A review. *Cerebellum* 4:218–223.

Jaskolski, D. J., Laing, R. J., Jarratt, J. A., and Jukubowski, J. 1990. Pre- and postoperative motor conduction times, measured using magnetic stimulation, in patients with cervical spondylosis. *British Journal of Neurosurgery* 4:187–192.

Jones, S. M., Streletz, L. J., Raab, V. E., Knobler, R. L., and Lublin, F. D. 1991. Lower extremity motor evoked-potentials in multiple-sclerosis. *Archives of Neurology* 48:944–948.

Kallmann, B. A., Fackelmann, S., Toyka, K. V., Rieckmann, P., and Reiners, K. 2006. Early abnormalities of evoked potentials and future disability in patients with multiple sclerosis. *Multiple Sclerosis Journal* 12:58–65.

Kameyama, O., Shibano, K., Kawakita, H., and Ogawa, R. 1995. Transcranial magnetic stimulation of the motor cortex in cervical spondylosis and spinal canal stenosis. *Spine* 20:1004–1010.

Kandler, R. H., Jarratt, J. A., Gumpert, E. J., Davies-Jones, G. A., Venables, G. S., and Sagar, H. J. 1991. The role of magnetic stimulation in the diagnosis of multiple-sclerosis. *Journal of Neurological Science* 106:25–30.

Kaneko, K., Kawai, S., Taguchi, T., Fuchigami, Y., and Shiraishi, G. 1997. Coexisting peripheral nerve and cervical cord compression. *Spine* 22:636–640.

Kanovsky, P., Bares, M., Streitova, H., Klajblova, H., Daniel, P., and Rektor, I. 2003. Abnormalities of cortical excitability and cortical inhibition in cervical dystonia. Evidence from somatosensory evoked potentials and paired transcranial magnetic stimulation recordings. *Journal of Neurology* 250:42–50.

Kessler, K. R., Schnitzler, A., Classen, J., and Benecke, R. 2002. Reduced inhibition within primary motor cortex in patients with poststroke focal motor seizures. *Neurology* 59:1028–1033.

Kidd, D., Thompson, P. D., Day, B. L. et al. 1998. Central motor conduction time in progressive multiple sclerosis—Correlations with MRI and disease activity. *Brain* 121:1109–1116.

Kimura, J. 1983. The F wave. In: *Electrodiagnosis in Disease of Nerve and Muscle: Principle and Practice*, ed. J. Kimura, pp. 353–377. Philadelphia, PA: Davies Publisher.

Kimura, J., Machida, M., Ishida, T. et al. 1986. Relation between size of compound sensory or muscle action potentials, and length of nerve segment. *Neurology* 36:647–652.

Knutsson, E., Lindblom, U., and Martensson, A. 1974. Plasma and cerebrospinal fluid levels of baclofen (Lioresal) at optimal therapeutic responses in spastic paresis. *Journal of Neurological Science* 23:473–484.

Koch, G., Fernandez del Olmo, M., Cheeran, B. et al. 2007. Focal stimulation of the posterior parietal cortex increases the excitability of the ipsilateral motor cortex. *Journal of Neuroscience* 27:6815–6822.

Koch, G., Fernandez del Olmo, M., Cheeran, B. et al. 2008a. Functional interplay between posterior parietal and ipsilateral motor cortex revealed by twin-coil transcranial magnetic stimulation during reach planning toward contralateral space. *Journal of Neuroscience* 28:5944–5953.

Koch, G., Franca, M., Del Olmo, M. F. et al. 2006. Time course of functional connectivity between dorsal premotor and contralateral motor cortex during movement selection. *Journal of Neuroscience* 26:7452–7459.

Koch, G., Mori, F., Marconi, B., Codecà, C., Pecchioli, C., and Salerno, S. 2008b. Changes in intracortical circuits of the human motor cortex following theta burst stimulation of the lateral cerebellum. *Clinical Neurophysiology* 119:2559–2569.

Koch, G., Ruge, D., Cheeran, B. et al. 2009. TMS activation of interhemispheric pathways between the posterior parietal cortex and the contralateral motor cortex. *Journal of Physiology* 587:4281–4292.

Korchounov, A., Ilic, T. V., and Ziemann, U. 2003. The alpha 2-adrenergic agonist guanfacine reduces excitability of human motor cortex through disfacilitation and increase of inhibition. *Clinical Neurophysiology* 114:1834–1840.

Kuhn, A. A., Grosse, P., Holtz, K., Brown, P., Meyer, B. U., and Kupsch, A. 2004. Patterns of abnormal motor cortex excitability in atypical parkinsonian syndromes. *Clinical Neurophysiology* 115:1786–1795.

Kujirai, T., Caramia, M. D., Rothwell, J. C. et al. 1993. Corticocortical inhibition in human motor cortex. *Journal of Physiology* 471:501–519.

Kukaswadia, S., Wagle-Shukla, A., Morgante, F., Gunraj, C., and Chen, R. 2005. Interactions between long latency afferent inhibition and interhemispheric inhibitions in the human motor cortex. *Journal of Physiology* 563:915–924.

Kwan, P., Sills, G. J., and Brodie, M. J. 2001. The mechanisms of action of commonly used antiepileptic drugs. *Pharmacology and Therapeutics* 90:21–34.

Lee, Y. C., Chen, J. T., Liao, K. K., Wu, Z. A., and Soong, B. W. 2003. Prolonged cortical relay time of long latency reflex and central motor conduction in patients with spinocerebellar ataxia type 6. *Clinical Neurophysiology* 114:458–462.

Lefaucheur, J. P. 2005. Excitability of the motor cortical representation of the external anal sphincter. *Experimental Brain Research* 160:268–272.

Lefaucheur, J. P., Drouot, X., Von Raison, F., Menard-Lefaucheur, I., Cesaro, P., and Nguyen, J. P. 2004. Improvement of motor performance and modulation of cortical excitability by repetitive transcranial magnetic stimulation of the motor cortex in Parkinson's disease. *Clinical Neurophysiology* 115:2530–2541.

Lefaucheur, J. P., Menard-Lefaucheur, I., Maison, P. et al. 2006. Electrophysiological deterioration over time in patients with Huntington's disease. *Movement Disorders* 21:1350–1354.

Leocani, L., Rovaris, M., Boneschi, F. M. et al. 2006. Multimodal evoked potentials to assess the evolution of multiple sclerosis: A longitudinal study. *Journal of Neurology, Neurosurgery and Psychiatry* 77:1030–1035.

Li, J-Y., Espay, A. J., Gunraj, C. A. et al. 2007. Interhemispheric and ipsilateral connections in Parkinson's disease: Relation to mirror movements. *Movement Disorders* 22:813–821.

Liepert, J., Bar, K. J., Meske, U., and Weiller, C. 2001. Motor cortex disinhibition in Alzheimer's disease. *Clinical Neurophysiology* 112:1436–1441.

Liepert, J., Restemeyer, C., Kucinski, T., Zittel, S., and Weiller, C. 2005. Motor strokes: Rhe lesion location determines motor excitability changes. *Stroke* 36:2648–2653.

Liepert, J., Storch, P., Fritsch, A., and Weiller, C. 2000. Motor cortex disinhibition in acute stroke. *Clinical Neurophysiology* 111:671–676.

Liepert, J., Wessel, K., Schwenkreis, P. et al. 1998. Reduced intracortical facilitation in patients with cerebellar degeneration. *Acta Neurologica Scandinavica* 98:318–323.

Lo, Y. L., Chan, L. L., Lim, W. et al. 2004. Systematic correlation of transcranial magnetic stimulation and magnetic resonance imaging in cervical spondylotic myelopathy. *Spine* 29:1137–1145.

Luft, A. R., Manto, M. U., and Taib, N. O. B. 2005. Modulation of motor cortex excitability by sustained peripheral stimulation: The interaction between the motor cortex and the cerebellum. *Cerebellum* 4:90–96.

Macdonell, R. A., King, M. A., Newton, M. R., Curatolo, J. M., Reutens, D. C., and Berkovic, S. F. 2001. Prolonged cortical silent period after transcranial magnetic stimulation in generalized epilepsy. *Neurology* 57:706–708.

Magistris, M. R. and Rösler, K. M. 2003. The triple stimulation technique to study corticospinal conduction. *Supplements to Clinical Neurophysiology* 56:24–32.

Magistris, M. R., Rosler, K. M., Truffert, A., Landis, T., and Hess, C. W. 1999. A clinical study of motor evoked potentials using a triple stimulation technique. *Brain* 122:265–279.

Magistris, M. R., Rosler, K. M., Truffert, A., and Myers, J. P. 1998. Transcranial stimulation excites virtually all motor neurons supplying the target muscle. A demonstration and a method improving the study of motor evoked potentials. *Brain* 121:437–450.

Manganotti, P., Bongiovanni, L. G., Zanette, G., and Fiaschi, A. 2000. Early and late intracortical inhibition in juvenile myoclonic epilepsy. *Epilepsia* 41:1129–1138.

Manganotti, P., Patuzzo, S., Cortese, F., Palermo, A., Smania, N., and Fiaschi, A. 2002. Motor disinhibition in affected and unaffected hemisphere in the early period of recovery after stroke. *Clinical Neurophysiology* 113:936–943.

Manganotti, P., Tamburin, S., Bongiovanni, L. G., Zanette, G., and Fiaschi, A. 2004. Motor responses to afferent stimulation in juvenile myoclonic epilepsy. *Epilepsia* 45:77–80.

Manganotti, P., Tamburin, S., Zanette, G., and Fiaschi, A. 2001. Hyperexcitable cortical responses in progressive myoclonic epilepsy: A TMS study. *Neurology* 57:1793–1799.

Manganelli, F., Vitale, C., Santangelo, G. et al. 2009. Functional involvement of central cholinergic circuits and visual hallucinations in Parkinson's disease. *Brain* 132:2350–2355.

Marangos, P. J., Post, R. M., Patel, J., Zander, A., Parma, K., and Weiss, S. 1983. Specific and potent interactions of carbamazepine with brain adenosine receptors. *European Journal of Pharmacology* 93:175–182.

Mariorenzi, R., Zarola, F., Caramia, M. D., Paradiso, C., and Rossini, P. M. 1991. Non-invasive evaluation of central motor tract excitability changes following peripheral nerve stimulation in healthy humans. *Electroencephalography and Clinical Neurophysiology* 81:90–101.

Mavroudakis, N., Caroyer, J. M., Brunko, E., and Zegers de Beyl, D. 1994. Effects of diphenylhydantoin on motor potentials evoked with magnetic stimulation. *Electroencephalography and Clinical Neurophysiology* 93:428–433.

Mavroudakis, N., Caroyer, J. M., Brunko, E., and Zegers de Beyl, D. 1997. Effects of vigabatrin on motor potentials evoked with magnetic stimulation. *Electroencephalography and Clinical Neurophysiology* 105:124–127.

Mayr, N., Baumgartner, C., Zeitlhofer, J., and Deecke, L. 1991. The sensitivity of transcranial cortical magnetic stimulation in detecting pyramidal tract lesions in clinically definite multiple sclerosis. *Neurology* 41:566–569.

Mazzocchio, R., Rothwell, J. C., Day, B. L., and Thompson, P. D. 1994. Effect of tonic voluntary activity on the excitability of human motor cortex. *Journal of Physiology* 474:261–267.

McCormick, D. A. 1992. Neurotransmitter actions in the thalamus and cerebral cortex. *Journal of Clinical Neurophysiology* 9:212–223.

Merton, P. A. and Morton, H. B. 1980. Stimulation of the cerebral cortex in the intact human subject. *Nature* 285:227.

Meyer, B. U., Röricht, S., Gräfin von Einsiedel, H., Kruggel, F., and Weindl, A. 1995. A inhibitory and excitatory interhemispheric transfers between motor cortical areas in normal humans and patients with abnormalities of the corpus callosum. *Brain* 118:429–440.

Meyer, B. U., Roricht, S., and Woiciechowsky, C. 1998. Topography of fibers in the human corpus callosum mediating interhemispheric inhibition between the motor cortices. *Annals of Neurology* 43:360–389.

Michels, R., Wessel, K., Klohn, S., and Kompf, D. 1993. Long-latency reflexes, somatosensory evoked potentials and transcranial magnetic stimulation: Relation of the three methods in multiple sclerosis. *Electroencephalography and Clinical Neurophysiology* 89:235–241.

Mills, K. R. 2003. The natural history of central motor abnormalities in amyotrophic lateral sclerosis. *Brain* 126:2558–2566.

Mills, K. R. and Murray, N. M. 1986. Electrical stimulation over the human vertebral column: Which neural elements are excited? *Electroencephalography and Clinical Neurophysiology* 63:582–589.

Mills, K. R. and Nithi, K. A. 1997. Corticomotor threshold is reduced in early sporadic amyotrophic lateral sclerosis. *Muscle & Nerve* 20:1137–1141.

Mochizuki, H., Huang, Y. Z., and Rothwell, J. C. 2004. Interhemispheric interaction between human dorsal premotor and contralateral primary motor cortex. *Journal of Physiology* 561:331–338.

Modugno, N., Nakamura, Y., Bestmann, S., Currà, A., Berardelli, A., and Rothwell, J. C. 2002. Neurophysiological investigations in patients with primary writing tremor. *Movement Disorders* 17:1336–1340.

Morgante, F., Dattola, V., Crupi, D. et al. 2011. Is central fatigue in multiple sclerosis a disorder of movement preparation? *Journal of Neurology* 258:263–272.

Murray, N. M. F., Hess, C. W., Mills, K. R., Schriefer, T. N., and Smith, S. J. M. 1987. Proximal facial nerve conduction using magnetic stimulation. *Electroencephalography and Clinical Neurophysiology* 66:S71.

Nakamura, H., Kitagawa, H., Kawaguchi, Y., and Tsuji, H. 1997. Intracortical facilitation and inhibition after transcranial magnetic stimulation in conscious humans. *Journal of Physiology* 498:817–823.

Nardone, R., Florio, I., Lochner, P., and Tezzon, F. 2005. Cholinergic cortical circuits in Parkinson's disease and in progressive supranuclear palsy: A transcranial magnetic stimulation study. *Experimental Brain Research* 163:128–131.

Netz, J., Lammers, T., and Homberg, V. 1997. Reorganization of motor output in the nonaffected hemisphere after stroke. *Brain* 120:1579–1586.

Ni, Z., Gunraj, C., Nelson, A. J., Yeh, I. J., Castillo, G., Hoque, T., and Chen, R. 2009. Two phases of interhemispheric inhibition between motor related cortical areas and the primary motor cortex in human. *Cerebral Cortex* 19:1654–1665.

Niehaus, L., Alt-Stutterheim, K., Roricht, S., and Meyer, B. U. 2001. Abnormal postexcitatory and interhemispheric motor cortex inhibition in writer's cramp. *Journal of Neurology* 248:51–56.

Niehaus, L., Bajbouj, M., and Meyer, B. U. 2003. Impact of interhemispheric inhibition on excitability of the non-lesioned motor cortex after acute stroke. *Supplements to Clinical Neurophysiology* 56:181–186.

Oliveri, M., Koch, G., Torriero, S., and Caltagirone, C. 2005. Increased facilitation of the motor cortex following 1 Hz repetitive transcranial magnetic stimulation of the contralateral cerebellum in normal humans. *Neuroscience Letters* 376:188–193.

Olivier, E., Baker, S. N., and Lemon, R. N. 2002. Comparison of direct and indirect measurements of the central motor conduction time in the monkey. *Clinical Neurophysiology* 113:469–477.

Orth, M., Snijders, A. H., and Rothwell, J. C. 2003. The variability of intracortical inhibition and facilitation. *Clinical Neurophysiology* 114:2362–2369.

Ozturk, V., Cakmur, R., Donmez, B., Yener, G. G., Kursad, F., and Idiman, F. 2002. Comparison of cortical excitability in chronic migraine (transformed migraine) and migraine without aura. A transcranial magnetic stimulation study. *Journal of Neurology* 249:1268–1271.

Palmieri, M. G., Iani, C., Scalise, A. et al. 1999. The effect of benzodiazepines and flumazenil on motor cortical excitability in the human brain. *Brain Research* 815:192–199.

Paradiso, G. O., Cunic, D. I., Gunraj, C. A., and Chen, R. 2005. Representation of facial muscles in human motor cortex. *Journal of Physiology* 567:323–336.

Pennisi, G., Alagona, G., Ferri, R., Greco, S., Santonocito, D., Pappalardo, A. et al. 2002. Motor cortex excitability in Alzheimer disease: One year followup study. *Neuroscience Letters* 329:293–296.

Pepin, J. L., Bogacz, D., De, P. V., and Delwaide, P. J. 1999. Motor cortex inhibition is not impaired in patients with Alzheimer's disease: Evidence from paired transcranial magnetic stimulation. *Journal of Neurological Science* 170:119–123.

Pierantozzi, M., Palmieri, M. G., Marciani, M. G., Bernardi, G., Giacomini, P., and Stanzione, P. 2001. Effect of apomorphine on cortical inhibition in Parkinson's disease patients: A transcranial magnetic stimulation study. *Experimental Brain Research* 141:52–62.

Pinto, A.D. and Chen, R. 2001. Suppression of the motor cortex by magnetic stimulation of the cerebellum. *Experimental Brain Research* 140:505–510.

Popa, T., Russo, M., and Meunier, S. 2010. Long-lasting inhibition of cerebellar output. *Brain Stimulation* 3:161–169.

Pouget, J., Trefouret, S., and Attarian, S. 2000. Transcranial magnetic stimulation (TMS): Compared sensitivity of different motor response parameters in ALS. *Amyotrophic Lateral Sclerosis and Other Motor Neuron Disorders* 1 (Suppl 2):S45–S49.

Priori, A., Berardelli, A., Inghilleri, M., Accornero, N., and Manfredi, M. 1994a. Motor cortical inhibition and the dopaminergic system. Pharmacological changes in the silent period after transcranial magnetic brain stimulation in normal subjects, patients with Parkinson's disease and drug induced parkinsonism. *Brain* 117:317–323.

Priori, A., Berardelli, A., Inghilleri, M., Polidori, L., and Manfredi, M. 1994b. Electromyographic silent period after transcranial brain-stimulation in Huntington's disease. *Movement Disorders* 9:178–182.

Prout, A. J. and Eisen, A. A. 1994. The cortical silent period and amyotrophic lateral sclerosis. *Muscle & Nerve* 17:217–223.

Quartarone, A., Rizzo, V., Terranova, C. et al. 2006a. Abnormal sensorimotor plasticity in organic but not in psychogenic dystonia. *Brain* 132:2871–2877.

Quartarone, A., Rizzo, V., Terranova, C. et al. 2009. Abnormal sensorimotor plasticity in organic but not in psychogenic dystonia. *Brain* 132:2871–2877.

Quartarone, A., Siebner, H. R., and Rothwell, J. C. 2006b. Task-specific hand dystonia: Can too much plasticity be bad for you? *Trends in Neurosciences* 29:192–199.

Ravnborg, M., Christiansen, P., Larsson, H., and Sorensen, P. S. 1992. The diagnostic reliability of magnetically evoked motor potentials in multiple-sclerosis. *Neurology* 42:1296–1301.

Restivo, D. A., Giuffrida, S., Rapisarda, G. et al. 2000. Central motor conduction to lower limb after transcranial magnetic stimulation in spinocerebellar ataxia type 2 (SCA2). *Clinical Neurophysiology* 111:630–635.

Restivo, D. A., Lanza, S., Saponara, R., Rapisarda, G., Giuffrida, S., and Palmeri, A. 2002. Changes of cortical excitability of human motor cortex in spinocerebellar ataxia type 2. A study with paired transcranial magnetic stimulation. *Journal of Neurological Science* 198:87–92.

Reutens, D. C. and Berkovic, S. F. 1992. Increased cortical excitability in generalised epilepsy demonstrated with transcranial magnetic stimulation. *Lancet* 339:362–363.

Reutens, D. C., Berkovic, S. F., Macdonell, R. A. L., and Baldin, P. F. 1993a. Magnetic stimulation of the brain in generalized epilepsy: Reversal of cortical hyperexcitability by anticonvulsants. *Annals of Neurology* 34:351–355.

Reutens, D. C., Puce, A., and Berkovic, S. F. 1993b. Cortical hyperexcitability in progressive myoclonus epilepsy: A study with transcranial magnetic stimulation. *Neurology* 43:186–192.

Reynolds, C. and Ashby, P. 1999. Inhibition in the human motor cortex is reduced just before a voluntary contraction. *Neurology* 53:730–735.

Richardson, M. P. and Lopes da Silva, F. H. 2011. TMS studies of preictal cortical excitability change *Epilepsy Research* 97:273–277.

Ridding, M. C., Inzelberg, R., and Rothwell, J. C. 1995a. Changes in excitability of motor cortical circuitry in patients with Parkinson's disease. *Annals of Neurology* 37:181–188.

Ridding, M. C. and Rothwell, J. C. 1997. Stimulus/response curves as a method of measuring motor cortical excitability in man. *Electroencephalography and Clinical Neurophysiology* 105:340–344.

Ridding, M. C., Sheean, G., Rothwell, J. C., Inzelberg, R., and Kujirai, T. 1995b. Changes in the balance between motor cortical excitation and inhibition in focal, task specific dystonia. *Journal of Neurology, Neurosurgery and Psychiatry* 39:493–498.

Ridding, M. C., Taylor, J. L., and Rothwell, J. C. 1995c. The effect of voluntary contraction on cortico-cortical inhibition in human motor cortex. *Journal of Physiology* 487:541–548.

Rimpilainen, I., Karma, P., Eskola, H., and Hakkinen, V. 1992. Magnetic facial nerve stimulation in normal subjects. Three groups of responses. *Acta Oto-Laryngologica Supplementum* 492:99–102.

Roick, H., von Giesen, H. J., and Benecke, R. 1993. On the origin of the postexcitatory inhibition seen after transcranial magnetic brain stimulation in awake human subjects. *Experimental Brain Research* 94:489–498.

Romeo, S., Berardelli, A., Pedace, F., Inghilleri, M., Giovannelli, M., and Manfredi, M. 1998. Cortical excitability in patients with essential tremor. *Muscle & Nerve* 21:1304–1308.

Romeo, S., Gilio, F., Pedace, F. et al. 2000. A changes in the cortical silent period after repetitive magnetic stimulation of cortical motor areas. *Experimental Brain Research* 135:504–510.

Rosenkranz, K. and Rothwell, JC. 2004. The effect of sensory input and attention on the sensorimotor organization of the hand area of the human motor cortex. *Journal of Physiology* 561:307–320.

Roshan, L., Paradiso, G. O., and Chen, R. 2003. Two phases of short-interval intracortical inhibition. *Experimental Brain Research* 151:330–337.

Rosler, K. M., Hess, C. W., and Schmid, U. D. 1989. Investigation of facial motor pathways by electrical and magnetic stimulation: Sites and mechanisms of excitation. *Journal of Neurology, Neurosurgery and Psychiatry* 52:1149–1156.

Rosler, K. M. and Magistris, M. R. 2004. Triple stimulation technique (TST) in amyotrophic lateral sclerosis. *Clinical Neurophysiology* 115:1715.

Rosler, K. M., Magistris, M. R., Glocker, F. X., Kohler, A., Deuschl, G., and Hess, C. W. 1995. Electrophysiological characteristics of lesions in facial palsies of different etiologies. A study using electrical and magnetic stimulation techniques. *Electroencephalography and Clinical Neurophysiology* 97:355–368.

Rosler, K. M., Petrow, E., Mathis, J., Aranyi, Z., Hess, C. W., and Magistris, M. R. 2002. Effect of discharge desynchronization on the size of motor evoked potentials: An analysis. *Clinical Neurophysiology* 113:1680–1687.

Rosler, K. M., Truffert, A., Hess, C. W., and Magistris, M. R. 2000. Quantification of upper motor neuron loss in amyotrophic lateral sclerosis. *Clinical Neurophysiology* 111:2208–2218.

Rossini, P. M., Barker, A. T., Berardelli, A. et al. 1994. Non-invasive electrical and magnetic stimulation of the brain, spinal cord and roots: Basic principles and procedures for routine clinical application. Report of an IFCN committee. *Electroencephalography and Clinical Neurophysiology* 91:79–92.

Rossini, P. M., Caramia, M. D., and Zarola, F. 1987. Mechanisms of nervous propagation along central motor pathways: Noninvasive evaluation in healthy subjects and in patients with neurological disease. *Neurosurgery* 20:183–191.

Rossini, P. M. and Rossi, S. 1998. Clinical applications of motor evoked potentials. *Electroencephalography and Clinical Neurophysiology* 106:180–194.

Rossini, P. M., Zarola, F., Floris, R. et al. 1989. Sensory (VEP, BAEP, SEP) and motor-evoked potentials, liquoral and magnetic resonance findings in multiple sclerosis. *European Journal of Neurology* 29:41–47.

Roth, G. and Magistris, M. R. 1989. Identification of motor conduction block despite desynchronisation. A method. *Electromyography and Clinical Neurophysiology* 29:305–313.

Roth, B. J., Saypol, J. M., Hallett, M., and Cohen, L. G. 1991. A theoretical calculation of the electric field induced in the cortex during magnetic stimulation. *Electroencephalography and Clinical Neurophysiology* 81:47–56.

Rothwell, J. C. 2010. Using transcranial magnetic stimulation methods to probe connectivity between motor areas of the brain. *Human Movement Science* 30:906–915.

Roy, F. D. and Gorassini, M. A. 2008. Peripheral sensory activation of cortical circuits in the leg motor cortex of man. *Journal of Physiology* 586:4091–4105.

Ruge, D., Cif, L., Limousin, P. et al. 2011. Shaping reversibility? Long-term deep brain stimulation in dystonia: The relationship between effects on electrophysiology and clinical symptoms. *Brain* 134:2106–2115.

Sailer, A., Cunic, D. I., Paradiso, G. O. et al. 2007. Subthalamic deep brain stimulation modulates afferent inhibition in Parkinson's disease. *Neurology* 68:356–364.

Sailer, A., Molnar, G. F., Paradiso, G., Gunraj, C. A., Lang, A. E., and Chen, R. 2003. Short and long latency afferent inhibition in Parkinson's disease. *Brain* 126:1883–1894.

Salerno, A., Thomas, E., Olive, P., Blotman, F., Picot, M. C., and Georgesco, M. 2000. Motor cortical dysfunction disclosed by single and double magnetic stimulation in patients with fibromyalgia. *Clinical Neurophysiology* 111:994–1001.

Schafer, M., Biesecker, J. C., Schulze-Bonhage, A., and Ferbert, A. 1997. Transcranial magnetic double stimulation: Influence of the intensity of the conditioning stimulus. *Electroencephalography and Clinical Neurophysiology* 105:462–469.

Schieppati, M., Trompetto, C., and Abbruzzese, G. 1996. Selective facilitation of responses to cortical stimulation of proximal and distal arm muscles by precision tasks in man. *Journal of Physiology* 491:551–562.

Schmid, U. D., Moller, A. R., and Schmid, J. 1991. Transcranial magnetic stimulation excites the labyrinthine segment of the facial nerve: An intraoperative electrophysiological study in man. *Neuroscience Letters* 124:273–276.

Schmid, U. D., Moller, A. R., and Schmid, J. 1992. Transcranial magnetic stimulation of the facial nerve: Intraoperative study on the effect of stimulus parameters on the excitation site in man. *Muscle & Nerve* 15:829–836.

Schmierer, K., Irlbacher, K., Grosse, P., Roricht, S., and Meyer, B. U. 2002. Correlates of disability in multiple sclerosis detected by transcranial magnetic stimulation. *Neurology* 59:1218–1224.

Schmierer, K., Niehaus, L., Roricht, S., and Meyer, B. U. 2000. Conduction deficits of callosal fibres in early multiple sclerosis. *Journal of Neurology, Neurosurgery and Psychiatry* 68:633–638.

Schriefer, T. N., Mills, K. R., Murray, N. M., and Hess, C. W. 1988. Evaluation of proximal facial nerve conduction by transcranial magnetic stimulation. *Journal of Neurology, Neurosurgery and Psychiatry* 51:60–66.

Schulze-Bonhage, A., Knott, H., and Ferbert, A. 1996. Effects of carbamazepine on cortical excitatory and inhibitory phenomena: A study with paired transcranial magnetic stimulation. *Electroencephalography and Clinical Neurophysiology* 99:267–273.

Schwenkreis, P., Janssen, F., Rommel, O. et al. 2003. Bilateral motor cortex disinhibition in complex regional pain syndrome (CRPS) type I of the hand. *Neurology* 61:515–519.

Schwenkreis, P., Liepert, J., Witscher, K. et al. 2000. Riluzole suppresses motor cortex facilitation in correlation to its plasma level. A study using transcranial magnetic stimulation. *Experimental Brain Research* 135:293–299.

Schwenkreis, P., Tegenthoff, M., Witscher, K. et al. 2002. Motor cortex activation by transcranial magnetic stimulation in ataxia patients depends on the genetic defect. *Brain* 125:301–309.

Schwenkreis, P., Witscher, K., Janssen, F. et al. 1999. Influence of the N-methyl-D-aspartate antagonist memantine on human motor cortex excitability. *Neuroscience Letters* 270:137–140.

Shahani, B. T. and Young, R. R. 1973. Studies of the normal human silent period. In: *New Developments in Electromyography and Clinical Neurophysiologyogy*. ed. J. Desmedt, pp. 589–602. Basel, Switzerland: Karger.

Shimizu, T., Hosaki, A., Hino, T. et al. 2002. Motor cortical disinhibition in the unaffected hemisphere after unilateral cortical stroke. *Brain* 125:1896–1907.

Siebner, H. R., Dressnandt, J., Auer, C., and Conrad, B. 1998. Continuous intrathecal baclofen infusions induced a marked increase of the transcranially evoked silent period in a patient with generalized dystonia. *Muscle & Nerve* 21:1209–1215.

Siebner, H. R., Mentschel, C., Auer, C., Lehner, C., and Conrad, B. 2000. Repetitive transcranial magnetic stimulation causes a short-term increase in the duration of the cortical silent period in patients with Parkinson's disease. *Neuroscience Letters* 284:147–150.

Smith, M. J., Keel, J. C., Greenberg, B. D. et al. 1999. Menstrual cycle effects on cortical excitability. *Neurology* 53:2069–2072.

Sohn, Y. H. and Hallett, M. 2004. Disturbed surround inhibition in focal hand dystonia. *Annals of Neurology* 56:595–599.

Sohn, Y. H., Wiltz, K., and Hallett, M. 2002. Effect of volitional inhibition on cortical inhibitory mechanisms. *Journal of Neurophysiology* 88:333–338.

Sommer, M., Ruge, D., Tergau, F., Beuche, W., Altenmuller, E., and Paulus, W. 2002. Intracortical excitability in the hand motor representation in hand dystonia and blepharospasm. *Movement Disorders* 17:1017–1025.

Stinear, C. M. and Byblow, W. D. 2003. Role of intracortical inhibition in selective hand muscle activation. *Journal of Neurophysiology* 89:2014–2020.

Stinear, C. M. and Byblow, W. D. 2004. Elevated threshold for intracortical inhibition in focal hand dystonia. *Movement Disorders* 19:1312–1317.

Strafella, A., Ashby, P., Lozano, A., and Lang, A. E. 1997. Pallidotomy increases cortical inhibition in Parkinson's disease. *Canadian Journal of Neurological Science* 24:133–136.

Tamburin, S., Fiaschi, A., Andreoli, A., Marani, S., Manganotti, P., and Zanette, G. 2004. Stimulus–response properties of motor system in patients with cerebellar ataxia. *Clinical Neurophysiology* 115:348–355.

Tassinari, C. A., Cincotta, M., Zaccara, G., and Michelucci, R. 2003. Transcranial magnetic stimulation and epilepsy. *Clinical Neurophysiology* 114:777–798.

Tegenthoff, M., Vorgerd, M., Juskowiak, F., Roos, V., and Malin, J. P. 1996. Postexcitatory inhibition after transcranial magnetic single and double brain stimulation in Huntington's disease. *Electroencephalography and Clinical Neurophysiology* 101:298–303.

Tergau, F., Wanschura, V., Canelo, M. et al. 1999. Complete suppression of voluntary motor drive during the silent period after transcranial magnetic stimulation. *Experimental Brain Research* 124:447–454.

Thickbroom, G. W., Byrnes, M. L., Archer, S. A., and Mastaglia, F. L. 2004. Motor outcome after subcortical stroke correlates with the degree of cortical reorganization. *Clinical Neurophysiology* 115:2144–2150.

Thompson, S. M. and Gahwiler, B. H. 1992. Effects of the GABA uptake inhibitor tiagabine on inhibitory synaptic potentials in rat hippocampal slice cultures. *Journal of Neurophysiology* 67:1698–1701.

Tokimura, H., Di Lazzaro, V., Tokimura, Y. et al. 2000. Short latency inhibition of human hand motor cortex by somatosensory input from the hand. *Journal of Physiology* 523:503–513.

Tokimura, H., Ridding, M. C., Tokimura, Y., Amassian, V. E., and Rothwell, J. C. 1996. Short latency facilitation between pairs of threshold magnetic stimuli applied to human motor cortex. *Electroencephalography and Clinical Neurophysiology* 103:263–272.

Tomberg, C. and Caramia, M. D. 1991. Prime mover muscle in finger lift or finger flexion reaction times: Identification with transcranial magnetic stimulation. *Electroencephalography and Clinical Neurophysiology* 81:319–322.

Torriero, S., Oliveri, M., Koch, G. et al. 2011. Changes in cerebello-motor connectivity during procedural learning by actual execution and observation. *Journal of Cognitive Neuroscience* 23:338–348.

Travlos, A., Pant, B., and Eisen, A. 1992. Transcranial magnetic stimulation for detection of preclinical cervical spondylotic myelopathy. *Archives of Physical Medicine and Rehabilitation* 73:442–446.

Triggs, W. J., Cros, D., Macdonell, R. A. L., Chiappa, K. H., Fang, J., and Day, B. J. 1993. Cortical and spinal motor excitability during the transcranial magnetic stimulation silent period in humans. *Brain Research* 628:39–48.

Triggs, W. J. and Edgara M. A. 1995. Case records of the Massachusetts General Hospital. Weekly clinicopathological exercises. Case 36–1995. A 61-year-old man with increasing weakness and atrophy of all extremities. *New England Journal of Medicine* 333:1406–1412.

Triggs, W. J., Menkes, D., Onorato, J. et al. 1999. Transcranial magnetic stimulation identifies upper motor neuron involvement in motor neuron disease. *Neurology* 53:605–611.

Trompetto, C., Assini, A., Buccolieri, A., Marchese, R., and Abbruzzese, G. 2000. Motor recovery following stroke: A transcranial magnetic stimulation study. *Clinical Neurophysiology* 111:1860–1867.

Trompetto, C., Bove, M., Marinelli, L., Avanzino, L., Buccolieri, A., and Abbruzzese, G. 2004. Suppression of the transcallosal motor output: A transcranial magnetic stimulation study in healthy subjects. *Experimental Brain Research* 158:133–140.

Trompetto, C., Buccolieri, A., Marchese, R., Marinelli, L., Michelozzi, G., Abbruzzese, G. 2003. Impairment of transcallosal inhibition in patients with corticobasal degeneration. *Clinical Neurophysiology* 114:2181–2187.

Trompetto, C., Caponnetto, C., Buccolieri, A., Marchese, R., and Abbruzzese, G. 1998 Responses of masseter muscles to transcranial magnetic stimulation in patients with amyotrophic lateral sclerosis. *Electroencephalography and Clinical Neurophysiology* 109:309–314.

Truffert, A., Rosler, K. M., and Magistris, M. R. 2000. Amyotrophic lateral sclerosis versus cervical spondylotic myelopathy: A study using transcranial magnetic stimulation with recordings from the trapezius and limb muscles. *Clinical Neurophysiology* 111:1031–1038.

Turk, U., Rosler, K. M., Mathis, J., Mullbacher, W., and Hess, C. W. 1994. Assessment of motor pathways to masticatory muscles: An examination technique using electrical and magnetic stimulation. *Muscle & Nerve* 17:1271–1277.

Turton, A., Wroe, S., Trepte, N., Fraser, C., and Lemon, R. N. 1996. Contralateral and ipsilateral EMG responses to transcranial magnetic stimulation during recovery of arm and hand function after stroke. Electroencephalogr *Clinical Neurophysiology* 101:316–328.

Ugawa, Y., Day, B. L., Rothwell, J. C., Thompson, P. D., Merton, P. A., and Marsden, C. D. 1991. Modulation of motor cortical excitability by electrical stimulation over the cerebellum in man. *Journal of Physiology* 441:57–72.

Ugawa, Y., Hanajima, R., and Kanazawa, I. 1994. Motor cortex inhibition in patients with ataxia. *Electroencephalography and Clinical Neurophysiology* 93:225–229.

Ugawa, Y., Terao, Y., Hanajima, R. et al. 1997. Magnetic stimulation over the cerebellum in patients with ataxia. *Electroencephalography and Clinical Neurophysiology* 104:453–458.

Ugawa, Y., Uesaka, Y., Terao, Y., Hanajima, R., and Kanazawa, I. 1995. Magnetic stimulation over the cerebellum in humans. *Annals of Neurology* 37:703–713.

Urban, P. P. 2003. Transcranial magnetic stimulation in brainstem lesions and lesions of the cranial nerves. *Supplements to Clinical Neurophysiology* 56:341–357.

Urban, P. P., Vogt, T., and Hopf, H. C. 1998. Corticobulbar tract involvement in amyotrophic lateral sclerosis. A transcranial magnetic stimulation study. *Brain* 121:1099–1108.

Valls-Sole, J., Pascual-Leone, A., Brasil-Neto, J. P., Cammarota, A., McShane, L., and Hallett, M. 1994. Abnormal facilitation of the response to transcranial magnetic stimulation in patients with Parkinson's disease. *Neurology* 44:735–741.

Valls-Sole, J., Pascual-Leone, A., Wassermann, E. M., and Hallett, M. 1992. Human motor evoked responses to paired transcranial magnetic stimuli. *Electroencephalography and Clinical Neurophysiology* 85:355–364.

Valzania, F., Quatrale, R., Strafella, A. P., Bombardi, R., Santangelo, M., Tassinari, C. A., and De Grandis, D. 1994. Pattern of motor evoked response to repetitive transcranial magnetic stimulation. *Electroencephalography and Clinical Neurophysiology* 93:312–317.

Valzania, F., Strafella, A. P., Quatrale, R. et al. 1997. Motor evoked responses to paired cortical magnetic stimulation in Parkinson's disease. *Electroencephalography and Clinical Neurophysiology* 105:37–43.

Valzania, F., Strafella, A. P., Tropeani, A., Rubboli, G., Nassetti, S. A., and Tassinari, C. A. 1999. Facilitation of rhythmic events in progressive myoclonus epilepsy: A transcranial magnetic stimulation study. *Clinical Neurophysiology* 110:152–157.

Van Der Kamp, W., Maertens de Noordhout, A., Thompson, P. D., Rothwell, J. C., Day, B. L., and Marsden, C. D. 1991. Correlation of phasic muscle strength and corticomotoneuron conduction time in multiple sclerosis. *Annals of Neurology* 29:6–12.

Wassermann, E. M., Fuhr, P., Cohen, L. G., and Hallett, M. 1991. Effects of transcranial magnetic stimulation on ipsilateral muscles. *Neurology* 41:1795–1799.

Wassermann, E. M., Greenberg, B. D., Nguyen, M. B., and Murphy, D. L. 2001. Motor cortex excitability correlates with an anxiety-related personality trait. *Biological Psychiatry* 50:377–382.

Wassermann, E. M., Pascual-Leone, A., Valls-Sole, J., Toro, C., Cohen, L. G., and Hallett, M. 1993. Topography of the inhibitory and excitatory responses to transcranial magnetic stimulation in a hand muscle. *Electroencephalography and Clinical Neurophysiology* 89:424–433.

Weber, M., Eisen, A., Stewart, H., and Hirota, N. 2000. The split hand in ALS has a cortical basis. *Journal of Neurological Science* 180:66–70.

Weise, D., Schramm A., Stefan K. et al. 2006. The two sides of associative plasticity in writer's cramp. *Brain* 129:2709–2721.

Werhahn, K. J., Kunesch, E., Noachtar, S., Benecke, R., and Classen, J. 1999. Differential effects on motorcortical inhibition induced by blockade of GABA uptake in humans. *Journal of Physiology* 517:591–597.

Werhahn, K. J., Taylor, J., Ridding, M., Meyer, B. U., and Rothwell, J. C. 1996. Effect of transcranial magnetic stimulation over the cerebellum on the excitability of human motor cortex. *Electroencephalography and Clinical Neurophysiology* 101:58–66.

Werhahn, K. J., Wiseman, K., Herzog, J., Forderreuther, S., Dichgans, M., and Straube, A. 2000. Motor cortex excitability in patients with migraine with aura and hemiplegic migraine. *Cephalalgia* 20:45–50.

Wright, M. A., Orth, M., Patsalos, P. N., Smith, S. J., and Richardson, M. P. 2006. Cortical excitability predicts seizures in acutely drug-reduced temporal lobe epilepsy patients. *Neurology* 67:1646–1651.

Yokota, T., Yoshino, A., Inaba, A., and Saito, Y. 1996. Double cortical stimulation in amyotrophic lateral sclerosis. *Journal of Neurology, Neurosurgery and Psychiatry* 61:596–600.

Young, W. B., Oshinsky, M. L., Shechter, A. L., Gebeline-Myers, C., Bradley, K. C., and Wassermann, E. M. 2004. Consecutive transcranial magnetic stimulation: Phosphene thresholds in migraineurs and controls. *Headache* 44:131–135.

Young, M. S., Triggs, W. J., Bowers, D., Greer, M., and Friedman, W. A. 1997. Stereotactic pallidotomy lengthens the transcranial magnetic cortical stimulation silent period in Parkinson's disease. *Neurology* 49:1278–1283.

Zanette, G., Tamburin, S., Manganotti, P., Refatti, N., Forgione, A., and Rizzuto, N. 2002. Changes in motor cortex inhibition over time in patients with amyotrophic lateral sclerosis. *Journal of Neurology* 249:1723–1728.

Ziemann, U. 2004. TMS and drugs. *Clinical Neurophysiology* 115:1717–1729.

Ziemann,U., Bruns, D., and Paulus, W. 1996a. Enhancement of human motor cortex inhibition by the dopamine receptor agonist pergolide: Evidence from transcranial magnetic stimulation. *Neuroscience Letters* 208:187–190.

Ziemann, U., Chen, R., Cohen, L. G., and Hallett, M. 1998a. Dextromethorphan decreases the excitability of the human motor cortex. *Neurology* 51:1320–1324.

Ziemann, U., Hallett, M., and Cohen, L. G. 1998b. Mechanisms of deafferentation-induced plasticity in human motor cortex. *Journal of Neuroscience* 18:7000–7007.

Ziemann, U., Lonnecker, S., and Paulus, W. 1995. Inhibition of human motor cortex by ethanol. A transcranial magnetic stimulation study. *Brain* 118:1437–1446.

Ziemann, U., Lönnecker, S., Steinhoff, B. J., and Paulus, W. 1996b. Effects of antiepileptic drugs on motor cortex excitability in humans: A transcranial magnetic stimulation study. *Annals of Neurology* 40:367–378.

Ziemann, U., Lonnecker, S., Steinhoff, B. J., and Paulus, W. 1996c. The effect of lorazepam on the motor cortex excitability in man. *Experimental Brain Research* 109:127–135.

Ziemann, U., Rothwell, J. C., and Ridding, M. C. 1996d. Interaction between intracortical inhibition and facilitation in human motor cortex. *Journal of Physiology* 496:873–881.

Ziemann, U., Steinhoff, B. J., Tergau, F., and Paulus, W. 1998c. Transcranial magnetic stimulation: Its current role in epilepsy research. *Epilepsy Research* 30:11–30.

Ziemann, U., Tergau, F., Bruns, D., Baudewig, J., and Paulus, W. 1997a. Changes in human motor cortex excitability induced by dopaminergic and antidopaminergic drugs. *Electroencephalography and Clinical Neurophysiology* 105:430–437.

Ziemann, U., Tergau, F., Wassermann, E. M., Wischer, S., Hildebrandt, J., and Paulus, W. 1998d. Demonstration of facilitatory I wave interaction in the human motor cortex by paired transcranial magnetic stimulation. *Journal of Physiology* 511:181–190.

Ziemann, U., Tergau, F., Wischer, S., Hildebrandt, J., and Paulus, W. 1998e. Pharmacological control of facilitatory I-wave interaction in the human motor cortex. A paired transcranial magnetic stimulation study. *Electroencephalography and Clinical Neurophysiology* 109:321–330.

Ziemann, U., Winter, M., Reimers, C. D., Reimers, K., Tergau, F., and Paulus, W. 1997b. Impaired motor cortex inhibition in patients with amyotrophic lateral sclerosis. Evidence from paired transcranial magnetic stimulation. *Neurology* 49:1292–1298.

Zoghi, M., Pearce, S. L., and Nordstrom, M. A. 2003. Differential modulation of intracortical inhibition in human motor cortex during selective activation of an intrinsic hand muscle. *Journal of Physiology* 550:933–946.

8 Neurophysiological (Mainly Transcranial Magnetic Stimulation) Techniques to Test Functional Neuroanatomy of Cortico-Cortical Connectivity

Paolo M. Rossini and Michael Charles Ridding

CONTENTS

Scientific evidences collected along a continuous research track in the last 150 years suggest that human behavior is characterized by engagement of functional distributed networks within the brain. Such network activation is especially apparent for higher functions including memory, action planning, and abstract reasoning. These networks dynamically connect adjacent and/or remote cortical neuronal assemblies via cortico-cortical connections.

In the most evolved nervous systems, cortico-cortical connectivity can be described with the *connectome* term in strict analogy with the genome concept. In the human brain, this connectome concept is strongly linked with the functional definition of which neuronal populations can interact with each other and on the directness and strength of their connections. Within this theoretical framework, cerebral networks can be classified into three organizational levels: (1) the individual neuron with its synaptic connections (microscale); (2) neuronal populations and assemblies and their interconnections (mesoscale); (3) anatomically distinct brain regions and their interregional connections (macroscale), defined as interregional connections between cortical regions. Within this context, techniques useful for examining network connectivity require a spatial resolution of several millimeters to several centimeters.

The main aim of this chapter is to describe the neurophysiological techniques available to investigate the level of connectivity of the brain within the macroscale scenario after having provided a brief introduction to some general ideas about network modeling, a useful basis on which to consider real neurophysiological data as well as future contributions from such technical possibilities.

A complex brain network can be represented by a graph G with N nodes and K connections, where nodes are modeling brain regions and connections are the interconnectivities linking such regions. Graphs are undirected (no preferential directional connectivity) and unweighted (or binary if every graph connection has a weight equal to 1). The graph also has a clustering coefficient (Cp) and individual interconnectional path length (Lp), which describe fundamentals characterizing its local architecture and the global network. The Cp is somehow reflecting the efficiency of the local information transfer (Latora and Marchiori, 2001) while the Lp quantifies the efficiency of information on parallel propagation within the network (Latora and Marchiori, 2001). By means of these two parameters, three types of networks can be distinguished: (a) regular with both high Cp and Lp due to lack of long-distance connectivity,

(b) random with a low Cp and a small Lp, and (c) small world in which not only nodes are more locally interconnected than in (b) type, but also showing a smaller minimal distance separating each node couple than in (a) type. Altogether, the vast majority of the neural networks follows the power's law, with a large number of nodes having very few connections and few nodes (hubs) having a large number of links. This represents a scale-free class network, which is quite resistant in case of damage of random nodes, while it is vulnerable when highly connected nodes are destroyed.

8.1   NEUROANATOMY (STRUCTURAL AND FUNCTIONAL) OF BRAIN CONNECTIVITY

Schulz and Braitenberg showed that there are three categories of cortico-cortical connections in the human brain: (1) intracortical connections (which represent the majority, are on the order of 0.1–0.5 cm, and involve collateral axonal connections that do not enter the cerebral white matter); (2) "U"-shaped myelinated fibers (the majority of the cerebral white matter that connect cortical gyri and sulci and are on the order of 0.3–3 cm); and (3) deeply located long-distance fiber systems (which are fasciculi with connections from 3 to 15 cm that represent approximately 4% of the cerebral white matter). This anatomical scaffold supports two different types of connectivities in the "working brain":

- Functional connectivity, which mainly represents a statistical concept based on the patterned deviation from the statistical independence among distributed neuronal units that are often spatially remote. Statistical interdependence can be measured on the basis of correlation, or covariance, or phase-locking of the activity of neuronal assemblies (either in their metabolic/blood flow fluctuations or in their rhythmic firing producing electroencephalographic [EEG] signals). In contrast to anatomical connectivity, this type is strictly time dependent.
- Effective connectivity, which represents a liaison between the previous types (anatomical and functional) and describes the causative interrelations between two neuronal systems activities. Here a causative model obtainable via experimental perturbations and/or the observation of the temporal order of the neuronal events is required.

The structural/anatomical and input/output connections of a given brain region are also the main constraints for its functional properties. Meanwhile, functional interactions contribute to modify the underlying structural/anatomical substrate by modifying the synaptic connections (enlarging, forming new synapses, pruning preexisting synapses).

Due to the distributed networks underlying sensory systems, sensory processing (e.g., object representation) requires a high degree of integration of information flow arriving from several, different, and often remote brain sources. Such intramodal integration must also be enriched by intermodal integrations combining visual, acoustic, olfactive, and tactile properties characterizing individual objects; the whole process is orchestrated by attention, which puts in order the individual sensory elements in a context-dependent hierarchical order. Finally (for instance, when considering the flexibility of sensory-motor coordination), a dynamic "binding" and "unbinding" between cooperating neuronal assemblies (i.e., sensory and motor cortical areas) is also required. Within this framework, transient synchronization of neuronal firing has been proposed as one of the most significant mechanisms for determining such rapidly time-varying binding/unbinding phenomenon, which can orchestrate the functional and dynamic linkage of separate and widely distributed neuronal assemblies within a unique and functionally coherent group.

Specific methods have been developed to identify transient functional coupling of distributed neuronal assemblies via EEG/MEG recordings, some of which can evaluate synchrony independently from the amplitude of the signal and differ from measurements of coherence, which determines the covariance of signal amplitude recorded at different sites for various frequency bands (Vecchio et al., 2010).

The investigation of network connectivity has the capacity to provide valuable new insights into brain function. While the theoretical approach described earlier provides a useful setting in which to explore these ideas, the provision of real data is a critical step to gaining greater understanding.

8.2   NEUROPHYSIOLOGICAL METHODS FOR CONNECTIVITY INVESTIGATION

Recent technical advances have led to the development of a number of neurophysiological approaches that noninvasively probe connectivity within, and between, cortical regions in conscious human subjects. These approaches provide an opportunity to gain a better understanding of functional brain connectivity in behavioral settings and involve the use of single- and double-pulse transcranial magnetic stimulation (TMS) either in isolation or in combination with EEG recordings. The measures obtained with these techniques, which are dependent upon the underlying structural connectivity, have provided novel insights into functional connectivity during behavioral engagement. Such approaches provide information on the functional network connectivity both locally (mesoscale) and between more distant cortical target areas or nodes (macroscale).

8.2.1   TESTING CORTICO-CORTICAL CONNECTIVITY WITH TMS

The connectivity within, and between, various cortical regions can be probed using TMS techniques. Both single-pulse TMS and trains of TMS (rTMS) can be used to temporarily modify activity in a targeted cortical region. The effects of this modified activity can be examined by investigating function dependent on the targeted, or an interconnected, cortical region. For example, Huang et al. (2005) demonstrated that an inhibitory rTMS paradigm (continuous theta burst stimulation) applied to the motor cortex lengthened simple reaction time for a task performed with the contralateral hand. In addition, connectivity can be investigated by combining TMS and functional imaging techniques. Using this approach, it has been shown that TMS applied at one cortical site can modify activity across a wide cortical and subcortical network (Fox et al., 1997; Paus et al., 1997), an effect reflecting network connectivity.

As discussed in following text, the level of excitability within a network can be used as an indicator of connectivity and short-term changes in excitability are largely indicative of changes in synaptic efficacy, which can effectively modify connectivity (Rothwell, 2011). Over the last 20 years or so, a number of TMS techniques have been developed to probe the excitability of cortico-cortical connections. Using these approaches, it is possible to examine cortico-cortical connections between both inhibitory and excitatory networks and the output cells of the primary motor cortex under a number of different conditions (Di Lazzaro et al., 2008). These techniques have provided novel and important insights into the functioning of these circuits and their role in normal motor control as well as highlighting impairments in their function in a range of neurological conditions (Berardelli et al., 2008; Di Lazzaro et al., 2008).

We will now outline briefly the main single- and double-pulse TMS techniques available to examine cortico-cortical connectivity in human subjects. This review is not designed to be comprehensive but to give a brief overview of each technique and identify some key studies.

8.2.2 PROBING LOCAL INTRACORTICAL CONNECTIVITY

8.2.2.1 Single-Pulse Paradigms

A supra-threshold transcranial magnetic stimulus to the motor cortex activates a complex network of excitatory and inhibitory elements. The resultant-evoked descending volley in the pyramidal tract is complex and reflects the activation and connectivity of this cortical network. The descending volley comprises a series of waves. These waves have been termed, according to the experimental studies by Patton and Amassian (1954), I-waves (indirect waves) (Day et al., 1987; Di Lazzaro et al., 1998a). In addition, with higher stimulus intensities, it is possible to record an earlier component termed the direct wave (D-wave) (Di Lazzaro et al., 1998a). There is good evidence that the D-wave reflects direct activation, and the later I-waves trans-synaptic activation, of the corticospinal neurons (see Di Lazzaro et al., 2008). Although the exact circuitry generating I-waves has not been determined, it is likely to predominantly involve horizontally oriented networks of excitatory interneurons with synapses onto corticospinal neurons. In addition, the earliest I-wave (I1) probably is due to different mechanisms than the later I-waves (see Di Lazzaro et al., 2008). If the descending volley evoked by a transcranial magnetic stimulus is composed of an appropriate number of D- and I-waves of sufficient amplitude, a population of spinal alpha-motoneurons will be excited and their synchronous firing will result in a motor-evoked potential (MEP). Therefore, it should be always kept in mind that the presence and amplitude of an MEP are also influenced by the excitability of the spinal motoneurons targeted by the descending volley and the related intraspinal circuits (Rossini et al., 1985, 1987). As the descending volley critically influences the amplitude of the MEP, changes in synaptic efficacy within cortical circuits responsible for the generation of the descending volley will, therefore, have a major influence on the MEP. Therefore, it is possible to obtain a measure of cortico-cortical connectivity from MEP characteristics. The MEP has proven to be a very sensitive marker of changes in the excitability of the cortical network responsible for its generation. Usually, changes in connectivity (or excitability) are investigated using a single TMS test intensity that either evokes baseline MEPs of fixed amplitude or is set relative to a particular threshold (e.g., resting motor threshold). However, it is also useful to construct stimulus response curves where the amplitude of the MEP is investigated across a range of stimulus intensities (from below motor threshold to supra-threshold or even maximal stimulator's output (Ridding and Rothwell, 1997). This type of data can be analyzed using curve-fitting techniques (Devanne et al., 1997). While there are significant differences in the strength of intracortical connections between subjects (reflected by differences in the input/output characteristics of the stimulus response curve), changes to an individual's input/output curve will reflect modifications in the strength of connections within the cortical network generating the MEP. Many conditions result in changes in MEP amplitude. For example, motor training can result in significant increases in MEP amplitude in the trained muscles (Pascual-Leone et al., 1995; Muellbacher et al., 2001), a change that reflects use-dependent strengthening of synapses within the motor cortex. A number of pharmacological agents also influence MEP amplitude. For example, benzodiazepines (Boroojerdi et al., 2001; Kimiskidis et al., 2006) and barbiturates (Inghilleri et al., 1996), which are positive allosteric modulators of the $GABA_A$ receptor, decrease MEP amplitude. Also norepinephrine (NE) antagonists (Korchounov et al., 2003) can reduce the amplitude of larger MEPs.

The strength of connectivity within inhibitory circuits of M1 can be assessed by recording the cortical silent period (CSP), which is a TMS-induced interruption (lasting up to 200–300 ms) in the ongoing electromyography (EMG) seen during a tonic contraction of the target muscle. The later part of the CSP (>75 ms) reflects recruitment of a cortical inhibitory network that projects onto the corticospinal output cells (Ziemann et al., 1993). The strength of connectivity both within the inhibitory network and also between the inhibitory and output networks will be reflected in the duration of the CSP. The duration of the CSP increases in a linear fashion with stimulus intensity (Cantello et al., 1992) but is not strongly related to the amplitude of the MEP (Inghilleri et al., 1993; Triggs et al., 1993) or the level of background EMG (Inghilleri et al., 1993). The mechanisms responsible for the CSP are complex with evidence that both $GABA_A$ and $GABA_B$ receptors are involved and the relative balance of these two systems is dictated by the intensity of stimulation (see Paulus et al., 2008). The functional significance of the CSP is not completely understood. However, the inhibition reflected by this measure may have an important role in inter-limb movement coordination and in limiting inter-limb interferences (Sohn et al., 2005). Changes have been reported in the CSP in a number of conditions characterized by abnormal movement control (Valls-Sole et al., 1994; Siebner et al., 2000).

It is possible to examine connectivity in pathways that convey afferent input to the motor cortex by measuring short-latency afferent inhibition (SAI). Mariorenzi and colleagues (1991) demonstrated that an electrical stimulus to the median nerve at the wrist (or elbow) could inhibit MEPs in thumb flexor muscles when the interstimulus interval (ISI) between the peripheral stimulus and the TMS was 20–22 ms. Similarly, an electrical stimulus to the median nerve at the wrist inhibits TMS-evoked MEPs in the first dorsal interosseous muscle if the conditioning test interval is approximately 19–21 ms (Tokimura et al., 2000). SAI is due in large part to the inhibition of I-waves (Tokimura et al., 2000). Pharmacological studies have indicated that $GABA_A$ receptors are involved, albeit a different population than those responsible for short-interval intracortical inhibition (SICI; see Paulus et al., 2008). In addition, there is evidence that cholinergic mechanisms are also involved in SAI, with SAI being reduced by blockade of cholinergic receptors (Di Lazzaro et al., 2000). Although little reported, afferent input can also facilitate MEPs at appropriate conditioning test intervals. Mariorenzi et al. (1991) demonstrated that median nerve stimulation also facilitated MEPs in thumb flexor muscles when the ISI was approximately 23–25 ms. This type of facilitatory interaction might be important in the genesis of the long-lasting MEP enhancement seen with the paired-associative plasticity inducing paradigm (Stefan et al., 2000). SAI can also interact with circuitry responsible for SICI. Appropriately timed peripheral stimulation can reduce SICI and, also, appropriately timed SICI conditioning stimuli reduce SAI (Alle et al., 2009). It is likely that SAI and SICI involve two distinct, but reciprocally connected subtypes of GABAergic networks, which project onto corticospinal output neurons.

8.2.2.2 Paired-Pulse Paradigms

As discussed in the following text, a number of paired-pulse TMS paradigms have been developed in recent years that have provided important insights into the functioning of specific excitatory (Kujirai et al., 1993; Tokimura et al., 1996; Ziemann et al., 1998b) and inhibitory (Kujirai et al., 1993; Valls-Sole et al., 2002) networks within the cortex. The excitability of these networks can be used as a marker of connectivity at both a local level and also between more distant cortical regions.

8.2.2.3 Paired-Pulse Paradigms: Short Interval Intracortical Inhibition

Kujirai et al. (1993) were the first to describe the use of paired magnetic stimuli to test the excitability of specific intracortical elements within the primary motor cortex. They demonstrated that a sub-motor threshold conditioning stimulus could inhibit the response to a subsequent test TMS pulse with an ISI of 1–5 ms, an effect that has been shown to be due to inhibition of I-waves (Di Lazzaro et al., 1998b). In particular, at least with conventional coil orientations, the conditioning stimulus largely exerts its inhibitory effect by reducing the later I3 wave (Di Lazzaro et al., 1998b), which occurs approximately 3 ms later than the I1 wave and is thought to be due to activation of more complex cortical circuitry (Di Lazzaro et al., 2011). The inhibitory circuits responsible for SICI are dependent on the major cortical inhibitory neurotransmitter γ-aminobutyric acid (GABA) (Ziemann et al., 1996b).

Several pieces of evidence suggest that the strength of connectivity within SICI circuits is functionally important in that there are clear behaviorally induced changes in SICI. For example, there is a reduction in SICI in the cortical representation of a voluntarily contracted muscle (Ridding et al., 1995c). Also during a focal contraction, SICI is not changed, or even increased, in the cortical representation of adjacent muscles that have to be maintained in a relaxed state (Stinear and Byblow, 2003; Zoghi et al., 2003). In addition, there are abnormalities in SICI in a number of pathological conditions associated with abnormal movements (e.g., Ridding et al., 1995a,b; Strafella et al., 2000; Bares et al., 2003) as well as in epilepsy and stroke (Caramia et al., 1996; Cicinelli et al., 2003). Therefore, SICI probably reflects the excitability (and connectivity) of an inhibitory network within M1, which is important for focusing muscle activation during task performance.

8.2.2.4   Paired-Pulse Paradigms: Intracortical Facilitation

Using the same paired-pulse paradigm as described earlier for SICI, a facilitation of test MEPs is seen with longer ISIs (Kujirai et al., 1993), which has been termed intracortical facilitation (ICF). The mechanism of ICF is not well understood, but it involves a different cortical network than that responsible for SICI (Ziemann et al., 1996c). Several pharmacological agents influence ICF. For example, $GABA_A$ receptor agonists (Ziemann et al., 1996b), anti-glutamatergic agents (Liepert et al., 1997; Ziemann et al., 1998a), and NE antagonists (Ilic et al., 2002a; Korchounov et al., 2003) decrease ICF, while NE agonists (Boroojerdi et al., 2001) increase ICF. In summary, the pharmacological data suggest that ICF reflects a net cortical facilitation due to an overlapping weaker inhibition and stronger facilitation with GABA and glutamate being the key neurotransmitters involved (Paulus et al., 2008).

8.2.2.5   Paired-Pulse Paradigms: SICI and ICF in Non-Motor Regions

SICI/ICF has most frequently been studied in primary motor cortex. However, these techniques can also be used to investigate functioning of non-motor areas. For example, Oliveri et al. (2000a) demonstrated that paired TMS applied to the parietal cortex could reveal inhibitory and facilitatory influences on behavior. Using a sensory perception task in the contralateral hand, these authors demonstrated that paired magnetic stimuli at short (1 ms) and long (5 ms) interval produced worsening and improvement respectively in perception. A similar approach was also used to examine tactile extinction in a group of brain-damaged patients (Oliveri et al., 2000b). Paired magnetic stimuli applied at ISIs of 1 and 10 ms were applied over the posterior parietal and frontal cortices in the unaffected hemisphere during a bimanual tactile discrimination task. With posterior parietal stimulation, paired stimuli at 1 ms improved perception of stimuli in the contralesional hand while paired stimuli at 10 ms worsened perception in the contralesional hand. Similar results were seen with frontal stimulation, but the time course of effect was slightly delayed.

8.2.2.6   Paired-Pulse Paradigms: Long Interval Intracortical Inhibition

Using modified intensities and longer ISIs, it is possible to test connectivity between a second inhibitory network and the cortical output cells within M1. Valls-Sole et al. (1992) first reported that a supra-threshold conditioning stimulus could inhibit the response to a test stimulus applied 60–200 ms later. Recordings of the descending volleys have provided evidence of a cortical origin for this inhibition with later I-waves being inhibited by the conditioning stimulus at ISIs of 100 and 150 ms (Nakamura et al., 1997; Di Lazzaro et al., 2002). Given the duration of this inhibition, it was proposed that slow IPSPs mediated by $GABA_B$ receptors are responsible for LICI, and this has received support from pharmacological studies in which $GABA_B$ receptor agonists have been shown to increase LICI (McDonnell et al., 2006). The functional role of LICI is not well understood, but it decreases with increasing levels of voluntary contraction in the target muscle and is likely to be important for focusing muscle activation during task performance (Hammond and Vallence, 2007). Abnormalities in LICI have been reported in a range of conditions including Huntington's disease (Tegenthoff et al., 1996), dystonia (Espay et al., 2006), and Parkinson's disease (Berardelli et al., 1996).

8.2.2.7   Paired-Pulse Paradigms: Short-Interval Intracortical Facilitation

The excitability of intracortical facilitatory circuits within the motor cortex can be investigated using a modification of the paired-pulse technique described earlier for SICI. Paired stimuli at just threshold intensities (Tokimura et al., 1996), or a pair of stimuli where the first is supra-threshold and the second just subthreshold (Ziemann et al., 1998b), result in marked facilitation at certain short ISIs (approximately 1.3, 2.5, and 4.3 ms). Epidural (Di Lazzaro et al., 1999) and single motor unit (Ilic et al., 2002b) recordings both indicate that these facilitatory peaks reflect I-wave interaction. Short-interval intracortical facilitation (SICF) is influenced by inhibitory networks as it is reduced by GABAergic drugs (Ziemann et al., 1996a). In summary, SICF is likely due to non-synaptic facilitation at the initial axon segment of interneurons in a high-threshold local excitatory pathway within M1 (Ilic et al., 2002b).

8.2.3   Probing Interregional Connectivity

As well as using paired-pulse TMS techniques to examine connectivity on a local level within M1, it is possible to examine functional connectivity between different regions of cortex using twin-coil approaches. These studies usually involve applying test stimuli to M1 (to evoke an MEP) and conditioning stimuli to remote but connected cortical regions.

8.2.3.1   Premotor Cortex: Motor Cortex Connectivity

Civardi and colleagues (2001) first reported that it is possible to test the excitability of inputs from premotor cortical regions to primary motor cortex using paired-pulse TMS techniques. Sub-motor threshold (measured for M1) conditioning stimuli applied at a site 3–5 cm anterior to the motor cortex hotspot, or 6 cm anterior to the vertex, inhibited MEPs evoked from M1 when the ISI was 4–6 ms. By contrast, MEPs were facilitated when the intensity of conditioning stimuli was increased to approximately 120% of active motor threshold. They suggested that these interactions reflected projections from anterior dorsal premotor (PMd) and supplementary motor (SMA) regions, respectively. The inhibition was reduced during a voluntary contraction although the authors cautioned about placing too much weight on this finding due to technical considerations.

Baumer and colleagues (2009) reported that conditioning stimuli at an intensity of 80% AMT applied over ventral premotor cortex (PMv) facilitated test responses evoked from M1 at ISIs of 4 or 6 ms. By contrast, conditioning stimulation at 90% RMT resulted in inhibition. Davare et al. (2008) also examined the influence of grasping behavior on the projection from PMv to M1. Interestingly, it was found that during a power grip the inhibition seen at rest disappeared, and during a precision grip the inhibition seen at rest was replaced by facilitation. These results suggest that the net influence of premotor area on primary motor cortex results from a balance of inhibitory and facilitatory projections. In the resting state, the inhibition reflects the predominant influence of inhibitory connections while the facilitation seen during precision grip reflects the greater influence of facilitatory projections.

8.2.3.2 Parietal Cortex: Primary Motor Cortex

Using a twin-coil approach, it is also possible to test connectivity between parietal cortex and primary motor cortex both in the same hemisphere and between hemispheres (see the following text). Koch et al. (2007) demonstrated that a conditioning pulse over the posterior parietal cortex (PPC) applied 4–6 or 15 ms prior to a test stimulus to M1 could facilitate the test response. The authors proposed that the early (4–6 ms) peak of facilitation reflected activation of direct projections while the later (15 ms) facilitatory peak reflected activation of polysynaptic circuits. This effect was sensitive to the conditioning stimulus intensity. Facilitatory effects were seen with a conditioning stimulus intensity of 90% RMT but not with lower or higher intensities. There are dense connections between PPC and M1, which are likely to be important for relaying information crucial for optimizing movement plans. Indeed, this interaction between parietal cortex and M1 has been shown to be influenced by specific motor plans. Using a choice reaction time task involving arm pointing to the left or right, Koch et al. (2008) demonstrated pre-movement facilitatory interactions between PPC and M1 when the arm contralateral to the stimulated hemisphere was moved but not when the ipsilateral limb was moved. This finding provides evidence that PPC–M1 connectivity is important in the early stages of movement planning.

8.2.4 Probing Interhemispheric Connectivity

8.2.4.1 Connectivity between Primary Motor Areas

Following pioneering studies utilizing transcranial electrical stimulation (Amassian and Cracco, 1987) in which a transcallosal nervous propagation was recorded from the scalp, Ferbert and colleagues (1992) reported that a conditioning TMS pulse applied to the motor cortex of one hemisphere could inhibit the response to a test TMS pulse applied to the contralateral motor cortex when the ISI was 6–30 ms. This inhibition was termed interhemispheric inhibition (IHI) and likely reflects activity in the transcallosal projections to the test hemisphere (Ferbert et al., 1992) with demonstrated abnormalities of IHI in patients with agenesis, or abnormalities, of the corpus callosum (Meyer et al., 1995). The effects are usually inhibitory, but with low-intensity conditioning stimuli and short ISIs, facilitatory interactions have been reported, which are, again, thought to reflect transcallosally mediated effects (Ugawa et al., 1993; Hanajima et al., 2001).

Ridding et al. (2000) demonstrated that there was stronger IHI between primary motor cortices in professional musicians. This suggests that the connectivity in this pathway may be influenced by training (although it is possible that such a finding reflects a genetic trait that predisposes to high-level musical achievement). Most studies have examined IHI between motor cortices in the relaxed condition. However, IHI can be modified by behavioral engagement of the conditioned motor cortex. When a hand muscle contralateral to the conditioned hemisphere is activated, the level of IHI onto the test hemisphere is increased (Ferbert et al., 1992; Vercauteren et al., 2008). This increase in IHI may have a role in reducing the likelihood of inappropriate recruitment of the test hemisphere during task performance. Also the magnitude of IHI changes during, and up to, a movement. Murase et al. (2004) examined IHI during a reaction time task performed with the right hand. During movement preparation, inhibition from the right (conditioned) hemisphere on to the left hemisphere gradually decreased and was replaced by facilitation at movement onset. The authors hypothesized that this change from IHI to interhemispheric facilitation might be useful for recruitment of the left M1. By contrast, in a group of patients with chronic subcortical strokes, the change from inhibition to facilitation during the development of a movement was not apparent.

8.2.4.2 Connectivity between Premotor and Primary Motor Areas

Although as originally described IHI was employed to examine interhemispheric connectivity between the primary motor cortices, this approach can also been used to examine transcallosal connectivity between premotor areas in one hemisphere and primary motor areas in the contralateral hemisphere. For example, a conditioning pulse at 90% or 110% RMT over the right PMd inhibits the response to a test TMS over the left M1 if the conditioning-test ISI is 8–10 ms (Mochizuki et al., 2004). Interestingly, the threshold for IHI was lower for PMd–M1 interactions than for M1–M1 interactions. Again, facilitatory effects were seen with lower intensity conditioning stimuli (80% AMT) and short ISIs (8 ms). Baumer et al. (2006) investigated such interhemispheric facilitatory effects when conditioning stimuli were applied to both M1 and PMd and test stimuli to the contralateral M1. Facilitatory effects were seen with conditioning of both PMd and M1 and provided evidence that these effects were due to separate anatomical pathways to the contralateral M1. In addition, by examining the effects with different test coil orientations, they suggested that the projections from PMd and M1 exerted their facilitatory effect by influencing different interneuron populations in the test M1. Behavioral engagement of the test M1 also differentially influenced interhemispheric M1–M1 and PMd–M1 effects; interhemispheric facilitatory effects were still seen with M1 conditioning but not with PMd conditioning. This finding suggests a functional segregation of premotor and motor influences, with M1 interhemispheric effects being important for the control of corticospinal output while interhemispheric PMd–M1 inputs might be more important for other aspects of motor control (e.g., visuomotor interaction).

Koch et al. (2006) investigated changes in transcallosal PMd–M1 interactions during a choice reaction time task consisting of a rapid isometric squeezing contraction with the thumb and index fingers of either hand. Conditioning stimuli were applied to the left PMd and test stimuli to the right M1. Both inhibitory and facilitatory effects were studied by using, respectively, a higher or lower conditioning stimulus intensity. During most of the reaction time period, inhibitory and facilitatory effects were reduced. However, the facilitatory pathway was active 75 ms after the cue (go) tone for movements of the left hand and the inhibitory pathway was active at 100 ms when movements involved the right hand, suggesting that PMd–M1 inputs are important for facilitating cued movements as well as inhibiting planned but non-executed movements.

In summary, there are multiple single- and double-pulse TMS techniques available to investigate the connectivity within local cortical circuits as well as between more remote cortical regions. These approaches have provided novel and important insights into the functional relationships within, and between, cortical areas.

In particular, probing the modifications in connectivity of these circuits during behavioral engagement has already provided important findings and is likely to be an approach that will provide additional and important insights into brain function in the future.

8.2.5 Contribution of TMS/EEG Techniques

Recent technical developments have allowed the recording of EEG responses to TMS of a given scalp site with millisecond resolution. Combining TMS with EEG provides a direct, noninvasive method to examine cortical reactivity and connectivity (Ilmoniemi et al., 1997). In contrast to TMS/TMS techniques in which either psychophysical measurements or the amplitude of MEPs (an indirect marker of cortico-spinal tract excitability and even less of M1 cortex excitability) are probed when "conditioning" stimuli of a given brain area are applied, TMS/EEG approaches record simultaneously the cortical activity from the whole scalp via a number of electrodes following single TMS on an individual brain site. A network of neuronal connections is engaged when TMS-evoked activation extends from a stimulation site to other parts of the brain and the summation of synaptic potentials produces deflections in the scalp EEG. These responses start a few milliseconds after the stimulus and last about 300 ms, first in the form of rapid oscillations and then as lower-frequency deflections of alternating polarity (Ilmoniemi and Karhu, 2008). The amplitude, latency, and scalp topography of single-pulse TMS-evoked EEG responses have been clearly described (Komssi et al., 2004). The characteristics of these responses are thought to depend on the stimulation intensity and functional state of both the stimulated cortex and the overall brain. In particular, it has been suggested that the very first part of the TMS-evoked EEG response reflects the reactivity, i.e., the functional state, of the stimulated cortex while its spatiotemporal distribution over the scalp and the spread of activation to other cortical areas via intra, and interhemispheric, cortico–cortical connections as well as to subcortical structures and spinal cord via projection fibers. The spatiotemporal distribution of the evoked activity therefore provides a marker of the stimulated area's effective connectivity (Lee et al., 2003; Komssi and Kähkönen, 2006). The correlates of SICI and ICF of the TMS-related EEG responses as well as their relationships with MEP modulation have yet to be clearly demonstrated (Paus et al., 2000; Komssi et al., 2004; Daskalakis et al., 2008). A recent study by Ferreri et al. (2011), utilizing EEG navigated-paired-pulse TMS (ppTMS) coregistration, characterized the neuronal circuits underlying human M1 connectivity. In particular, this study aimed to evaluate SICI and ICF directly from the cortex using EEG and to investigate whether EEG measures of SICI and ICF are related to the same mechanisms underlying the conventional EMG measures of SICI and ICF. This study adds new insight into previously delineated (Komssi et al., 2004) functional behavior of human brain as investigated by EEG oscillations evoked by TMS to M1 and confirmed that ppTMS can modulate early and late EEG evoked responses as well as MEPs (Paus et al., 2001; Bonato et al., 2006). This suggests that for some TMS-induced EEG peaks, the measures of SICI and ICF are somewhat related to the same mechanisms mediating EMG (MEP) measures of SICI and ICF. Additionally, it implies that intracortical inhibition and facilitation, which until now could only be indirectly demonstrated in the M1 by measuring MEP amplitude modulation, could be directly evaluated in all cortical area by measuring EEG responses to ppTMS (Daskalakis et al., 2008). TMS-evoked brain responses are generated by the temporal and spatial summation in the superficial cortical layers of the electrical currents caused by the slow postsynaptic excitatory (EPSP) and/or inhibitory potentials (IPSP) with little or no contribution from the brief action potentials. On the other hand, the MEP is produced from brief, indirect-induced descending action potentials generated by the activation of pyramidal neurons via presynaptic IPSP and/or EPSP (Rossini et al., 1994). This activation is influenced by both the intrinsic proprieties of corticospinal tract and inhibitory–excitatory circuit interactions in different proportions depending on the timing, on the "stimulus intensity/evoked impulse synchronicity" and direction of induced current. Since SICI and ICF are thought to represent separate intracortical phenomena (Ziemann et al., 1996c), starting from the fifth millisecond from the conditioning stimulus, a significant amount of data about the M1 connectivity and functional state can be collected. On the other hand, the clear modulation of the amplitude of the evoked EEG peaks as well as their global mean field power by different TMS intensities and paradigms provides information about the functional state of M1 cortex and its large-scale connected networks, via different interneuronal circuit activation and/or modulation (Ilmoniemi and Karhu, 2008). Thus it is reasonable to propose that the EEG–TMS evoked activity should not be seen as the result of single process and the conventionally negative or positive peaks should not be considered as reflecting inhibitory or excitatory circuit activities but probably a balance of both, as suggested by the study of Ferreri et al. (2011), where a modulation of the peak amplitude was often seen at ISI 3 as well as at ISI 11 even in the same direction.

8.2.6   POSSIBLE MOLECULAR ORIGIN OF THE EEG–TMS EVOKED RESPONSES: INSIGHT INTO M1 FUNCTIONAL STATE FROM ppTMS EXPERIMENT

The firing of cortical neurons enveloped in the EEG signal is associated with the activation of both fast and slow excitatory postsynaptic potentials (fEPSPs and sEPSPs respectively) and fast and slow inhibitory postsynaptic potentials (fIPSPs and sIPSPs Rosenthal et al., 1967). fEPSPs are mediated by non-NMDA, AMPA/kainate receptors with a rise time of 0.5–1.9 ms. Given this short rise time, the contribution of fEPSPs to TMS-evoked EEG responses is often obscured by the TMS artifact. On the other hand, sEPSPs are mediated by N-methyl-D-aspartate (NMDA) receptors with a rise time of 4–9 ms and are possibly involved in the generation and/or modulation of the N7 component, while fIPSPs are mediated by $GABA_A$ postsynaptic receptors lasting approximately 20–30 ms (Davies et al., 1990; Deisz, 1999) and are possibly involved in the generation and/or modulation of the P13, N18, P30, and N44 components. Finally, sIPSPs are related to presynaptic and postsynaptic $GABA_B$ receptors with an inhibition that peaks around 100–200 ms starting around 50 ms and lasting up to a few hundred milliseconds, being possibly involved in the genesis of the P60, N100, P180, and N280 potentials (Tamás et al., 2003; McDonnel et al., 2006).

On this basis, SICI as evaluated by means of MEP amplitude modulation at an ISI of 3 ms (Fisher et al., 2002) is thought to explore the net effect of the activation of inhibitory $GABA_A$ circuits in M1. This activation occurs through the summation at the pyramidal neurons, of presynaptic low-threshold $GABA_A$ fIPSPs elicited by the sub-threshold TMS pulse as well as high-threshold non-NMDA fEPSPs elicited by the supra-threshold pulse. This means that SICI produces MEP inhibition by reducing the late indirect waves (Hanajima et al., 1998; Di Lazzaro et al., 1999), and that in the EEG evoked activity these hyperpolarizing currents should be hidden and their 20–30 ms lasting effects on the initial waveform evoked: effectively a modulation at ISI 3 is observable at P13 and slightly at N18 and P30. By contrast, the modulations seen in components P60–N100, P180, and N280 could be related to the activity of $GABA_B$ receptors. Regarding ICF, there is consensus for a cortical (rather than spinal) site of action of the subthreshold pulse, although emerging data suggest complex mechanisms (Di Lazzaro et al., 2002, 2006). Converging evidence support the idea that ICF can interact with inhibitory circuits and is the result of a net facilitation consisting of prevailing facilitation and weaker inhibition. The inhibition probably comes from $GABA_A$ fIPSP, while the facilitation is due to NMDA-dependent sEPSPs with a rise time of 4–9 ms (Ziemann et al., 2003). Indeed, according to this study, a modulation at ISI 11 is clearly observable at N7. Moreover, in line with previous results, a clear modulation of N44 (and slightly of P30) at ISI 11 was found. It has been suggested (Paus et al., 2001) that this modulation is related to a disruption of the rhythmic activity of an ongoing M1 pacemaker, which is selectively induced by ISI 11 ppTMS and strictly correlated with beta and gamma oscillatory activity in M1 (Van Der Werf and Paus, 2006). Notably, generation of gamma oscillations, similar to ICF, has been associated with both $GABA_A$ and NMDA receptor activity. Regarding the modulation of later waves in respect to supra-threshold single pulse, it seems to reflect quite exactly what seen at ISI 3, suggesting also here an involvement of $GABA_B$ receptor activity.

8.2.6.1   M1 Connectivity as Evaluated by Means of Spatial Distribution of the EEG–TMS-Evoked Activity: EEG–MEP Modulation Correlations

It has been suggested that TMS to M1 predominantly affects the site of stimulation, although, according to an emerging principle, this effect should also produce mutual interaction with other brain areas (Matsumoto et al., 2007). In fact, it is known that M1 receives input from, and projects output bilaterally to, primary motor, premotor, SMA, somatosensory cortices, as well as to the thalamus. Recent fMRI and PET studies, which however suffer from a poor temporal resolution (for review see Bestmann et al., 2008), have demonstrated some spread of activation from M1 to remote brain areas. Indeed, as described in the earlier sections, TMS to M1 can influence distant cortical regions, and stimulation of remote motor and non-motor areas can affect excitability within the target M1.

The scalp topography of the first wave detected, i.e., *N7*, suggests the engagement of the ipsilateral non-primary motor cortices, possibly with the prevalent involvement of the premotor cortex, from the stimulated M1 via cortico-cortical projections. Animal data suggested strong connections from primary and non-primary motor cortices, both facilitatory and inhibitory with a predominance of the inhibitory ones at rest. Indeed, as described earlier, recent ppTMS studies (Davare et al., 2008) revealed the clear existence of functional interactions between motor and premotor areas in humans at similar latencies (6–8 ms).

Wave *P13* shows a topographical map with a clear engagement of the contralateral cortices, possibly with the prevalent involvement of the non-primary motor ones. This is consistent with the transcallosal conduction time of 12–15 ms found in studies using both electrical stimulation and TMS (Amassian and Cracco, 1987; Ferbert et al., 1992; Meyer et al., 1998). It was demonstrated that transcallosal projections are mainly excitatory, synapsing onto local inhibitory circuits mediating SICI and LICI within the target hemisphere (Daskalakis et al., 2002). Recently, homologous connections and their preferential inhibitory effects were indirectly observed in humans at consistent latencies (10 ms, Mochizuki et al., 2004). The correlation observed with the MEP amplitude during ppTMS at ISI 11 suggests an involvement of this peak and these cortices in modulating EMG measures of ICF. Moreover, if P13 modulation is related to IPSPs, it could be inferred that at this latency, the contralateral ICF could be mediated by $GABA_A$ receptors.

Wave *N18* appears as a focal negativity in electrode P3 (that possibly corresponds to the ipsilateral PPC, Civardi et al., 2001) and could reflect sustained and reverberant activation extending from the stimulated M1 via cortico-cortical projections. Pioneering studies (Meynert, 1865) revealed that PPC is anatomically interconnected with motor, premotor, and more frontal areas through distinct white matter tracts forming the superior longitudinal fasciculus. Indeed, ppTMS studies have provided evidence to support the existence of functional interactions between the PPC and the ipsilateral M1 (Koch et al., 2007) at a consistent timing (about 15 ms). Moreover, it was proposed that also parieto-premotor projections, which are known to be more numerous than direct parieto-motor ones transferring information for visuomotor plan and transformations, could be involved in this pathway and modulated by SICI mechanism (Koch et al., 2007). Previous studies suggested a complex source structure for the wave *P30* (Van Der Werf et al., 2006), linked to a strong activation of the contralateral hemisphere, which is slightly increased following ppTMS at ISI 11. This wave might be due to further interhemispheric spread of transcallosal activation or via a subcortical pathway (via thalamic nuclei and/or basal ganglia) projecting back diffusely to the cortex or both. The thalamus is in fact believed to play a minor role in the generation of early EEG responses to TMS but could be involved in reverberating circuits at longer poststimulus latencies also explaining the large scalp distribution of the later waves (Ziemann and Rothwell, 2000). The correlation observed with the MEP amplitude both after single pulse and after ISI 11 in contralateral hemisphere suggests an involvement of this peak and these brain structures in determining M1 net output and modulating EMG measures of ICF; moreover, if P30 modulation is related to fIPSPs, it could be inferred that at this latency the contralateral ICF could be mediated by $GABA_A$ receptors.

Wave *N44* ms is widely distributed with clear spatial and amplitude modulations. It shows high amplitude in the stimulated area in each condition, while in the parieto-posterior regions of the non-stimulated hemisphere it was observed at suprathreshold single pulse and at ISI 3, whereas it was virtually absent at ISI 11. Although it was previously proposed that N44 and the MEP should be considered separate markers of M1 functional properties, as the N44 can arise following sub-threshold TMS (Van Der Werf and Paus, 2006), the correlation of the N44 amplitude in the stimulated hemisphere with the MEP amplitude both after single pulse and after ISI 3 suggests a role of this peak and these brain structures in determining M1 net output and modulating EMG measures of SICI (Figure 8B of Ferreri et al. [2011]). N44 was possibly related to somatosensory-evoked potentials generated as feedback from the TMS-induced muscle twitch (Bonato et al., 2006) even if this view is contradicted by the fact that the same component, in some brain areas even stronger than during supra-threshold single-pulse session, was also recorded during ISI 3 session despite remarkably depressed MEP.

Wave *P60* ms shows a wide distribution with clear spatial and amplitude modulations. Interestingly enough, this wave is in fact virtually absent in the stimulated area after supra-threshold single-pulse TMS and ISI 11 and specifically reversed in polarity at ISI 3.

Wave *N100*, considered the dominant peak in TMS-evoked EEG activity, shows a consistent lateralization to the stimulated side, being stronger both after ISI 3 and after ISI 11 than after supra-threshold single-pulse TMS. In preliminary experiments, it has attributed as a brain response to the coil "click" and to bone-conducted sound, although later studies have partially excluded such a contamination (Nikouline et al., 1999; Bender et al., 2005; Kicic et al., 2008), suggesting that this component reflects, at least in part, TMS-induced cortical inhibitory process.

The topography of *P190* ms shows a clear and circumscribed involvement of the contralateral hemisphere. This component, like N100, has been associated with the auditory N1–P2 complex in previous studies (latency 100–200 ms, Bonato et al., 2006). However, the lateralization found by others in the hemisphere opposite to the stimulation and the audio-masking procedure adopted does not support this hypothesis. Instead its long latency and wide distribution could suggest the engagement of a reverberant cortico-subcortical circuit.

Finally, wave *N280* peaked in the premotor areas of the stimulated hemisphere. As for P190, its long latency and wide distribution could suggest the involvement of a reverberant cortico-subcortical circuit.

The data of Ferreri et al. (2011) and others demonstrate that EEG–ppTMS is a promising tool to characterize the neuronal circuits underlying human cortical effective connectivity as well as the neural mechanisms regulating the balance between inhibition and facilitation within the cortices and the corticospinal pathway. It was in fact proved that ppTMS modulates MEPs, as well as EEG early and late evoked responses and that for some peaks the EEG variability is partly linked with, and therefore might partially explain, MEP variability. Future studies parsing EEG measures of SICI and ICF into their component frequencies are needed to further characterize their physiology. Several animal studies, computer simulations, and finally a human investigation (Farzan et al., 2009) have implicated glutamatergic and GABAergic neurotransmissions in both the generation and the modulation of theta, beta, and gamma frequency oscillations in the cortex (Amzica and Steriade, 1995; Traub et al., 1998; Mann and Paulsen, 2007). Finally, it should be stressed that the correlations observed between EEG and EMG measures do not necessarily imply that the mechanisms subtending EEG response modulation in the central nervous system are causally linked with those subtending MEP amplitude modulations in SICI and ICF. However, it is likely that future studies using glutamate, $GABA_A$, as well as $GABA_B$ receptor agonists or antagonists in healthy subjects could establish this relationship more directly.

8.3 SUMMARY

We have described a number of neurophysiological approaches employing neurophysiological—mainly TMS—techniques that have been utilized to provide insights into cortical connectivity. These techniques offer a number of advantages over other approaches to examine connectivity such as functional MRI including their greater temporal resolution and the ability to test connectivity in a wide range of behavioral conditions and environments. However, while these neurophysiological approaches provide powerful tools to investigate cortical connectivity, it must be remembered that they also have some limitations. For example, these techniques involve the stimulation of, and recording from, large populations of cortical neurons with resultant limited spatial specificity. Related to this is that the stimulation with TMS usually involves activation of both excitatory and inhibitory cortical elements. However, a number of methodological approaches, including manipulation of stimulus intensities and the use of paired-pulse technology with various ISIs, allow preferential targeting of excitatory or inhibitory circuits. By combining TMS and EEG techniques, it has recently become possible to more directly test cortical connectivity. While these TMS/EEG approaches present significant technical challenges, they do offer an opportunity to gain additional insights into network connectivity in a variety of behavioral conditions. It is likely that data gained from neurophysiological studies employing TMS and TMS/EEG techniques will continue to provide information that can be used to validate and develop our understanding of human cortical network connectivity in behaviorally relevant conditions.

REFERENCES

Alle H, Heidegger T, Krivanekova L, and Ziemann U. (2009). Interactions between short-interval intracortical inhibition and short-latency afferent inhibition in human motor cortex. *J Physiol* 587, 5163–5176.

Amassian VE and Cracco RQ. (1987). Human cerebral cortical responses to contralateral transcranial magnetic stimulation. *Neurosurgery* 20, 148–155.

Amzica F and Steriade M. (1995). Short- and long-range neuronal synchronization of the slow (<1 Hz) cortical oscillation. *J Neurophysiol* 73(1), 20–38.

Bares M, Kanovsky P, Klajblova H, and Rektor I. (2003). Intracortical inhibition and facilitation are impaired in patients with early Parkinson's disease: A paired TMS study. *Eur J Neurol* 10, 385–389.

Baumer T, Bock F, Koch G, Lange R, Rothwell JC, Siebner HR, and Munchau A. (2006). Magnetic stimulation of human premotor or motor cortex produces interhemispheric facilitation through distinct pathways. *J Physiol* 572, 857–868.

Baumer T, Schippling S, Kroeger J, Zittel S, Koch G, Thomalla G, Rothwell JC, Siebner HR, Orth M, and Munchau A. (2009). Inhibitory and facilitatory connectivity from ventral premotor to primary motor cortex in healthy humans at rest—A bifocal TMS study. *Clin Neurophysiol* 120, 1724–1731.

Bender S, Basseler K, Sebastian I, Resch F, Kammer T, Oelkers-Ax R, and Weisbrod M. (2005). Electroencephalographic response to transcranial magnetic stimulation in children: Evidence for giant inhibitory potentials. *Ann Neurol* 58(1), 58–67.

Berardelli A, Abbruzzese G, Chen R, Orth M, Ridding MC, Stinear C, Suppa A, Trompetto C, and Thompson PD. (2008). Consensus paper on short-interval intracortical inhibition and other transcranial magnetic stimulation intracortical paradigms in movement disorders. *Brain Stimul* 1, 183–191

Berardelli A, Rona S, Inghilleri M, and Manfredi M. (1996). Cortical inhibition in Parkinson's disease. A study with paired magnetic stimulation. *Brain* 119(Pt 1), 71–77.

Bestmann S, Ruff CC, Blankenburg F, Weiskopf N, Driver J, and Rothwell JC. (2008). Mapping causal interregional influences with concurrent TMS-fMRI. *Exp Brain Res* 191(4), 383–402. Review.

Bonato C, Miniussi C, and Rossini PM. (2006). Transcranial magnetic stimulation and cortical evoked potentials: A TMS/EE Gco-registration study. *Clin Neurophysiol* 117(8), 1699–1707.

Boroojerdi B, Battaglia F, Muellbacher W, and Cohen LG. (2001). Mechanisms influencing stimulus-response properties of the human corticospinal system. *Clin Neurophysiol* 112, 931–937.

Cantello R, Gianelli M, Civardi C, and Mutani R. (1992). Magnetic brain stimulation: The silent period after the motor evoked potential. *Neurology* 42, 1951–1959.

Caramia MD, Gigli G, Iani C, Desiato MT, Diomedi M, Palmieri MG, and Bernardi G. (1996). Distinguishing forms of generalized epilepsy using magnetic brain stimulation. *Electroencephalogr Clin Neurophysiol* 98, 14–19.

Cicinelli P, Pasqualetti P, Zaccagnini M, Traversa R, Oliveri M, and Rossini PM. (2003). Interhemispheric asymmetries of motor cortex excitability in the postacute stroke stage: A paired-pulse transcranial magnetic stimulation study. *Stroke* 34, 2653–2658.

Civardi C, Cantello R, Asselman P, and Rothwell JC. (2001). Transcranial magnetic stimulation can be used to test connections to primary motor areas from frontal and medial cortex in humans. *Neuroimage* 14, 1444–1453.

Daskalakis ZJ, Christensen BK, Fitzgerald PB, Roshan L, and Chen R. (2002). The mechanisms of interhemispheric inhibition in the human motor cortex. *J Physiol* 15, 543(Pt 1), 317–326.

Daskalakis ZJ, Farzan F, Barr MS, Maller JJ, Chen R, and Fitzgerald PB. (2008). Long-interval cortical inhibition from the dorsolateral prefrontal cortex: A TMS–EEG study. *Neuropsychopharmacology* 33(12), 2860–2869.

Davare M, Lemon R, and Olivier E. (2008). Selective modulation of interactions between ventral premotor cortex and primary motor cortex during precision grasping in humans. *J Physiol* 586(Pt 11), 2735–2742.

Davies CH, Davies SN, and Collingridge GL. (1990). Paired-pulse depression of monosynaptic GABA-mediated inhibitory postsynaptic responses in rat hippocampus. *J Physiol* 424, 513–531.

Day BL, Rothwell JC, Thompson PD, Dick JP, Cowan JM, Berardelli A, and Marsden CD. (1987). Motor cortex stimulation in intact man. 2. Multiple descending volleys. *Brain* 110, 1191–1209.

Deisz RA. (1999). GABA(B) receptor-mediated effects in human and rat neocortical neurons in vitro. *Neuropharmacology* 38(11), 1755–1766.

Devanne H, Lavoie BA, and Capaday C. (1997). Input-output properties and gain changes in the human corticospinal pathway. *Exp Brain Res* 114, 338.

Di Lazzaro V, Oliviero A, Mazzone P, Pilato F, Saturno E, Insola A, Visocchi M, Colosimo C, Tonali PA, and Rothwell JC. (2002). Direct demonstration of long latency cortico-cortical inhibition in normal subjects and in a patient with vascular parkinsonism. *Clin Neurophysiol* 113, 1673–1679.

Di Lazzaro V, Oliviero A, Profice P, Pennisi MA, Di Giovanni S, Zito G, Tonali P, and Rothwell JC. (2000). Muscarinic receptor blockade has differential effects on the excitability of intracortical circuits in the human motor cortex. *Exp Brain Res* 135, 455–461.

Di Lazzaro V, Oliviero A, Profice P, Saturno E, Pilato F, Insola A, Mazzone P, Tonali P, and Rothwell JC. (1998a). Comparison of descending volleys evoked by transcranial magnetic and electric stimulation in conscious humans. *Electroencephalogr Clin Neurophysiol* 109, 397–401.

Di Lazzaro V, Pilato F, Oliviero A, Dileone M, Saturno E, Mazzone P, Insola A, Profice P, Ranieri F, Capone F, Tonali PA, and Rothwell JC. (2006). Origin of facilitation of motor-evoked potentials after paired magnetic stimulation: Direct recording of epidural activity in conscious humans. *J Neurophysiol* 96(4), 1765–1771.

Di Lazzaro V, Profice P, Ranieri F, Capone F, Dileone M, Oliviero A, and Pilato F. (2011). I wave origin and modulation. *Brain Stimul* [Epub. ahead of print].

Di Lazzaro V, Restuccia D, Oliviero A, Profice P, Ferrara L, Insola A, Mazzone P, Tonali P, and Rothwell JC. (1998b). Magnetic transcranial stimulation at intensities below active motor threshold activates intracortical inhibitory circuits. *Exp Brain Res* 119, 265–268.

Di Lazzaro V, Rothwell JC, Oliviero A, Profice P, Insola A, Mazzone P, and Tonali P. (1999). Intracortical origin of the short latency facilitation produced by pairs of threshold magnetic stimuli applied to human motor cortex. *Exp Brain Res* 129, 494–499.

Di Lazzaro V, Ziemann U, and Lemon RN. (2008). State of the art: Physiology of transcranial motor cortex stimulation. *Brain Stimul* 1, 345–362.

Espay AJ, Morgante F, Purzner J, Gunraj CA, Lang AE, and Chen R. (2006). Cortical and spinal abnormalities in psychogenic dystonia. *Ann Neurol* 59, 825–834.

Farzan F, Barr MS, Wong W, Chen R, Fitzgerald PB, and Daskalakis ZJ. (2009). Suppression of gamma oscillations in the dorsolateral prefrontal cortex following long interval cortical inhibition: ATMS-EEG study. *Neuropsychopharmacology* 34(6), 1543–1551.

Ferbert A, Priori A, Rothwell JC, Day BL, Colebatch JG, and Marsden CD. (1992). Interhemispheric inhibition of the human motor cortex. *J Physiol* 453, 525–546.

Ferrari F, Pasqualetti P, Määttä S, Ponzo D, Ferrarelli F, Tononi G, Mervaala E, Miniussi C, and Rossini PM. (2011). Human brain connectivity during single and paired pulse transcranial magnetic stimulation. *Neuroimage* 54, 90–102.

Fisher RJ, Nakamura Y, Bestmann S, Rothwell JC, and Bostock H. (2002). Two phases of intracortical inhibition revealed by transcranial magnetic threshold tracking. *Exp Brain Res* 143(2), 240–248.

Fox P, Ingham R, George MS, Mayberg H, Ingham J, Roby J, Martin C, and Jerabek P. (1997). Imaging human intra-cerebral connectivity by PET during TMS. *Neuroreport* 8, 2787–2791.

Hammond G and Vallence AM. (2007). Modulation of long-interval intracortical inhibition and the silent period by voluntary contraction. *Brain Res* 1158, 63–70.

Hanajima R, Ugawa Y, Machii K, Mochizuki H, Terao Y, Enomoto H, Furubayashi T, Shiio Y, Uesugi H, and Kanazawa I. (2001). Interhemispheric facilitation of the hand motor area in humans. *J Physiol* 531, 849–859.

Hanajima R, Ugawa Y, Terao Y, Sakai K, Furubayashi T, Machii K, and Kanazawa I. (1998). Paired-pulse magnetic stimulation of the human motorcortex: Differences among I waves. *J Physiol* 509(Pt 2), 607–618.

Huang YZ, Edwards MJ, Rounis E, Bhatia KP, and Rothwell JC. (2005). Theta burst stimulation of the human motor cortex. *Neuron* 45, 201–206.

Ilic TV, Korchounov A, and Ziemann U. (2002a). Complex modulation of human motor cortex excitability by the specific serotonin re-uptake inhibitor sertraline. *Neurosci Lett* 319, 116–120.

Ilic TV, Meintzschel F, Cleff U, Ruge D, Kessler KR, and Ziemann U. (2002b). Short-interval paired-pulse inhibition and facilitation of human motor cortex: The dimension of stimulus intensity. *J Physiol* 545, 153–167.

Ilmoniemi RJ and Karhu J. (2008). TMS and electroencephalography: Methods and current advances. In *The Oxford Handbook of Transcranial Stimulation* (eds. Wassermann E, Epstein C, Ziemann U, Walsh V, Paus T, and Lisanby S). Oxford University Press, Oxford, U.K., pp. 593–608.

Ilmoniemi RJ, Virtanen J, Ruohonen J, Karhu J, Aronen HJ, Näätänen R, and Katila T. (1997). Neuronal responses to magnetic stimulation reveal cortical reactivity and connectivity. *Neuroreport* 8, 3537–3540.

Inghilleri M, Berardelli A, Cruccu G, and Manfredi M. (1993). Silent period evoked by transcranial stimulation of the human cortex and cervicomedullary junction. *J Physiol* 466, 521–534.

Inghilleri M, Berardelli A, Marchetti P, and Manfredi M. (1996). Effects of diazepam, baclofen and thiopental on the silent period evoked by transcranial magnetic stimulation in humans. *Exp Brain Res* 109, 467–472.

Kicic D, Lioumis P, Ilmoniemi RJ, and Nikulin VV. (2008). Bilateral changes in excitability of sensorimotor cortices during unilateral movement: Combined electroencephalographic and transcranial magnetic stimulation study. *Neuroscience* 152, 1119–1129.

Kimiskidis VK, Papagiannopoulos S, Kazis DA, Sotirakoglou K, Vasiliadis G, Zara F, Kazis A, and Mills KR. (2006). Lorazepam-induced effects on silent period and corticomotor excitability. *Exp Brain Res* 173, 603–611.

Koch G, Fernandez Del Olmo M, Cheeran B, Ruge D, Schippling S, Caltagirone C, and Rothwell JC. (2007). Focal stimulation of the posterior parietal cortex increases the excitability of the ipsilateral motor cortex. *J Neurosci* 27(25), 6815–6822.

Koch G, Fernandez Del Olmo M, Cheeran B, Schippling S, Caltagirone C, Driver J, and Rothwell JC. (2008). Functional interplay between posterior parietal and ipsilateral motor cortex revealed by twin-coil transcranial magnetic stimulation during reach planning toward contralateral space. *J Neurosci* 28, 5944–5953.

Koch G, Franca M, Del Olmo MF, Cheeran B, Milton R, Alvarez SM, and Rothwell JC. (2006). Time course of functional connectivity between dorsal premotor and contralateral motor cortex during movement selection. *J Neurosci* 26, 7452–7459.

Komssi S and Kähkönen S. (2006). The novelty value of the combined use of electroencephalography and transcranial magnetic stimulation for neuroscience research. *Brain Res Rev* 52(1), 183–192. Review.

Komssi S, Kähkönen S, and Ilmoniemi RJ. (2004). The effect of stimulus intensity on brain responses evoked by transcranial magnetic stimulation. *Hum Brain Mapp* 21, 154–164.

Korchounov A, Ilic TV, and Ziemann U. (2003). The alpha2-adrenergic agonist guanfacine reduces excitability of human motor cortex through disfacilitation and increase of inhibition. *Clin Neurophysiol* 114, 1834–1840.

Kujirai T, Caramia MD, Rothwell JC, Day BL, Thompson PD, Ferbert A, Wroe S, Asselman P, and Marsden CD. (1993). Corticocortical inhibition in human motor cortex. *J Physiol* 471, 501–519.

Lee L, Harrison LM, and Mechelli A. (2003). A report of the functional connectivity workshop, Dusseldorf. *Neuroimage* 19(2 Pt 1), 457–465.

Liepert J, Schwenkreis P, Tegenthoff M, and Malin JP. (1997). The glutamate antagonist riluzole suppresses intracortical facilitation. *J Neural Transm* 104, 1207–1214.

Mann EO and Paulsen O. (2007). Role of GABAergic inhibition in hippocampal network oscillations. *Trends Neurosci* 30, 343–349.

Mariorenzi R, Zarola F, Caramia MD, Paradiso C, and Rossini PM. (1991). Non-invasive evaluation of central motor tract excitability changes following peripheral nerve stimulation in healthy humans. *Electroencephalogr Clin Neurophysiol* 81, 90–101.

Matsumoto R, Nair DR, LaPresto E, Bingaman W, Shibasaki H, and Lüders HO. (2007). Functional connectivity in human cortical motor system: A cortico-cortical evoked potential study. *Brain* 130(Pt 1), 181–197.

McDonnell MN, Orekhov Y, and Ziemann U. (2006). The role of GABA(B) receptors in intracortical inhibition in the human motor cortex. *Exp Brain Res* 173(1), 86–93.

Meyer BU, Roricht S, Grafin von Einsiedel H, Kruggel F, and Weindl A. (1995). Inhibitory and excitatory interhemispheric transfers between motor cortical areas in normal humans and patients with abnormalities of the corpus callosum. *Brain* 118(Pt 2), 429–440.

Meyer BU, Röricht S, and Woiciechowsky C. (1998). Topography of fibres in the human corpus callosum mediating interhemispheric inhibition between the motor cortices. *Ann Neurol*, 4, 360–369.

Meynert T. (1865). Anatomie der hirnrinde und ihre verbindungsbahnen mit den empfindenden oberflachen und den bewegenden massen. In: *Leidesdorf's lehrbuch der phychischen krankheiten*. Erlangen, Germany.

Mochizuki H, Huang YZ, and Rothwell JC. (2004). Interhemispheric interaction between human dorsal premotor and contralateral primary motor cortex. *J Physiol* 15, 561(Pt 1):331–338.

Muellbacher W, Ziemann U, Boroojerdi B, Cohen L, and Hallett M. (2001). Role of the human motor cortex in rapid motor learning. *Exp Brain Res* 136, 431–438.

Murase N, Duque J, Mazzocchio R, and Cohen LG. (2004). Influence of interhemispheric interactions on motor function in chronic stroke. *Ann Neurol* 55, 400–409.

Nakamura H, Kitagawa H, Kawaguchi Y, and Tsuji H. (1997). Intracortical facilitation and inhibition after transcranial magnetic stimulation in conscious humans. *J Physiol* 498(Pt 3), 817–823.

Nikouline V, Ruohonen J, and Ilmoniemi RJ. (1999). The role of the coil click in TMS assessed with simultaneous EEG. *Clin Neurophysiol* 110(8), 1325–1328.

Oliveri M, Caltagirone C, Filippi MM, Traversa R, Cicinelli P, Pasqualetti P, and Rossini PM. (2000a). Paired transcranial magnetic stimulation protocols reveal a pattern of inhibition and facilitation in the human parietal cortex. *J Physiol* 529(Pt 2), 461–468.

Oliveri M, Rossini PM, Filippi MM, Traversa R, Cicinelli P, Palmieri MG, Pasqualetti P, and Caltagirone C. (2000b). Time-dependent activation of parieto-frontal networks for directing attention to tactile space. A study with paired transcranial magnetic stimulation pulses in right-brain-damaged patients with extinction. *Brain* 123(Pt 9), 1939–1947.

Pascual-Leone A, Nguyet D, Cohen LG, Brasil-Neto JP, Cammarota A, and Hallett M. (1995). Modulation of muscle responses evoked by transcranial magnetic stimulation during the acquisition of new fine motor skills. *J Neurophysiol* 74, 1037–1045.

Patton HD and Amassian VE. (1954). Single- and multiple-unit analysis of cortical stage of pyramidal tract activation. *J Neurophysiol* 17, 345–363.

Paulus W, Classen J, Cohen LG, Large CH, Di Lazzaro V, Nitsche M, Pascual-Leone A, Rosenow F, Rothwell JC, and Ziemann U. (2008). State of the art: Pharmacologic effects on cortical excitability measures tested by transcranial magnetic stimulation. *Brain Stimul* 1, 151–163.

Paus T, Jech R, Thompson CJ, Comeau R, Peters T, and Evans AC. (1997). Transcranial magnetic stimulation during positron emission tomography: A new method for studying connectivity of the human cerebral cortex. *J Neurosci* 17, 3178–3184.

Paus T, Sipilä PK, and Strafella AP. (2001). Synchronization of neuronal activity in the human primary motor cortex by transcranial magnetic stimulation: An EEG study. *J Neurophysiol* 86, 1983–1990.

Ridding MC, Brouwer B, and Nordstrom MA. (2000). Reduced interhemispheric inhibition in musicians. *Exp Brain Res* 133, 249–253.

Ridding MC, Inzelberg R, and Rothwell JC. (1995a). Changes in excitability of motor cortical circuitry in patients with Parkinson's disease. *Ann Neurol* 37, 181–188.

Ridding MC and Rothwell JC. (1997). Stimulus/response curves as a method of measuring motor cortical excitability in man. *Electroencephalogr Clin Neurophysiol* 105, 340–344.

Ridding MC, Sheean G, Rothwell JC, Inzelberg R, and Kujirai T. (1995b). Changes in the balance between motor cortical excitation and inhibition in focal, task specific dystonia. *J Neurol Neurosurg Psychiatry* 59, 493–498.

Ridding MC, Taylor JL, and Rothwell JC. (1995c). The effect of voluntary contraction on cortico-cortical inhibition in human motor cortex. *J Physiol (Lond)* 487, 541–548.

Rosenthal J, Waller HJ, and Amassian VE. (1967). An analysis of the activation of motor cortical neurons by surface stimulation. *J Neurophysiol* 30(4), 844–858.

Rossini PM, Barker AT, Berardelli A, Caramia MD, Caruso G et al. (1994). Non-invasive electrical and magnetic stimulation of the brain, spinal cord and roots: Basic principles and procedures for routine clinical application. *Electroencephalogr Clin Neurophysiol* 91, 79–92.

Rossini PM, Caramia M, and Zarola F. (1987). Central motor tract propagation in man: Studies with non-invasive, unifocal, scalp stimulation. *Brain Res* 415, 211–225.

Rossini PM, Marciani MG, Caramia M, Roma V, and Zarola F. (1985). Nervous propagation along 'central' motor pathways in intact man: Characteristics of motor responses to 'bifocal' and 'unifocal' spine and scalp non-invasive stimulation. *Electroencephalogr Clin Neurophysiol* 61, 272–286.

Rothwell JC. (2011). Using transcranial magnetic stimulation methods to probe connectivity between motor areas of the brain. *Hum Mov Sci* 30, 906–915.

Siebner HR, Mentschel C, Auer C, Lehner C, and Conrad B. (2000). Repetitive transcranial magnetic stimulation causes a short-term increase in the duration of the cortical silent period in patients with Parkinson's disease. *Neurosci Lett* 284, 147–150.

Sohn YH, Kang SY, and Hallett M. (2005). Corticospinal disinhibition during dual action. *Exp Brain Res* 162, 95–99.

Stefan K, Kunesch E, Cohen LG, Benecke R, and Classen J. (2000). Induction of plasticity in the human motor cortex by paired associative stimulation. *Brain* 123(Pt 3), 572–584.

Stinear CM and Byblow WD. (2003). Role of intracortical inhibition in selective hand muscle activation. *J Neurophysiol* 89, 2014–2020.

Strafella AP, Valzania F, Nassetti SA, Tropeani A, Bisulli A, Santangelo M, and Tassinari CA. (2000). Effects of chronic levodopa and pergolide treatment on cortical excitability in patients with Parkinson's disease: A transcranial magnetic stimulation study. *Clin Neurophysiol* 111, 1198–1202.

Tamás G, Lorincz A, Simon A, and Szabadics J. (2003). Identified sources and targets of slow inhibition in the neocortex. *Science* 299, 1902–1905.

Tegenthoff M, Vorgerd M, Juskowiak F, Roos V, and Malin JP. (1996). Postexcitatory inhibition after transcranial magnetic single and double brain stimulation in Huntington's disease. *Electroencephalogr Clin Neurophysiol* 101, 298–303.

Tokimura H, Di LV, Tokimura Y, Oliviero A, Profice P, Insola A, Mazzone P, Tonali P, and Rothwell JC. (2000). Short latency inhibition of human hand motor cortex by somatosensory input from the hand. *J Physiol (Lond)* 523(Pt 2), 503–513.

Tokimura H, Ridding MC, Tokimura Y, Amassian VE, and Rothwell JC. (1996). Short latency facilitation between pairs of threshold magnetic stimuli applied to human motor cortex. *Electroencephalogr Clin Neurophysiol* 101, 263–272.

Traub RD, Spruston N, Soltesz I, Konnerth A, Whittington MA, and Jefferys GR. (1998). Gamma-frequency oscillations: A neuronal population phenomenon, regulated by synaptic and intrinsic cellular processes, and inducing synaptic plasticity. *Prog Neurobiol* 55, 563–575.

Triggs WJ, Cros D, Macdonell RA, Chiappa KH, Fang J, and Day BJ. (1993). Cortical and spinal motor excitability during the transcranial magnetic stimulation silent period in humans. *Brain Res* 628, 39–48.

Ugawa Y, Hanajima R, and Kanazawa I. (1993). Interhemispheric facilitation of the hand area of the human motor cortex. *Neurosci Lett* 160, 153–155.

Valls-Sole J, Pascual-Leone A, Brasil-Neto JP, Cammarota A, McShane L, and Hallett M. (1994). Abnormal facilitation of the response to transcranial magnetic stimulation in patients with Parkinson's disease. *Neurology* 44, 735–741.

Valls-Sole J, Pascual-Leone A, Wassermann EM, and Hallett M. (1992). Human motor evoked responses to paired transcranial magnetic stimuli. *Electroencephalogr Clin Neurophysiol* 85, 355–364.

Van Der Werf YD and Paus T. (2006). The neural response to transcranial magnetic stimulation of the human motor cortex. I. Intracortical and cortico-cortical contributions. *Exp Brain Res* 175(2), 231–245.

Van Der Werf YD Sadikot AF, Strafella AP, and Paus T. (2006). The neural response to transcranial magnetic stimulation of the human motor cortex. II. Thalamocortical contributions. *Exp Brain Res* 175(2), 246–255.

Vecchio F, Babiloni C, Ferreri F, Buffo P, Cibelli G, Curcio G, van Dijkman S, Melgari JM, Giambattistelli F, and Rossini PM. (2010). Mobile phone emission modulates interhemispheric functional coupling of EEG alpha rhythms in elderly compared to young subjects. *Clin Neurophysiol* 121, 163–171.

Vercauteren K, Pleysier T, Van Belle L, Swinnen SP, and Wenderoth N. (2008). Unimanual muscle activation increases interhemispheric inhibition from the active to the resting hemisphere. *Neurosci Lett* 445, 209–213.

Ziemann U. (2003). Pharmacology of TMS. *Suppl Clin Neurophysiol* 56, 226–231. Review.

Ziemann U, Chen R, Cohen LG, and Hallett M. (1998a). Dextromethorphan decreases the excitability of the human motor cortex. *Neurology* 51, 1320–1324.

Ziemann U, Lönnecker S, Steinhoff BJ, and Paulus W. (1996a). Effects of antiepileptic drugs on motor cortex excitability in humans: A transcranial magnetic stimulation study [see comments]. *Ann Neurol* 40, 367–378.

Ziemann U, Lonnecker S, Steinhoff BJ, and Paulus W. (1996b). The effect of lorazepam on the motor cortical excitability in man. *Exp Brain Res* 109, 127–135.

Ziemann U, Netz J, Szelenyi A, and Homberg V. (1993). Spinal and supraspinal mechanisms contribute to the silent period in the contracting soleus muscle after transcranial magnetic stimulation of human motor cortex. *Neurosci Lett* 156, 167–171.

Ziemann U and Rothwell JC. (2000). I-waves in motor cortex. *J Clin Neurophysiol* 17(4), 397–405.

Ziemann U, Rothwell JC, and Ridding MC. (1996c). Interaction between intracortical inhibition and facilitation in human motor cortex. *J Physiol* 496, 873–881.

Ziemann U, Tergau F, Wassermann EM, Wischer S, Hildebrandt J, and Paulus W. (1998b). Demonstration of facilitatory I wave interaction in the human motor cortex by paired transcranial magnetic stimulation. *J Physiol (Lond)* 511, 181–190.

Zoghi M, Pearce SL, and Nordstrom MA. (2003). Differential modulation of intracortical inhibition in human motor cortex during selective activation of an intrinsic hand muscle. *J Physiol* 550, 933–946.

9 Understanding Homeostatic Metaplasticity

Anke Karabanov, Ulf Ziemann, Joseph Classen, and Hartwig Roman Siebner

CONTENTS

Throughout life the brain maintains a remarkable potential to reorganize in response to experience and environmental changes and to compensate for brain damage or neurological disorders (Sanes and Donoghue, 2000). To achieve this level of flexibility, the brain needs to be able to up- and down-regulate synaptic activity while keeping a relatively stable equilibrium of activity over time. The term synaptic plasticity refers to up- and down-regulation in direct response to external stimuli whereas *homeostatic plasticity* describes regulatory processes guaranteeing stable levels of synaptic activity over time. Metaplasticity refers to the malleability of plasticity (synaptic or non-synaptic) by prior activity (Abraham, 2008).

At the synaptic level, long-term potentiation (LTP) and long-term depression (LTD) have been identified as key mechanisms for synaptic up- and down-regulation. LTP and LTD cause lasting modifications in synaptic strength and, hereby, enable neurons to dynamically adapt the functional weight of specific synaptic inputs. There is ample evidence that both LTP and LTD are, at least partially, N-methyl-D-aspartate (NMDA) receptor mediated (Tsumoto, 1992; Turrigiano and Nelson, 2000). The function of NMDA receptors can be conceptualized as "coincidence detectors," which allow neurons to modulate the strength of a given synapse by relating it to other inputs that the postsynaptic cell receives at the same time (Shouval et al., 2002). Coincidence detection can be achieved by removing the Mg^{2+} block of the NMDA receptor when both the pre- and postsynaptic cells are depolarized simultaneously.

The removal of the Mg^{2+} block triggers the influx of calcium in the postsynaptic neuron. This activity-induced increase in postsynaptic Ca^{2+} levels triggers LTP or LTD (Tsumoto, 1992; Abbott and Nelson, 2000). Whether LTP or LTD is triggered depends on the amount and slope of postsynaptic Ca^{2+} influx. A fast, large increase in calcium selectively triggers LTP while a slow, modest increase triggers LTD (Lisman, 1989; Artola and Singer, 1993; Yang et al., 1999). The concept of a "threshold" for LTP and LTD induction that depends on the dynamics of postsynaptic Ca^{2+} influx is also supported by experiments in rat visual cortex: the same tetanic stimulation protocol could induce either LTP or LTD depending on the level of postsynaptic depolarization, which was pharmacologically manipulated by the gamma-aminobutyric acid A ($GABA_a$) receptor antagonist bicuculline (Artola et al., 1990). LTD was induced only when the postsynaptic depolarization exceeded a critical level, but stayed below the threshold for LTP induction. These observations imply that the successful induction of synaptic plasticity (LTD and LTD) does depend not only on a certain stimulation protocol but also on the excitability level of the postsynaptic neuron at the time of stimulation by influencing the magnitude and temporal dynamics of the stimulation-induced postsynaptic Ca^{2+} signal.

LTP or LTD is expressed in the primary motor cortex (M1) under physiological conditions (Rioult-Pedotti et al., 1998, 2000): in rats that had been trained for 5 days in a skilled reaching task, the amplitude of field potentials evoked by stimulation of rat M1 horizontal connections in the motor forelimb region contralateral to the trained limb (i.e., the trained M1) was significantly increased relative to the contralateral "untrained" M1. The trained M1 also expressed less LTP but more LTD as compared with control rats without training. This seminal work has two important implications. First, the results showed that in rats, motor skill learning shares common mechanisms with LTP. Second, these studies demonstrated that the ability to induce LTP and LTD is modulated by previous learning experience.

9.1 HOMEOSTATIC PLASTICITY

LTP and LTD of synaptic efficacy play a crucial role for learning and memory (Malenka, 1994; Bergmann et al., 2008; Ziemann and Siebner, 2008), but the positive feedback nature of synaptic plasticity threatens the stability of neural networks (Abbott and Nelson, 2000; Turrigiano and Nelson, 2000, 2004): Without additional safeguard mechanisms, LTP would reinforce itself promoting further LTP induction. Likewise, LTD would tend to trigger further LTD. "Unsupervised" LTP or LTD would eventually result in extreme neural states, resulting in excessive firing (in the case of excessive LTP) or silencing (in the case of excessive LTD). This "runaway" problem compromises information storage and transfer in neural networks. An extensive body of research has shown that the expression of LTP and LTD is effectively controlled by a multitude of homeostatic regulatory mechanisms, which stabilize the synaptic excitability level within a physiologically safe range through modulation of synaptic efficacy and membrane excitability (Turrigiano, 2000; Turrigiano and Nelson, 2000, 2004; Bear, 2003; Abraham, 2008).

Whenever mechanisms counteract the destabilizing effects of self-enhancing LTP or LTD, they are referred to as homeostatic plasticity. Homeostatic plasticity represents a form of "metaplasticity" as it is triggered by a lasting change in synaptic efficiency that changes the level of postsynaptic activity (Turrigiano, 2000; Turrigiano and Nelson, 2000, 2004; Bear, 2003; Abraham, 2008). Therefore, homeostatic interactions have to be investigated in a setting of metaplasticity requiring consecutive application of interventions. One intervention causing a change in synaptic efficiency (i.e., the priming intervention) is followed by a second intervention, which probes the induced homeostatic response. The priming intervention induces a lasting change in the tendency to express LTP and LTD, which can be probed by the second intervention. The second intervention is critical to reveal a homeostatic modulation of the threshold to induce LTP or LTD depending on the priming protocol. This "metaplastic setting" to trace prolonged homeostatic changes in the ability to induce LTP and LTD has to be distinguished from "associative protocols," which apply two plasticity-inducing interventions concurrently and exploit the acute interactions between the two protocols in terms of inducing lasting changes in synaptic strength (Turrigiano, 2000; Turrigiano and Nelson, 2000, 2004; Bear, 2003; Abraham, 2008).

While most homeostatic interactions are investigated in a setting of metaplasticity with consecutive application of interventions, the success rate for inducing synaptic plasticity can also be modulated acutely by manipulating the excitability level of the stimulated neurons at the time of conditioning, for instance, by application of CNS-active drugs or hormones. This *concurrent* manipulation of the neural excitability threshold at the time of LTP or LTD induction in order to influence the efficacy of the LTP or LTD induction protocol is referred to as "gating" (Quartarone et al., 2006; Hallett, 2007; Ziemann and Siebner, 2008). For instance, the efficacy of an LTP-inducing protocol can be markedly increased by manipulations that concurrently weaken the excitability of intracortical inhibitory circuits during the administration of an LTP-inducing protocol. It is critical to bear in mind that gating does not invoke metaplasticity and, thus, needs to be distinguished from homeostatic plasticity (Quartarone et al., 2006; Hallett, 2007; Ziemann and Siebner, 2008).

Homeostatic plasticity can be both homo- and heterosynaptic in nature. In homosynaptic plasticity, the changeability of synaptic excitability is altered by prior activity of the same synaptic pathway. In heterosynaptic plasticity, the changeability of synaptic excitability is altered by prior activity of neighboring synapses. One of the most influential conceptual models of heterosynaptic homeostatic plasticity is the Bienenstock–Cooper–Munro (BCM) theory of a "sliding threshold" of bidirectional synaptic plasticity (Bienenstock et al., 1982). The core postulate of the BCM theory is that the threshold for LTP and LTD induction is not fixed but varies as a function of the integrated level of previous synaptic activity. According to the BCM theory, a history of low synaptic activity will lower the modification threshold favoring LTP induction, whereas a history of high synaptic activity will shift the modification threshold favoring the induction of LTD (Figure 9.1). Several lines of experiments in both humans and animals provided experimental evidence that the threshold for LTP and LTD induction indeed can slide up and down depending on the level of previous postsynaptic activity (Kirkwood et al., 1996; Wang and Wagner, 1999; Hamada et al., 2009; Hamada and Ugawa, 2010). Homeostatic and synaptic plasticity may share the same molecular substrates (Turrigiano and Nelson, 2000). Like LTP/LTD, homeostatic plasticity is NMDA receptor dependent, involving a shift in the regulatory

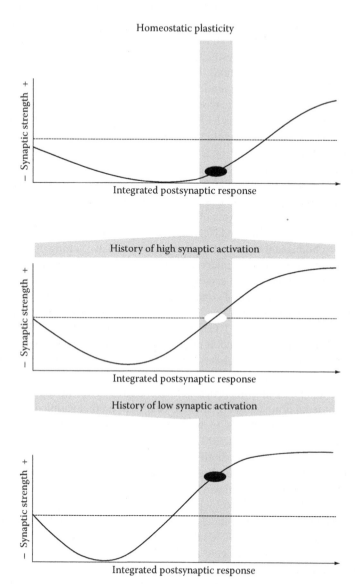

FIGURE 9.1   Change in synaptic strength as a function of the integrated postsynaptic response at baseline (middle graph) and in response to a history of high (upper diagram) or low (lower diagram) synaptic activation. At baseline, a given stimulation (indicated by the gray bar) does not result in any overt change in synaptic strength (shown by the white circle). Lower diagram: When the same stimulation is applied after a period of low synaptic activity (e.g., after 1 Hz TMS or iTBS), the LTP induction threshold slides to the left and the stimulation leads to LTP (shown by the black circle). Upper diagram: When the same stimulation is applied after a period of high synaptic activity (e.g., after 10 Hz TMS or cTBS), the LTP induction threshold slides to the right and the stimulation leads now to LTD (shown by the gray circle).

subunit expression of the NMDA receptors: low levels of postsynaptic activity cause the 2A subunit of the receptor to be replaced with the 2B subunit. NMDA receptors containing a 2B subunit allow longer current influx and can carry more Ca^{2+} per current unit. This results in stronger intracellular Ca^+ and in turn stronger postsynaptic activation and increased likelihood of LTP induction (Bear, 2003; Philpot et al., 2007; Yashiro and Philpot, 2008).

9.2   PLASTICITY-INDUCING INTERVENTIONS IN HUMANS

In animal preparations, LTD is often induced by low-frequency (≤ 1 Hz) synaptic stimulation whereas LTP is induced by higher-frequency stimulation (≥ 20 Hz) (Cooke and Bliss, 2006; Massey and Bashir, 2007) In human cortex, transcranial brain stimulation (TMS) is an effective means to induce LTP-like and LTD-like plasticity, using stimulation parameters similar to those found effective in slice preparations: high-frequency repetitive magnetic stimulation (rTMS; ≥ 5 Hz) (Takano et al., 2004), intermittent theta burst stimulation (iTBS) (Huang et al., 2005), anodal transcranial direct current stimulation (aTDCS) (Nitsche and Paulus, 2000; Nitsche et al., 2003), paired associative stimulation at an interstimulus interval (ISI) of around 25 ms (PAS_{25ms}) (Stefan et al., 2000), and quadripulse stimulation with and interstimulus interval of 1.5–10 ms (QPS_{short}) (Hamada et al., 2007) increase excitability in the stimulated area (Quartarone et al., 2006; Hallett, 2007; Ziemann and Siebner, 2008). In all these protocols, the excitability increase outlasts the stimulation time. Magnitude and duration of this effect depend on several factors such as stimulation intensity, stimulation duration, and protocol type, but also sex, age, CNS-active drugs, and single-nucleotide polymorphisms of important regulating molecules such as brain-derived neurotrophic factor (BDNF) and catechol-O-methyltransferase (COMT) (Ridding and Ziemann, 2010). All the aforementioned protocols can also decrease cortical excitability if their parameters are slightly changed: rTMS with a frequency of around 1 Hz (Chen et al., 1997), continuous theta burst stimulation (cTBS) (Huang et al., 2005), cathodal transcranial direct current stimulation (cTDCS) (Nitsche and Paulus, 2000), PAS at ISIs of around 10 ms (PAS_{10ms}) (Stefan et al., 2000), as well as quadripulse stimulation with ISIs of ≥ 30 ms (QPS_{long}) (Hamada et al., 2007) decrease excitability, outlasting stimulation time (Quartarone et al., 2006; Hallett, 2007; Ziemann and Siebner, 2008). In addition to these protocols, which induce a rather artificial pattern of neural stimulation, physiological patterns of neural activity associated with motor learning can also alter cortical plasticity. For instance, a relatively short period of motor sequence learning can modulate the cortical output maps of the trained muscle (Pascual-Leone et al., 1995) and increase the excitability of the motor cortex (Classen et al., 1998). Several studies have provided evidence that LTP-like processes underlie learning-induced cortical plasticity in humans, since motor learning modulates subsequent LTP and LTD induction in a similar fashion to Rioult-Pedotti's seminal animal experiments (Ziemann et al., 2004; Stefan et al., 2006; Rosenkranz et al., 2007; Jung and Ziemann, 2009). The evidence for the LTP-like nature of motor learning in humans will be discussed in greater detail in the next section.

9.3 INVESTIGATING HUMAN HOMEOSTATIC PLASTICITY

Homeostatic plasticity following the BCM theory predicts that high levels of prior activity favor the induction of LTD while low levels of prior activity favor LTP. Homeostatic patterns can be tested in the human cortex by pairing excitatory and inhibitory TMS protocols (Iyer et al., 2003; Lang et al., 2004; Muller et al., 2007; Nitsche et al., 2007; Todd et al., 2009): one of the first studies applied facilitatory aTDCS, inhibitory cTDCS, or sham stimulation prior to a 15 min treatment session of low-intensity 1 Hz TMS (Siebner et al., 2004). Facilitatory preconditioning caused the subsequent 1 Hz stimulation to reduce cortical excitability whereas inhibitory preconditioning caused the subsequent 1 Hz stimulation to increase cortical excitability. When preconditioned by sham TDCS, the 1 Hz protocol did not have an effect on cortical excitability (Siebner et al., 2004). The differential effects of preconditioning stimulation strongly suggest the existence of homeostatic mechanisms in the human cortex. Other studies showed similar homeostatic effects when pairing a "priming" TMS or TDCS protocol with a "treatment" TMS or TDCS protocol (Iyer et al., 2003; Lang et al., 2004; Muller et al., 2007; Nitsche et al., 2007; Todd et al., 2009). Even though these studies used different stimulation protocols, they provide convergent evidence that priming enhances the effect of treatment stimulation if the prime has the opposite effect on excitability as the treatment stimulation. If both priming and treatment have the same effect on excitability, the prime often reverses the effect normally seen after the treatment. It may be important to note that treatment effect reversal by the priming procedure is not necessarily embraced by the concept of homeostatic plasticity, because homeostasis of network excitability would be achieved by attenuating (up to completely blocking) the efficacy of the treatment. A reversal of the excitability has also been observed when giving the same protocol twice (Muller et al., 2007; Nitsche et al., 2007, 2008; Todd et al., 2009), when changing the duration of stimulation, or, in case of 5 Hz TMS, when omitting breaks between stimulation trains (Gentner et al., 2008; Gamboa et al., 2010; Rothkegel et al., 2010).

The timing with which priming and treatment stimulation are paired is of crucial importance, and only few studies have examined the time course of homeostatic plasticity (Fricke et al., 2011). Fricke and coworkers paired two identical 5 min sessions of aTDCS or cTDCS. Priming and treatment TDCS were administered with a 30, 3, or 0 min intervals. With the 0 min interval, the effects of both aTDCS and cTDCS were prolonged, while there was no priming effect when the sessions were separated by 30 min. However, at the interval of 3 min, the two TDCS protocols interacted in a homeostatic manner. This finding suggests that a homeostatic response pattern may be triggered by transcranial priming protocols within several minutes. However, the critical time window during which homeostatic metaplasticity can be observed may differ considerably depending on which priming protocol is applied. This important issue remains to be addressed in future studies.

Quadripulse stimulation (QPS) has been introduced in 2007 as an effective burst TMS protocol that can induce lasting changes in corticomotor excitability (Hamada et al., 2007, 2008). QPS consists of 360 four-pulse bursts with an inter-burst interval of 5 s. Depending on the ISI that separates the stimuli of a single train, QPS induces either an LTP-like increase in corticomotor excitability (i.e., at short

ISIs of 1.5 or 5 ms) or an LTD-like decrease in corticomotor excitability (i.e., at long ISIs of 30 or 100 ms) (Hamada et al., 2007, 2008). An LTP–LTD induction curve was derived from the results, which plotted the induced LTP- or LTD-like effects (y-axis) against the stimulation frequency of QPS (i.e., the reciprocal of the ISI of QPS, x-axis). A priming QPS protocol induced bidirectional shifts in the ability to induce LTP- or LTD-like effects with a subsequent QPS test protocol, and the priming effects closely resembled the predictions of the BCM model (Hamada et al., 2007, 2008): 10 min of priming QPS at a short ISI of 5 ms resulted in switches of LTP-like to LTD-like plasticity for most of the short QPS testing protocols, causing a rightward shift of the LTP–LTD induction curve. Only QPS at very short ISIs were still effective in producing LTP-like facilitation. A reciprocal effect was induced by priming QPS at a long ISI of 50 ms. Now test QPS at various ISIs revealed a leftward shift of the LTP–LTD induction curve. Priming with long 50 ms QPS resulted in switches of LTD-like to LTP-like plasticity for the long QPS testing protocols. This priming-induced bidirectional shift of the LTP–LTD induction curve strongly resembles the homeostatic properties as predicted by the BCM theory with the demonstration of a sliding crossover point from LTD- to LTP-like plasticity (for review also see: Hamada and Ugawa, 2010). The results of QPS priming and its resemblance to the BCM curves are depicted in Figure 9.2.

There is evidence that homeostatic interactions can be triggered in human M1 when the priming protocol is given over secondary motor areas: Potter-Nerger and coworkers showed that rTMS priming over the dorsal premotor cortex (dPMC) had a classical homeostatic effect on subsequent M1 excitability: combining an "inhibitory" prime over dPMC with an inhibitory treatment protocol over M1 resulted in an increase in M1 excitability whereas combining an excitatory prime over PMC with and excitatory treatment over M1 resulted in a decrease in M1 excitability (Potter-Nerger et al., 2009). Homeostatic modulations were also demonstrated in M1 using QPS test protocols at various ISIs after QPS priming of the supplementary motor area (Hamada et al., 2009). Likewise, priming 1 Hz rTMS of M1 influenced the LTP-like effects of iTBS in the opposite M1 in a homeostatic fashion (Ragert et al., 2009). These findings indicate that homeostatic interactions can be elicited through different input channels in the intact human M1.

9.4 HOMEOSTATIC PLASTICITY AND MOTOR LEARNING

Stimulation-induced plasticity has also been shown to interact homeostatically with motor learning: Ziemann and coworkers showed that a simple motor learning task could act as a "primer" to subsequent PAS protocols: motor learning prevented subsequent induction of LTP-like PAS effects and enhanced the induction of subsequent LTD-like PAS effects (Ziemann et al., 2004). The effect that motor learning has on subsequent PAS stimulation is dependent on the learning phase: training a novel motor task reversed the effect of subsequent facilitatory PAS whereas training of a well-practiced task did not have a significant modulating effect on subsequent PAS (Rosenkranz et al., 2007).

These studies show that motor learning has an impact on LTP- and LTD-like plasticity induced by subsequent brain stimulation, but can prior neuronal activity also

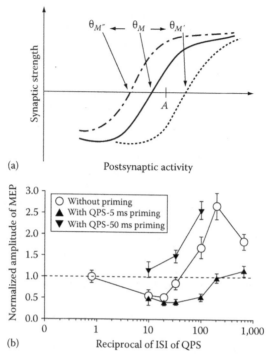

FIGURE 9.2 Shows the bidirectional shift of the LTP–LTD induction curve predicted by the BCM theory (a) and induced by QPS priming (b). (a) In BCM theory, the LTD–LTP cross-over point (θ_M) slides to the right on the x-axis if the preceding neuronal activity is high ($\theta_{M'}$), and to the left if preceding activity is low ($\theta_{M''}$). (b) QPS with priming over $M1$. The normalized amplitudes of MEP at 30 min post conditioning as a function of the reciprocal of ISI of QPS (in Hertz) with and without priming over $M1$. QPS-5 ms priming over $M1$ resulted in a rightward shift, whereas QPS-50 ms priming produced a leftward shift. The x-axis is logarithmically scaled. (Reprinted from Hamada, M. and Ugawa, Y., *Restor. Neurol. Neurosci.*, 28, 419, 2010. With permission from IOS Press and the original authors.)

modify subsequent motor learning? According to the rules of homeostatic plasticity, it would be plausible to expect that an excitability decreasing stimulation protocol could improve subsequent motor learning. Few studies have investigated this question, and the results are complex. When a 30 min training session that required rapid thumb abduction movements was primed with PAS stimulation at a 0 min delay, both excitability-decreasing and excitability-increasing PAS enhanced motor learning (Jung and Ziemann, 2009). However, if the priming PAS protocol was given 90 min prior to learning, excitability-decreasing PAS still had a beneficial effect on motor learning, but excitability-increasing PAS disrupted motor learning, indicating a full homeostatic interaction of brain stimulation and motor learning. These results suggest that non-homeostatic mechanisms may play a role when the interval between stimulation protocol and motor training is short, whereas homeostatic interactions with motor learning do occur at longer delays between priming and learning.

Studies of homeostatic mechanisms in the context of motor learning are more complicated because other mechanisms than synaptic strengthening may play a role.

A recent study reported that a priming iTBS enhanced performance in a subsequent ballistic motor learning task (Teo et al., 2011). This beneficial priming effect of iTBS on subsequent motor learning was blocked by nicotine. Behavioral analysis and modeling suggested that iTBS might have facilitated performance by increasing motor output variability. The authors hypothesized that nicotine blocked this effect presumably by increasing the signal-to-noise ratio in cerebral cortex (Teo et al., 2011). This may explain why other studies, which assessed the priming effects of brain stimulation on motor learning, failed to reveal homeostatic effects (Kuo et al., 2008). To further exploit the possibility of enhancing motor learning by homeostatic mechanisms, it is important to better understand through which mechanisms brain stimulation modulates motor learning and how these mechanisms are influenced by priming stimulation protocols.

TMS has been successfully used to enhance learning when transcranial stimulation was given *concurrently* with the motor learning task (Nitsche et al., 2003; Antal et al., 2004; Reis et al., 2009; Reis and Fritsch, 2011; Schambra et al., 2011; Stagg et al., 2011). It is important to stress that these learning-enhancing effects cannot be attributed to homeostatic plasticity as brain stimulation and learning were not separated in time. The most likely mechanism that triggered the improvement in motor learning is "gating," which is a stimulation-induced increase in net calcium influx into the targeted cortical neurons, thereby inducing a transient increase in excitability of the stimulated cortex and facilitating LTP-like plasticity during learning. Please note that gating is a non-homeostatic phenomenon, as it does not shift the LTP–LTD induction curve of the stimulated neurons (Quartarone et al., 2006; Hallett, 2007; Ziemann and Siebner, 2008). Gating rather facilitates the learning-induced calcium influx. Thus, a stronger intracellular signal can more easily induce LTP in the cortical neurons involved in motor learning without changing the threshold for inducing LTP or LTD. A recent pharmacological TMS study provided mechanistic support to the relevance of the magnitude of calcium signaling for the induction of LTP- and LTD-like phenomena in humans (Wankerl et al., 2010): The LTP-like effects of cTBS on corticomotor excitability were reversed when healthy volunteers were treated with nimodipine, an L-type voltage-gated Ca^{2+} channel antagonist. Pharmacological blockade of the NMDA receptor by dextromethorphan abolished both the LTD-like effect of cTBS caused by nimodipine and the normal LTP-like effect of cTBS alone in M1.

9.5   NON-HOMEOSTATIC METAPLASTICITY

There is ample evidence for homeostatic interactions in the human M1, but other forms of metaplasticity may also be expressed when a priming and a test stimulation protocol are subsequently applied: a priming protocol consisting of very low-frequency (0.1 Hz) rTMS given to M1 abolished the ability to induce LTP- and LTD-like plasticity in the primed M1 with subsequent PAS (Delvendahl et al., 2010; Siebner, 2010). When given alone, the 0.1 Hz rTMS protocol did not alter corticospinal excitability as measured by motor-evoked potential (MEP) amplitude, but increased short-interval and long-interval intracortical inhibition in the stimulated M1. It was proposed that the lasting increase in intracortical inhibitory circuits caused by the priming 0.1 Hz rTMS protocol blocked the ability of subsequent PAS to induce LTP- or LTD-like changes in corticospinal excitability, presumably by reducing the calcium influx into the corticospinal neurons during PAS. This mechanism does not invoke a homeostatic counter-regulation; it rather invokes a sort of "anti-gating" effect, which is a reduction of activity-induced calcium influx caused by increased excitability of intracortical inhibitory circuits (Delvendahl et al., 2010; Siebner, 2010).

A related but separate phenomenon is depotentiation (or de-depression), which erases LTP (or LTD) *after* LTD (or LTP) has been induced: a recent study showed that LTP-like effect induced by facilitatory iTBS was abolished (depotentiated), if followed by a short train of inhibitory cTBS. Vice versa, the LTD-like effect induced by cTBS protocol was abolished (de-depressed) if followed by a short train of facilitatory iTBS. If the short trains of TBS were given alone, they did not change corticomotor excitability (Huang et al., 2010). There is ample evidence for depotentiation and de-depression in the animal literature (Larson et al., 1993; Kulla and Manahan-Vaughan, 2000; Huang et al., 2001), and depotentiation has been implicated as a key factor in forgetting and learning reversal (Zhou and Poo, 2004). The 0–15 min period in which depotentiation is possible has been suggested to be a "grace period for the neurons to correct whatever mistakes they might have made before the mistakes are stabilized" (Zhou and Poo, 2004). Future studies need to explore the interplay between these non-homeostatic forms of cortical plasticity and homeostatic plasticity.

9.6 IMPAIRED HOMEOSTATIC PLASTICITY

The study of homeostatic plasticity with transcranial stimulation of M1 can be used to gain valuable insights into the mechanisms underlying the development of detrimental cortical plasticity in neurological and psychiatric diseases. For instance, it has been shown that homeostatic metaplasticity is dysfunctional in task-specific hand dystonia (Quartarone et al., 2005). In patients with writer's cramp and eight healthy age-matched controls, low-frequency 1 Hz rTMS was given after a priming transcranial direct current stimulation (tDCS) session. In accordance with the study by Siebner, anodal tDCS enhanced the inhibitory effect of subsequent 1 Hz rTMS on corticospinal excitability (Siebner et al., 2004; Quartarone et al., 2005). Conversely, cathodal TDCS flipped the aftereffect of 1 Hz rTMS, which now increased corticospinal excitability (Quartarone et al., 2005). In patients with writer's cramp, 1 Hz rTMS failed to induce consistent changes in corticospinal excitability regardless of the type of priming TDCS (Quartarone et al., 2005). Further, the normal inhibitory effect of cathodal TDCS was absent (Quartarone et al., 2005). Together, these results indicate dysfunctional homeostatic plasticity in writer's cramp, which might contribute to the development of maladaptive motor plasticity during skilled manual tasks. A more recent study showed a similar lack of homeostatic modulation of practice-dependent plasticity: Kang et al. (2011) observed that improvement in a motor training task was suppressed in healthy controls, when the M1 was primed by facilitatory PAS. No such modulation was observed in a group of dystonic patients. The lack of suppression correlated with the clinical severity of dystonia, suggesting that deficient homeostatic regulation in focal dystonia indeed impacts on motor learning (Quartarone et al., 2006).

The work in focal hand dystonia illustrates that the priming effects of transcranial cortical stimulation provide a powerful tool to detect alterations in homeostatic plasticity. We anticipate that studies of homeostatic plasticity will reveal valuable insights into the pathophysiology of other neuropsychiatric disorders associated with alterations in cortical excitability, such as epilepsy, migraine, or Parkinson's disease (Siniatchkin et al., 2011).

9.7 CONCLUDING REMARKS

In this chapter, we have reviewed how a wide array of brain stimulation protocols and motor learning tasks can influence brain plasticity. When two noninvasive brain stimulation protocols or a brain stimulation protocol and a motor learning task are applied sequentially, homeostatic effects can often be observed. They are caused by a lasting effect of the priming protocol on postsynaptic activity. If such mechanisms maintain a physiological level of neural activity, they are termed homeostatic. Homeostatic metaplasticity modulates the threshold required for LTP and LTD induction in the cortex and thereby can switch the direction of an intervention from an LTP- to an LTD-inducing protocol and vice versa. The interval that separates the priming intervention and the test intervention seems to be crucial for inducing homeostatic effects.

In most experiments on cortical homeostatic metaplasticity, the mean MEP amplitude is usually used to probe changes in corticomotor excitability. However, homeostatic metaplasticity is also expressed in other cortical areas such as the primary somatosensory cortex or the visual cortex and may be probed by recording the somatosensory-evoked potential or visual-evoked potential as a measure of cortical excitability (Bliem et al., 2008). Proton magnetic resonance spectroscopy offers another interesting option to probe homeostatic metaplasticity in the human cortex by monitoring changes in the regional GABA and glutamate content (Siniatchkin et al., 2011).

In addition to homeostatic metaplasticity, there are many other determinants of cortical plasticity (Ridding and Ziemann, 2010): genetic factors such as the BDNF and COMT single-nucleotide polymorphisms (Kleim et al., 2006; Witte et al., 2012), attention (Stefan et al., 2004), age (Muller-Dahlhaus et al., 2008), sex (Kuo et al., 2006; Sale et al., 2007; Fumagalli et al., 2010), time of day (Sale et al., 2007), and endogenous brain oscillations (Marshall et al., 2006). Future research needs to address how these factors impact on the expression of homeostatic metaplasticity in the human cortex.

To establish a more solid physiological framework that can be used to predict the effect of priming interventions on plasticity in the human cortex, it will be crucial to closely examine how different stimulation protocols, behavioral interventions, and pharmacological interventions interact and to keep the possibility of a whole array of other influencing variables in mind. It will also be of importance to further study the effect of repeated interventional sessions on plasticity and to investigate how these potentially long-term interactions could be used in a therapeutic setting to help ease the symptoms of various neurological and psychiatric diseases.

REFERENCES

Abbott LF and Nelson SB (2000) Synaptic plasticity: Taming the beast. *Nat Neurosci* 3 (Suppl):1178–1183.

Abraham WC (2008) Metaplasticity: Tuning synapses and networks for plasticity. *Nat Rev Neurosci* 9:387.

Antal A, Nitsche MA, Kincses TZ, Kruse W, Hoffmann KP, and Paulus W (2004) Facilitation of visuo-motor learning by transcranial direct current stimulation of the motor and extrastriate visual areas in humans. *Eur J Neurosci* 19:2888–2892.

Artola A, Brocher S, and Singer W (1990) Different voltage-dependent thresholds for inducing long-term depression and long-term potentiation in slices of rat visual cortex. *Nature* 347:69–72.

Artola A and Singer W (1993) Long-term depression of excitatory synaptic transmission and its relationship to long-term potentiation. *Trends Neurosci* 16:480–487.

Bear MF (2003) Bidirectional synaptic plasticity: From theory to reality. *Philos Trans R Soc Lond B Biol Sci* 358:649–655.

Bergmann TO, Molle M, Marshall L, Kaya-Yildiz L, Born J, and Roman Siebner H (2008) A local signature of LTP- and LTD-like plasticity in human NREM sleep. *Eur J Neurosci* 27:2241–2249.

Bienenstock EL, Cooper LN, and Munro PW (1982) Theory for the development of neuron selectivity: Orientation specificity and binocular interaction in visual cortex. *J Neurosci* 2:32–48.

Bliem B, Muller-Dahlhaus JF, Dinse HR, and Ziemann U (2008) Homeostatic metaplasticity in the human somatosensory cortex. *J Cogn Neurosci* 20:1517–1528.

Chen R, Classen J, Gerloff C, Celnik P, Wassermann EM, Hallett M, and Cohen LG (1997) Depression of motor cortex excitability by low-frequency transcranial magnetic stimulation. *Neurology* 48:1398–1403.

Classen J, Gerloff C, Honda M, and Hallett M (1998) Integrative visuomotor behavior is associated with interregionally coherent oscillations in the human brain. *J Neurophysiol* 79:1567–1573.

Cooke SF and Bliss TV (2006) Plasticity in the human central nervous system. *Brain* 129:1659–1673.

Delvendahl I, Jung NH, Mainberger F, Kuhnke NG, Cronjaeger M, and Mall V (2010) Occlusion of bidirectional plasticity by preceding low-frequency stimulation in the human motor cortex. *Clin Neurophysiol* 121:594–602.

Fricke K, Seeber AA, Thirugnanasambandam N, Paulus W, Nitsche MA, and Rothwell JC (2011) Time course of the induction of homeostatic plasticity generated by repeated transcranial direct current stimulation of the human motor cortex. *J Neurophysiol* 105:1141–1149.

Fumagalli M, Vergari M, Pasqualetti P, Marceglia S, Mameli F, Ferrucci R, Mrakic-Sposta S et al. (2010) Brain switches utilitarian behavior: Does gender make the difference? *PLoS One* 5:e8865.

Gamboa OL, Antal A, Moliadze V, and Paulus W (2010) Simply longer is not better: Reversal of theta burst after-effect with prolonged stimulation. *Exp Brain Res* 204:181–187.

Gentner R, Wankerl K, Reinsberger C, Zeller D, and Classen J (2008) Depression of human corticospinal excitability induced by magnetic theta-burst stimulation: Evidence of rapid polarity-reversing metaplasticity. *Cereb Cortex* 18:2046–2053.

Hallett M (2007) Transcranial magnetic stimulation: A primer. *Neuron* 55:187–199.

Hamada M, Hanajima R, Terao Y, Arai N, Furubayashi T, Inomata-Terada S, Yugeta A, Matsumoto H, Shirota Y, and Ugawa Y (2007) Quadro-pulse stimulation is more effective than paired-pulse stimulation for plasticity induction of the human motor cortex. *Clin Neurophysiol* 118:2672–2682.

Hamada M, Hanajima R, Terao Y, Okabe S, Nakatani-Enomoto S, Furubayashi T, Matsumoto H, Shirota Y, Ohminami S, and Ugawa Y (2009) Primary motor cortical metaplasticity induced by priming over the supplementary motor area. *J Physiol* 587:4845–4862.

Hamada M, Terao Y, Hanajima R, Shirota Y, Nakatani-Enomoto S, Furubayashi T, Matsumoto H, and Ugawa Y (2008) Bidirectional long-term motor cortical plasticity and metaplasticity induced by quadripulse transcranial magnetic stimulation. *J Physiol* 586:3927–3947.

Hamada M and Ugawa Y (2010) Quadripulse stimulation—A new patterned rTMS. *Restor Neurol Neurosci* 28:419–424.

Huang YZ, Edwards MJ, Rounis E, Bhatia KP, and Rothwell JC (2005) Theta burst stimulation of the human motor cortex. *Neuron* 45:201–206.

Huang CC, Liang YC, and Hsu KS (2001) Characterization of the mechanism underlying the reversal of long term potentiation by low frequency stimulation at hippocampal CA1 synapses. *J Biol Chem* 276:48108–48117.

Huang YZ, Rothwell JC, Lu CS, Chuang WL, Lin WY, and Chen RS (2010) Reversal of plasticity-like effects in the human motor cortex. *J Physiol* 588:3683–3693.

Iyer MB, Schleper N, and Wassermann EM (2003) Priming stimulation enhances the depressant effect of low-frequency repetitive transcranial magnetic stimulation. *J Neurosci* 23:10867–10872.

Jung P and Ziemann U (2009) Homeostatic and nonhomeostatic modulation of learning in human motor cortex. *J Neurosci* 29:5597–5604.

Kang JS, Terranova C, Hilker R, Quartarone A, and Ziemann U (2011) Deficient homeostatic regulation of practice-dependent plasticity in writer's cramp. *Cereb Cortex* 21:1203–1212.

Kirkwood A, Rioult MC, and Bear MF (1996) Experience-dependent modification of synaptic plasticity in visual cortex. *Nature* 381:526–528.

Kleim JA, Chan S, Pringle E, Schallert K, Procaccio V, Jimenez R, and Cramer SC (2006) BDNF val66met polymorphism is associated with modified experience-dependent plasticity in human motor cortex. *Nat Neurosci* 9:735–737.

Kulla A and Manahan-Vaughan D (2000) Depotentiation in the dentate gyrus of freely moving rats is modulated by D1/D5 dopamine receptors. *Cereb Cortex* 10:614–620.

Kuo MF, Paulus W, and Nitsche MA (2006) Sex differences in cortical neuroplasticity in humans. *Neuroreport* 17:1703–1707.

Kuo MF, Unger M, Liebetanz D, Lang N, Tergau F, Paulus W, and Nitsche MA (2008) Limited impact of homeostatic plasticity on motor learning in humans. *Neuropsychologia* 46:2122–2128.

Lang N, Siebner HR, Ernst D, Nitsche MA, Paulus W, Lemon RN, and Rothwell JC (2004) Preconditioning with transcranial direct current stimulation sensitizes the motor cortex to rapid-rate transcranial magnetic stimulation and controls the direction of after-effects. *Biol Psychiatry* 56:634–639.

Larson J, Xiao P, and Lynch G (1993) Reversal of LTP by theta frequency stimulation. *Brain Res* 600:97–102.

Lisman J (1989) A mechanism for the Hebb and the anti-Hebb processes underlying learning and memory. *Proc Natl Acad Sci USA* 86:9574–9578.

Malenka RC (1994) Synaptic plasticity in the hippocampus: LTP and LTD. *Cell* 78:535–538.

Marshall L, Helgadottir H, Molle M, and Born J (2006) Boosting slow oscillations during sleep potentiates memory. *Nature* 444:610–613.

Massey PV and Bashir ZI (2007) Long-term depression: Multiple forms and implications for brain function. *Trends Neurosci* 30:176–184.

Muller JF, Orekhov Y, Liu Y, and Ziemann U (2007) Homeostatic plasticity in human motor cortex demonstrated by two consecutive sessions of paired associative stimulation. *Eur J Neurosci* 25:3461–3468.

Muller-Dahlhaus JF, Orekhov Y, Liu Y, and Ziemann U (2008) Interindividual variability and age-dependency of motor cortical plasticity induced by paired associative stimulation. *Exp Brain Res* 187:467–475.

Nitsche MA, Cohen LG, Wassermann EM, Priori A, Lang N, Antal A, Paulus W et al. (2008) Transcranial direct current stimulation: State of the art 2008. *Brain Stimul* 1:206–223.

Nitsche MA and Paulus W (2000) Excitability changes induced in the human motor cortex by weak transcranial direct current stimulation. *J Physiol* 527 (Pt 3):633–639.

Nitsche MA, Roth A, Kuo MF, Fischer AK, Liebetanz D, Lang N, Tergau F, and Paulus W (2007) Timing-dependent modulation of associative plasticity by general network excitability in the human motor cortex. *J Neurosci* 27:3807–3812.

Nitsche MA, Schauenburg A, Lang N, Liebetanz D, Exner C, Paulus W, and Tergau F (2003) Facilitation of implicit motor learning by weak transcranial direct current stimulation of the primary motor cortex in the human. *J Cogn Neurosci* 15:619–626.

Pascual-Leone A, Nguyet D, Cohen LG, Brasil-Neto JP, Cammarota A, and Hallett M (1995) Modulation of muscle responses evoked by transcranial magnetic stimulation during the acquisition of new fine motor skills. *J Neurophysiol* 74:1037–1045.

Philpot BD, Cho KK, and Bear MF (2007) Obligatory role of NR2A for metaplasticity in visual cortex. *Neuron* 53:495–502.

Potter-Nerger M, Fischer S, Mastroeni C, Groppa S, Deuschl G, Volkmann J, Quartarone A, Munchau A, and Siebner HR (2009) Inducing homeostatic-like plasticity in human motor cortex through converging corticocortical inputs. *J Neurophysiol* 102:3180–3190.

Quartarone A, Rizzo V, Bagnato S, Morgante F, Sant'Angelo A, Romano M, Crupi D, Girlanda P, Rothwell JC, and Siebner HR (2005) Homeostatic-like plasticity of the primary motor hand area is impaired in focal hand dystonia. *Brain* 128:1943–1950.

Quartarone A, Siebner HR, and Rothwell JC (2006) Task-specific hand dystonia: Can too much plasticity be bad for you? *Trends Neurosci* 29:192–199.

Ragert P, Camus M, Vandermeeren Y, Dimyan MA, and Cohen LG (2009) Modulation of effects of intermittent theta burst stimulation applied over primary motor cortex (M1) by conditioning stimulation of the opposite M1. *J Neurophysiol* 102:766–773.

Reis J and Fritsch B (2011) Modulation of motor performance and motor learning by transcranial direct current stimulation. *Curr Opin Neurol* 24:590–596.

Reis J, Schambra HM, Cohen LG, Buch ER, Fritsch B, Zarahn E, Celnik PA, and Krakauer JW (2009) Noninvasive cortical stimulation enhances motor skill acquisition over multiple days through an effect on consolidation. *Proc Natl Acad Sci USA* 106:1590–1595.

Ridding MC and Ziemann U (2010) Determinants of the induction of cortical plasticity by non-invasive brain stimulation in healthy subjects. *J Physiol* 588:2291–2304.

Rioult-Pedotti MS, Friedman D, and Donoghue JP (2000) Learning-induced LTP in neocortex. *Science* 290:533–536.

Rioult-Pedotti MS, Friedman D, Hess G, and Donoghue JP (1998) Strengthening of horizontal cortical connections following skill learning. *Nat Neurosci* 1:230–234.

Rosenkranz K, Kacar A, and Rothwell JC (2007) Differential modulation of motor cortical plasticity and excitability in early and late phases of human motor learning. *J Neurosci* 27:12058–12066.

Rothkegel H, Sommer M, and Paulus W (2010) Breaks during 5 Hz rTMS are essential for facilitatory after effects. *Clin Neurophysiol* 121:426–430.

Sale MV, Ridding MC, and Nordstrom MA (2007) Factors influencing the magnitude and reproducibility of corticomotor excitability changes induced by paired associative stimulation. *Exp Brain Res* 181:615–626.

Sanes JN and Donoghue JP (2000) Plasticity and primary motor cortex. *Annu Rev Neurosci* 23:393–415.

Schambra HM, Abe M, Luckenbaugh DA, Reis J, Krakauer JW, and Cohen LG (2011) Probing for hemispheric specialization for motor skill learning: A transcranial direct current stimulation study. *J Neurophysiol* 106:652–661.

Shouval HZ, Bear MF, and Cooper LN (2002) A unified model of NMDA receptor-dependent bidirectional synaptic plasticity. *Proc Natl Acad Sci USA* 99:10831–10836.

Siebner HR (2010) A primer on priming the human motor cortex. *Clin Neurophysiol* 121:461–463.

Siebner HR, Lang N, Rizzo V, Nitsche MA, Paulus W, Lemon RN, and Rothwell JC (2004) Preconditioning of low-frequency repetitive transcranial magnetic stimulation with transcranial direct current stimulation: Evidence for homeostatic plasticity in the human motor cortex. *J Neurosci* 24:3379–3385.

Siniatchkin M, Sendacki M, Moeller F, Wolff S, Jansen O, Siebner H, and Stephani U (2011) Abnormal changes of synaptic excitability in migraine with aura. *Clin Neurophysiol* 122:2475–2481.

Stagg CJ, Jayaram G, Pastor D, Kincses ZT, Matthews PM, and Johansen-Berg H (2011) Polarity and timing-dependent effects of transcranial direct current stimulation in explicit motor learning. *Neuropsychologia* 49:800–804.

Stefan K, Kunesch E, Cohen LG, Benecke R, and Classen J (2000) Induction of plasticity in the human motor cortex by paired associative stimulation. *Brain* 123 (Pt 3):572–584.

Stefan K, Wycislo M, and Classen J (2004) Modulation of associative human motor cortical plasticity by attention. *J Neurophysiol* 92:66–72.

Stefan K, Wycislo M, Gentner R, Schramm A, Naumann M, Reiners K, and Classen J (2006) Temporary occlusion of associative motor cortical plasticity by prior dynamic motor training. *Cereb Cortex* 16:376–385.

Takano B, Drzezga A, Peller M, Sax I, Schwaiger M, Lee L, and Siebner HR (2004) Short-term modulation of regional excitability and blood flow in human motor cortex following rapid-rate transcranial magnetic stimulation. *Neuroimage* 23:849–859.

Teo JT, Swayne OB, Cheeran B, Greenwood RJ, and Rothwell JC (2011) Human theta burst stimulation enhances subsequent motor learning and increases performance variability. *Cereb Cortex* 21:1627–1638.

Todd G, Flavel SC, and Ridding MC (2009) Priming theta-burst repetitive transcranial magnetic stimulation with low- and high-frequency stimulation. *Exp Brain Res* 195:307–315.

Tsumoto T (1992) Long-term potentiation and long-term depression in the neocortex. *Prog Neurobiol* 39:209–228.

Turrigiano GG (2000) AMPA receptors unbound: Membrane cycling and synaptic plasticity. *Neuron* 26:5–8.

Turrigiano GG and Nelson SB (2000) Hebb and homeostasis in neuronal plasticity. *Curr Opin Neurobiol* 10:358–364.

Turrigiano GG and Nelson SB (2004) Homeostatic plasticity in the developing nervous system. *Nat Rev Neurosci* 5:97–107.

Wang H and Wagner JJ (1999) Priming-induced shift in synaptic plasticity in the rat hippocampus. *J Neurophysiol* 82:2024–2028.

Wankerl K, Weise D, Gentner R, Rumpf JJ, and Classen J (2010) L-type voltage-gated Ca^{2+} channels: A single molecular switch for long-term potentiation/long-term depression-like plasticity and activity-dependent metaplasticity in humans. *J Neurosci* 30:6197–6204.

Witte AV, Kurten J, Jansen S, Schirmacher A, Brand E, Sommer J, and Floel A (2012) Interaction of BDNF and COMT polymorphisms on paired-associative stimulation-induced cortical plasticity. *J Neurosci* 32:4553–4561.

Yang SN, Tang YG, and Zucker RS (1999) Selective induction of LTP and LTD by postsynaptic $[Ca^{2+}]i$ elevation. *J Neurophysiol* 81:781–787.

Yashiro K and Philpot BD (2008) Regulation of NMDA receptor subunit expression and its implications for LTD, LTP, and metaplasticity. *Neuropharmacology* 55:1081–1094.

Zhou Q and Poo MM (2004) Reversal and consolidation of activity-induced synaptic modifications. *Trends Neurosci* 27:378–383.

Ziemann U, Ilic TV, Pauli C, Meintzschel F, and Ruge D (2004) Learning modifies subsequent induction of long-term potentiation-like and long-term depression-like plasticity in human motor cortex. *J Neurosci* 24:1666–1672.

Ziemann U and Siebner HR (2008) Modifying motor learning through gating and homeostatic metaplasticity. *Brain Stimul* 1:60–66.

10 Methodological Aspects of Transcranial Magnetic Stimulation Combined with Neuroimaging

Hanna Mäki and Risto J. Ilmoniemi

CONTENTS

Neuroimaging can be helpful or even essential for transcranial magnetic stimulation (TMS) studies in several ways. First, targeting of TMS can be made accurate and reliable by using navigation techniques based on anatomical imaging such as magnetic resonance imaging (MRI) as the "map" according to which one navigates. Second, functional neuroimaging can be used to measure the distribution and timing of the activity elicited by TMS. Third, the results of TMS mapping studies are often best presented as maps drawn on anatomical brain images. Fourth, neuroimaging may guide the researcher in deciding what area of the brain to stimulate and in interpreting the data.

We will start by discussing the basis of targeting TMS pulses accurately, because this is the starting point of combining TMS with neuroimaging techniques. First, we will introduce the concept of activating function; then, principles that govern the focusing of TMS; and then, navigated TMS (nTMS). The rest of the chapter will cover the recording of TMS-evoked activity with electromyography (EMG), electroencephalography (EEG), functional MRI (fMRI), positron emission tomography (PET), and near-infrared spectroscopy (NIRS).

10.1 TARGETING OF TMS

10.1.1 ACTIVATION FUNCTION

It is of great importance to know which neurons and which parts of neurons are first or most readily excited by the electric field induced by TMS. Since action potentials are triggered when the membrane voltage exceeds a threshold value, we need a function that is proportional to local membrane potential changes. This function is called the activating function and it forms the basis for nTMS: Maximal neuronal responses are expected at sites where the activation function has its peak value. In the case of a straight axon in a homogeneous medium, as has been concluded based on the cable equation, the driving force for membrane potential change is the rate of change of the induced electric field in the direction of the axon, $\partial E_z/\partial z$, where z is the coordinate along the axon. If the stimulation current changes slowly compared to the membrane time constant and is uniform on the scale of the length constant, the change in membrane voltage is equal to $\lambda^2 \partial E_z/\partial z$ (Basser and Roth 2000).

The situation is more complicated if the neuron is bent, if there is a constriction in it, if there are extracellular conductivity inhomogeneities next to it, or if the end of the axon is considered. For a bent axon, the activating function is approximately the same as that for the straight axon if the coordinate z is the considered distance along the bent axon. In such a case, $\partial E_z/\partial z$ is nonzero even if the E-field is uniform. At and near a constriction where the fiber diameter changes, the activating function includes a term proportional to the E-field. The effect of external inhomogeneities is more complicated (Maccabee et al. 1991).

TMS pulses are typically short compared to the membrane time constant. It turns out that in such a case the length constant is reduced because of the capacitance-caused reduction of membrane impedance (Ilmoniemi et al. 2011).

In the real brain, the microscopic details of the arrangement of the neurons are unknown and very complicated. Furthermore, even the macroscopic E-field is usually not well known because of irregular conductivity structure and tissue anisotropy. However, experimental studies have shown that E-field is more effective when it is oriented perpendicularly with respect to cortical surface and when the current is directed from the surface of the cortex toward white matter. One explanation for this finding is the predominant perpendicular-to-cortex orientation of pyramidal cells in the gray matter. Fox et al. have argued that the effective activation function in the cortex is proportional to the cosine of the angle between the surface normal and the E-field (Fox et al. 2004). For small angles, this may be a valid approximation.

10.1.2 FOCUSING OF THE TMS EFFECT

Ideally, the investigator using TMS should be able to control the location, direction, intensity, and focality of the E-field. With fixed coils, only the location, direction, and intensity can be easily adjusted although the focality can also be adjusted by changing the distance of the coil from the head. In general, the closer to the head the coil is, the more focused is the E-field. The location of the E-field maximum and its orientation can be determined based on the measured location and orientation of the coil with respect to the head. This is best done by a stereotactic or nTMS system (e.g., Ruohonen and Karhu 2010), where the location and orientation of the coil with respect to the head is continuously monitored with millimeter precision and the induced E-field is computed in real time on the basis of an individual MRI-based model of the head shape, ideally taking into account irregularities in the conductivity structure formed by the cerebrospinal fluid, gray and white matter, and the anisotropy of tissue.

The electric field induced in the brain by a changing current in a TMS coil is roughly a blurred mirror image of the current pattern in the coil, the direction of the induced current being opposite to the direction of the rate of change of the current in the coil. Thus, a circular coil with, say, a diameter of 100 mm placed over the vertex of the head with current increasing in the clockwise direction induces current in the head under the wiring in the anticlockwise direction. The current pattern of a round coil being circular, the electric field is also circular, there being no single focal spot of the E-field. However, when the induced current is anticlockwise, the current in the right hemisphere is postero–anterior whereas it is antero–posterior in the left hemisphere. In such a case, the right motor cortex is activated about 50 mm from the vertex, because current flowing from the surface of cortex toward white matter stimulates the cortex optimally. Similarly, if the induced current is clockwise, the left motor cortex is activated.

A more elegant focusing method was introduced by Ueno et al. (1988), who proposed the so-called figure-of-eight coil. Here, two circular loops of wire are placed side by side, with their windings in opposite directions so that the current in the adjoining wire segments flow in the same direction. In such a case, the currents induced by the two loops are superposed and peak under the center of the coil. It has been demonstrated that the focality of figure-of-eight loops can be improved by making the coils smaller, but there is a limit to this; even an idealized, infinitely small figure-of-eight coil would have a relatively widely distributed E-field.

Ruohonen and Ilmoniemi (1998) published a theoretical study to investigate the focusing of TMS by using multiple coils. They showed that if the currents in an array of a large number of coils are properly adjusted, the focality of the E-field can be improved and, simultaneously, one can target the E-field to any desired location on the brain surface. Although an arbitrarily good focus can in principle be obtained by increasing the number of coils in the array, the currents in each coil would need to be increased indefinitely so as to impose a practical limit to the number of coils and to the obtained focality. So far, large TMS arrays have not been constructed.

In the preceding text, we have been talking about focusing on the surface of the brain. In fact, this is the best we can do. Namely, Heller and van Hulsteyn (1992) showed mathematically that focusing in depth is impossible: the maximal induced E-field, at least in the spherical model approximation, is always at the surface of the brain.

As was discussed earlier, the effective activating function depends not only on the electric field but also on the orientation of the sulcus with respect to the induced field.

Thanks to the fact that the optimal E-field is oriented perpendicularly to the cortical surface, improved physiological focusing can be obtained. For example, when the omega-shaped knob of the motor cortex is stimulated, one can dramatically change the elicited muscle activity by simply rotating the coil.

10.1.3 MRI-Based Navigation

nTMS, i.e., the real-time computation and display of the predicted TMS-induced electric field on anatomical brain images (such as MRI or CT), was suggested in the patent application of Ilmoniemi and Grandori (1993). Although the magnetic field passes the scalp and skull freely, the electric field generated inside the head depends on the conductivity geometry, especially on the local curvature of the head. Formulas that take into account the spherical structure of the head in computing the induced field were published by Ruohonen and Ilmoniemi (1996, 1998). Early experiments using simple MRI-based targeting were reported by Krings et al. (1997, 1998, 2001), but in their experiments, the induced field or its maximum was not calculated; only the approximate location of the peak field was shown.

nTMS with E-field computation, also called navigated brain stimulation (NBS), was developed and commercialized by Nexstim Ltd. (Ruohonen and Karhu 2010). This system is based on continuous optical measurement of the 3D location and orientation of the TMS coil with respect to the head. From this measurement, the precise configuration of the coil wiring, coil current waveform, and the individual MRI registered to the head, one can calculate the induced field in the brain (Figure 10.1). NBS systems have been approved by FDA for locating the motor cortex; the system is in active use in presurgical localization after it was demonstrated that its accuracy and reliability are at par with direct cortical stimulation (Picht et al. 2009, 2011a,b).

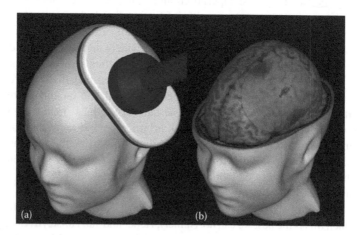

FIGURE 10.1 **(See color insert.)** Navigated brain stimulation. The coil (a) and the cortically induced electric field (b) can be displayed with respect to the subject's anatomical MR image. The coloring of the cortex indicates the strength of the calculated E-field; the maximum is shown by the small red area and the arrows show the direction of the induced current.

10.2 MEASURING TMS-EVOKED ACTIVITY

10.2.1 RECORDING OF PERIPHERAL RESPONSES

A basic TMS-based neuroimaging procedure is the mapping of representation areas of the cortex by observing the effects of TMS. The peripherally measurable effects of TMS range from evoked motor activity and the perception of phosphenes (Meyer et al. 1991) to interference with complex motor (e.g., Ruspantini et al. 2011) and cognitive functions such as speech (Pascual-Leone et al. 1991). The most easily measurable peripheral response to TMS, and also the most extensively studied one, is the motor-evoked potential (MEP) appearing in the surface EMG signal recorded over a target muscle as a result of the stimulation of the motor cortex. The MEP amplitudes are highly variable; thus, averaging is required to obtain a reliable MEP amplitude measure. When combined with a neuronavigation system, TMS-evoked MEPs can also be recorded in preoperative motor cortex mapping. The so-obtained motor map agrees well with that acquired with direct cortical stimulation, the gold standard in preoperative mapping (Picht et al. 2011a).

Because of the easily measurable MEP response, TMS has been traditionally applied mostly on the primary motor cortex (M1). Along with the emergence of TMS-compatible neuroimaging techniques such as TMS–EEG and TMS–fMRI as well as the therapeutic applications, TMS is increasingly applied to cortical areas other than M1. Nevertheless, because of the lack of an easily determinable excitability measure to stimulation of most areas, the stimulation intensity is often chosen based on the excitability of M1, as a certain percentage of the motor threshold (MT). Adjusting the stimulation intensity according to the MT may not, however, be optimal since the apparent excitability of cortical sites depends also, e.g., on the scalp-to-cortex distance and on differences in cortical folding and neuronal organization. An intensity measure that is independent of motor cortex excitability or scalp-to-cortex distance is the E-field amplitude, which is routinely computed by nTMS systems.

10.2.2 TMS AND EEG

The first TMS-evoked EEG recordings were performed by Cracco et al. (1989), who reported responses contralateral to the stimulated hemisphere with an onset latency of about 10 ms from the pulse. Large electromagnetic stimulus artifacts made the experiments difficult. In the same laboratory, Amassian et al. (1992) recorded EEG responses from the interaural line after TMS delivered to the cerebellum. In their study, TMS-induced artifacts were reduced by adjusting the placement of the coil and the electrodes and by placing a scalp-grounded metal strip between them. However, reliable measurements were still difficult.

The subsequent introduction of TMS-compatible EEG devices has greatly expanded the possibilities to measure TMS-evoked reactions of the brain. As the postsynaptic potentials generated through the TMS-induced synaptic activation are reflected in the EEG signals, the combined TMS–EEG allows measuring the electrical activity evoked by stimulation of any cortical area, not just motor, and tracking the spread of activation to connected brain regions at a millisecond time resolution (Ilmoniemi et al. 1997). The reaction of each neuron stimulated directly or trans-synaptically depends on neuronal excitability and on the state of the synapses along the pathway. The evoked signals thus reflect the functional state of the brain at the time of the stimulation. It has been shown that the TMS-evoked EEG response is highly repeatable between measurement sessions (Lioumis et al. 2009, Casarotto et al. 2010), yet sensitive to changes in the stimulation parameters such as the intensity, stimulation site, and direction of the induced current (Iramina et al. 2003, Komssi et al. 2004, Iwahashi et al. 2008, Casarotto et al. 2010). The repeatability and sensitivity make TMS–EEG a reliable tool for studying modulations in the state of the brain. This approach enables studying, e.g., the neuronal networks involved in different tasks and their functional modulation and possibly also changes caused by diseases affecting the central nervous system. In addition, it can probe the effects of rTMS on the neuronal state, which is beneficial, e.g., considering the growing interest in therapeutic applications of TMS. In case the stimulation has behavioral effects, their connection with the activated neuronal network can be assessed.

The electromagnetic interference problems related to combining the EEG measurement with the powerful TMS pulse have been largely overcome, but this combination of methods is still challenging, especially because of the muscle artifacts that may contaminate the signal. Careful preparation of the electrode contacts to decrease their impedance, signal separation or filtering methods, and control measurements are needed to reduce these artifacts and to make sure that the evoked responses originate in the brain.

10.2.2.1 TMS-Compatible Instrumentation

Combining TMS and EEG has been challenging because of the strong electromotive force induced by the magnetic pulse in the loops comprising the electrode leads, the amplifiers, and the head. If a standard EEG system is combined with TMS, the amplifiers may saturate and it takes up to 100 ms for them to recover. This problem has been dealt with by EEG instrumentation made TMS compatible either by eliminating the artifact with gain-control and sample-and-hold circuits or by electronics that do not become saturated during the pulse and recover quickly to measure the evoked brain activity. In addition, the systems utilize electrodes that do not heat up excessively due to the eddy currents.

10.2.2.1.1 Amplifiers

The TMS-induced artifact can be effectively blocked by gain-control and sample-and-hold circuits (Virtanen et al. 1999, Iramina et al. 2003). In such systems, the gain of the amplifier is reduced and the signal level is latched during a gating period, typically a few milliseconds, after which EEG signals free of the induced artifacts can be recorded. Another way to prevent the saturation of the amplifiers is to use a preamplifier with a limited slew rate (Thut et al. 2005, Ives et al. 2006). Because of the low slew rate, the circuit does not respond to the high slew rate of the TMS pulse, which reduces the artifact and prevents the amplifiers from saturating. The amplifiers recover from the pulse within 30 ms. If constant, the remaining artifact can be eliminated by subtracting the responses obtained in two conditions from each other (Thut et al. 2005). The low slew rate limits the bandwidth typically to frequencies below 100 Hz. Yet another way of preventing the saturation is to use amplifiers with adjustable sensitivity and operational range; the system of Veniero et al. (2009) was reported to record artifact-free data about 5 ms after the pulse.

10.2.2.1.2   Electrode Design

The problems with recording TMS-evoked EEG with conventional electrodes are the heating and electrode movement due to the TMS-induced eddy currents. Excessive heating causes skin burns and electrode movement produces artifacts, which is why electrode design is a crucial step in TMS-compatible EEG system design. As the rise of temperature in the electrode is proportional to the conductivity and the square of the diameter of the electrode (Roth et al. 1992), it can be limited by reducing the conductivity and the current-loop area of the electrode. Small low-conductivity Ag/AgCl pellet electrodes are therefore used in many TMS-compatible EEG systems. Also conductive plastic electrodes covered with silver epoxy have been found suitable (Ives et al. 2006).

10.2.2.2   Measures of Excitability and Connectivity

As TMS-evoked EEG responses are generated trans-synaptically, they respond to changes in the functional state of the brain. Modulation of the responses has been shown, e.g., during motor preparation (Nikulin et al. 2003, Bender et al. 2005, Kičić et al. 2008, Bonnard et al. 2009), after a conditioning TMS pulse (Daskalakis et al. 2008, Fitzgerald et al. 2009) or rTMS (Esser et al. 2006), following cutaneous stimulation (Bikmullina et al. 2009), after the intake of alcohol (Kähkönen and Wilenius 2007), along with spontaneous changes in cortical excitability (Mäki and Ilmoniemi 2010), and during sleep (Massimini et al. 2005).

Different measures of cortical excitability have been derived from the TMS-evoked EEG signals. The global mean field amplitude (GMFA; Lehmann and Skrandies 1980), i.e., the square root of the sum of squares of the average-referenced signals of the EEG channels, has been used as a measure of overall cortical excitability. Komssi et al. (2004) presented a model of the dependence of the TMS-evoked EEG response on stimulation intensity based on the assumed normal distribution of neuron membrane potentials. They found an agreement between the model and the experimental intensity–GMFA response curve. Esser et al. (2006) demonstrated long-term potentiation of the GMFA after 5 Hz stimulation to the motor cortex. Daskalakis et al. (2008) applied a more localized measure of cortical reactions, the mean area under rectified curve calculated from a channel above the stimulation site, to measure the long-interval intracortical inhibition (LICI) in the motor, parietal, and dorsolateral prefrontal cortices. To utilize both the spatial and the temporal information in the signals, in many studies, the analysis has concentrated on certain EEG deflections measured with channels in the region of interest, which gives more specific information about the cortical reactions to the stimulus. So far, the best-studied deflection is the negativity peaking around 100 ms (N100) recorded over the stimulated M1. The N100 decreases in amplitude during motor preparation (Nikulin et al. 2003, Bender et al. 2005) and increases when the subject is prepared to resist the TMS-evoked movement (Bonnard et al. 2009). Accordingly, the N100 is suggested to reflect the postsynaptic effects of the inhibitory interneurons (Nikulin et al. 2003). This view is supported by the overlap in time with the

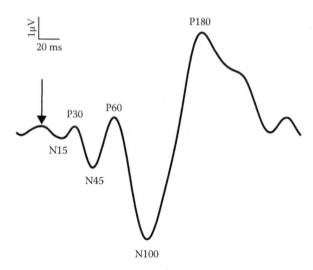

FIGURE 10.2   A typical averaged EEG response to TMS delivered to the left M1, measured between an electrode at the vertex and a reference behind the right ear. The arrow shows the time when the TMS pulse was delivered.

inhibitory postsynaptic potentials in intracellular recordings (Krnjević et al. 1966) and the LICI (Valls-Sole et al. 1992, Fitzgerald et al. 2009) as well as by the fact that inhibitory processes in the deeper cortical layers produce surface-negative potentials (Caspers et al. 1980). Unless proper hearing protection and/or noise masking is used, part of the N100 reflects the auditory response to the TMS coil click (see Section 10.2.2.3.4). The response following M1 stimulation consists of a typical sequence of deflections: N15, P30, N45, P55 (or P60), N100, and P180 (Paus et al. 2001, Komssi et al. 2002, 2004, 2007, Bonato et al. 2006, Esser et al. 2006, Figure 10.2). The origin of these components is not fully clarified, but a few studies have revealed some of their characteristics. The peak-to-peak amplitude of the N15–P30 complex has been shown to correlate with the evoked MEP with constant stimulation parameters and the subjects in rest, showing that the complex reflects spontaneous fluctuations in motor cortical excitability (Mäki and Ilmoniemi 2010). The N45 component appears to reflect activity at the stimulated M1: it correlates with the stimulation intensity and is inhibited by a conditioning pulse delivered 12 ms earlier (Paus et al. 2001). The response structure varies between individuals (Komssi et al. 2004) and stimulation parameters such as the coil orientation (Bonato et al. 2006, Casarotto et al. 2010) and the stimulation site (Rosanova et al. 2009, Casarotto et al. 2010).

The first milliseconds of the evoked response reflect activation of the stimulated cortical area, but later components also include signals from sites to where the activation propagates. TMS–EEG thus provides a way to probe the time-resolved effective connectivity in the human brain. TMS–EEG has been successfully applied to study the spread of activation from the stimulated motor and visual cortices to the other hemisphere (Ilmoniemi et al. 1997; Figure 10.3). The modulation of connectivity patterns has been shown, e.g., during sleep (Massimini et al. 2005) and as a result of alcohol intake (Kähkönen et al. 2001). The TMS-evoked

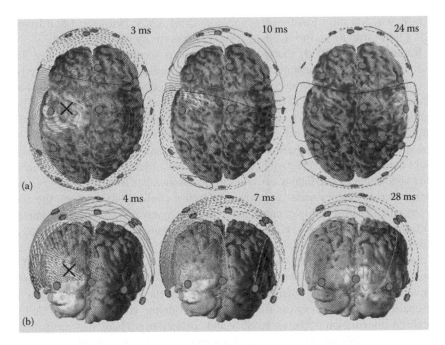

FIGURE 10.3   (**See color insert.**) Spread of TMS-evoked activation from the stimulated motor (a) or visual (b) cortex to the contralateral side. The color maps show the activation as calculated from the averaged TMS-evoked EEG with the minimum norm estimate. The contour lines show the potential patterns of the measured EEG signals (red, positive potential; blue, negative potential; contour spacing: 1 μV (a), 2 μV (b)). The cross indicates the stimulation site. (From Ilmoniemi, R.J. et al., *NeuroReport,* 8, 3537, 1997. With permission.)

brain activation can be modeled as dipoles (Scherg 1992) or as continuous current distributions with minimum norm estimation (Hämäläinen and Ilmoniemi 1994). Massimini et al. (2005) applied permutation statistics on the estimated source responses to consider only activations significantly different from the baseline. They showed that, as opposed to the waking state, during sleep the evoked activity was restricted to the vicinity of the stimulation site and died out quickly after the pulse. In addition to connectivity studies, source modeling is also beneficial in determining the local excitability.

In addition to the traditional assessment of evoked responses averaged over trials, frequency analysis of the signals provides interesting insights into connectivity and other properties of the stimulated networks. Paus et al. (2001) were the first to report effects of TMS on oscillatory activity. They detected synchronization of beta oscillations in single trials after stimulation of the motor cortex. Because the amplitude of these oscillations was not modulated by low-frequency rTMS, it was suggested that they are not induced by the pulse but the pulse resets the ongoing oscillations (van der Werf and Paus 2006). Further, Fuggetta et al. (2005) reported modulation of alpha- and beta-rhythm power as well as event-related coherence between EEG channels after motor cortex TMS, emphasizing the role of network dynamics in the formation of the measured response. Similarly, Brignani et al. (2008) reported increased power of alpha rhythm after low-frequency motor cortex TMS. However, when Iramina et al. (2002) stimulated the occipital cortex, alpha activity was suppressed.

Rosanova et al. (2009) showed that the frequency content of the evoked TMS response depends on the natural frequency of the stimulated corticothalamic system: stimulation of occipital areas produced a response predominantly in the alpha band, whereas parietal and frontal cortex stimulation evoked beta and beta/gamma band oscillations, respectively. The responses originating in these areas oscillated in their natural frequencies also when activated indirectly by stimulation of the other areas. The effects of TMS on oscillatory activity in distant areas provide information about the functional connectivity patterns. One has to keep in mind, however, that coherence between oscillatory activities in two brain areas (often termed functional connectivity) does not necessarily imply effective connectivity between the areas. In any case, TMS–EEG is gaining momentum as a new tool to perform controlled studies of brain rhythms, e.g., in the context of cognitive neuroscience (Thut and Miniussi 2009).

As a direct, noninvasive, and relatively focal method to activate the cortex, TMS combined with EEG can probe the functional state of the brain, its excitability, connectivity, and plastic changes in healthy subjects and in patients (Thut and Pascual-Leone 2010). In future studies, source modeling is needed to provide more localized information of the excitability and connectivity. There also seems to be a need for straightforward indices that characterize the response of the brain to TMS (Casali et al. 2010). Source localization is beneficial also in coherence analysis, which, when conducted in the signal space, suffers from artifactual correlations between electrodes that detect the same oscillatory sources. The studies may also benefit from the anatomical connectivity information obtained with diffusion tensor imaging (Le Bihan et al. 2001), a type of magnetic resonance imaging providing directional information of white matter tracts.

10.2.2.3 Dealing with EEG Artifacts

Even with a TMS-compatible EEG system, the TMS-evoked EEG signals need to be interpreted carefully to control for the various artifacts the signals may contain. These include the electrode polarization artifact, electrode movement artifacts, muscle artifacts due to stimulation of cranial or facial muscles, auditory evoked potentials as a response to the coil click, somatosensory evoked potentials (SEPs) as a response to scalp sensory nerve or muscle stimulation or target muscle contraction, and ocular artifacts.

10.2.2.3.1 Electrode Polarization

It has been suggested that a TMS pulse can produce a charge displacement over the electrode contact. Because the contact has capacitive properties, the charge distribution does not return to equilibrium immediately after the pulse, but the recovery may take up to hundreds of milliseconds. The exponentially diminishing polarization artifact can be approximated by fitting exponentially decaying functions to the signal, which can then be removed from the data (Litvak et al. 2007). Low contact impedance reduces the polarization artifact, which stresses the importance of good electrode contacts.

10.2.2.3.2   *Electrode Movement Artifacts*

The electrodes may move slightly if the coil touches them, if the muscles under the electrodes move, or if the currents induced in traditional electrodes give rise to a force. Movement may disturb the charge distribution over the electrode contact, causing the electrode movement artifact, which is actually a polarization artifact by nature. To avoid electrode movements due to coil vibrations or the investigator moving the coil, the coil should not touch the electrodes. This may require keeping the coil at a small distance from the head. However, the distance complicates keeping a constant coil position with respect to the head. Alternatively, the movement-artifact-contaminated channels may need to be excluded from the analysis.

Muscle movement makes both the electrode and the skin to move. The movement-related changes in skin potential produce another component to the movement artifact (Tam and Webster 1977), which can be reduced by preparing the skin carefully to reduce the impedance of the contact; in one study, mini-puncturing the skin, with the effect of short-circuiting the epithelium, decreased the artifact amplitudes to half (Julkunen et al. 2008).

10.2.2.3.3   *Muscle Artifacts*

The muscle artifacts caused by the stimulation of cranial or facial muscles are typically biphasic, peak at around 5 and 10 ms, after which the signal returns back to baseline in a few tens of milliseconds (Figure 10.4). Both the electrical activity of the muscles and the electrode movements are likely to contribute to these artifacts. Muscle artifacts do not generally cause problems in the TMS–EEG recordings when relatively central areas far away from the cranial muscles are stimulated. For example, EEG signals without or with only moderate muscle artifacts have been reported to stimulation of M1 hand area (Ilmoniemi et al. 1997, Paus et al. 2001, Komssi et al. 2002, 2004, 2007, Nikulin et al. 2003, Esser et al. 2006, Kičić et al. 2008, Bikmullina et al. 2009), S1 hand area (Raij et al. 2008), dorsal premotor cortex (Massimini et al. 2005, Rosanova et al. 2009), dorsolateral prefrontal cortex (Kähkönen et al. 2004, 2005a,b), parietal cortex (Rosanova et al. 2009), and the

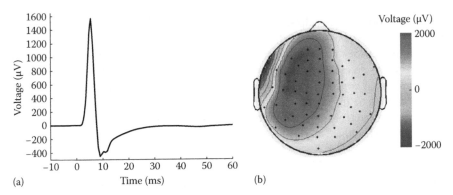

FIGURE 10.4   (See color insert.) A typical biphasic muscle artifact recorded at the vertex (reference behind right ear) after the stimulation of Broca's area (left hemisphere) at time $t = 0$ (a) and the corresponding topography at the time of the signal maximum (6 ms; b).

occipital Brodmann area 19 (Rosanova et al. 2009). However, because of individual differences in head size, anatomy, and cortical excitability, the activation of the cranial muscles varies between subjects. For example, for some subjects, stimulation of the M1 hand area at 100% of the MT produces seemingly artifact-free data, while stimulation of other subjects evokes muscle artifacts of tens of microvolts. The possibility of muscle artifacts has to be taken into consideration also when stimulating relatively central sites, especially if the early components of the evoked EEG are of interest. Within the limits of the study paradigm, the stimulation site and other parameters, such as the induced current direction, coil tilting, and intensity, should be chosen so that the muscle artifacts are minimized.

The largest muscle artifacts are generated by stimulation of lateral brain areas. For example, stimulation of Broca's area can produce muscle artifacts of hundreds to thousands of µV (Mäki and Ilmoniemi 2011). If the early ($t < 60$ ms) evoked signals are of interest, these large artifacts need to be dealt with by signal-processing methods such as independent component analysis (Korhonen et al. 2011) or signal-space projection (SSP; Mäki and Ilmoniemi 2011).

10.2.2.3.4   Auditory Artifacts

If the subject can hear the stimulation coil click, an auditory evoked potential with pronounced N100 and P180 deflections is produced in the EEG (Tiitinen et al. 1999). The auditory artifacts can be reduced by hearing protection, which is, however, usually not sufficient to completely block the perception of the click. Instead, or in addition to it, the subject can listen to noise masking the stimulation sound. Both white noise (Paus et al. 2001, Fuggetta et al. 2005) and noise whose spectrum approximates that of the coil click (Massimini et al. 2005) have been used. If the spectrum of the click can be reliably reproduced, smaller masking sound volumes compared to white noise are needed.

The click sound is perceived not only through air-conducted sound waves, but also through vibrations in the skull (Nikouline et al. 1999). The vibrations of the coil transmitted via the skull can be reduced by placing a thin piece of plastic foam between the coil and the head (Massimini et al. 2005).

10.2.2.3.5   Somatosensory Artifacts

The stimulation of the scalp sensory nerves or muscles and contraction of the target muscle (when M1 is stimulated) produces SEPs in the EEG recording. The effects of both the sensory nerve stimulation (Nikouline et al. 1999, Paus et al. 2001) and the target muscle contraction (Paus et al. 2001, Nikulin et al. 2003) on the evoked EEG have been concluded to be small, but the possibly larger contribution of SEPs due to cranial or facial muscle stimulation needs further studies. In addition, the possibility of intersensory facilitation or cross-modality suppression has to be taken into account (Kadunce et al. 1997).

10.2.2.3.6   Ocular Artifacts

Eye movements cause large artifacts in the EEG signals, because there is a voltage of several millivolts over the eyeball. Depending on the paradigm, a fixation cross can help preventing eye movements. The subjects may also blink because of the startle effect caused by the stimulation or because of stimulation of nerves innervating eye muscles or brain areas controlling eye movements. The ocular artifacts can be controlled with an electrooculogram recording to exclude the trials with eye movements. In addition, if blinks cannot be avoided, SSP can be applied to project the signal vector(s) related to eye movements out of the data (Huotilainen et al. 1993).

10.2.3 TMS COMBINED WITH PET, fMRI, AND NIRS

TMS combined with hemodynamics-based neuroimaging, such as PET (Fox et al. 1997, Paus et al. 1997), fMRI (Bohning et al. 1998), or NIRS (Oliviero et al. 1999), provides another way to study the local and remote effects of the cortical modulation. As opposed to TMS–EEG, the temporal resolution of these methods is typically limited to seconds because of the inherent properties of the hemodynamic response. Because of the necessary data collection period, the temporal resolution of PET is even poorer (from tens of seconds to minutes). However, all of these methods measure slightly different parameters and provide complementary information about brain activity due to the stimulation: fMRI, measuring the blood-oxygen-level-dependent signal, has a superior spatial resolution of the order of millimeters and thus might be the method of choice when accurate spatial activation patterns are of interest. PET, in turn, measures changes in regional cerebral blood flow (rCBF), regional cerebral metabolic rate of glucose (rCMRglc), or binding of other molecules of interest, e.g., neurotransmitters, depending on the radioactive tracer used. As opposed to fMRI, PET measures the changes in absolute, quantitative units. With NIRS, it is possible to measure changes in both oxygenated (HbO_2) and deoxygenated (HbR) hemoglobin concentrations. In addition, NIRS measurements are not disturbed by the TMS pulses, which makes this combination of methods attractive. Each of the methods has its challenges as well: In TMS–PET experiments, one has to make sure that the PET detectors are not disturbed by the TMS pulses; also the attenuation of radiation caused by the TMS coil needs to be taken into account (Fox et al. 1997, Paus et al. 1997). In addition, subject exposure to radiation limits the use of PET. TMS–fMRI measurements are technically more challenging, requiring special stimulation strategies, such as interleaved TMS pulse–fMRI acquisition sequences, to prevent signal losses and image distortion (Bohning et al. 1998). NIRS signals, in turn, include contributions from superficial (skin) circulation, which obscures the brain signals (Tachtsidis et al. 2009, Näsi et al. 2011). In addition, the stimulus-related somatosensory and auditory sensations discussed in Section 10.2.2.3 also affect the hemodynamic measurements. Accordingly, the resulting artifacts need to be prevented or controlled for.

Similar to TMS–EEG, stimulation combined with hemodynamic neuroimaging can be applied to study the activity at the stimulated and interconnected areas, thus allowing the assessment of the functional state of the brain and its modulation in a multitude of situations. In addition, the combination of controlled neuronal modulation achieved by TMS and hemodynamic measurement may provide a means to study neurovascular coupling.

10.2.3.1 TMS–fMRI

The magnetic fields generated by the TMS pulse interfere with the fMRI measurement, which is why combining these two methods is challenging. Typically, the problem is solved by separating the TMS pulses and echo planar imaging (EPI) sequences in time: the TMS stimuli can be administered in between acquired volumes or slices. In addition to stimulation during the radiofrequency excitation period or data acquisition, which obviously causes signal loss and image distortions, one should avoid stimulation just before the EPI sequence (100 ms or less), because artifacts may be generated due to residual currents or coil vibrations (Bohning et al. 1998). Moreover, the introduction of the stimulation coil into the scanner increases the noise level and reduces sensitivity (Bohning et al. 1998).

Also the MRI environment poses several restrictions to TMS. Naturally, a non-ferromagnetic stimulation coil should be used. Even then, image distortions close to the coil can occur because of the resulting field inhomogeneity, although these appear insignificant at the scalp–cortex distance (Baudewig et al. 2000). Additionally, the constricted space inside the scanner restricts the positioning of the coil.

These issues taken into account, several TMS–fMRI studies have been successfully conducted (see Bestmann et al. 2008 for a review). Because of the excellent spatial resolution of fMRI, it provides a method to accurately examine the networks activated by the TMS pulse (see Figure 10.5 for an example of fMRI responses to TMS), which in turn can be timed precisely and targeted relatively focally.

10.2.3.2 TMS–PET

PET measures the distribution of molecules tagged with positron-emitting tracers in the tissue; electron–positron annihilation produces pairs of gamma-ray photons that are detected by the PET detector. Commonly, tagged water molecules ($H_2{}^{15}O$) are used to assess rCBF. Tagging fluorodeoxyglucose (FDG) sugar with ^{18}F enables tracing the rCMRglc. The data need to be integrated over time, which causes the compromised temporal resolution of PET: $H_2{}^{15}O$-PET requires a data collection period of tens of seconds to sample enough photon pairs, while in FDG-PET, a sampling time of several minutes is needed.

Combining TMS with PET is relatively straightforward: the major technical challenge is the effect of the magnetic pulse on the photomultiplier tubes (PMTs). While the effect does not appear significant in some measurement systems (Fox et al. 1997), mu-metal shielding of the photomultipliers is needed in others (Paus et al. 1997). The shielding attenuates the radiation slightly (about 20%; Paus et al. 1997). Also the stimulation coil causes attenuation, which has to be measured by a separate scan with the coil in the desired position and taken into account in the analysis (Paus et al. 1997). Alternatively, the coil needs to be placed outside the field of view of PET (Fox et al. 1997).

An example of TMS–PET rCMRglc responses to motor cortex stimulation is shown in Figure 10.6.

10.2.3.3 TMS–NIRS

NIRS measures the attenuation of near-infrared light (typically in the range 650–950 nm) in the tissue. Changes in attenuation result mainly from varying HbO_2 and HbR concentrations, which can be calculated from NIRS signals recorded with two different wavelengths (at which the two hemoglobin types are absorbed in different proportions). Combining the optical NIRS recording with TMS is technically

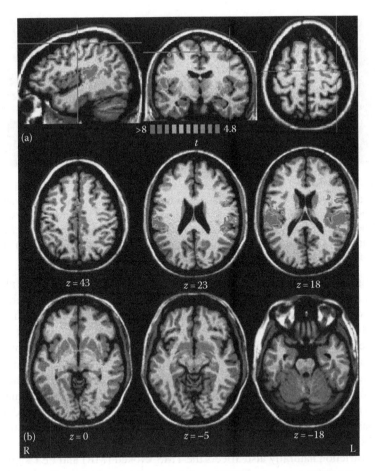

FIGURE 10.5   (**See color insert.**) Local and distant fMRI responses to TMS delivered to the left dorsal premotor cortex (indicated with crosshair). The colors mark areas with statistically significant activation according to a group analysis: left dorsal premotor cortex (a), cingulate gyrus, ventral premotor cortex, auditory cortex, caudate nucleus, left posterior temporal lobe, medial geniculate nucleus, and cerebellum (b). R, right, L, left. (From Bestmann, S. et al., *NeuroImage*, 28, 22, 2005. With permission.)

easy because the stimulation and the measurement do not interfere; this requires placing the PMTs at a distance from the stimulation site and, therefore, guiding the detected light to the PMTs via optical fibers. Prism fiber terminals (Näsi et al. 2011, Virtanen et al. 2011), reflecting the light so that the optical fibers can be placed tangentially against the head, enable positioning the stimulation coil close to the head above the NIRS fibers. In addition to being easily TMS compatible, NIRS is relatively cheap, mobile, and safe. In contrast to the compromised spatial resolution of centimeters, the temporal resolution is good enough to obtain the shape of the hemodynamic response (around 1 s depending on the system and fiber configuration).

Because all the light going to the brain travels through surface tissue, NIRS is prone to extracerebral interference due to, e.g., stimulus-related arousal and the

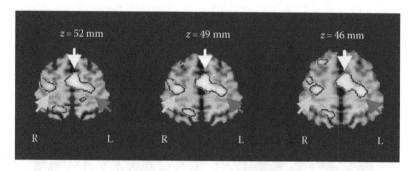

FIGURE 10.6 **(See color insert.)** PET responses showing rCMRglc increases on the group level as a response to 5 Hz TMS delivered to the left M1 (red arrow). Yellow and white arrows point to activations at contralateral M1 and supplementary motor area, respectively. The responses are superimposed on axial sections of stereotactically normalized MRIs and the Talairach z-coordinates are given on top of the sections. R, right, L, left. (Data published in Siebner, H.R. et al., *Neurology*, 54, 956, 2000. With permission.)

resulting changes in surface circulation. These artifacts are common to all NIRS studies, not just those measuring the effects of TMS. Different signal-processing methods have been presented to overcome this problem. These methods utilize the fact that global changes appear in all the measurement channels (Zhang et al. 2005, Kohno et al. 2007, Virtanen et al. 2009) or that the ratio of surface and brain contributions in the signals depends on the distance between NIRS source and detector (Saager et al. 2005, Gregg et al. 2010). Additionally, it was recently shown that TMS generates local vasoconstriction near the stimulation site, which masks the hemodynamic response measured with NIRS (Näsi et al. 2011). At which depth this vasoconstriction is significant still remains to be clarified; if occurring in the brain, it masks the hemodynamic response at the stimulation site. If this is the case, the vasoconstriction naturally also affects the TMS-evoked fMRI and PET signals.

REFERENCES

Amassian, V. E., R. Q. Cracco, P. J. Maccabee et al. 1992. Cerebello-frontal cortical projections in humans studied with the magnetic coil. *Electroencephalography and Clinical Neurophysiology* 85:265–272.

Basser, P. J. and B. J. Roth. 2000. New currents in electrical stimulation of excitable tissues. *Annual Review of Biomedical Engineering* 2:377–397.

Baudewig, J., W. Paulus, and J. Frahm. 2000. Artifacts caused by transcranial magnetic stimulation coils and EEG electrodes in T2*-weighted echo-planar imaging. *Magnetic Resonance Imaging* 18:479–484.

Bender, S., K. Basseler, I. Sebastian et al. 2005. Transcranial magnetic stimulation evokes giant inhibitory potentials in children. *Annals of Neurology* 58:58–67.

Bestmann, S., J. Baudewig, H. R. Siebner et al. 2005. BOLD MRI responses to repetitive TMS over human dorsal premotor cortex. *NeuroImage* 28:22–29.

Bestmann, S., C. C. Ruff, F. Blankenburg et al. 2008. Mapping causal interregional influences with concurrent TMS–fMRI. *Experimental Brain Research* 191:383–402.

Bikmullina, R., D. Kičić, S. Carlson et al. 2009. Electrophysiological correlates of short-latency afferent inhibition: A combined EEG and TMS study. *Experimental Brain Research* 194:517–526.

Bohning, D. E., A. Shastri, Z. Nahas et al. 1998. Echoplanar BOLD fMRI of brain activation induced by concurrent transcranial magnetic stimulation. *Investigative Radiology* 33:336–340.

Bonato, C., C. Miniussi, and P. M. Rossini. 2006. Transcranial magnetic stimulation and cortical evoked potentials: A TMS/EEG co-registration study. *Clinical Neurophysiology* 117:1699–1707.

Bonnard, M., L. Spieser, H. B. Meziane et al. 2009. Prior intention can locally tune inhibitory processes in the primary motor cortex: Direct evidence from combined TMS–EEG. *European Journal of Neuroscience* 30:913–923.

Brignani, D., P. Manganotti, P. M. Rossini et al. 2008. Modulation of cortical oscillatory activity during transcranial magnetic stimulation. *Human Brain Mapping* 29:603–612.

Casali, A. G., S. Casarotto, M. Rosanova et al. 2010. General indices to characterize the electrical response of the cerebral cortex to TMS. *NeuroImage* 49:1459–1468.

Casarotto, S., L. J. Romero Lauro, V. Bellina et al. 2010. EEG responses to TMS are sensitive to changes in the perturbation parameters and repeatable over time. *PLoS One* 5:e10281.

Caspers, H., E. J. Speckmann, and A. Lehmenkühler. 1980. Electrogenesis of cortical DC potentials. *Progress in Brain Research* 54:3–15.

Cracco, R. Q., V. E. Amassian, P. J. Maccabee et al. 1989. Comparison of human transcallosal responses evoked by magnetic coil and electrical stimulation. *Electroencephalography and Clinical Neurophysiology* 74:417–424.

Daskalakis, Z. J., F. Farzan, M. S. Barr et al. 2008. Long-interval cortical inhibition from the dorsolateral prefrontal cortex: A TMS–EEG study. *Neuropsychopharmacology* 33:2860–2869.

Esser, S. K., R. Huber, M. Massimini et al. 2006. A direct demonstration of cortical LTP in humans: A combined TMS/EEG study. *Brain Research Bulletin* 69:86–94.

Fitzgerald, P. B., J. J. Maller, K. Hoy et al. 2009. GABA and cortical inhibition in motor and nonmotor regions using combined TMS–EEG: A time analysis. *Clinical Neurophysiology* 120:1706–1710.

Fox, P., R. Ingham, M. S. George et al. 1997. Imaging human intra-cerebral connectivity by PET during TMS. *NeuroReport* 8:2787–2791.

Fox, P. T., S. Narayana, N. Tandon et al. 2004. Column-based model of electric field excitation of cerebral cortex. *Human Brain Mapping* 22:1–14.

Fuggetta, G., A. Fiaschi, and P. Manganotti. 2005. Modulation of cortical oscillatory activities induced by varying single-pulse transcranial magnetic stimulation intensity over the left primary motor area: A combined EEG and TMS study. *NeuroImage* 27:896–908.

Gregg, N. M., B. R. White, B. W. Zeff et al. 2010. Brain specificity of diffuse optical imaging: improvements from superficial signal regression and tomography. *Frontiers in Neuroenergetics* 2:14.

Hämäläinen, M. S. and R. J. Ilmoniemi. 1994. Interpreting magnetic fields of the brain: Minimum norm estimates. *Medical and Biological Engineering and Computing* 32:35–42.

Heller, L. and D. B. van Hulsteyn. 1992. Brain stimulation using electromagnetic sources: Theoretical aspects. *Biophysical Journal* 63:129–138.

Huotilainen, M., R. J. Ilmoniemi, H. Tiitinen et al. 1993. Eye-blink removal for multichannel MEG measurements. In *Abstracts of International Conference on Biomagnetism*, Vienna, Austria, 14–20 August, eds. L. Deecke, C. Baumgartner, G. Stroink, and S.J. Williamson, pp. 209–210.

Ilmoniemi, R. and F. Grandori. 1993. A programmable applicator for an electromagnetic field especially for stimulation of the central and peripheral nervous systems and for tissue therapy and hyperthermia applications. Finnish Patent Application No. 934511 (October 13).

Ilmoniemi, R. J., H. Mäki, and J. Saari. 2011. The length constant is frequency-dependent: implications for transcranial magnetic stimulation. In *Society for Neuroscience Meeting*, Washington, DC, November.

Ilmoniemi, R. J., J. Virtanen, J. Ruohonen et al. 1997. Neuronal responses to magnetic stimulation reveal cortical reactivity and connectivity. *NeuroReport* 8:3537–3540.

Iramina, K., T. Maeno, Y. Kowatari et al. 2002. Effects of transcranial magnetic stimulation on EEG activity. *IEEE Transactions on Magnetics* 38:3347–3349.

Iramina, K., T. Maeno, Y. Nonaka et al. 2003. Measurement of evoked electroencephalography induced by transcranial magnetic stimulation. *Journal of Applied Physics* 93:6718–6720.

Ives, J. R., A. Rotenberg, R. Poma et al. 2006. Electroencephalographic recording during transcranial magnetic stimulation in humans and animals. *Clinical Neurophysiology* 117:1870–1875.

Iwahashi, M., T. Arimatsu, S. Ueno et al. 2008. Differences in evoked EEG by transcranial magnetic stimulation at various stimulus points on the head. *Conference Proceedings of IEEE Engineering in Medicine and Biology Society* 2008:2570–2573.

Julkunen, P., A. Pääkkönen, T. Hukkanen et al. 2008. Efficient reduction of stimulus artefact in TMS–EEG by epithelial short-circuiting by mini-punctures. *Clinical Neurophysiology* 119:475–481.

Kadunce, D. C., J. W. Vaughan, M. T. Wallace et al. 1997. Mechanisms of within- and cross-modality suppression in the superior colliculus. *Journal of Neurophysiology* 78:2834–2847.

Kähkönen, S., M. Kesäniemi, V. V. Nikouline et al. 2001. Ethanol modulates cortical activity: Direct evidence with combined TMS and EEG. *NeuroImage* 14:322–328.

Kähkönen, S., S. Komssi, J. Wilenius et al. 2005a. Prefrontal transcranial magnetic stimulation produces intensity-dependent EEG responses in humans. *NeuroImage* 24:955–960.

Kähkönen, S., S. Komssi, J. Wilenius et al. 2005b. Prefrontal TMS produces smaller EEG responses than motor-cortex TMS: Implications for rTMS treatment in depression. *Psychopharmacology* 181:16–20.

Kähkönen, S. and J. Wilenius. 2007. Effects of alcohol on TMS-evoked N100 responses. *Journal of Neuroscience Methods* 166:104–108.

Kähkönen, S., J. Wilenius, S. Komssi et al. 2004. Distinct differences in cortical reactivity of motor and prefrontal cortices to magnetic stimulation. *Clinical Neurophysiology* 115:583–588.

Kičić, D., P. Lioumis, R. J. Ilmoniemi et al. 2008. Bilateral changes in excitability of sensorimotor cortices during unilateral movement: Combined electroencephalographic and transcranial magnetic stimulation study. *Neuroscience* 152:1119–1129.

Kohno, S., I. Miyai, A. Seiyama et al. 2007. Removal of the skin blood flow artifact in functional near-infrared spectroscopic imaging data through independent component analysis. *Journal of Biomedical Optics* 12:062111.

Komssi, S., H. J. Aronen, J. Huttunen et al. 2002. Ipsi- and contralateral EEG reactions to transcranial magnetic stimulation. *Clinical Neurophysiology* 113:175–184.

Komssi, S., S. Kähkönen, and R. J. Ilmoniemi. 2004. The effect of stimulus intensity on brain responses evoked by transcranial magnetic stimulation. *Human Brain Mapping* 21:154–164.

Komssi, S., P. Savolainen, J. Heiskala et al. 2007. Excitation threshold of the motor cortex estimated with transcranial magnetic stimulation electroencephalography. *NeuroReport* 18:13–16.

Korhonen, R. J., J. C. Hernandez Pavon, J. Metsomaa et al. 2011. Removal of large muscle artifacts from transcranial magnetic stimulation-evoked EEG by independent component analysis. *Medical and Biological Engineering and Computing* 49:397–407.

Krings, T., B. R. Buchbinder, W. E. Butler et al. 1997. Stereotactic transcranial magnetic stimulation: Correlation with direct electrical cortical stimulation. *Neurosurgery* 41:1319–1325.

Krings, T., C. Naujokat, and D. G. von Keyserlingk. 1998. Representation of cortical motor function as revealed by stereotactic transcranial magnetic stimulation.
Electroencephalography and Clinical Neurophysiology 109:85–93.

Krings, T., M. Schreckenberger, V. Rohde et al. 2001. Metabolic and electrophysiological
validation of functional MRI. *Journal of Neurology, Neurosurgery and Psychiatry*
71:762–771.

Krnjević, K., M. Randić, and D. W. Straughan. 1966. An inhibitory process in the cerebral
cortex. *Journal of Physiology* 184:16–48.

Le Bihan, D., J. F. Mangin, C. Poupon et al. 2001. Diffusion tensor imaging: Concepts and
applications. *Journal of Magnetic Resonance Imaging* 13:534–546.

Lehmann, D. and W. Skrandies. 1980. Reference-free identification of components of check-
erboard-evoked multichannel potential fields. *Electroencephalography and Clinical
Neurophysiology* 48:609–621.

Lioumis, P., D. Kičić, P. Savolainen et al. 2009. Reproducibility of TMS-evoked EEG
responses. *Human Brain Mapping* 30:1387–1396.

Litvak, V., S. Komssi, M. Scherg et al. 2007. Artifact correction and source analysis of early
electroencephalographic responses evoked by transcranial magnetic stimulation over
primary motor cortex. *NeuroImage* 37:56–70.

Maccabee, P. J., V. E. Amassian, L. P. Eberle et al. 1991. Measurement of the electric field
induced into inhomogeneous volume conductors by magnetic coils: Application to
human spinal neurogeometry. *Electroencephalography and Clinical Neurophysiology*
81:224–237.

Mäki, H. and R. J. Ilmoniemi. 2010. The relationship between peripheral and early cortical activation induced by transcranial magnetic stimulation. *Neuroscience Letters* 478:24–28.

Mäki, H. and R. J. Ilmoniemi. 2011. Projecting out muscle artifacts from TMS-evoked EEG.
NeuroImage 54:2706–2710.

Massimini, M., F. Ferrarelli, R. Huber et al. 2005. Breakdown of cortical effective connectivity
during sleep. *Science* 309:2228–2232.

Meyer, B. U., R. Diehl, H. Steinmetz et al. 1991. Magnetic stimuli applied over motor and
visual cortex: Influence of coil position and field polarity on motor responses, phosphenes, and eye movements. *Electroencephalography and Clinical Neurophysiology*
43:121–134.

Näsi, T., H. Mäki, K. Kotilahti et al. 2011. Magnetic-stimulation-related physiological artifacts in hemodynamic near-infrared spectroscopy signals. *PLoS ONE* 6:e24002.

Nikouline, V., J. Ruohonen, and R. J. Ilmoniemi. 1999. The role of the coil click in TMS
assessed with simultaneous EEG. *Clinical Neurophysiology* 110:1325–1328.

Nikulin, V. V., D. Kičić, S. Kähkönen et al. 2003. Modulation of electroencephalographic
responses to transcranial magnetic stimulation: Evidence for changes in cortical excitability related to movement. *European Journal of Neuroscience* 18:1206–1212.

Oliviero, A., V. Di Lazzaro, O. Piazza et al. 1999. Cerebral blood flow and metabolic
changes produced by repetitive magnetic brain stimulation. *Journal of Neurology*
246:1164–1168.

Pascual-Leone, A., J. R. Gates, and A. Dhuna. 1991. Induction of speech arrest and counting
errors with rapid-rate transcranial magnetic stimulation. *Neurology* 41:697–702.

Paus, T., R. Jech, C. J. Thompson et al. 1997. Transcranial magnetic stimulation during positron emission tomography: A new method for studying connectivity of the human cerebral cortex. *Journal of Neuroscience* 17:3178–3184.

Paus, T., P. K. Sipilä, and A. P. Strafella. 2001. Synchronization of neuronal activity in the
human primary motor cortex by transcranial magnetic stimulation: An EEG study.
Journal of Neurophysiology 86:1983–1990.

Picht, T., S. Mularski, B. Kuehn et al. 2009. Navigated transcranial magnetic stimulation for
preoperative functional diagnostics in brain tumor surgery. *Neurosurgery* 65:93–98.

Picht, T., S. Schmidt, S. Brandt et al. 2011a. Preoperative functional mapping for rolandic brain tumor surgery: Comparison of navigated transcranial magnetic stimulation to direct cortical stimulation. *Neurosurgery* 69:581–588.

Picht, T., J. Schulz, M. Hanna et al. 2011b. Assessment of the influence of navigated transcranial magnetic stimulation on surgical planning for tumors in or near the motor cortex. *Neurosurgery* 70:1248–1256.

Raij, T., J. Karhu, D. Kičić et al. 2008. Parallel input makes the brain run faster. *NeuroImage* 40:1792–1797.

Rosanova, M., A. Casali, V. Bellina et al. 2009. Natural frequencies of human corticothalamic circuits. *Journal of Neuroscience* 29:7679–7685.

Roth, B. J., A. Pascual-Leone, L. G. Cohen et al. 1992. The heating of metal electrodes during rapid-rate magnetic stimulation: A possible safety hazard. *Electroencephalography and Clinical Neurophysiology* 85:116–123.

Ruohonen, J. and R. J. Ilmoniemi. 1996. Multichannel magnetic stimulation: Improved stimulus targeting. In *Advances in Magnetic Stimulation: Mathematical Modeling and Clinical Applications, Advances in Occupational Medicine and Rehabilitation 2*, eds. J. Nilsson, M. Panizza, and F. Grandori. Maugeri Foundation, Pavia, Italy, pp. 55–64.

Ruohonen J. and R. J. Ilmoniemi. 1998. Focusing and targeting of magnetic brain stimulation using multiple coils. *Medical and Biological Engineering and Computing* 36:297–301.

Ruohonen, J. and J. Karhu. 2010. Navigated transcranial magnetic stimulation. *Neurophysiologie Clinique* 40:7–17.

Ruspantini, I., H. Mäki, R. Korhonen et al. 2011. The functional role of the ventral premotor cortex in a visually paced finger tapping task: A TMS study. *Behavioural Brain Research* 220:325–330.

Saager, R. B. and A. J. Berger. 2005. Direct characterization and removal of interfering absorption trends in two-layer turbid media. *Journal of the Optical Society of America A: Optics, Image Science, and Vision* 22:1874–1882.

Scherg, M. 1992. Functional imaging and localization of electromagnetic brain activity. *Brain Topography* 5:103–111.

Siebner, H. R., M. Peller, F. Willoch et al. 2000. Lasting cortical activation after repetitive TMS of the motor cortex: A glucose metabolic study. *Neurology* 54:956–963.

Tachtsidis, I., T. S. Leung, A. Chopra et al. 2009. False positives in functional near-infrared topography. *Advances in Experimental Medicine and Biology* 645: 307–314.

Tam, H. W. and J. G. Webster. 1977. Minimizing electrode motion artifact by skin abrasion. *IEEE Transactions on Biomedical Engineering* 24:134–139.

Thut, G., J. R. Ives, F. Kampmann et al. 2005. A new device and protocol for combining TMS and online recordings of EEG and evoked potentials. *Journal of Neuroscience Methods* 141:207–217.

Thut, G. and C. Miniussi. 2009. New insights into rhythmic brain activity from TMS–EEG studies. *Trends in Cognitive Sciences* 13:182–189.

Thut, G. and A. Pascual-Leone. 2010. A Review of combined TMS–EEG studies to characterize lasting effects of repetitive TMS and assess their usefulness in cognitive and clinical neuroscience. *Brain Topography* 22:219–232.

Tiitinen, H., J. Virtanen, R. J. Ilmoniemi et al. 1999. Separation of contamination caused by coil clicks from responses elicited by transcranial magnetic stimulation. *Clinical Neurophysiology* 110:982–985.

Ueno, S., T. Tashiro, and K. Harada. 1988. Localized stimulation of neural tissues in the brain by means of a paired configuration of time-varying magnetic fields. *Journal of Applied Physics* 64:5862–5864.

Valls-Solé, J., A. Pascual-Leone, E. M. Wassermann et al. 1992. Human motor evoked responses to paired transcranial magnetic stimuli. *Electroencephalography and Clinical Neurophysiology* 85:355–364.

Van Der Werf, Y. D. and T. Paus. 2006. The neural response to transcranial magnetic stimulation of the human motor cortex. I. Intracortical and cortico-cortical contributions. *Experimental Brain Research* 175:231–245.

Veniero, D., M. Bortoletto, and C. Miniussi. 2009. TMS–EEG co-registration: On TMS-induced artifact. *Clinical Neurophysiology* 120:1329–1399.

Virtanen, J., T. Noponen, K. Kotilahti et al. 2011. Accelerometer-based method for correcting signal baseline changes caused by motion artifacts in medical near-infrared spectroscopy. *Journal of Biomedical Optics* 16:087005.

Virtanen, J., T. Noponen, and P. Meriläinen. 2009. Comparison of principal and independent component analysis in removing extracerebral interference from near-infrared spectroscopy signals. *Journal of Biomedical Optics* 14:054032.

Virtanen, J., J. Ruohonen, R. Näätänen et al. 1999. Instrumentation for the measurement of electric brain responses to transcranial magnetic stimulation. *Medical and Biological Engineering and Computing* 37:322–326.

Zhang, Y., D. H. Brooks, M. A. Franceschini et al. 2005. Eigenvector-based spatial filtering for reduction of physiological interference in diffuse optical imaging. *Journal of Biomedical Optics* 10:11014.

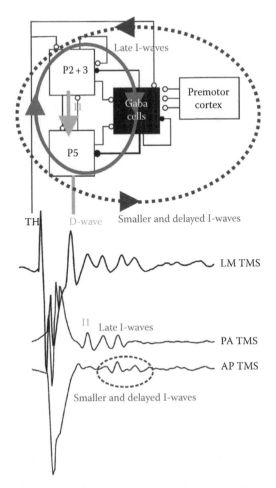

FIGURE 2.1 A schematic view of the model of corticospinal volley generation based on canonical cortical circuit proposed by Douglas et al. (1989). This model includes the superficial population of excitatory pyramidal neurons of layers II and III (P2–P3), the large PTNs in layer V (P5), and the inhibitory GABA cells and the thalamocortical projections. (Modified from Sheperd, G. *The Synaptic Organization of the Brain*, Cambridge University Press, New York, Figure 1.14, 2004.) Magnetic stimulation with a PA induced current in the brain produces monosynaptic activation of P5 cells by the axons of superficial pyramidal neurons (green arrow) evoking the I1 wave; at high intensities, it also produces recurrent activity in a circuit composed of the layer II and III and layer V pyramidal neurons together with their connections with local GABAergic interneurons (red ellipse and arrows) evoking later I-waves, which include the same cortical elements. Magnetic stimulation with an LM induced current in the brain produces a direct activation of the axons of corticospinal cells (blue line) evoking the D-wave followed, at high intensities, by I-waves. Magnetic stimulation with an AP induced current in the brain recruits smaller and delayed descending volleys with slightly different peak latencies and longer duration than those seen after PA magnetic stimulation. It is proposed that this more dispersed descending activity is produced by a more complex circuit (violet-dotted ellipse and arrows) that might include cortico-cortical fibers originating from the premotor cortex and projecting upon the M1 circuits generating the I-waves.

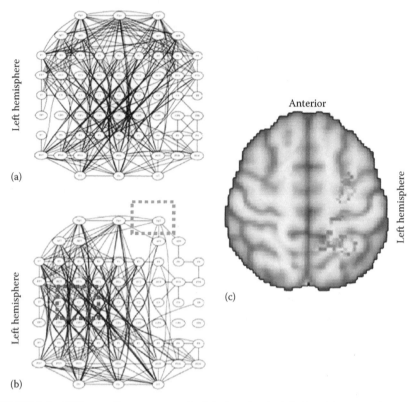

(a)

(b)

(c)

Anterior

Left hemisphere

Left hemisphere

Left hemisphere

FIGURE 4.4 EEG recordings were acquired during simple right-hand finger tapping before (a) and after (b) the application of 10 min anodal tDCS over the left primary motor cortex. After the end of stimulation, functional connectivity significantly increased in the left hemisphere in the high-gamma band (60–90 Hz), whereas the number of inter-hemispheric significantly decreased. Panel (c) shows regions that significantly increased the functional coupling with the left sensorimotor cortex following anodal stimulation of the left primary motor cortex as measured by BOLD-fMRI during resting state (axial image is displayed according to radiological convention, left is right).

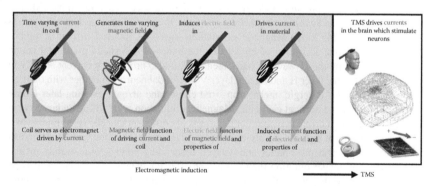

Time varying current in coil | Generates time varying magnetic field | Induces electric field in | Drives current in material | TMS drives currents in the brain which stimulate neurons

Coil serves as electromagnet driven by current | Magnetic field function of driving current and coil | Electric field function of magnetic field and properties of | Induced current function of electric field and properties of

Electromagnetic induction ————▶ TMS

FIGURE 5.3 Basics of TMS. (Partially adapted from Wagner, T. et al., *Cortex*, 45, 1025, 2009.)

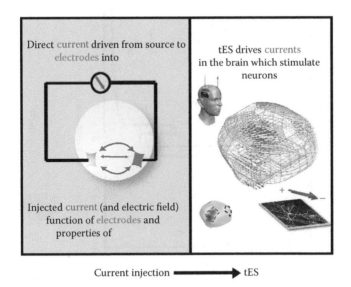

Direct current driven from source to electrodes into

tES drives currents in the brain which stimulate neurons

Injected current (and electric field) function of electrodes and properties of

Current injection ⟶ tES

FIGURE 5.4 Basics of tES.

Stimulation location, focality, and orientation

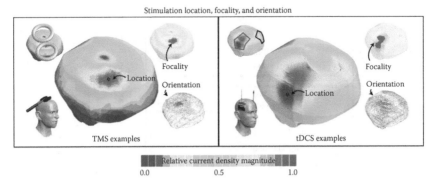

Focality

Location

Orientation

TMS examples

Focality

Location

Orientation

tDCS examples

Relative current density magnitude

0.0   0.5   1.0

FIGURE 5.5 TMS and tDCS field modeling examples of focality, orientation, and location. (Partially adapted from Wagner, T. et al., *Cortex*, 45, 1025, 2009.)

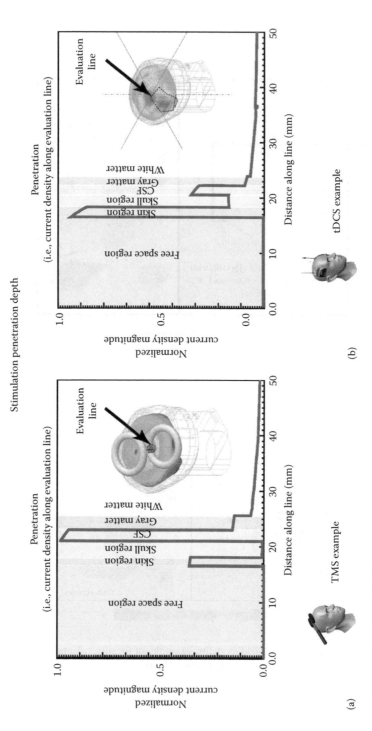

FIGURE 5.6 (a) TMS and (b) tDCS field modeling example of depth of stimulation. (Adapted from Wagner, T. et al., *Annu. Rev. Biomed. Eng.*, 9, 527, 2007b.)

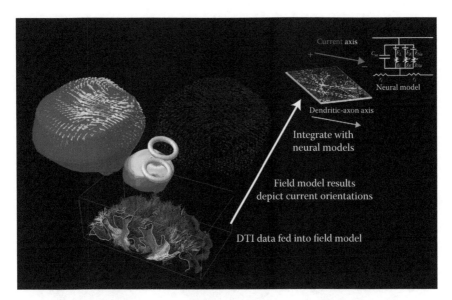

FIGURE 5.7 From DTI to a neural model (a method for field localization and tracking). (From Wagner, T. et al., Transcranial magnetic stimulation: High resolution tracking of the induced current density in the individual human brain, *12th Annual Meeting of the Organization for Human Brain Mapping*, Florence, Italy, 2006b.)

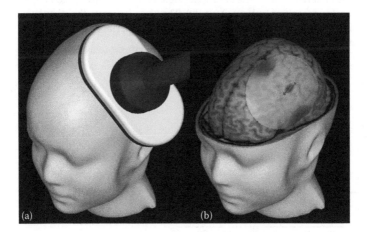

FIGURE 10.1 Navigated brain stimulation. The coil (a) and the cortically induced electric field (b) can be displayed with respect to the subject's anatomical MR image. The coloring of the cortex indicates the strength of the calculated E-field; the maximum is shown by the small red area and the arrows show the direction of the induced current.

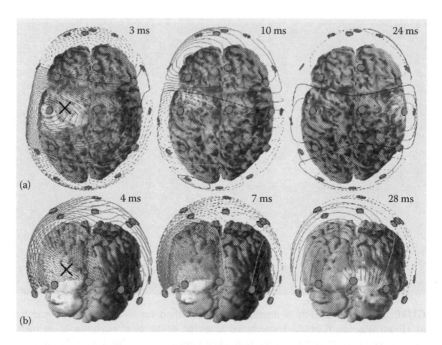

FIGURE 10.3 Spread of TMS-evoked activation from the stimulated motor (a) or visual (b) cortex to the contralateral side. The color maps show the activation as calculated from the averaged TMS-evoked EEG with the minimum norm estimate. The contour lines show the potential patterns of the measured EEG signals (red, positive potential; blue, negative potential; contour spacing: 1 μV (a), 2 μV (b)). The cross indicates the stimulation site. (From Ilmoniemi, R.J. et al., *NeuroReport,* 8, 3537, 1997. With permission.)

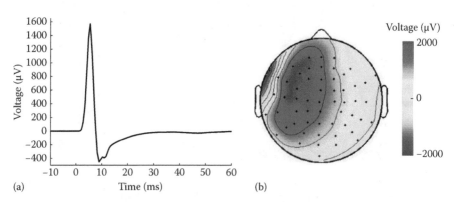

FIGURE 10.4 A typical biphasic muscle artifact recorded at the vertex (reference behind right ear) after the stimulation of Broca's area (left hemisphere) at time $t = 0$ (a) and the corresponding topography at the time of the signal maximum (6 ms; b).

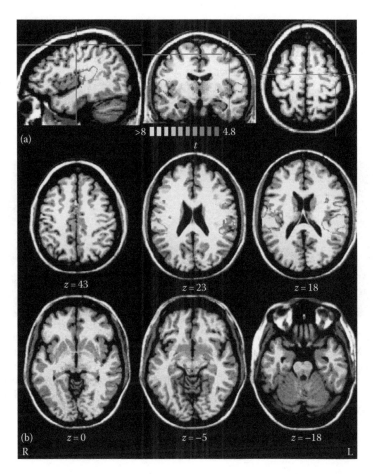

FIGURE 10.5 Local and distant fMRI responses to TMS delivered to the left dorsal premotor cortex (indicated with crosshair). The colors mark areas with statistically significant activation according to a group analysis: left dorsal premotor cortex (a), cingulate gyrus, ventral premotor cortex, auditory cortex, caudate nucleus, left posterior temporal lobe, medial geniculate nucleus, and cerebellum (b). R, right, L, left. (From Bestmann, S. et al., *NeuroImage,* 28, 22, 2005. With permission.)

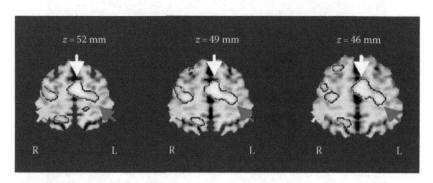

FIGURE 10.6 PET responses showing rCMRglc increases on the group level as a response to 5 Hz TMS delivered to the left M1 (red arrow). Yellow and white arrows point to activations at contralateral M1 and supplementary motor area, respectively. The responses are superimposed on axial sections of stereotactically normalized MRIs and the Talairach z-coordinates are given on top of the sections. R, right, L, left. (Data published in Siebner, H.R. et al., *Neurology,* 54, 956, 2000. With permission.)

Part II

Biology

11 Molecular Mechanisms of Brain Stimulation

Luisella Bocchio-Chiavetto,
James T.H. Teo, and Binith Cheeran

CONTENTS

ABBREVIATIONS

ACTH	adrenocorticotropic hormone
cAMP	cyclic adenosine monophosphate
APP	amyloid precursor protein
ATP	adenosine triphosphate
BDNF	brain-derived neurotrophic factor
CCK	cholecystokinin
COX-2	cyclooxygenase-2
CREB	cAMP-responsive element binding protein
CSF	cerebrospinal fluid
DA	dopamine
DAT	dopamine transporter
DEX/CRT	dexamethasone/corticotropin-releasing hormone

DLPFC	dorsolateral prefrontal cortex
DOPA	dihydroxyphenylalanine
DRG	dorsal root ganglia
ES	electrical stimulation
tES	transcranial electrical stimulation
GABA	gamma-aminobutyric acid
GAD 65/67	glutamic acid decarboxylase 65/67
GAP-43	growth-associated protein 43
GAT-1	presynaptic GABA transporter 1
GFAP	glial fibrillary acidic protein
GluR1	glutamate AMPA receptor subunit 1
HD	Huntington's disease
5-HIAA	5-hydroxyindoleacetic acid
HSP70	heat-shock protein 70
5-HT	5-hydroxytryptamine or serotonin
$5\text{-HT}_{1A/1B/2/2A}$	serotonin receptor $_{1A/1B/2/2A}$
HVA	homovanillic acid
M1	primary motor cortex
MAP-2	microtubule-associated protein-2
MD	major depression
MHPG	3-methoxy-4-hydroxyphenylglycol
NE	norepinephrine
NET	norepinephrine transporter
NFM	medium-molecular-weight neurofilament
NMDAR	N-methyl-D-aspartate receptor
PD	Parkinson's disease
PFC	prefrontal cortex
PV	parvalbumin
SERT	serotonin transporter
SOD-1	superoxide dismutase 1
cTBS	continuous theta burst stimulation
iTBS	intermittent theta burst stimulation
TH	tyrosine hydroxylase
rTMS	repetitive transcranial magnetic stimulation
TNF-α	tumor necrosis factor-α
TRD	treatment-resistant depression
TSH	thyroid-stimulating hormone

11.1 INTRODUCTION

In the last two decades, noninvasive brain stimulation (NIBS) has been tested in a number of neurological and neuropsychiatric conditions for potential clinical applications. NIBS paradigms such as repetitive transcranial magnetic stimulation (rTMS) or transcranial electrical stimulation (tES) have been shown in small pilot studies to improve symptoms in a number of neurological and many neuropsychiatric disorders with relatively fewer side effects. However, research and clinical applications of

NIBS have largely preceded the understanding of the neurobiological mechanisms of NIBS. It is increasingly accepted that NIBS involve long-term potentiation (LTP) and depression (LTD)-like processes, as well as the inhibitory processes involving gamma-aminobutyric acid (GABA)-ergic activity, but it is likely that many other mechanisms are also involved. This chapter will review the literature on what is known about the molecular mechanisms.

Several strategies have been utilized to dissect the molecular mechanisms of NIBS. Pharmacological manipulation of the effects of NIBS has been an important tool for the study of the mechanisms. Another route for understanding the molecular pathways is with molecular neuroimaging techniques such as magnetic resonance spectroscopy (MRS), single-photon emission computed tomography (SPECT), and others. An alternative route is using animal or cell models of NIBS to assay molecular parameters and pathways. Finally, one could study the effect of genetic polymorphisms or genetic disease models on NIBS in humans.

11.2   CLUES FROM PHARMACOLOGICAL MANIPULATION

To date, the primary method of investigating the molecular mechanisms of NIBS has been pharmacological manipulation. Modulation or blockade of the effects of a NIBS paradigm by drugs is used to determine if that particular NIBS protocol involves pathways affected by the drug. Often, to provide information on the physiological properties of a novel TMS measure, the effect of a drug with a single (or narrow) mode of action is tested.

The main parameter measured with noninvasive stimulation is the motor-evoked potential (MEP) that is considered broadly analogous to cortical excitability. With TMS, a single stimulus delivered at strengths above the resting motor threshold (RMT) produces a synchronous depolarization of axons of cortical pyramidal neurons and cortical interneurons of the corticospinal tract. As such, the RMT reflects the propensity of cortical pyramidal neurons to generate an action potential and thus the membrane potential of these neurons. This relation with the membrane potential is supported by the fact that membrane-stabilizing drugs (such as lamotrigine, phenytoin, and carbamazepine, which act by blocking voltage-gated sodium channels) increase RMT (Mavroudakis et al., 1994; Ziemann et al., 1996).

However, there is some evidence that drugs that act on glutamate receptors also affect the RMT. Di Lazzaro and colleagues found that ketamine, an antagonist of the NMDA receptor (*N*-methyl-D-aspartate receptor, an ionotropic glutamate receptor), reduced RMT and enhanced MEP amplitude with TMS but not tES (Di Lazzaro et al., 2003). This could be explained by ketamine also enhancing glutamatergic transmission by action on non-NMDA glutamatergic receptors such as α-amino-3-hydroxy-5-methyl-4-isoxazolepropionic acid (AMPA) and kainate receptors (Moghaddam et al., 1997) rather than the NMDA receptor antagonism causing lowering of RMT. This is supported by the lack of effect on RMT of alternative NMDA antagonists, such as dextromethorphan (Ziemann et al., 1998a), D-cycloserine (Nitsche et al., 2004b; Teo et al., 2007), and memantine (Schwenkreis et al., 2005; Huang et al., 2007).

However, the membrane potential model for RMT does not explain why inhibition of glutamatergic synaptic transmission should affect RMT. Levetiracetam (an inhibitor of synaptic vesicle release of glutamate) raises the RMT (Reis et al., 2004; Epstein et al., 2008; Solinas et al., 2008) while talampanel, an antagonist of the ionotropic glutamatergic AMPA receptor, elevated RMT by TMS (Danielsson et al., 2004). One explanation could be that a small component of the RMT constitutes background trans-synaptic activity onto the AMPA receptors of pyramidal interneurons (i.e., background subthreshold excitatory postsynaptic potentials). Alternatively, TMS produces 700 Hz repetitive discharges of neurons (Di Lazzaro et al., 1998), so it is conceivable that these repetitive discharges may involve neurotransmission via AMPA receptors and therefore RMT.

In addition to an effect on RMT, both talampanel and levetiracetam also reduced MEP amplitude (Sohn et al., 2001; Danielsson et al., 2004; Reis et al., 2004, Epstein et al., 2008; Solinas et al., 2008); this can be explained if MEP amplitude measurement also comprises two components: cortical glutamatergic interneuron excitability (i.e., active motor threshold) and the subsequent trans-synaptic glutamatergic activation of pyramidal neurons by AMPA receptors (Cheeran et al., 2010). The model of RMT predominantly reflecting membrane potential (rather than synaptic neurotransmission) would also be consistent with a lack of effect on RMT by other drugs acting on other neurotransmitter receptors such as GABA (Ziemann et al., 1996) or dopamine (Ziemann et al., 1997).

On the other hand, NIBS that induces effects outlasting the period of stimulation, such as rTMS, paired associative stimulation (PAS), and transcranial direct current stimulation (tDCS), involves glutamatergic synaptic transmission. rTMS, tDCS, and PAS induce facilitation or inhibition of MEP amplitude (and with high-frequency rTMS (HF-rTMS) even lower RMT), and this effect is blocked by NMDA receptor antagonists (Ziemann et al., 1998a; Stefan et al., 2002) and benzodiazepines (Ziemann et al., 1998b). In addition, NMDA partial agonist D-cycloserine modulates the effects of theta burst rTMS and tDCS (Nitsche et al., 2004b; Teo et al., 2007). This association with the NMDA receptor has led to the effects of NIBS being termed "cortical plasticity," and the association with motor learning has led it to be compared to glutamatergic synaptic plasticity as measured on animal neocortical slices. These "cortical plasticity" effects are modulated by dopaminergic drugs (Monte-Silva et al., 2009, 2011; Nitsche et al., 2009), noradrenergic drugs (Nitsche et al., 2004a), and cholinergic drugs (Kuo et al., 2008).

Other aspects of cortical physiology can also be assessed with paired-pulse TMS paradigms. In these paradigms, a conditioning stimulus modulates the amplitude of the MEP of a subsequent stimulus. By varying the interstimulus interval, the intensity of the conditioning stimulus and even the type of the stimulus, and various intracortical and corticocortical circuits can be studied. The most established measure is short-interval intracortical inhibition (SICI): this is the inhibition of MEP amplitude by a subthreshold conditioning stimulus 1–5 ms before the test stimulus and is thought to represent GABAergic inhibitory interneurons acting through $GABA_A$ receptors on cortical excitatory interneurons (Ziemann et al., 1996; Di Lazzaro et al., 2007). Other paired-pulse measures involve other neurotransmitter systems: long-interval intracortical inhibition (LICI) is modulated by $GABA_B$ agonist receptors (McDonnell et al., 2007) while short-interval afferent inhibition (SAI) is reduced by agonism of

$GABA_A$ receptors lacking the $GABA_{A\alpha1}$ subunit (Di Lazzaro et al., 2007; Teo et al., 2009) and are enhanced by cholinesterase inhibitors such as rivastigmine (Di Lazzaro et al., 2004), donepezil (Nardone et al., 2012), and tacrine (Korchounov et al., 2005).

Pharmacological studies have important limitations: (1) most drugs have multiple mechanisms of action; (2) chronic and acute administration effects can vary (e.g., acute administration of carbamazepine has no effect on SICI, but chronic administration increases SICI); and (3) effects of drugs may be dose dependent with few NIBS studies exploring dose-response effects. Finally, intersubject variability has not been systematically studied, and genetic factors could account for substantial differences (Missitzi et al., 2011).

11.3 CLUES FROM NEUROIMAGING STUDIES

Molecular neuroimaging provides another route to studying the molecular pathways underlying NIBS. MRS is a noninvasive imaging technique, which, by varying echo times, can quantify concentrations of small molecules of interest in a brain voxel (Burlina et al., 2000). This is typically normalized to a reference peak of another small molecule not expected to change during stimulation (e.g., N-acetyl-aspartate) to avoid confounds such as edema. As such, MRS can be used to assess glutamate and GABA levels in a voxel, but cannot distinguish between intracellular (i.e., in synaptic vesicles) or extracellular (synaptic) molecules, nor can it measure receptors.

Michael et al. (2003) described an increase in glutamate levels in stimulated dorsolateral prefrontal cortex (DLPFC) in normal individuals after one or five sessions of 20 Hz rTMS (Michael et al., 2003). This was replicated in patients with major depression (MD) after 10 sessions of 20 Hz rTMS, where an increase in the levels of glutamate was observed, which correlated with stimulation intensity (Luborzewski et al., 2007). Interestingly, both studies showed that the amount of glutamate increase was dependent on the pre-stimulation glutamate level (the lower the pre-stimulation level, the higher the increase observed).

More recently, other NIBS protocols have also been studied with MRS: continuous theta burst stimulation (cTBS, a high-frequency burst patterned TMS paradigm), which is inhibitory on cortical excitability, leads to an increase in GABA concentration within the stimulated area but not glutamate concentrations (Stagg et al., 2009b). It is not possible to definitively identify in which GABA pool this increase occurs, but it seems likely that the increase in GABA concentration is secondary to changes in glutamic acid decarboxylase (GAD) activity as there was no significant change in glutamate levels (the synthetic precursor of GABA).

However, MRS measurements of another inhibitory transcranial stimulation paradigm, cathodal tDCS (which exerts very similar effects on net cortical excitability as cTBS), reduced glutamate levels and a secondary decrease in GABA (Stagg et al., 2009a), suggesting that different NIBS paradigms, with very similar outcomes using MEP as a summary variable, have distinct effects and therefore may involve different molecular processes and different interneuronal populations.

The mechanisms of NIBS can also be studied using SPECT or positron emission tomography (PET). Many studies using these techniques looked at the effects of rTMS on cerebral blood flow or glucose metabolism (Catafau et al., 2001;

Siebner et al., 2001a,b; Okabe et al., 2003; Speer et al., 2003; Ohnishi et al., 2004), and although helpful in understanding cortical connectivity, this provides little information on the molecular mechanisms. However, the development of specific radioligands allows the molecular neuroimaging of neurotransmitter receptors. Using the radioligand ^{11}C-raclopride (a radiolabeled D2-antagonist), PET has shown a reduction in binding in the striatum after 10 Hz rTMS to the motor cortex in healthy subjects indicating an increase in striatal dopaminergic neurotransmission after rTMS (Strafella et al., 2001, 2003). The same effect was also seen in SPECT studies using the D2-receptor radioligand ^{123}I-iodobenzamide to study patients with depression treated with 10 Hz rTMS (Pogarell et al., 2006, 2007).

The serotonergic system has also recently been studied with this approach. The radioligand R91150 (radiolabeled 5-HT$_{2A}$ antagonist) had reduced binding in both DLPFCs and increased binding in the right hippocampus after 10 Hz rTMS to the left DLPFC in patients with MD (Baeken et al., 2011). This indicates that rTMS induces increased serotonergic neurotransmission in both DLPFCs, and the correlation with treatment response suggests that this might be the mechanism of action in the treatment of MD.

11.4 CLUES FROM GENETIC DETERMINANTS OF RESPONSE TO NIBS

Recent advances in genetics together with the growing data from clinical studies now permit the study of genetic variants influencing the action of NIBS. One method would be to study disease models caused by single gene mutations, where the function of the gene is well characterized. For example, patients with an inherited generalized epilepsy (due to the R43Q missense mutation in the GABAA-γ2 receptor subunit) and found an impairment of SICI but normal RMT (Fedi et al., 2008). This is as predicted from pharmacological studies and lends support to the idea of RMT representing membrane potential and SICI representing GABAergic synaptic transmission.

Fragile X syndrome is a disorder where a loss of function of the fragile-X mental retardation protein (FMRP) (either from a deletion or a trinucleotide repeat in the FMR1 gene) causes abnormal learning disability and autism (Lugenbeel et al., 1995). Patients with fragile X syndrome have normal SICI and LICI, impaired response to cTBS, and excessive response to intermittent TBS (iTBS) (which causes an increase in cortical excitability as measured by MEPs) (Oberman et al., 2010a,b), and repeated testing also demonstrated altered metaplasticity. FMRP is an mRNA-binding protein that regulates synaptic protein synthesis triggered by metabotropic glutamate receptors (mGluR1/5) (Auerbach and Bear, 2010; Auerbach et al., 2011). This lends credence to the hypothesis that the cortical excitability changes induced by NIBS are a measure of synaptic plasticity.

NIBS has been used to study many other monogenic diseases such as DYT1 dystonia (Edwards et al., 2003, 2006), Huntington's disease (Orth et al., 2009), and spinocerebellar ataxias (Schwenkreis et al., 2002; Teo et al., 2008); however, as the function of the gene products are unknown currently, these studies add little to the molecular understanding of the mechanisms of NIBS.

An alternative method is to study the effect of common (non-disease) polymorphisms in genes with well-characterized function on NIBS. Cheeran and colleagues, who undertook this approach with studies of the common variation in the brain-derived neurotrophic factor (BDNF) gene Val66Met (rs6265) on cortical neuroplasticity using NIBS, have demonstrated the effects of this genetic variation on human neuroplasticity (Cheeran et al., 2008). BDNF has been shown to modulate NMDA-dependent LTP (Figurov et al., 1996; Aicardi et al., 2004) and LTD (Aicardi et al., 2004; Woo et al., 2005) in animals. Most neurotrophins are secreted constitutively, but BDNF secretion from central synapses is activity-dependent (Lu, 2003), possibly by virtue of a "sorting motif" in the "pro" region of its precursor molecule pro-BDNF. In humans, the "pro" region of BDNF is critical for intracellular trafficking and activity-dependent secretion and a common single-nucleotide polymorphism (SNP)—BDNF Val66Met—affects intracellular trafficking of pro-BDNF and activity-dependent secretion of mature BDNF. This SNP has known functional consequences in healthy human subjects including reduced hippocampal volume and episodic memory (Egan et al., 2003; Hariri et al., 2003; Pezawas et al., 2004).

Cheeran et al. (2008) explored whether various NIBS paradigms would also be influenced by the Val66Met polymorphism in the BDNF gene. The authors studied three NMDA-dependent plasticity-inducing protocols: (1) cTBS and iTBS to study LTP-like and LTD-like plasticity (Huang et al., 2005); (2) PAS, which engages spike-timing-dependent plasticity by pairing electrical stimulation of the contralateral median nerve with single-pulse TMS of the primary motor cortex at a critical interstimulus interval (Stefan et al., 2000; Wolters et al., 2003), to study associative plasticity; and (3) a meta-plasticity protocol in which 10 min tDCS of M1 primary motor cortex precedes 1 Hz rTMS of the same region (Siebner et al., 2004). Healthy Met allele carriers (Met/Val or Met/Met) were compared with homozygous Val/Val carriers. The results showed that the aftereffects of both iTBS (n = 18) and cTBS (n = 18) are reduced or absent in subjects carrying the Met allele. Although there was no significant effect of the BDNF polymorphism on PAS (n = 18)-induced plasticity in the homotopic muscle, there was less spread of LTP-like change to heterotopic muscles in those carrying a Met allele. This suggests that individuals carrying the Met allele have a reduced response to NMDA-dependent plasticity induction by rTMS paradigms. Finally, Met allele carriers had reduced or absent metaplasticity (n = 16); cathodal tDCS produced the same amount of LTD-like suppression of corticospinal activity in all subjects, regardless of genotype; there was a lack of the expected metaplastic effect on subsequent 1 Hz rTMS in subjects carrying the Met allele. Thus, the response of Met allele carriers differed significantly in all protocols compared with the response in Val/Val individuals. The implication is that the polymorphism of BDNF influences the susceptibility of synapses to undergo LTP/LTD-like change in response to NIBS. This finding has been replicated with the study of another independent group that analyzed retrospectively collected data with new genotype data and confirmed that the response of the BDNF Val/Met allele carriers to iTBS stimulation differed when compared with the response of Val/Val individuals (Antal et al., 2010). The additional finding in this retrospective analysis was that carriers of the Val/Met allele had prolonged responses to facilitatory tDCS (24 subjects, 10 heterozygotes) and inhibitory tDCS (19 subjects, 8 heterozygotes) compared with Val/Val allele. This paradoxical effect for tDCS suggests that the effect of tDCS operates via a different (but overlapping) molecular pathway, which is in line with the results of MRS neuroimaging studies.

Using this approach, other SNPs in other genes have been linked with response to NIBS; testing of selected NMDA subunit SNPs showed that an SNP in GRIN2B was associated with alterations in response to TBS (Mori et al., 2011). An SNP in the TRPV-1 gene coding a vanilloid receptor was associated with increased short-interval intracortical facilitation, a more subtle measure of glutamatergic transmission (Mori et al., 2012). However, standard statistical genetic stringency was not used, and neither of these studies have been replicated yet.

Another approach is to use gene expression studies in animal models of rTMS to elucidate the effects. In mice, 20 days of rTMS treatment downregulated the mRNA expression of the serotonin (5-HT) transporter (SERT) and increased dopamine and norepinephrine (NE) transporter mRNA levels (Ikeda et al., 2005). The 5-HT transporter is coded by the SLC6A4 gene, and a 44-base pair common variation in the promoter region (5-HTTLPR) affects SLC6A4 transcription mRNA expression. The short (S) allele is associated with a decreased transcription of the SLC6A4 gene (Heils et al., 1996) and variations in 5-HTTLPR is associated with individual differences in brain morphological and functional variability (Canli et al., 2005; Serretti et al., 2006; Young, 2007). The individuals who are homozygote with the long (L) allele have been shown to have responded differently to citalopram (selective serotonin reuptake inhibitor, SSRI) compared with LS or SS genotypes with enhancement of SICI (Eichhammer et al., 2003) in line with a meta-analysis showing a difference in SSRI treatment response depending on genotype (Serretti et al., 2007).

As rTMS is used to treat depression, it would be expected that these genotypes would have an effect on response to rTMS treatment. Bocchio-Chiavetto and colleagues (2008) showed that BDNF Val66Met and 5-HTTLPR polymorphisms had an effect on the response to TMS (1 or 17 Hz) of MD patients: response rates were increased (average improvement >30%) in 5-HTTLPR LL and BDNF Val/Val homozygote subjects. The effects of the two polymorphisms were independent, and no other demographic and clinical variable was found to affect rTMS response.

A previous study that analyzed the 5-HTTLPR polymorphism as well as the HTR1A gene (coding the $5HT_{1A}$ receptor) showed slightly different results: and also the 5-HTTLPR polymorphism had an effect in the sham stimulation group (Zanardi et al., 2007). The 5HTR1A promoter polymorphism (−1019C/G) influenced treatment response: C/C homozygotes showed a greater response between active and sham stimulation (Zanardi et al., 2007). However, more recent work by the same group replicated the result in 5-HTTLPR and 5HTR1A (Malaguti et al., 2011).

Recently, gene–gene interactions have been reported in the BDNF and COMT genes using PAS: the BDNF Val66Met polymorphism having an effect only conditional on the COMT Met/Met homozygotes (Witte et al., 2012). This indication of epistasis underlines the complexity of the molecular mechanisms of NIBS.

In summary, the earlier studies suggest that response to NIBS is a complex trait with contribution from many genes (and their molecular pathways); therefore, it is only a matter of time before NIBS researchers start to adopt techniques used in other genetic studies of complex traits such as quantitative trait loci techniques or genome-wide association studies.

11.5 MOLECULAR CONSEQUENCES OF BRAIN STIMULATIONS: STUDIES IN ANIMALS, NEURONAL CELL MODELS, AND HUMAN PERIPHERAL TISSUES

Animal and cell models provide the most direct way of measuring the molecular consequences of NIBS. Several aspects of NIBS have been studied using this approach: rTMS effects on neurotransmitter systems (neurotransmitter release and effects on function of their receptors and transporters), potential neuroprotective effects of rTMS in animal models of brain injury, and neurodegeneration have been the focus of some investigations and non-CNS effects of rTMS (e.g., in peripheral blood). Here we review only positive findings on molecular effects induced by brain stimulations.

11.5.1 Neurotransmitter Systems (Dopaminergic, Noradrenergic, Serotonergic Systems, Excitatory and Inhibitory Amino Acids, and Endogenous Opioids)

A single 25 Hz rTMS session has been reported to increase dopamine turnover rates in rat frontal cortex, to decrease the rates in striatum and hippocampus, to decrease dopamine concentration in cortex, and to increase it in other brain regions (Ben-Shachar et al., 1999). Chronic 20 Hz rTMS in rat is also associated with an increase in the transcript levels of the dopamine transporter (DAT), together with an augmented DA transport rate (Ikeda et al., 2005). The most conclusive evidence of rTMS inducing dopamine release comes from direct measurements of dopamine levels using cerebral microdialysis showing that acute 20 Hz rTMS increased dopamine levels in the hippocampus (Keck et al., 2000b) and nucleus accumbens (Erhardt et al., 2004) and that 25 Hz rTMS also produced dopamine increase in the striatum (Kanno et al., 2004).

The regulation of the dopaminergic release has been studied in neuroblastoma cell cultures showing a decrease in intracellular DA and L-3,4-dihydroxyphenylalanine (L-DOPA) and a diminished activity of tyrosine hydroxylase (TH) after acute 3 Hz magnetic stimulation (MS), whereas 9 Hz MS showed opposite effects, increasing the TH activity (Shaul et al., 2003). The same study reported effects at 3 and 9 Hz, downregulating and upregulating the NE intracellular content, respectively. On the other hand, chronic 20 Hz rTMS was able to upregulate the transcription of the norepinephrine transporter (NET), the NE uptake, and binding in the mouse brain. Furthermore, even experiments in PC12 cell cultures treated with MS for 15 days showed the same modifications on the NET expression (Ikeda et al., 2005). In addition, also β-adrenergic receptor changes have been reported after chronic rTMS but with conflicting results in the cortex: 15 Hz rTMS induced downregulation of β-adrenergic receptors (Fleischmann et al., 1996) and 25 Hz rTMS induced upregulation of β-adrenergic receptors (Ben-Shachar et al., 1999).

Chronic rTMS has been found, by two independent studies, to modulate the binding of β-adrenergic receptors in the cortex, but the direction of these effects is inconsistent between studies. Fleischmann et al. (1996) showed a downregulation of cortical β-adrenergic receptors after rTMS (Fleischmann et al., 1996), whereas Ben-Shachar et al. (1999) reported an upregulation (Ben-Shachar et al., 1999).

Both studies applied chronic stimulation paradigms but used a different stimulation frequency (15 vs. 25 Hz, respectively). The latter study also reported a reduction in β-adrenergic binding in striatum (Ben-Shachar et al., 1999).

Molecular measurements in humans also support an effect of NIBS on dopaminergic and noradrenergic systems. An elevation of serum dopamine was found in Parkinson's disease (PD) patients treated with 25 Hz rTMS to both motor cortices for 6 sessions (Khedr et al., 2007). On the other hand, low-frequency rTMS treatment on the motor cortex in PD patients decreased homovanillic acid levels in the cerebrospinal fluid (CSF) (Shimamoto et al., 2001). Another study reported a decrease in the plasma levels of the 3-methoxy-4-hydroxyphenylglycol (MHPG), the main metabolite of NE after 10 sessions of 20 Hz rTMS in drug-resistant MD patients (Yukimasa et al., 2006). tES has also been shown to induce an increase in urinary MHPG in treating patients with chronic pain (Capel et al., 2003).

There are also evidences for the effects of acute and chronic rTMS on the serotonergic system. An increase in hippocampal levels of 5-HT and its main metabolite 5-HIAA has been observed in rat brains treated with acute rTMS (Ben-Shachar et al., 1997). *In vivo* microdialysis studies in rat showed that both chronic and acute rTMS treatments are able to suppress the 5-HT increase in prefrontal cortex induced by the stress associated with sham treatment (handling procedures and rTMS sound) and anxiety generated by elevated maze test (Kanno et al., 2003a,b), whereas tES of the medial prefrontal cortex produced current-dependent increases in limbic 5-HT content in both anesthetized and behaving rats (Juckel et al., 1999). Furthermore, rTMS has effects on the regulation of 5-HT presynaptic autoreceptors, which control 5-HT synthesis and output. HF-rTMS has also been reported to increase the density of 5-HT_{1A} binding sites in different cortical areas and in the anterior olfactory nucleus (Kole et al., 1999).

In parallel, chronic administration of rTMS leads to a reduction in the sensitivity of both types of serotonin receptors 5-HT_{1A} and 5-HT_{1B} in prefrontal cortex (Gur et al., 2000). Decrease in 5-HT_2 postsynaptic receptors in rat frontal cortex (but not in striatum and hippocampus) chronically treated with rTMS was also observed (Ben-Shachar et al., 1999). Finally, a downregulation of SERT mRNA levels and a subsequent decrease in serotonin uptake and binding were observed after chronic rTMS, while acute treatment induced a transient increase in SERT after 1 h (Ikeda et al., 2005).

Recently, several studies investigated the molecular modifications induced by different rTMS protocols on glutamate and GABA regulatory pathways in animals with studies of rTMS-induced changes in the expression of the glutamic acid decarboxylase (GAD) enzyme isoforms (the synaptically located GAD65 and the more uniformly distributed GAD67) and the presynaptic transmembranous GABA transporter GAT-1, a marker of GABA turnover (Patel et al., 2006; Trippe et al., 2009). Acutely, 1 Hz rTMS, iTBS, and cTBS reduced the GAD67 expression in frontal, motor, somatosensory, and visual cortex whereas increased GAD65 and GAT-1, with iTBS having the strongest effect. Time course studies on chronic effects showed that GAD65 and GAT-1 expression reversed after 1 day for iTBS and cTBS while 1 Hz rTMS induced a stable increase for 7 days, suggesting possible contributions of different neuronal subsystems. A further study confirmed the reduction of GAD67 expression after acute iTBS and associated this molecular effect with an improvement in learning performance in animals (Mix et al., 2010), while cTBS induced an increase in GAD65 synthesis.

Concentrations of glutamate and GABA in the brain of rats subjected to chronic 0.5 Hz rTMS have been reported to be modulated in different directions in different brain regions: in the hippocampus and striatum, levels of both the neurotransmitters were increased, whereas in the hypothalamus both were decreased (Yue et al., 2009). A later study was conducted to evaluate the effects of chronic rTMS at 1 and 20 Hz on the expression of the glutamate AMPA receptor subunit GluR1 in awake and anesthetized rats. The high-frequency treatment was reported to induce a long-lasting increase in GluR1 and its phosphorylated form levels in the hippocampus and prelimbic cortex in awake animals, whereas opposite effects, consisting in a down-regulation, were observed in the same areas in anesthetized animals (Gersner et al., 2011). In addition, the levels of other excitatory amino acids as taurine, serine, and aspartate were significantly increased in the hypothalamic paraventricular nucleus after acute rTMS treatment (Keck et al., 2000b).

A double-blind placebo-controlled study of patients with back pain reported that tES was associated with increased serum β-endorphin levels and reduced pain intensity (Gabis et al., 2003). This increase in serum β-endorphin level was also noted after rTMS in patients with phantom limb pain (Ahmed et al., 2011) without any correlation with treatment response.

11.5.2 NEUROENDOCRINE EFFECTS

Effects of rTMS on the hypothalamic–pituitary–adrenal axis in rats have been reported: 20 Hz rTMS stimulation decreased the hypothalamic release of the neuro-hypophysial hormone vasopressin, determined through *in vivo* microdialysis (Keck et al., 2000b), while chronic treatment induced a decrease in stress-induced elevation of plasma corticotropin and corticosterone (Keck et al., 2000a). The same reduction in stress-related hormones was observed in rodents selectively bred for high-anxiety-related behavior subjected to chronic 20 Hz rTMS (Keck et al., 2001).

Studies in humans have also replicated this: decrease in serum cortisol levels after left-sided 10–20 Hz rTMS at 90% RMT has been found in healthy individuals (Evers et al., 2001). Downregulation of saliva cortisol levels has been observed in melancholic MD and bulimia patients after a single session of 10 Hz rTMS applied to the left DLPFC (Baeken et al., 2009; Claudino et al., 2010). Finally, in a group of MD patients treated with HF-rTMS, a significant correlation was detected between the reduction of depressive symptoms and the decrease in post-dexamethasone cortisol levels (Zwanzger et al., 2003). Diminished salivary cortisol levels have been observed also after tES that reduced chronic pain intensity in patients with spinal cord injuries (Capel et al., 2003). Anodal tDCS reduces blood cortisol and adrenocorticotropic hormone (ACTH) in healthy subjects using MRS (Binkofski et al., 2011).

Contrasting results have been reported on the effect of rTMS on thyroid-stimulating hormone (TSH) levels. Whereas two studies observed no differences or a decrease in plasma TSH after acute rTMS (Pascual-Leone et al., 1993; Evers et al., 2001), other groups have observed a slight increase in plasma TSH levels in healthy volunteers and in MD patients after HF-rTMS (5–20 Hz) (George et al., 1996; Cohrs et al., 2001; Szuba et al., 2001). Recently, studies on MD patients have reported a significant increase in plasma TSH also after 1 Hz rTMS treatments (Kito et al., 2010;

Trojak et al., 2011). Interestingly, the augmentation of TSH levels correlated with the therapeutic efficacy of rTMS treatment on depressive symptoms.

11.5.3 NEUROTROPHINS, NEUROPROTECTORS, AND NEURONAL REPAIR

As discussed earlier, the role of the BDNF polymorphism in NIBS has been extensively studied in humans. However, the direct molecular evidence of this in animals has only relatively recently been forthcoming. A recent study by Fritsch et al. (2010) showed for the first time that tDCS in mouse neocortex promotes LTP with effects outlasting the stimulation period and goes on to characterize the effect by demonstrating a key role for the NMDA receptor and BDNF-TrkB (tyrosine kinase) signaling in the induction of synaptic plasticity of tDCS in slices of mouse cortex.

Increased BDNF mRNA and immunoreactivity in rat cortex has been shown after chronic daily 20 Hz rTMS (Muller et al., 2000) and chronic daily 0.5 Hz rTMS (Zhang et al., 2007). A recent study observed that chronic 20 Hz rTMS induced a long-lasting increase in BDNF levels (as measured by immunohistochemistry) in awake animals but a decrease in anesthetized animals (Gersner et al., 2011). These results provide evidence to support the notion that rTMS effects are strongly dependent on the brain state at the time of stimulation (Siebner et al., 2009).

Studies of peripheral nervous tissue with tES showed that after the cut and suture of rat femoral nerve, application of 20 Hz (100 μs, 3 V) tES (for 1, 8, and 48 h), an upregulation of BDNF and its receptor TrkB mRNA levels was noted in motoneurons, and these changes correlated to nerve regeneration (Al-Majed et al., 2000). A wider study of the expression of other genes implicated in neuronal regeneration showed that Tα1-tubulin and growth-associated protein-43 (GAP-43) mRNA levels significantly increased after tES, while the marker of neuronal damage medium molecular weight neurofilament (NFM) mRNA was reduced (Al-Majed et al., 2004). These findings were also seen in sensory neurons of the dorsal root ganglia using a similar protocol with an elevation in BDNF and GAP-43 mRNA levels (Geremia et al., 2007).

Effects of rTMS on BDNF plasma levels were observed also in humans. Angelucci et al. (2004) reported that chronic 1 Hz rTMS was able to decrease the levels of this neurotrophin in the plasma of healthy volunteers (Angelucci et al., 2004), measured 48 h after the last session, while Zanardini et al. (2006) reported an increase in serum BDNF in patients with depression after five daily sessions of 1 or 17 Hz rTMS, and this is confirmed in another study with 20 Hz rTMS (Yukimasa et al., 2006). More recently, higher BDNF plasma levels and an increased TrkB–NMDAR interaction in lymphocytes were reported in healthy subjects after a 5 Hz treatment (Wang et al., 2011).

11.5.4 NEURODEGENERATION, INFLAMMATION, AND APOPTOSIS

Some studies report that rTMS may have effects on anti-neurodegenerative, anti-inflammatory, and anti-apoptotic pathways. In a 3-nitropropionic acid rat model of Huntington's disease, chronic 60 Hz rTMS reduced cell loss in the striatum (Tunez et al., 2006), while Yang et al. (2010) described that chronic 0.5 Hz rTMS treatment in a 6-OHDA lesion rat model of Parkinson's disease was able to decrease the expression of the inflammation markers tumor necrosis factor-α (TNF-α) and cyclooxygenase-2 (COX–2) in the substantia nigra (Yang et al., 2010). This suggested that rTMS was able to alleviate the lesion presumably by rescuing diseased neurons.

Acute and chronic 20 Hz stimulation was reported to increase the concentration of amyloid precursor protein in rat CSF and mouse hippocampal HT22 cells (Post et al., 1999). In another study employing rats subjected to middle cerebral artery occlusion, chronic 20 Hz rTMS appeared to have anti-apoptotic effects by increasing glucose metabolism in affected areas and decreasing the caspase-3 cell positivity (Gao et al., 2010).

11.5.5 EXPRESSION OF OTHER GENES

Transcription factors such as CREB and c-fos, which are strongly associated with LTP induction, have been linked with NIBS in animal studies. An increase in c-fos, c-jun, and CREB expression was observed in rats after acute 25 Hz rTMS, in the thalamic paraventricular nucleus and other regions involved in the regulation of circadian rhythms (Ji et al., 1998). An elevation of c-fos expression in the parietal cortex and hippocampus of rats subjected to chronic 20 Hz rTMS has also been reported (Hausmann et al., 2000), while in another group, a similar effect was observed, exerted in several brain regions, most notably cortex and thalamus, by a 20 Hz treatment conducted for 2 to 6 sessions, with a more definite effect induced by the 6-session stimulation (Doi et al., 2001). The level of c-fos was increased in the cortex of rats with cerebral infarction subjected to a chronic treatment at 0.5 Hz over 7–28 days (Zhang et al., 2007) in all cortical area of rats stimulated with rTMS at 1 and 10 Hz and iTBS (Aydin-Abidin et al., 2008). Increased expression of c-fos in multiple brain regions has also been shown in tES to rat sensorimotor cortex exposed to anodal tDCS (Islam et al., 1995).

Expression of another transcription factor, nerve-growth-factor-induced protein A (NGFI-A or zif268), increased in all cortical areas after iTBS, and in primary and sensory cortices after 10 Hz rTMS (Aydin-Abidin et al., 2008).

In a neuroblastoma cell line differentiated to a neuronal phenotype by retinoic acid treatment, a 5 Hz MS treatment was reported to increase intracellular cyclic adenosine monophosphate (cAMP) levels and CREB phosphorylation (Hellmann et al., 2011). This is contrasted against another report of a downregulation of the cAMP-generating system in the cortex of rats after 12 sessions of chronic 10 Hz rTMS (Vetulani et al., 2010).

Cortical cells positive for the calcium-binding protein calbindin, a molecule in the LTP pathway, has also been shown to be decreased in rats after acute 1 Hz rTMS treatment or cTBS (Benali et al., 2011), while iTBS lowered the number of cells positive for parvalbumin (another calcium-binding protein). The result obtained for acute iTBS confirmed previous evidence from a study in the same animal model, in which this treatment was associated with a reduced cortical expression of parvalbumin (Mix et al., 2010).

Other isolated reports include the following: increased MAP-2 expression after 5 or 20 Hz rTMS in rats after cerebral ischemia-reperfusion injury (Feng et al., 2008); increased glial fibrillary acidic protein mRNA in the hippocampal dentate gyrus of mice acutely after 25 Hz rTMS (Fujiki and Steward, 1997); and an increased cholecystokinin expression in hippocampal and cortical regions of rats after chronic rTMS (Muller et al., 2000).

11.6 DIFFERENCES AND OVERLAPS BETWEEN MAGNETIC AND ELECTRICAL STIMULATION

The combined studies provided earlier indicate that TMS and rTMS affect multiple monoamine neurotransmitter systems (particularly potentiating dopaminergic neurotransmission by the modulation of the dopamine release, turnover, and transport). Isolated reports suggest that rTMS might activate or facilitate repair mechanisms in substantia nigra and peripheral nerve tissue and possibly even have neurotrophic and anti-apoptotic effects. The TBS form of rTMS also appears to be distinct from other forms of rTMS, but the nature of the distinction remains to be determined. Studies of neuroendocrine effects of NIBS suggest a reduction of stress-related hormones, which may partially explain the antidepressant effects.

On the other hand, molecular neuroimaging using MRS provides evidence that tES and rTMS, while apparently producing similar effects on MEP, can produce quite different effects on a molecular level and involve different mechanisms. Studies of humans with genetic polymorphisms also indicate that although common molecules are involved (i.e., NMDA, BDNF, dopamine), the different NIBS produce distinct effects, and the response to NIBS should constitute a complex trait.

11.7 CONCLUSIONS

The study of the molecular mechanisms and consequences of NIBS is an emerging science, and it is worthwhile remembering that a significant proportion of molecular studies of NIBS have been conducted by a few groups—reproducing these results will reinforce their validity.

The growing number of studies on NIBS indicate a plethora of molecular effects of these techniques on neurotransmitter systems, neuroplasticity effectors, neurodegenerative and repair mechanisms, together with secondary effects on hypothalamic–pituitary–adrenal and hypothalamic–pituitary–thyroid axes to mention a few. Furthermore, recent studies in humans with neuroimaging, genomic, and molecular techniques suggest that the enhanced response to newer innovative NIBS paradigms (such as patterned stimulation) and the enhanced (or diminished) response of certain patient subgroups to particular NIBS paradigms may have a molecular basis.

In the last years, based at least in part on these results, NIBS have been widely studied for potential clinical applications in neurodegenerative and neuropsychiatric disorders with promising results. With this in mind, the clarification of complex molecular consequences of different noninvasive stimulation techniques will provide the foundation for the implementation of innovative therapeutic strategies. Concerns regarding the safety of NIBS protocols (particularly the safety of chronic administration) may also drive further growth in the study of the molecular effects of NIBS. The growth in future applications of noninvasive neurostimulation will therefore be dependent in no small part on further characterization of its molecular mechanisms and consequences.

At the end of this chapter, we have summarized all the studies looking at the molecular mechanisms of NIBS: on rTMS in animal and cell models (Table 11.1) and in humans (Table 11.2); and on tES in animal and cell models (Table 11.3) and in humans (Table 11.4).

TABLE 11.1
Studies of the Molecular Effects of rTMS in Animal and Cell Models

rTMS in Animal and Cell Models

Target		Stimulation Protocol	Model	Effect	Reference
Neurotransmitter Systems					
Dopaminergic system	DA	Acute 25 Hz rTMS	Rats	DA levels reduced in frontal cortex and increased in striatum and hippocampus; DA turnover rates increased in frontal cortex and decreased in striatum and hippocampus	Ben-Shachar et al. (1997)
	DA	Acute 20 Hz rTMS	Rat *in vivo* microdialysis	Increase in DA release in hippocampus	Keck et al. (2000a)
	DA, L-DOPA, TH	Acute/chronic MS 3 and 9 Hz	Neuroblastoma SH-SY5Y cells	Intracellular DA and L-DOPA and TH activity decreased by 3 Hz and increased by 9 Hz stimulation	Shaul et al. (2003)
	DA	Acute 20 Hz rTMS	Morphine-sensitized rat *in vivo* microdialysis	Increased concentration of DA in nucleus accumbens	Erhardt et al. (2004)
	DA	Acute 25 Hz rTMS	Rat *in vivo* microdialysis	Increase in extracellular DA in dorsolateral striatum	Kanno et al. (2004)
	DAT	Chronic rTMS, 20 days	Mice	Increase in DAT mRNA and DA transport rate in synaptosomes of whole brain	Ikeda et al. (2005)
	DA	Chronic 0.5 Hz rTMS, 42 days	Rat model of PD (6-OHDA lesions)	Increase in striatal DA levels	Yang et al. (2010)
Noradrenergic system	β-adrenoceptor	Chronic 25 Hz rTMS, 9 days	Rats	Reduction in β-adrenoceptor binding in cortex	Fleischmann et al. (1996)

(continued)

TABLE 11.1 (continued)

Studies of the Molecular Effects of rTMS in Animal and Cell Models

rTMS in Animal and Cell Models

Target		Stimulation Protocol	Model	Effect	Reference
	β-adrenoceptor	Chronic 15 Hz rTMS, 10 days	Rats	β-adrenoceptor binding increased in frontal cortex and decreased in striatum	Ben-Shachar et al. (1999)
	NE	Acute magnetic stimulation 3 Hz	Neuroblastoma SH-SY5Y cells	Increase in intracellular NE	Shaul et al. (2003)
		Acute magnetic stimulation 9 Hz		Decrease in intracellular NE	
	NET	Chronic 20 Hz rTMS, 20 days	Mice	Increase in NET mRNA and NE uptake and binding	Ikeda et al. (2005)
		Acute rTMS	PC12 cells	After 1 h, NET mRNA decreased	
		rTMS-like treatment, 15 days		After 15 days, increase in NET promoter luciferase activity and in NET mRNA levels	
Serotonergic system	5-HT 5-HIAA	Acute 25 Hz rTMS	Rats	5-HT and 5-HIAA levels increased in hippocampus	Ben-Shachar et al. (1997)
	5-HT$_2$	Chronic 15 Hz rTMS, 10 days	Rats	5-HT$_2$ binding and affinity decreased in frontal cortex	Ben-Shachar et al. (1999)
	5-HT$_{1A}$ 5-HT$_{1B}$	Acute 20 Hz rTMS	Rats	Increased density of 5-HT$_{1A}$ binding sites in cortical areas and olfactory nucleus	Kole et al. (1999)
	5-HT$_{1A}$ 5-HT$_{1B}$	Chronic 15 Hz rTMS, 10 days	Rat *in vivo* microdialysis	Reduction in sensitivity of 5-HT$_{1A}$ and 5-HT$_{1B}$ receptors in the frontal cortex	Gur et al. (2000)

	5-HT	Acute 25 Hz rTMS	Rat *in vivo* microdialysis	rTMS eliminated the increase in extracellular 5-HT levels observed in the sham group	Kanno et al. (2003a)
	5-HT	Chronic 25 Hz rTMS	Rat *in vivo* microdialysis	rTMS suppressed the increase in 5-HT induced by the plus-maze test	Kanno et al. (2003b)
	SERT	Chronic 20 Hz rTMS, 20 days Acute rTMS	Mice	Decrease in SERT mRNA, 5-HT uptake, and binding After 1 h, transient increase in SERT mRNA	Ikeda et al. (2005)
Excitatory and inhibitory amino acids	Taurine serine aspartate	Acute rTMS	Rat *in vivo* microdialysis	Increase in taurine, serine, aspartate in hypothalamus	Keck et al. (2000b)
	GAD67 GAD65 GAT-1	Acute rTMS (iTBS, cTBS, or 1 Hz) 2 h	Rats	All protocols reduced the expression of GAD67 in cortex and increased GAD65 and GAT-1 with iTBS having the strongest effect	Trippe et al. (2009)
		Chronic rTMS (iTBS, cTBS, or 1 Hz) 1–7 days		1 Hz rTMS, iTBS, and cTBS increased GAD67 expression after initial decreases on the first day; 1 Hz rTMS also produced sustained increases in GAD65 and GAT-1	
	Glutamate GABA	Chronic 0.5 Hz rTMS, 15 sessions	Rats	Glutamate and GABA increased in hippocampus and striatum and decreased in hypothalamus	Yue et al. (2009)
	GAD67 GAD65	iTBS cTBS	Rats	iTBS reduced cortical GAD67 and improved learning cTBS increased the expression of GAD65	Mix et al. (2010)

(continued)

TABLE 11.1 (continued)
Studies of the Molecular Effects of rTMS in Animal and Cell Models

rTMS in Animal and Cell Models

Target	Stimulation Protocol	Model	Effect	Reference
GluR1	Chronic rTMS 20 Hz and 1 Hz, 10 sessions	Awake rats	20 Hz rTMS induced long-lasting increase in GluR1 and phosphorylated GluR1 expression in hippocampus and prelimbic cortex	Gersner et al. (2011)
		Anesthetized rats	20 Hz rTMS induced in hippocampus and prelimbic cortex downregulation of GluR1 and phosphorylated GluR1 expression in hippocampus and prelimbic cortex	
Hypothalamic–Pituitary–Adrenal Axis				
Arginine vasopressin	Acute 20 Hz rTMS	Rat *in vivo* microdialysis	Decrease in arginine vasopressin release in hypothalamus	Keck et al. (2000a)
Corticotropin corticosterone	Chronic rTMS 20 Hz, 40 sessions	Rats under stress exposure	Attenuation of corticotropin and corticosterone increase in plasma after stress exposure	Keck et al. (2000b)
Corticotropin corticosterone	Chronic 20 Hz rTMS, six sessions	Rats with anxiety behavior	Decrease in stress-induced elevation of plasma corticotropin and corticosterone	Keck et al. (2001)
Neurotrophins, Neuroprotectors, and Neuronal Repair				
APP	Acute and 55 sessions of 20 Hz rTMS	Rats	Increased APP in CSF after acute and chronic rTMS	Post et al. (1999)

Target	Protocol	Model	Effect	Reference
BDNF CCK	rTMS analogous treatment *in vitro*	HT22 cells	Increased APP release *in vitro* 6 h after treatment	Muller et al. (2000)
	Chronic rTMS 20 Hz, 55 sessions	Rats	Increase in BDNF and CCK mRNA in several hippocampus and cortex regions	
BDNF	Chronic rTMS 0.5 Hz, 14–42 sessions	Rat stroke model	rTMS promoted the expression of BDNF in cortex	Zhang et al. (2007)
BDNF	Chronic rTMS 20 Hz and 1 Hz, 10 sessions	Awake/anesthetized rats	20 Hz rTMS induced a long-lasting increase in BDNF expression in hippocampus and prelimbic cortex of the awake rat, while downregulated BDNF in the same areas in anesthetized animals	Gersner et al. (2011)
BDNF-TrkB signaling	Chronic rTMS 5 Hz, 5 sessions	Rats	rTMS increased TrkB activation and TrkB–NMDAR interaction in prefrontal cortex and in lymphocytes, while increasing plasma BDNF	Wang et al. (2011)
Neurodegeneration, Inflammation, and Apoptosis				
Oxidative stress markers and neurodegeneration	Chronic 60 Hz rTMS 16 sessions	Rat model for HD	rTMS prevented neurodegeneration and oxidative stress	Tunez et al. (2006)
Caspase-3 Bcl-2/Bax ratio	Chronic 20 Hz rTMS, 7 sessions	Rat model of transient cerebral ischemia	Increased glucose metabolism in lesioned areas; Inhibition of apoptosis (decrease in caspase-3, increase in Bcl-2/Bax ratio)	Gao et al. (2010)
TNF-α COX-2	Chronic 0.5 Hz rTMS, 28 sessions	Rat model for PD (6-OHDA lesions)	Decreased TNF-α and COX-2 expression in the substantia nigra	Yang et al. (2010)

(continued)

Transcranial Brain Stimulation — page 290

TABLE 11.1 (continued)
Studies of the Molecular Effects of rTMS in Animal and Cell Models

rTMS in Animal and Cell Models

Target	Stimulation Protocol	Model	Effect	Reference
Expression of Other Genes				
GFAP	Acute 25 Hz rTMS	Mice	Increase in GFAP mRNA in dentate gyrus	Fujiki and Steward (1997)
c-fos c-jun CREB	Acute 25 Hz rTMS	Rats	Increased expression of c-fos, c-jun, CREB in paraventricular nucleus of the thalamus and specific cortical regions and in regions controlling circadian rhythms	Ji et al. (1998)
c-fos	Chronic 20 Hz rTMS, 14 sessions	Rats	c-fos mRNA increased in parietal cortex and in hippocampus	Hausmann et al. (2000)
c-fos	Chronic 20 Hz rTMS, two to six sessions	Rats	Increased c-fos expression in several brain regions (cortex and thalamus)	Doi et al. (2001)
kf-1	Chronic rTMS 20 Hz, 10 days	Rats	Increased mRNA levels of kf-1 in frontal cortex and hippocampus	Kudo et al. (2005)
c-fos	Chronic rTMS 0.5 Hz, 14–42 sessions	Rats with cerebral infarction	rTMS promoted the expression of c-fos in cortex	Zhang et al. (2007)

Target	Stimulation	Model	Effect	Reference
c-fos zif268	Acute rTMS (10 Hz, 1 Hz)	Rats	Enhanced c-fos expression in all cortical areas, while Zif268 expression increased only after 10 Hz in primary motor and sensory cortices	Aydin-Abidin et al. (2008)
	Acute rTMS (iTBS)		Enhanced c-fos expression in limbic cortices and enhanced Zif268 expression in all cortical areas	
cAMP	Chronic 10 Hz rTMS, 12 sessions	Rats	Downregulation of cAMP-generating system	Vetulani et al. (2010)
Parvalbumin GAD	Acute iTBS, cTBS	Rats	iTBS reduced cortical expression of parvalbumin and GAD_{67} and improved learning; cTBS increased cortical expression of GAD_{65}	Mix et al. (2010)
Parvalbumin calbindin	Acute iTBS, cTBS, 1 Hz rTMS	Rats	iTBS reduced the number of PV-positive cells and PV protein levels; cTBS and 1 Hz rTMS reduced the number of calbindin-positive cells	Benali et al. (2011)
ATP MAP-2	5, 20 Hz rTMS	Rat cerebral ischemia-reperfusion injury	Increase in ATP and MAP-2 expression in injured areas	Feng et al. (2008)
cAMP CREB	5 Hz rTMS	Neuroblastoma cells	Increased intracellular cAMP levels Increased CREB phosphorylation	Hellmann et al. (2011)

TABLE 11.2

Studies of the Molecular Effects of rTMS in Humans

rTMS in Humans

Target		Stimulation Protocol	Subjects	Effect	Reference
Neurotransmitter Systems					
Dopaminergic	HVA	Chronic 0.2 Hz rTMS, Once weekly for 3–4 months	17 PD patients	CSF HVA levels showed a significant decrease	Shimamoto et al. (2001)
	DA	Chronic 25 Hz rTMS right and left motor cortices, six sessions	20 PD patients	Increased concentration of serum DA, correlated to illness symptomatology	Khedr et al. (2007)
Noradrenergic	MHPG	Chronic 20 Hz rTMS left prefrontal cortex, 10 sessions	26 MD patients	MHPG plasma levels were decreased	Yukimasa et al. (2006)
Hypothalamic–Pituitary–Adrenal Axis					
	Cortisol, TSH	Acute rTMS (10, 20 Hz, sub-, supra-threshold, placebo)	Healthy subjects	Serum cortisol and TSH decreased after infrathreshold rTMS	Evers et al. (2001)
	Cortisol	10 Hz rTMS on DLPFC, 13 sessions	37 MD patients tested with DEX/CRH test	A positive correlation was detected between the reduction of depressive symptoms and the reduction of post-DEX basal cortisol levels	Zwanzger et al. (2003)

Cortisol	10 Hz rTMS applied on the left DLPFC, one application	20 MD patients	Salivary cortisol concentrations decreased significantly immediately and 30 min after the treatment	Baeken et al. (2009)
Cortisol	10 Hz rTMS on DLPFC	22 patients with bulimic disorder	Decreased salivary cortisol levels	Claudino et al. (2010)
Hypothalamic–Pituitary–Thyroid Axis				
TSH	5 Hz rTMS over prefrontal cortex	MD patients and healthy volunteers	Increase in plasma TSH	George et al. (1996)
TSH	20 Hz rTMS	Healthy volunteers	Supra-threshold rTMS increased TSH plasma levels 10–60 min poststimulation	Cohrs et al. (2001)
TSH	10 Hz rTMS over left PFC, 10 sessions	MD patients	Increased serum TSH levels after one single rTMS session	Szuba et al. (2001)
TSH	1 Hz rTMS, 12 sessions	TRD patients	TSH serum levels significantly increased after the treatment in responder patients. Negative correlation between TSH levels at pretreatment and decrease in the depression score	Kito et al. (2010)
TSH	1 Hz rTMS over right DLPFC with venlafaxine over 6 weeks	Case report of TRD patient	TSH plasma levels rapidly increased after the treatment and remained elevated during treatment. When rTMS was stopped plasma TSH returned to normal levels	Trojak et al. (2011)

(continued)

TABLE 11.2 (continued)
Studies of the Molecular Effects of rTMS in Humans

rTMS in Humans

Target	Stimulation Protocol	Subjects	Effect	Reference
Neurotrophins, Neuroprotectors, and Neuronal Repair				
BDNF	1 Hz rTMS, eight sessions	10 healthy volunteers	BDNF plasma levels were decreased 48 h after the last treatment	Angelucci et al. (2004)
BDNF	20 Hz rTMS over left prefrontal, 10 sessions	26 MD patients	Plasma BDNF increased in responders and partial responders, but not in nonresponders	Yukimasa et al. (2006)
BDNF	1 Hz or 17 Hz rTMS, five sessions	16 MD patients	Serum BDNF increased after the treatment and correlated with symptom improvement	Zanardini et al. (2006)
BDNF-TrkB	5 Hz rTMS, five sessions	8 healthy volunteers	TrkB signaling and TrkB–NMDA interaction were activated in peripheral lymphocytes and increase in plasma BDNF levels	Wang et al. (2011)

TABLE 11.3

Studies of the Molecular Effects of ES, tES, and tDCS in Animal and Cell Models

tES in Animal and Cell Models

Target		Stimulation Protocol	Model	Effect	Reference
Neurotrophins, Neuroprotectors, and Neuronal Repair					
Serotonergic system	5-HT	tES 1 s trains of 5 ms stimuli (60 Hz, 100 and 150 mA), every 5 s for 20 min	Rat medial prefrontal cortex	Increase in limbic 5-HT release	Juckel et al. (1999)
BDNF-TrkB system	BDNF TrkB	ES 20 Hz, 100 μs, 3 V, 1 h	Rat femoral motoneurons	BDNF and TrkB mRNAs were upregulated 8 h and 2 days after ES	Al-Majed et al. (2000)
	BDNF GAP-43	ES 20 Hz, 100 μs, 3 V, 1 h	Adult rat motoneurons and DRG sensory neurons	ES enhanced motor axonal regeneration and elevated the cell body expression of GAP-43 and BDNF in DRG sensory neurons 2 days after stimulation	Geremia et al. (2007)
	BDNF TrkB	tDCS, 10 μA	Mice M1 slices	tDCS induced synaptic activation enhancing BDNF secretion and TrkB activation	Fritsch et al. (2010)
Others	c-fos	Anodal tDCS	Rats	Increase in c-fos immunoreactive neurons in cingulate, piriform, frontoparietal cortices, hippocampus	Islam et al. (1995)
	Tα1-tubulin GAP-43 NFM	ES 20 Hz, 100 μs, 3 V, 1 h	Rat femoral motor neurons	Tα1-tubular and GAP-43 mRNA were significantly elevated while NFM mRNA was reduced by 2 days after tES stimulation	Al-Majed et al. (2004)

TABLE 11.4
Studies of the Molecular Effects of tES and tDCS in Humans

tES in Humans				
Target	**Stimulation Protocol**	**Subjects**	**Effect**	**Reference**
Neurotransmitter Systems				
MHPG	tES 12 μA, 53 min, 10 cps, two treatments daily for 4 days	30 subjects with spinal cord injury	Urinary MHPG was higher after tES	Capel et al. (2003)
β-endorphin	tES, forehead, 4 mA, 30 min, 77 Hz, one treatment session	20 patients with cervical/back pain	Serum β-endorphin levels showed a trend for increase after tES	Gabis et al. (2003)
Hypothalamic–Pituitary–Adrenal Axis				
Cortisol	tES 12 μA, 53 min, 10 cps, two treatments daily for 4 days	30 subjects with spinal cord injury	Salivary cortisol was lower after tES	Capel et al. (2003)
Cortisol ACTH	Anodal tDCS primary motor cortex, 1 mA, for 20 min	15 healthy male volunteers	Cortisol and ACTH blood levels decreased after anodal tDCS	Binkofski et al. (2011)

REFERENCES

Ahmed, M. A., Mohamed, S. A., and Sayed, D. (2011) Long-term antalgic effects of repetitive transcranial magnetic stimulation of motor cortex and serum beta-endorphin in patients with phantom pain. *Neurol Res*, 33, 953–958.

Aicardi, G., Argilli, E., Cappello, S. et al. (2004) Induction of long-term potentiation and depression is reflected by corresponding changes in secretion of endogenous brain-derived neurotrophic factor. *Proc Natl Acad Sci USA*, 101, 15788–15792.

Al-Majed, A. A., Brushart, T. M., and Gordon, T. (2000) Electrical stimulation accelerates and increases expression of BDNF and trkB mRNA in regenerating rat femoral motoneurons. *Eur J Neurosci*, 12, 4381–4390.

Al-Majed, A. A., Tam, S. L., and Gordon, T. (2004) Electrical stimulation accelerates and enhances expression of regeneration-associated genes in regenerating rat femoral motoneurons. *Cell Mol Neurobiol*, 24, 379–402.

Angelucci, F., Oliviero, A., Pilato, F. et al. (2004) Transcranial magnetic stimulation and BDNF plasma levels in amyotrophic lateral sclerosis. *Neuroreport*, 15, 717–720.

Antal, A., Chaieb, L., Moliadze, V. et al. (2010) Brain-derived neurotrophic factor (BDNF) gene polymorphisms shape cortical plasticity in humans. *Brain Stimul*, 3, 230–237.

Auerbach, B. D. and Bear, M. F. (2010) Loss of the fragile X mental retardation protein decouples metabotropic glutamate receptor dependent priming of long-term potentiation from protein synthesis. *J Neurophysiol*, 104, 1047–1051.

Auerbach, B. D., Osterweil, E. K., and Bear, M. F. (2011) Mutations causing syndromic autism define an axis of synaptic pathophysiology. *Nature*, 480, 63–68.

Aydin-Abidin, S., Trippe, J., Funke, K. et al. (2008) High- and low-frequency repetitive transcranial magnetic stimulation differentially activates c-Fos and zif268 protein expression in the rat brain. *Exp Brain Res*, 188, 249–261.

Baeken, C., De Raedt, R., Bossuyt, A. et al. (2011) The impact of HF-rTMS treatment on serotonin(2A) receptors in unipolar melancholic depression. *Brain Stimul*, 4, 104–111.

Baeken, C., De Raedt, R., Vanderhasselt, M. A. et al. (2009) A "hypersensitive" hypothalamic-pituitary-adrenal system could be indicative for a negative clinical high-frequency repetitive transcranial magnetic stimulation outcome in melancholic depressed patients. *Brain Stimul*, 3, 54–57.

Benali, A., Trippe, J., Weiler, E. et al. (2011) Theta-burst transcranial magnetic stimulation alters cortical inhibition. *J Neurosci*, 31, 1193–1203.

Ben-Shachar, D., Belmaker, R. H., Grisaru, N. et al. (1997) Transcranial magnetic stimulation induces alterations in brain monoamines. *J Neural Transm*, 104, 191–197.

Ben-Shachar, D., Gazawi, H., Riboyad-Levin, J. et al. (1999) Chronic repetitive transcranial magnetic stimulation alters beta-adrenergic and 5-HT2 receptor characteristics in rat brain. *Brain Res*, 816, 78–83.

Binkofski, F., Loebig, M., Jauch-Chara, K. et al. (2011) Brain energy consumption induced by electrical stimulation promotes systemic glucose uptake. *Biol Psychiatry*, 70, 690–695.

Bocchio-Chiavetto, L., Miniussi, C., Zanardini, R. et al. (2008) 5-HTTLPR and BDNF Val66Met polymorphisms and response to rTMS treatment in drug resistant depression. *Neurosci Lett*, 437, 130–134.

Burlina, A. P., Aureli, T., Bracco, F. et al. (2000) MR spectroscopy: A powerful tool for investigating brain function and neurological diseases. *Neurochem Res*, 25, 1365–1372.

Canli, T., Omura, K., Haas, B. W. et al. (2005) Beyond affect: A role for genetic variation of the serotonin transporter in neural activation during a cognitive attention task. *Proc Natl Acad Sci USA*, 102, 12224–12229.

Capel, I. D., Dorrell, H. M., Spencer, E. P. et al. (2003) The amelioration of the suffering associated with spinal cord injury with subperception transcranial electrical stimulation. *Spinal Cord*, 41, 109–117.

Catafau, A. M., Perez, V., Gironell, A. et al. (2001) SPECT mapping of cerebral activity changes induced by repetitive transcranial magnetic stimulation in depressed patients. A pilot study. *Psychiatry Res*, 106, 151–160.

Cheeran, B., Koch, G., Stagg, C. J. et al. (2010) Transcranial magnetic stimulation: From neurophysiology to pharmacology, molecular biology and genomics. *Neuroscientist*, 16, 210–221.

Cheeran, B., Talelli, P., Mori, F. et al. (2008) A common polymorphism in the brain-derived neurotrophic factor gene (BDNF) modulates human cortical plasticity and the response to rTMS. *J Physiol*, 586, 5717–5725.

Claudino, A. M., Van Den Eynde, F., Stahl, D. et al. (2010) Repetitive transcranial magnetic stimulation reduces cortisol concentrations in bulimic disorders. *Psychol Med*, 1–8.

Cohrs, S., Tergau, F., Korn, J. et al. (2001) Suprathreshold repetitive transcranial magnetic stimulation elevates thyroid-stimulating hormone in healthy male subjects. *J Nerv Ment Dis*, 189, 393–397.

Danielsson, I., Su, K. G., Kauer, L. et al. (2004) Talampanel and human cortical excitability: EEG and TMS. *Epilepsia*, 45, 120.

Di Lazzaro, V., Oliviero, A., Pilato, F. et al. (2004) Motor cortex hyperexcitability to transcranial magnetic stimulation in Alzheimer's disease. *J Neurol Neurosurg Psychiatry*, 75, 555–559.

Di Lazzaro, V., Oliviero, A., Profice, P. et al. (1998) Comparison of descending volleys evoked by transcranial magnetic and electric stimulation in conscious humans. *Electroencephalogr Clin Neurophysiol*, 109, 397–401.

Di Lazzaro, V., Oliviero, A., Profice, P. et al. (2003) Ketamine increases human motor cortex excitability to transcranial magnetic stimulation. *J Physiol*, 547, 485–496.

Di Lazzaro, V., Pilato, F., Dileone, M. et al. (2007) Segregating two inhibitory circuits in human motor cortex at the level of GABAA receptor subtypes: A TMS study. *Clin Neurophysiol*, 118, 2207–2214.

Doi, W., Sato, D., Fukuzako, H. et al. (2001) c-Fos expression in rat brain after repetitive transcranial magnetic stimulation. *Neuroreport*, 12, 1307–1310.

Edwards, M. J., Huang, Y. Z., Mir, P. et al. (2006) Abnormalities in motor cortical plasticity differentiate manifesting and nonmanifesting DYT1 carriers. *Mov Disord*, 21, 2181–2186.

Edwards, M. J., Huang, Y. Z., Wood, N. W. et al. (2003) Different patterns of electrophysiological deficits in manifesting and non-manifesting carriers of the DYT1 gene mutation. *Brain*, 126, 2074–2080.

Egan, M. F., Kojima, M., Callicott, J. H. et al. (2003) The BDNF val66met polymorphism affects activity-dependent secretion of BDNF and human memory and hippocampal function. *Cell*, 112, 257–269.

Eichhammer, P., Langguth, B., Wiegand, R. et al. (2003) Allelic variation in the serotonin transporter promoter affects neuromodulatory effects of a selective serotonin transporter reuptake inhibitor (SSRI). *Psychopharmacology (Berl)*, 166, 294–297.

Epstein, C. M., Girard-Siqueira, L., and Ehrenberg, J. A. (2008) Prolonged neurophysiologic effects of levetiracetam after oral administration in humans. *Epilepsia*, 49, 1169–1173.

Erhardt, A., Sillaber, I., Welt, T. et al. (2004) Repetitive transcranial magnetic stimulation increases the release of dopamine in the nucleus accumbens shell of morphine-sensitized rats during abstinence. *Neuropsychopharmacology*, 29, 2074–2080.

Evers, S., Hengst, K., and Pecuch, P. W. (2001) The impact of repetitive transcranial magnetic stimulation on pituitary hormone levels and cortisol in healthy subjects. *J Affect Disord*, 66, 83–88.

Fedi, M., Berkovic, S. F., Macdonell, R. A. et al. (2008) Intracortical hyperexcitability in humans with a GABAA receptor mutation. *Cereb Cortex*, 18, 664–669.

Feng, H. L., Yan, L., and Cui, L. Y. (2008) Effects of repetitive transcranial magnetic stimulation on adenosine triphosphate content and microtubule associated protein-2 expression after cerebral ischemia-reperfusion injury in rat brain. *Chin Med J (Engl)*, 121, 1307–1312.

Figurov, A., Pozzo-Miller, L. D., Olafsson, P. et al. (1996) Regulation of synaptic responses to high-frequency stimulation and LTP by neurotrophins in the hippocampus. *Nature*, 381, 706–709.

Fleischmann, A., Sternheim, A., Etgen, A. M. et al. (1996) Transcranial magnetic stimulation downregulates beta-adrenoreceptors in rat cortex. *J Neural Transm*, 103, 1361–1366.

Fritsch, B., Reis, J., Martinowich, K. et al. (2010) Direct current stimulation promotes BDNF-dependent synaptic plasticity: potential implications for motor learning. *Neuron*, 66, 198–204.

Fujiki, M. and Steward, O. (1997) High frequency transcranial magnetic stimulation mimics the effects of ECS in upregulating astroglial gene expression in the murine CNS. *Brain Res Mol Brain Res*, 44, 301–308.

Gabis, L., Shklar, B., and Geva, D. (2003) Immediate influence of transcranial electrostimulation on pain and beta-endorphin blood levels: An active placebo-controlled study. *Am J Phys Med Rehabil*, 82, 81–85.

Gao, F., Wang, S., Guo, Y. et al. (2010) Protective effects of repetitive transcranial magnetic stimulation in a rat model of transient cerebral ischaemia: A microPET study. *Eur J Nucl Med Mol Imaging*, 37, 954–961.

George, M. S., Wassermann, E. M., Williams, W. A. et al. (1996) Changes in mood and hormone levels after rapid-rate transcranial magnetic stimulation (rTMS) of the prefrontal cortex. *J Neuropsychiatry Clin Neurosci*, 8, 172–180.

Geremia, N. M., Gordon, T., Brushart, T. M. et al. (2007) Electrical stimulation promotes sensory neuron regeneration and growth-associated gene expression. *Exp Neurol*, 205, 347–359.

Gersner, R., Kravetz, E., Feil, J. et al. (2011) Long-term effects of repetitive transcranial magnetic stimulation on markers for neuroplasticity: Differential outcomes in anesthetized and awake animals. *J Neurosci*, 31, 7521–7526.

Gur, E., Lerer, B., Dremencov, E. et al. (2000) Chronic repetitive transcranial magnetic stimulation induces subsensitivity of presynaptic serotonergic autoreceptor activity in rat brain. *Neuroreport*, 11, 2925–2929.

Hariri, A. R., Goldberg, T. E., Mattay, V. S. et al. (2003) Brain-derived neurotrophic factor val-66met polymorphism affects human memory-related hippocampal activity and predicts memory performance. *J Neurosci*, 23, 6690–6694.

Hausmann, A., Weis, C., Marksteiner, J. et al. (2000) Chronic repetitive transcranial magnetic stimulation enhances c-fos in the parietal cortex and hippocampus. *Brain Res Mol Brain Res*, 76, 355–362.

Heils, A., Teufel, A., Petri, S. et al. (1996) Allelic variation of human serotonin transporter gene expression. *J Neurochem*, 66, 2621–2624.

Hellmann, J., Juttner, R., Roth, C. et al. (2011) Repetitive magnetic stimulation of human-derived neuron-like cells activates cAMP-CREB pathway. *Eur Arch Psychiatry Clin Neurosci*, 262, 87–91.

Huang, Y. Z., Chen, R. S., Rothwell, J. C. et al. (2007) The after-effect of human theta burst stimulation is NMDA receptor dependent. *Clin Neurophysiol*, 118, 1028–1032.

Huang, Y. Z., Edwards, M. J., Rounis, E. et al. (2005) Theta burst stimulation of the human motor cortex. *Neuron*, 45, 201–206.

Ikeda, T., Kurosawa, M., Uchikawa, C. et al. (2005) Modulation of monoamine transporter expression and function by repetitive transcranial magnetic stimulation. *Biochem Biophys Res Commun*, 327, 218–224.

Islam, N., Moriwaki, A., Hattori, Y. et al. (1995) c-Fos expression mediated by N-methyl-D-aspartate receptors following anodal polarization in the rat brain. *Exp Neurol*, 133, 25–31.

Ji, R. R., Schlaepfer, T. E., Aizenman, C. D. et al. (1998) Repetitive transcranial magnetic stimulation activates specific regions in rat brain. *Proc Natl Acad Sci USA*, 95, 15635–15640.

Juckel, G., Mendlin, A., and Jacobs, B. L. (1999) Electrical stimulation of rat medial prefrontal cortex enhances forebrain serotonin output: Implications for electroconvulsive therapy and transcranial magnetic stimulation in depression. *Neuropsychopharmacology*, 21, 391–398.

Kanno, M., Matsumoto, M., Togashi, H. et al. (2003a) Effects of acute repetitive transcranial magnetic stimulation on extracellular serotonin concentration in the rat prefrontal cortex. *J Pharmacol Sci*, 93, 451–457.

Kanno, M., Matsumoto, M., Togashi, H. et al. (2003b) Effects of repetitive transcranial magnetic stimulation on behavioral and neurochemical changes in rats during an elevated plus-maze test. *J Neurol Sci*, 211, 5–14.

Kanno, M., Matsumoto, M., Togashi, H. et al. (2004) Effects of acute repetitive transcranial magnetic stimulation on dopamine release in the rat dorsolateral striatum. *J Neurol Sci*, 217, 73–81.

Keck, M. E., Engelmann, M., Muller, M. B. et al. (2000a) Repetitive transcranial magnetic stimulation induces active coping strategies and attenuates the neuroendocrine stress response in rats. *J Psychiatr Res*, 34, 265–276.

Keck, M. E., Sillaber, I., Ebner, K. et al. (2000b) Acute transcranial magnetic stimulation of frontal brain regions selectively modulates the release of vasopressin, biogenic amines and amino acids in the rat brain. *Eur J Neurosci*, 12, 3713–3720.

Keck, M. E., Welt, T., Post, A. et al. (2001) Neuroendocrine and behavioral effects of repetitive transcranial magnetic stimulation in a psychopathological animal model are suggestive of antidepressant-like effects. *Neuropsychopharmacology*, 24, 337–349.

Khedr, E. M., Rothwell, J. C., Shawky, O. A. et al. (2007) Dopamine levels after repetitive transcranial magnetic stimulation of motor cortex in patients with Parkinson's disease: Preliminary results. *Mov Disord*, 22, 1046–1050.

Kito, S., Hasegawa, T., Fujita, K. et al. (2010) Changes in hypothalamic-pituitary-thyroid axis following successful treatment with low-frequency right prefrontal transcranial magnetic stimulation in treatment-resistant depression. *Psychiatry Res*, 175, 74–77.

Kole, M. H., Fuchs, E., Ziemann, U. et al. (1999) Changes in 5-HT1A and NMDA binding sites by a single rapid transcranial magnetic stimulation procedure in rats. *Brain Res*, 826, 309–312.

Korchounov, A., Ilic, T. V., Schwinge, T. et al. (2005) Modification of motor cortical excitability by an acetylcholinesterase inhibitor. *Exp Brain Res*, 164, 399–405.

Kudo, K., Yamada, M., Takahashi, K. et al. (2005) Repetitive transcranial magnetic stimulation induces kf-1 expression in the rat brain. *Life Sci*, 76, 2421–2429.

Kuo, M. F., Paulus, W., and Nitsche, M. A. (2008) Boosting focally-induced brain plasticity by dopamine. *Cereb Cortex*, 18, 648–651.

Lu, B. (2003) BDNF and activity-dependent synaptic modulation. *Learn Mem*, 10, 86–98.

Luborzewski, A., Schubert, F., Seifert, F. et al. (2007) Metabolic alterations in the dorsolateral prefrontal cortex after treatment with high-frequency repetitive transcranial magnetic stimulation in patients with unipolar major depression. *J Psychiatr Res*, 41, 606–615.

Lugenbeel, K. A., Peier, A. M., Carson, N. L. et al. (1995) Intragenic loss of function mutations demonstrate the primary role of FMR1 in fragile X syndrome. *Nat Genet*, 10, 483–485.

Malaguti, A., Rossini, D., Lucca, A. et al. (2011) Role of COMT, 5-HT(1A), and SERT genetic polymorphisms on antidepressant response to transcranial magnetic stimulation. *Depress Anxiety*, 28, 568–573.

Mavroudakis, N., Caroyer, J. M., Brunko, E. et al. (1994) Effects of diphenylhydantoin on motor potentials evoked with magnetic stimulation. *Electroencephalogr Clin Neurophysiol*, 93, 428–433.

Mcdonnell, M. N., Orekhov, Y., and Ziemann, U. (2007) Suppression of LTP-like plasticity in human motor cortex by the GABAB receptor agonist baclofen. *Exp Brain Res*, 180, 181–186.

Michael, N., Gosling, M., Reutemann, M. et al. (2003) Metabolic changes after repetitive transcranial magnetic stimulation (rTMS) of the left prefrontal cortex: A sham-controlled proton magnetic resonance spectroscopy (1H MRS) study of healthy brain. *Eur J Neurosci*, 17, 2462–2468.

Missitzi, J., Gentner, R., Geladas, N. et al. (2011) Plasticity in human motor cortex is in part genetically determined. *J Physiol*, 589, 297–306.

Mix, A., Benali, A., Eysel, U. T. et al. (2010) Continuous and intermittent transcranial magnetic theta burst stimulation modify tactile learning performance and cortical protein expression in the rat differently. *Eur J Neurosci*, 32, 1575–1586.

Moghaddam, B., Adams, B., Verma, A. et al. (1997) Activation of glutamatergic neurotransmission by ketamine: A novel step in the pathway from NMDA receptor blockade to dopaminergic and cognitive disruptions associated with the prefrontal cortex. *J Neurosci*, 17, 2921–2927.

Monte-Silva, K., Kuo, M. F., Thirugnanasambandam, N. et al. (2009) Dose-dependent inverted U-shaped effect of dopamine (D2-like) receptor activation on focal and nonfocal plasticity in humans. *J Neurosci*, 29, 6124–6131.

Monte-Silva, K., Ruge, D., Teo, J. T. et al. (2011) D2 receptor block abolishes theta burst stimulation-induced neuroplasticity in the human motor cortex. *Neuropsychopharmacology*, 36, 2097–2102.

Mori, F., Ribolsi, M., Kusayanagi, H. et al. (2011) Genetic variants of the NMDA receptor influence cortical excitability and plasticity in humans. *J Neurophysiol*, 106, 1637–1643.

Mori, F., Ribolsi, M., Kusayanagi, H. et al. (2012) TRPV1 channels regulate cortical excitability in humans. *J Neurosci*, 32, 873–879.

Muller, M. B., Toschi, N., Kresse, A. E. et al. (2000) Long-term repetitive transcranial magnetic stimulation increases the expression of brain-derived neurotrophic factor and cholecystokinin mRNA, but not neuropeptide tyrosine mRNA in specific areas of rat brain. *Neuropsychopharmacology*, 23, 205–215.

Nardone, R., Bergmann, J., Christova, M. et al. (2012) Short latency afferent inhibition differs among the subtypes of mild cognitive impairment. *J Neural Transm*, 119, 463–471.

Nitsche, M. A., Grundey, J., Liebetanz, D. et al. (2004a) Catecholaminergic consolidation of motor cortical neuroplasticity in humans. *Cereb Cortex*, 14, 1240–1245.

Nitsche, M. A., Jaussi, W., Liebetanz, D. et al. (2004b) Consolidation of human motor cortical neuroplasticity by D-cycloserine. *Neuropsychopharmacology*, 29, 1573–1578.

Nitsche, M. A., Kuo, M. F., Grosch, J. et al. (2009) D1-receptor impact on neuroplasticity in humans. *J Neurosci*, 29, 2648–2653.

Oberman, L. M., Horvath, J. C., and Pascual-Leone, A. (2010a) TMS: Using the theta-burst protocol to explore mechanism of plasticity in individuals with Fragile X syndrome and autism. *J Vis Exp* doi:10.3791/2272.

Oberman, L., Ifert-Miller, F., Najib, U. et al. (2010b) Transcranial magnetic stimulation provides means to assess cortical plasticity and excitability in humans with fragile x syndrome and autism spectrum disorder. *Front Synaptic Neurosci*, 2, 26.

Ohnishi, T., Matsuda, H., Imabayashi, E. et al. (2004) rCBF changes elicited by rTMS over DLPFC in humans. *Suppl Clin Neurophysiol*, 57, 715–720.

Okabe, S., Hanajima, R., Ohnishi, T. et al. (2003) Functional connectivity revealed by single-photon emission computed tomography (SPECT) during repetitive transcranial magnetic stimulation (rTMS) of the motor cortex. *Clin Neurophysiol*, 114, 450–457.

Orth, M., Schippling, S., Schneider, S. A. et al. (2009) Abnormal motor cortex plasticity in premanifest and very early manifest Huntington disease. *J Neurol Neurosurg Psychiatry*, 81, 267–270.

Pascual-Leone, A., Houser, C. M., Reese, K. et al. (1993) Safety of rapid-rate transcranial magnetic stimulation in normal volunteers. *Electroencephalogr Clin Neurophysiol*, 89, 120–130.

Patel, A. B., De Graaf, R. A., Martin, D. L. et al. (2006) Evidence that GAD65 mediates increased GABA synthesis during intense neuronal activity *in vivo*. *J Neurochem*, 97, 385–396.

Pezawas, L., Verchinski, B. A., Mattay, V. S. et al. (2004) The brain-derived neurotrophic factor val66met polymorphism and variation in human cortical morphology. *J Neurosci*, 24, 10099–10102.

Pogarell, O., Koch, W., Popperl, G. et al. (2006) Striatal dopamine release after prefrontal repetitive transcranial magnetic stimulation in major depression: Preliminary results of a dynamic [123I] IBZM SPECT study. *J Psychiatr Res*, 40, 307–314.

Pogarell, O., Koch, W., Popperl, G. et al. (2007) Acute prefrontal rTMS increases striatal dopamine to a similar degree as D-amphetamine. *Psychiatry Res*, 156, 251–255.

Post, A., Muller, M. B., Engelmann, M. et al. (1999) Repetitive transcranial magnetic stimulation in rats: Evidence for a neuroprotective effect *in vitro* and *in vivo*. *Eur J Neurosci*, 11, 3247–3254.

Reis, J., Wentrup, A., Hamer, H. M. et al. (2004) Levetiracetam influences human motor cortex excitability mainly by modulation of ion channel function—A TMS study. *Epilepsy Res*, 62, 41–51.

Schwenkreis, P., Tegenthoff, M., Witscher, K. et al. (2002) Motor cortex activation by transcranial magnetic stimulation in ataxia patients depends on the genetic defect. *Brain*, 125, 301–309.

Schwenkreis, P., Witscher, K., Pleger, B. et al. (2005) The NMDA antagonist memantine affects training induced motor cortex plasticity—A study using transcranial magnetic stimulation. *BMC Neurosci*, 6, 35.

Serretti, A., Calati, R., Mandelli, L. et al. (2006) Serotonin transporter gene variants and behavior: A comprehensive review. *Curr Drug Targets*, 7, 1659–1669.

Serretti, A., Kato, M., De Ronchi, D. et al. (2007) Meta-analysis of serotonin transporter gene promoter polymorphism (5-HTTLPR) association with selective serotonin reuptake inhibitor efficacy in depressed patients. *Mol Psychiatry*, 12, 247–257.

Shaul, U., Ben-Shachar, D., Karry, R. et al. (2003) Modulation of frequency and duration of repetitive magnetic stimulation affects catecholamine levels and tyrosine hydroxy-lase activity in human neuroblastoma cells: Implication for the antidepressant effect of rTMS. *Int J Neuropsychopharmacol*, 6, 233–241.

Shimamoto, H., Takasaki, K., Shigemori, M. et al. (2001) Therapeutic effect and mechanism of repetitive transcranial magnetic stimulation in Parkinson's disease. *J Neurol*, 248(Suppl 3), III48–III52.

Siebner, H. R., Hartwigsen, G., Kassuba, T. et al. (2009) How does transcranial magnetic stimulation modify neuronal activity in the brain? Implications for studies of cognition. *Cortex*, 45, 1035–1042.

Siebner, H. R., Lang, N., Rizzo, V. et al. (2004) Preconditioning of low-frequency repetitive transcranial magnetic stimulation with transcranial direct current stimulation: Evidence for homeostatic plasticity in the human motor cortex. *J Neurosci*, 24, 3379–3385.

Siebner, H., Peller, M., Bartenstein, P. et al. (2001a) Activation of frontal premotor areas during suprathreshold transcranial magnetic stimulation of the left primary sensorimotor cortex: A glucose metabolic PET study. *Hum Brain Mapp*, 12, 157–167.

Siebner, H. R., Takano, B., Peinemann, A. et al. (2001b) Continuous transcranial magnetic stimulation during positron emission tomography: A suitable tool for imaging regional excitability of the human cortex. *Neuroimage*, 14, 883–890.

Sohn, Y. H., Kaelin-Lang, A., Jung, H. Y. et al. (2001) Effect of levetiracetam on human corticospinal excitability. *Neurology*, 57, 858–863.

Solinas, C., Lee, Y. C., and Reutens, D. C. (2008) Effect of levetiracetam on cortical excitability: A transcranial magnetic stimulation study. *Eur J Neurol*, 15, 501–505.

Speer, A. M., Willis, M. W., Herscovitch, P. et al. (2003) Intensity-dependent regional cerebral blood flow during 1-Hz repetitive transcranial magnetic stimulation (rTMS) in healthy volunteers studied with H215O positron emission tomography: II. Effects of prefrontal cortex rTMS. *Biol Psychiatry*, 54, 826–832.

Stagg, C. J., Best, J. G., Stephenson, M. C. et al. (2009a) Polarity-sensitive modulation of cortical neurotransmitters by transcranial stimulation. *J Neurosci*, 29, 5202–5206.

Stagg, C. J., Wylezinska, M., Matthews, P. M. et al. (2009b) Neurochemical effects of theta burst stimulation as assessed by magnetic resonance spectroscopy. *J Neurophysiol*, 101, 2872–2877.

Stefan, K., Kunesch, E., Benecke, R. et al. (2002) Mechanisms of enhancement of human motor cortex excitability induced by interventional paired associative stimulation. *J Physiol*, 543, 699–708.

Stefan, K., Kunesch, E., Cohen, L. G. et al. (2000) Induction of plasticity in the human motor cortex by paired associative stimulation. *Brain*, 123 Pt 3, 572–584.

Strafella, A. P., Paus, T., Barrett, J. et al. (2001) Repetitive transcranial magnetic stimulation of the human prefrontal cortex induces dopamine release in the caudate nucleus. *J Neurosci*, 21, RC157.

Strafella, A. P., Paus, T., Fraraccio, M. et al. (2003) Striatal dopamine release induced by repetitive transcranial magnetic stimulation of the human motor cortex. *Brain*, 126, 2609–2615.

Szuba, M. P., O'Reardon, J. P., Rai, A. S. et al. (2001) Acute mood and thyroid stimulating hormone effects of transcranial magnetic stimulation in major depression. *Biol Psychiatry*, 50, 22–27.

Teo, J. T., Schneider, S. A., Cheeran, B. J. et al. (2008) Prolonged cortical silent period but normal sensorimotor plasticity in spinocerebellar ataxia 6. *Mov Disord*, 23, 378–385.

Teo, J. T., Swayne, O. B., and Rothwell, J. C. (2007) Further evidence for NMDA-dependence of the after-effects of human theta burst stimulation. *Clin Neurophysiol*, 118, 1649–1651.

Teo, J. T., Terranova, C., Swayne, O. et al. (2009) Differing effects of intracortical circuits on plasticity. *Exp Brain Res*, 193, 555–563.

Trippe, J., Mix, A., Aydin-Abidin, S. et al. (2009) Theta burst and conventional low-frequency rTMS differentially affect GABAergic neurotransmission in the rat cortex. *Exp Brain Res*, 199, 411–421.

Trojak, B., Chauvet-Gelinier, J. C., Verges, B. et al. (2011) Significant increase in plasma thyroid-stimulating hormone during low-frequency repetitive transcranial magnetic stimulation. *J Neuropsychiatry Clin Neurosci*, 23, E12.

Tunez, I., Drucker-Colin, R., Jimena, I. et al. (2006) Transcranial magnetic stimulation attenuates cell loss and oxidative damage in the striatum induced in the 3-nitropropionic model of Huntington's disease. *J Neurochem*, 97, 619–630.

Vetulani, J., Roman, A., Kowalska, M. et al. (2010) Paroxetine pretreatment does not change the effects induced in the rat cortical beta-adrenergic receptor system by repetitive transcranial magnetic stimulation and electroconvulsive shock. *Int J Neuropsychopharmacol*, 13, 737–746.

Wang, H. Y., Crupi, D., Liu, J. et al. (2011) Repetitive transcranial magnetic stimulation enhances BDNF-TrkB signaling in both brain and lymphocyte. *J Neurosci*, 31, 11044–11054.

Witte, A. V., Kurten, J., Jansen, S. et al. (2012) Interaction of BDNF and COMT polymorphisms on paired-associative stimulation-induced cortical plasticity. *J Neurosci*, 32, 4553–4561.

Wolters, A., Sandbrink, F., Schlottmann, A. et al. (2003) A temporally asymmetric Hebbian rule governing plasticity in the human motor cortex. *J Neurophysiol*, 89, 2339–2345.

Woo, N. H., Teng, H. K., Siao, C. J. et al. (2005) Activation of p75NTR by proBDNF facilitates hippocampal long-term depression. *Nat Neurosci*, 8, 1069–1077.

Yang, X., Song, L., and Liu, Z. (2010) The effect of repetitive transcranial magnetic stimulation on a model rat of Parkinson's disease. *Neuroreport*, 21, 268–272.

Young, S. N. (2007) How to increase serotonin in the human brain without drugs. *J Psychiatry Neurosci*, 32, 394–399.

Yue, L., Xiao-Lin, H., and Tao, S. (2009) The effects of chronic repetitive transcranial magnetic stimulation on glutamate and gamma-aminobutyric acid in rat brain. *Brain Res*.

Yukimasa, T., Yoshimura, R., Tamagawa, A. et al. (2006) High-frequency repetitive transcranial magnetic stimulation improves refractory depression by influencing catecholamine and brain-derived neurotrophic factors. *Pharmacopsychiatry*, 39, 52–59.

Zanardi, R., Magri, L., Rossini, D. et al. (2007) Role of serotonergic gene polymorphisms on response to transcranial magnetic stimulation in depression. *Eur Neuropsychopharmacol*, 17, 651–657.

Zanardini, R., Gazzoli, A., Ventriglia, M. et al. (2006) Effect of repetitive transcranial magnetic stimulation on serum brain derived neurotrophic factor in drug resistant depressed patients. *J Affect Disord*, 91, 83–86.

Zhang, X., Mei, Y., Liu, C. et al. (2007) Effect of transcranial magnetic stimulation on the expression of c-Fos and brain-derived neurotrophic factor of the cerebral cortex in rats with cerebral infarct. *J Huazhong Univ Sci Technolog Med Sci*, 27, 415–418.

Ziemann, U., Chen, R., Cohen, L. G. et al. (1998a) Dextromethorphan decreases the excitability of the human motor cortex. *Neurology*, 51, 1320–1324.

Ziemann, U., Hallett, M., and Cohen, L. G. (1998b) Mechanisms of deafferentation-induced plasticity in human motor cortex. *J Neurosci*, 18, 7000–7007.

Ziemann, U., Lonnecker, S., Steinhoff, B. J. et al. (1996) Effects of antiepileptic drugs on motor cortex excitability in humans: A transcranial magnetic stimulation study. *Ann Neurol*, 40, 367–378.

Ziemann, U., Tergau, F., Bruns, D. et al. (1997) Changes in human motor cortex excitability induced by dopaminergic and anti-dopaminergic drugs. *Electroencephalogr Clin Neurophysiol*, 105, 430–437.

Zwanzger, P., Baghai, T. C., Padberg, F. et al. (2003) The combined dexamethasone-CRH test before and after repetitive transcranial magnetic stimulation (rTMS) in major depression. *Psychoneuroendocrinology*, 28, 376–385.

Part III

Imaging-Brain Mapping

12 Noninvasive Brain Stimulation and Neuroimaging
Novel Ways of Assessing Causal Relationships in Brain Networks

Jacinta O'Shea, Gregor Thut, and Sven Bestmann

CONTENTS

The behavioral consequences of TMS offer potential insight into both the neural underpinnings of cognitive processes in the brain (Pascual-Leone et al., 2000; Cowey and Walsh, 2001; Miniussi et al., 2010) and the mechanisms of action of TMS itself (Bestmann et al., 2008b; Reithler et al., 2011). In this chapter, we review how the combination of TMS and neuroimaging has led to the formulation of novel and testable hypotheses about the functioning of distributed brain networks in vivo in the human brain. We focus on studies that have tested the effect of TMS on functional magnetic resonance imaging (fMRI) or electroencephalographic (EEG) correlates of behavior and discuss how TMS-induced functional changes may relate to behavioral changes induced by TMS. We outline examples of work that have led to specific predictions about TMS actions and then review how these predictions were tested by combining TMS with neuroimaging. Finally, we discuss how these "stimulate-and-record" approaches (Paus, 2005) have advanced our understanding of brain–behavior relationships. Our chapter illustrates how concurrent recordings of brain signals during TMS have afforded increasingly refined hypotheses about how TMS interacts with networks in the human brain. Such understanding is paramount for adequate interpretation of TMS effects, for testing novel hypotheses on how functional interactions among cortical and subcortical brain regions support and enable perception and cognition, and for optimizing potential clinical applications of neurostimulation. The latter is of particular importance in light of the vast number of TMS parameters, which can be potentially exploited for translational goals.

12.1   EVIDENCE FOR NETWORK EFFECTS ELICITED BY NEUROSTIMULATION

That noninvasive cortical stimulation elicits significant changes in neural activity outside the directly stimulated area has been clear since the earliest TMS studies. This fact was evidenced by the very first demonstrations that stimulation of primary motor cortex could elicit muscle activity, motor-evoked potentials (MEPs), from the distal muscles of the contralateral hand (Barker et al., 1985; Cracco, 1987; Mills et al., 1987; Rothwell et al., 1987a,b, Cracco and Cracco, 1999). MEPs are elicited at least two synapses away from the stimulation site, demonstrating that the effects of TMS over M1 could spread polysynaptically, at least to peripheral muscles. However, this effect is not confined to cortico-spinal projections: early work also demonstrated that the effects of single TMS pulses to M1 can spread to the homologous area of the opposite hemisphere, via transcallosal projections, and can cause significant functional changes in cortical excitability at this distant but connected site (Ferbert et al., 1992; Meyer et al., 1995). Similar remote effects of single-pulse TMS were demonstrated in other cortical systems. For example, stimulation of visual area V5 has been shown to significantly change the excitability of primary visual cortex, a measurable functional interaction with significant consequence for visual awareness (Pascual-Leone and Walsh, 2001). Such findings demonstrate that stimulation of a single cortical region also affects closely interconnected regions. Hence, the findings have been clear since even relatively early unimodal TMS studies, namely that *almost nothing goes on internally in one area without this activity being transmitted to at least one other area* (Mumford, 1992). The great potential of combined TMS/neuroimaging approaches lies in clarifying the functional implications of these connectional effects, thus promising to revolutionize our conceptual interpretation of TMS effects, which are still typically confined to a consideration of the cortex directly underneath the stimulating coil.

Combined TMS/neuroimaging approaches now make it possible to map and quantify induced network effects and to characterize the extent and contextual dependence of TMS-induced remote activity changes during or subsequent to cortical stimulation. Here we focus primarily on studies that have combined TMS with EEG (Virtanen et al., 1999; Paus et al., 2001a; Komssi et al., 2002; Kahkonen et al., 2005; Massimini et al., 2005; Fuggetta et al., 2006; Komssi and Kahkonen, 2006; Van Der Werf and Paus, 2006; Taylor et al., 2007a,b; Ilmoniemi and Karhu, 2008; Julkunen et al., 2008; Romei et al., 2008; Bonnard et al., 2009; Rosanova et al., 2009; Veniero et al., 2009;

Hamidi et al., 2010; Johnson et al., 2010; Taylor et al., 2010; Thut et al., 2011b; Reichenbach et al., 2011) or with positron emission tomography (PET) (Fox et al., 1997; Paus et al., 1997, 1998, 2001b; Strafella and Paus, 2001; Siebner et al., 1998, 1999, 2000; Speer et al., 2003; Paus, 2005, Paus and Wolforth, 1998) or with fMRI (Bohning et al., 1997, 1998, 2000, 2003b; Baudewig et al., 2001; Nahas et al., 2001; Bestmann et al., 2003, 2004, 2005, 2006, 2008a,b,c, 2010; Denslow et al., 2005, Ruff et al., 2006, 2008, 2009a; Sack et al., 2007; Blankenburg et al., 2008, 2010). The evidence for remote activity changes is what motivated the initial pioneering work that established the feasibility of various TMS and neuroimaging combinations. Interestingly, the combination with PET (Paus et al., 1997), EEG (Ilmoniemi et al., 1997), and fMRI (Bohning et al., 1997) emerged near-simultaneously. The inevitable technical demands for combining neuroimaging with TMS can be substantial, and following early demonstrations of feasibility, technical refinements are continuously being advanced (Bestmann et al., 2003; Thut et al., 2005; Ives et al., 2006; Litvak et al., 2007; Morbidi et al., 2007; Taylor et al., 2008; Moisa et al., 2009, 2010; Weiskopf et al., 2009; Hamidi et al., 2010; Ilmoniemi and Kicić, 2010; Bungert et al., 2012).

Feasibility established, early studies were motivated by the prediction, outlined earlier, that TMS should elicit activity changes not only locally at the stimulation site, but also distally in remote but putatively interconnected brain regions. To investigate this, researchers used so-called perturb-and-measure approaches (Paus, 2005), in which neuroimaging served to identify which brain regions were affected by the stimulation of a given cortical site. This approach made it possible to map the spatial topography of TMS-evoked activity changes, both locally and distally in remote brain regions. The appeal of this approach lies in the ability of TMS to enable direct causal tests of the role of a given brain area in a given behavior. This is because TMS bypasses normal routes of entry to cortex via the sensory input pathways, making it possible to directly manipulate cortical activity in a given area of interest. Classically, psychological research has sought to achieve this through behavioral tasks in which activity in the brain region of interest is likely to change through complex interactions with other regions. The combination with neuroimaging has opened complementary means of addressing such questions through direct manipulation of activity in a brain region of interest and the concurrent measurement of concomitant activity changes throughout the brain.

Collectively, these studies have consistently demonstrated TMS-induced activity changes in remote but interconnected brain regions, irrespective of the brain area stimulated or the specific TMS protocols that were used (Ilmoniemi et al., 1997, 1999; Bohning et al., 1999; Kimbrell et al., 2002; Bestmann et al., 2003, 2004; Chouinard et al., 2003; Speer et al., 2003; Denslow et al., 2005). For example, TMS applied during fMRI to either M1 or dorsal premotor cortex can elicit activity changes in remote regions of the motor system (Bestmann et al., 2003, 2004, 2005) that cannot be trivially explained by re-afferent feedback from activation of peripheral muscles (Bestmann et al., 2005, 2008c). Combined TMS–EEG studies during rest have established that the rapid propagation of the evoked TMS effects is not arbitrary, but closely maps onto the known connectivity patterns in the brain (Ilmoniemi et al., 1997; Massimini et al., 2005, 2007; Litvak et al., 2007; Iwahashi et al., 2008).

Since the earliest behavioral TMS studies, it has been clear that stimulation at a given cortical site induces remote activation changes elsewhere in the brain. One key contribution of early resting-state neuroimaging and TMS studies was the empirical confirmation that noninvasive cortical stimulation has the capacity to activate distributed cortical networks in the human brain, in an anatomically specific way. Prior to the combination with neuroimaging, such distributed effects could have been investigated only in a limited way.

12.2   STATE DEPENDENCE OF INTERREGIONAL TMS EFFECTS

As explained earlier, early work largely investigated how TMS affects activity at distant cortical and subcortical sites while volunteers are at rest. Resting-state approaches can characterize how stimulation of different cortical regions induces specific activity changes in brain networks, but they cannot reveal whether such remote effects have any causal role in supporting behavior: the impact of the same TMS protocol at rest may differ profoundly during changing perceptual or cognitive states.

There is ample evidence demonstrating that the impact of TMS varies profoundly if the activity level of the targeted region changes (cf Silvanto et al., 2008; Ruff et al., 2009b). For example, early work showed that the amplitude and the latency of evoked peripheral hand muscle responses (i.e., the size and the number of descending volleys) increase dramatically during voluntary contraction, as opposed to when the hand is at rest (Hufnagel et al., 1990; Mazzocchio et al., 1994; Ridding et al., 1995; Wilson et al., 1995; Fujiwara and Rothwell 2004). Further, the MEP threshold is markedly reduced by slight voluntary contraction of the target muscle (Mills et al., 1987). Subsequent studies have also shown that cognitive load, such as motor performance or motor learning, can induce activity-dependent excitability modulation of M1 (e.g., Bonato et al., 1994, 1996; Pascual-Leone et al., 1994, 1995).

Indirect, yet compelling, evidence for the cognitive state dependence of TMS effects was also present in early behavioral studies, which showed that transient disruption of a specific brain region only led to behavioral impairments if the stimulation was applied at the specific time, presumably when that brain region was critically involved (i.e., "active") in sustaining the relevant behavior. For example, V1 TMS only suppresses visual perception when applied at the time when sensory input reaches early visual cortex (Amassian et al., 1989). Moreover, the critical period of V1 in visual awareness of moving phosphenes both pre-dates and post-dates that of area V5/MT (Silvanto et al., 2005). Stimulation of V5/MT at the critical time for V1 has no effect, nor does stimulation of V1 at V5/MT's critical period. Another example is that contextual factors that are known to modulate neural excitability in visual cortex, such as spatial attention, can alter the intensities required for occipital TMS to elicit phosphenes (Bestmann et al., 2007). We have long known that the state dependence of TMS effects can be demonstrated by comparing a brain region's response to stimulation under different cognitive conditions, as evidenced, e.g., by an early study that demonstrated the criticality of right posterior parietal cortex for conjunction but not serial visual search tasks (Ashbridge et al., 1997).

Recent combined TMS and EEG work has now demonstrated that changes in "effective or functional connectivity," i.e., the influence one brain region exerts over another, strongly depend on the cognitive or sensory state of the targeted networks (Massimini et al., 2005; Taylor et al., 2007a). For example, stimulation of the right frontal eye fields has been shown to modulate visually evoked negativities at posterior electrodes, starting from around 200 ms after visual onset, and crucially depending on visual attention. This has been taken as evidence that activity within occipital or parieto-occipital visual regions is under causal influence and top-down control from frontal regions such as the frontal eye fields when attention is being allocated.

Combined EEG–TMS studies have also investigated whether the state of the targeted cortex, specifically, the pattern of ongoing intrinsic oscillatory cortical activity, determines the evoked physiological response elicited by TMS. In a recent study (Romei et al., 2008), phosphenes were induced by applying single TMS pulses over early visual cortex. Whether or not the same TMS intensity, which was near phosphene threshold, actually evoked a phosphene depended on the level of spontaneous posterior alpha-band activity. This suggests that moment-by-moment variations in the excitability of early visual cortex determine the impact of a single pulse of TMS applied to that region.

Just as for resting-state observations, recent data indicate that state-dependent TMS effects can also induce signal changes at distant but putatively interconnected sites. For example, Morishima and colleagues (2009) showed that a single TMS pulse applied over prefrontal electrode position FC2 during concurrent EEG recordings altered event-related potentials (ERPs) at occipito-temporal electrodes in a state-dependent way. More specifically, the effects on posterior ERP responses changed depending on the task (judge direction of a moving grating versus the gender of a face stimulus) and the pre-cuing of the tasks. Finally, these effects related to the speed of participants' responses, which suggests a causal influence of top-down prefrontal control over posterior cortex, which changes as a function of attentional state.

Complementary results come from combined TMS–fMRI experiments. In contrast to the TMS/EEG studies, which allow fine-grained temporal assessment of interregional influences and their dependence on activation state, fMRI studies enable mapping of the topographic specificity of TMS-evoked network changes. For example, Ruff and coworkers (Ruff et al., 2009a) could show that TMS to the cortical region proposed to be involved in top-down visual control, the right parietal eye fields in the intraparietal sulcus, induced activity increases in early visual cortex during the absence of visual stimulation, but influenced activity in the human motion complex (V5/MT+) only when moving visual input was presented simultaneously. No such changes were observed for left parietal TMS (Ruff et al., 2009a).

One intriguing question is whether changes in cognitive state may also change how TMS-evoked activity at the targeted site propagates throughout one or more functional networks? Given the assumption that the more excitable a given connection at the time of stimulation, the more likely it is to be affected by TMS (Mazzocchio et al., 1994; Ridding et al., 1995; Fujiwara and Rothwell, 2004), recent combined TMS–fMRI work has addressed this question using short bursts of TMS to left dorsal premotor cortex while subjects performed brief isometric grips of the left hand or maintained rest (Bestmann et al., 2008c). In this case, TMS was used to probe connectivity during the task or at rest, while concurrent fMRI measured the resulting activity changes in remote brain regions. The key finding was that left dorsal premotor cortex stimulation changed contralateral right premotor cortex and M1 activity in a state-dependent way: During the simple motor task, effective TMS over left dorsal premotor cortex increased activity in contralateral right premotor cortex and right M1, whereas, during rest, the same stimulation decreased activity in these regions. An additional analysis of interregional coupling furthermore suggested that coupling between the targeted left dorsal premotor cortex and right dorsal premotor cortex/M1 was stronger when high-intensity TMS was applied during the active left-hand grip task, compared with rest. Collectively, these results suggest that voluntary movements of the hand, such as simple grip behavior, change the interhemispheric interactions between left dorsal premotor cortex and contralateral dorsal premotor cortex and M1. They further suggest that these interactions are specific to the brain regions that are actually engaged in that behavior (rather than being widely distributed across all regions putatively interconnected to the stimulation site) and that TMS can preferentially activate pathways that, at the time of stimulation, show increased effective connectivity with the stimulation site.

The examples provided earlier demonstrate how combined neuroimaging and TMS work has addressed previously inaccessible questions about the state dependence of TMS-evoked network effects. The key insight is that the degree with which interconnected brain regions are modulated depends on both the activity state of the stimulated brain region and the activation state of those remote areas that form part of that functionally active network.

12.3 INTERPRETING "VIRTUAL LESION" EFFECTS: LOCAL VERSUS DISTRIBUTED CAUSATION

One question arising from these studies is whether the behavioral consequences that can be induced by TMS, and which allow for causal structure–function inferences, arise only from activity changes at the stimulation site or whether they also reflect functionally critical perturbations at more distant brain regions that are active during a specific task. Some evidence suggests that the latter may indeed be the case. For example, Sack and coworkers (2007) used combined TMS–fMRI to ask about task-specific causal interactions between parietal regions implicated in visuospatial processing and other regions known to be active during such tasks. The authors applied TMS over left or right parietal cortex while participants performed either a spatial task (angle judgments) or a nonspatial control task (color judgment). Only right parietal TMS increased reaction times. Interestingly, TMS-related BOLD signal decreases were found not only in the (stimulated) right parietal cortex, but also in parts of the right medial frontal gyrus. This effect was strongest during the spatial task. This finding of state-dependent (i.e., task-specific) effects of parietal TMS on frontal cortex raised the question whether this disruption contributed to the behavioral impairments observed. One indication that this may be the case was that the TMS-evoked activity changes in right frontal and parietal cortex correlated with the behavioral impairments.

Another example comes from Ruff et al. (2006), who tested whether the spatial pattern of BOLD effects induced in retinotopic visual areas V1–V4 by right FEF–TMS could relate to perception in a comparable way. In an initial experiment, right FEF–TMS increased BOLD activity in peripheral visual field representations. By contrast, BOLD activity was decreased in the central visual field. Based on this observation, the authors sought to test whether right FEF–TMS could enhance perceptual salience of peripheral relative to central visual stimuli. This prediction was confirmed in a subsequent psychophysical experiment, where perceived contrast was enhanced for peripheral relative to central Gabor gratings during right FEF–TMS (compared with control stimulation of the vertex). These results indicate that remote physiological effects of TMS can be used to generate new predictions for behavioral TMS effects. Such observations would not have been possible without the combination of neuroimaging and TMS and now open the interesting question whether TMS effects on perception and cognition always arise only from local perturbations of activity or may sometimes be caused by disruption of information processing in a distributed functional network.

We emphasize that such findings do not invalidate the "virtual lesion" inferential approach, which interprets behavioral TMS with respect to the tissue directly underneath the coil. Instead, by providing accompanying physiological information, they offer new insight into how distributed functional systems react to virtual lesions and may even reorganize in response (see, e.g., O'Shea et al., 2007, as discussed in the following text). Behavioral effects elicited by TMS provide clear evidence for some causal involvement of the stimulated cortical region. But in combination with neuroimaging approaches, one can additionally ask whether interactions between different parts of task-specific brain networks are critical for supporting cognition, perception, and behavior (O'Shea et al., 2007; Bestmann et al., 2008b,c; Ruff et al., 2009b).

For virtual lesion TMS studies, one can now ask to what degree the causal contribution of a given brain region reflects induced changes in local computations or instead arises from altered patterns of functional interaction with interconnected regions of that task network.

12.4   MANIPULATING SUBCORTICAL NUCLEI VIA CORTICAL STIMULATION

If cortical stimulation is capable of evoking activity changes in cortico-cortical networks, then one prediction is that specific subcortical changes should also occur, given the intimate cortico-basal ganglia-thalamic loops that define the connectivity patterns of distinct functional systems in the brain (Alexander et al., 1986). Consistent with this prediction, combined neuroimaging and TMS studies have confirmed specific patterns of subcortical activity change during, or following, cortical stimulation. For example, when short bursts of TMS are applied to primary motor or premotor cortex at rest, specific activity increases can indeed be observed in the corresponding sensorimotor nuclei of the thalamus, as well as in the putamen and caudate nucleus (Bestmann et al., 2004, 2005; Denslow et al., 2005). These findings are of interest, since it has long been reasoned that TMS might have therapeutic potential in neuropsychiatric disorders, which are often linked to specific abnormalities in subcortical structures.

Even more precise hypotheses have been formulated prior to these studies by using combined ligand-PET and TMS approaches: we know, e.g., that various parts of the prefrontal cortex receive rich dopaminergic projections and, furthermore, that the prefrontal cortex might play an important role in regulating the release of dopamine (DA) in subcortical structures. This has led to the prediction that prefrontal stimulation will not only change activity in subcortical structures, but more specifically may trigger the release of DA in the basal ganglia. Dysfunction in DA plays a role in a variety of psychiatric conditions such as Parkinson's disease, drug addiction, and schizophrenia. Hence, if true, this could indicate a mechanism through which noninvasive stimulation protocols could potentially exert therapeutic effects, analogous to more invasive clinically established procedures such as electroconvulsive therapy or deep-brain stimulation. A variety of combined TMS/neuroimaging protocols could then be used to assess the physiological efficacy of stimulation protocols over different cortical target sites.

Using different radio-ligands in combination with PET, Strafella and colleagues have put these ideas to the test (Ko et al., 2008; Ko and Strafella, 2012). This approach allows for indirectly quantifying synaptic changes in DA release in vivo, by comparing control (e.g., occipital TMS) versus active (e.g., prefrontal TMS) conditions. By inferring changes in synaptic DA concentration from changes in binding potential associated with the active condition, these studies (Strafella et al., 2001, 2003, 2005) have demonstrated significant and reproducible changes in endogenous DA release following a specific high-frequency repetitive TMS protocol. For example, such stimulation, when applied over the dorsolateral prefrontal cortex, was shown to increase DA release in the ipsilateral caudate nucleus (Strafella et al., 2001), while the same stimulation applied to M1 increased DA release in the ipsilateral putamen (Strafella et al., 2003). One appealing explanation for this specific effect is a TMS-induced activation of corticostriatal projections that leads to focal release in DA in the striatal projection site of the stimulated cortical target region. The observed dopaminergic effects were in striking correspondence with known anatomical data as well as previous physiological studies in nonhuman primates (Yeterian and Pandya, 1991; Takada et al., 1998a,b; Tachibana et al., 2004).

Importantly, observation and characterization of such subcortical changes are not accessible without the combination of TMS and neuroimaging. This is perhaps one of the strongest demonstrations why these combinations are important, particularly in light of the growing interest in using TMS for therapeutic interventions.

To summarize, TMS can alter activity in subcortical regions in an anatomically specific way. Combinations with ligand-based PET techniques suggest that such remote, subcortical effects can also involve alterations in neurotransmitter release, such as DA, in the basal ganglia. This may allow for the development of therapeutic stimulation protocols, designed to alter neurochemical and/or functional signaling in specific subcortical nuclei, which are frequently disordered in a variety of neuropsychiatric conditions.

12.5 MEASURING AND MANIPULATING FUNCTIONAL PLASTICITY

"Plasticity" refers to the brain's capacity to reorganize how it functions to support mental processes and behavior. Functional plasticity unfolds over the course of normal development, enables adaptive behavioral modification during learning, and can permit behavioral recovery from brain injury. Increasingly, combined TMS/fMRI approaches are being used to measure (Johansen-Berg et al., 2002), induce (O'Shea et al., 2007), and manipulate (Tegenthoff et al., 2005) neural processes of plastic functional change.

Learning-related changes in functional brain activity occur over a variety of timescales. Early TMS studies of cortical plasticity demonstrated motor cortical map enlargement during motor learning (Pascual-Leone et al., 1994) and confirmed the causal role of visual cortex activation in tactile perception in the blind (Cohen et al., 1997). More recent work, combining TMS with fMRI, has made it possible to assess the causal role of distinct nodes within a given cortical network supporting a specific cognitive function. In a clever paradigm, Wig et al. (2005) combined offline TMS with fMRI to test the causal status of frontal cortex BOLD signal changes for behavioral priming. When subjects are presented with a series of visual objects, they are quicker to recognize and classify objects that they have recently viewed, even if they do not consciously remember seeing the objects before. This repetition-based learning facilitation, "priming," is associated with reductions in BOLD signal in visual sensory regions and inferior frontal cortex. The authors aimed to test whether the inferior frontal cortex activity had a causal role in behavioral priming. Subjects performed the task and underwent fMRI twice, both before and after offline 10 Hz TMS to left inferior frontal cortex or motor cortex. In both sessions, subjects showed the typical behavioral and fMRI signatures of priming, except for trials with repeated objects that had been previously paired with frontal TMS. On those trials, behavioral priming was significantly reduced and the neural priming effect was abolished in frontal but not sensory visual cortex. Hence, the combination of TMS/fMRI demonstrated the causal link between neural and behavioral priming and revealed a functional dissociation between neural priming in sensory versus frontal cortex.

As detailed earlier, it is now well established that TMS induces cognitive state-dependent distal activation changes, but the functional status of remote activations has been largely unaddressed. A key question is whether TMS-induced remote activity changes are causally implicated in the observed behavioral performance or whether they are epiphenomenal. This is important, both for understanding how TMS changes behavior and for advancing knowledge on functional redundancy, reorganization, and causality in neural circuits. In a study of motor response selection (O'Shea et al., 2007), 1 Hz TMS to the left dorsal premotor cortex (PMd) was shown to induce functionally specific reorganization in the cortical network implementing response selection. Specifically, although behavior was unchanged, and activation levels under the coil remained stable, there was increased activity in the right PMd and medial motor regions whenever subjects performed response selection, as opposed to a control task of action execution. This suggested that the reorganized fMRI activity was causally contributing to intact performance. To test this, the authors applied a second TMS intervention to probe the causal role of reorganized activity in the right PMd. The results showed that, after the left PMd had been challenged by TMS, the reorganized activity in right PMd became task critical: perturbation of those signals led to performance breakdown. In this example, the combination of TMS/fMRI could demonstrate latent capacity within the action selection system: when a key node in the task network (left PMd) was challenged by TMS, functionally connected regions in both hemispheres showed either increased local activity or enhanced functional coupling with that node to enable performance to be maintained. The criticality of this newly emergent activation for maintaining behavior could be directly confirmed by showing that, when it was subjected to challenge, intact performance then broke down.

One domain in which combined TMS/neuroimaging approaches are being increasingly used is to test the functional significance of abnormal functional activation patterns in stroke-damaged brains. Using either whole-brain approaches, such as TMS/fMRI, or circuit-level approaches, such as dual-site TMS combined with MEP recordings, investigators have begun to disentangle adaptive versus maladaptive changes in brain functioning that occur subsequent to injury, such as stroke (e.g., Lotze et al., 2006; Koch et al., 2008; Bestmann et al., 2010). A pioneering study (Johansen-Berg et al., 2002) asked whether increased activity in ipsilateral motor regions post-stroke reflects merely system dysfunction or rather a process of adaptive functional reorganization—areas of increased activity helping to drive retained or recovered use of the hemiparetic hand. Single-pulse TMS to ipsilateral PMd, but not M1, led to a breakdown of hemiparetic hand performance. This effect was graded by symptom severity, such that those who were most impaired post-stroke showed the greatest reliance on the hemisphere contralateral to the stroke, while those with milder impairments showed less reliance and had activation patterns more closely resembling healthy brains.

This kind of work is set to become increasingly important now that TMS and its methodological relative, transcranial direct current stimulation (TDCS), are under development as adjuvant rehabilitation interventions post-stroke (for a review of therapeutic brain stimulation approaches, see Fregni and Pascual-Leone, 2007). Following a stroke, there are widespread changes in the functioning of cortical networks that implement the compromised behaviors. Many of these changes are dysfunctional and thus contribute to patient disability. Others may be advantageous, reflecting adaptive reorganization, and thus enabling partial recovery of lost functions previously carried out by the damaged tissue. A growing research effort aims to use brain stimulation to proactively manipulate such patterns of post-stroke neural activity change for functional benefit, e.g., by increasing excitability in the stroke-affected hemisphere (Hummel et al., 2005; Khedr et al., 2005) or decreasing excitability in the intact hemisphere (Fregni et al., 2006) or by pairing these stimulation approaches with occupational or physical therapy (Lindenberg et al., 2010; Flöel et al., 2011). Each of these approaches can bring about short-term behavioral gains, but patient heterogeneity is a challenge: variation in lesion site, pre-morbid functional status, and recovery rate suggests that no "one size fits all" approach is likely to be successful (Swayne et al., 2008), motivating the need for individualized therapy tailored on a patient-by-patient basis. Combined TMS/neuroimaging approaches offer a powerful means to identify potential biomarkers of therapeutic response (Stinear et al., 2007), to determine the causal status of abnormal activation patterns prior to attempted remedial intervention (Lotze et al., 2006; Koch et al., 2008), and to characterize induced functional activity changes following successful or unsuccessful experimental treatment attempts (Stagg et al., 2012).

Combined TMS/neuroimaging approaches offer significant potential for interrogating plasticity-induced changes in cortical network functioning, such as during normal learning or subsequent to brain injury. Imaging can be used to map how distributed networks change, while TMS can be used to probe the functional status of reorganized brain activity in supporting a given behavior. Increasingly, brain stimulation is being used to manipulate observed patterns of neural activation change—in an attempt to promote adaptive functional plasticity and thus enhance behavior.

12.6 MAPPING FACTORS LEADING TO INDIVIDUAL VARIABILITY IN RESPONSE TO TMS

Individual variation is a challenge, not only with brain damaged patients, but also with healthy volunteers. From the earliest work, it was clear that the impact of TMS is highly variable across individuals (Maeda et al., 2002), and the "nonresponder" is a common feature of many papers reporting behavioral interference effects. Protocol titration based on motor or phosphene threshold is a good practice, where possible, in an attempt to match TMS dose delivery and can be further refined by factoring in skull–cortex distance (Stokes et al., 2007). Nevertheless, a variety of other biological factors can influence the impact of the induced stimulation, and recent combined TMS/neuroimaging work has begun to identify some of those parameters— anatomical, genetic, neurochemical—that contribute to outcome variability.

It has long been clear, since early studies of motor cortex, that the orientation of the TMS coil strongly influences the physiological impact of the induced current— MEPs are more readily evoked when current is applied perpendicular to the central sulcus. Recent studies have begun to characterize how local gray and white matter anatomy affects the distribution of the induced electric field, by combining structural MR with realistic head models (see, e.g., Opitz et al., 2011; Thielscher et al., 2011). Given large interindividual variation in gyral folding patterns and white matter microstructure, a growing number of studies have also begun to address how such intrinsic variation may contribute to outcome variability in TMS studies. In a dual-coil TMS experiment of response selection, Boorman and colleagues (2007) showed that interindividual differences in the strength of task-specific interhemispheric PMd–M1 functional connectivity related to underlying variation in white matter microstructure connecting those two brain regions. Similar results have been found intra-hemispherically (Groppa et al., 2012) and for transcallosal connections between left and right M1 (Wahl et al., 2007). Variation in resting and active motor threshold has also been shown to relate to heterogeneity in cerebral white matter structure (Klöppel et al., 2008).

Genotypic variation influences not only brain structure and function, but also its response to modification, such as via drugs. Recently, there has been an upsurge of research interest within the TMS community in genotypic variation underlying brain-derived neurotropic factor (BDNF), a neural growth factor that is important for brain functions such as plasticity and repair. A single nucleotide polymorphism for this growth factor, val(66)met, is relatively common (frequency of 16%–28% in European Caucasians) and is associated with a reduction in activity-dependent BDNF release. This polymorphism has been shown to have (detrimental) effects on motor function, short-term plasticity, and learning. In a motor cortex TMS study, Kleim and colleagues (2006) showed that individuals with this polymorphism showed a significant reduction in the normal pattern of training-induced MEP increase and motor map reorganization that occurs during motor learning. Cheeran et al. (2008) showed that these individuals also show a reduction in or absence of the normal patterns of MEP change subsequent to inhibitory theta burst TMS of the motor cortex. In a mouse study, Fritsch et al. (2010) further showed that this polymorphism significantly reduced both the neural and the behavioral plasticity, which could be induced by TDCS. Hence, BDNF polymorphism affects behavior, responsiveness to stimulation, and also functional brain activity as measured by fMRI (McHughen et al., 2010).

As an excitability manipulation, it is not surprising that the magnitude of TMS effects is influenced by cross-subject variation in the baseline balance of excitation and inhibition in a targeted region of cortex. Using magnetic resonance spectroscopy (MRS), it is possible to measure volumetric concentrations of gaba-amino-butyric acid (GABA) and glutamate, the major inhibitory and excitatory cortical neurotransmitters, in circumscribed regions of cortex (but not yet routinely at whole-brain level). MRS measures of GABA and glutamate have been demonstrated to correlate with specific TMS measures of cortical excitation and inhibition (Stagg et al., 2009, 2011b). Further, using MRS, it has been shown that interindividual differences in regionally varying basal GABA concentration influence a variety of behavioral tasks known to depend on those brain regions. For example, subjects with higher GABA levels, or stronger inhibitory "tone," in the frontal eye fields showed lower levels of distractibility (Sumner et al., 2010); while individuals with higher GABA levels in the supplementary motor area showed less subconscious motor control (Boy et al., 2010). The responsiveness of the GABA system to motor cortex stimulation (in this case, using TDCS) has recently been shown to correlate significantly with individual differences in both motor learning rates and associated fMRI signal change (Stagg et al., 2011a). Those subjects who showed the greatest reduction in GABA following excitatory (anodal) TDCS also showed the greatest motor learning rates and the greatest change in learning-related BOLD signal in M1. Similar covariation between GABA, BOLD, and MEG measures of gamma oscillations has been reported (Muthukumaraswamy et al., 2009), suggesting that all of these three different measurement techniques are sensitive to common, underlying neural sources that give rise to genuine functional variation among individuals.

Interindividual variability in responsiveness is a common feature of TMS studies. Recent work, combining stimulation with anatomical, genetic, and neurochemical measures, offers a powerful new means to exploit and explain these sources of intrinsic variation across individuals. Incorporating measures of these covariates in standard TMS/neuroimaging protocols promises an increase in power and sensitivity to distinguish noise from meaningful neuro-behavioral variation. This may become particularly important in the context of clinical applications.

12.7 TARGETING BRAIN OSCILLATIONS THROUGH CORTICAL STIMULATION

So far, we have reviewed TMS-evoked network effects, spreading from the target area to anatomically connected sites. We conclude our review by considering TMS effects on oscillatory EEG or MEG activity, measurements that are thought to directly reflect neuronal network interactions. Brain oscillations are generated by synchronized activity of an ensemble of a large number of neurons. As such, brain oscillations are thought to reflect the functional grouping of neuronal elements in local or larger-scale neuronal networks as a function of the current state of the brain and the task that is currently being executed (e.g., Buzsáki and Draguhn, 2004). TMS can affect brain oscillations in a number of ways, from triggering oscillations via a single TMS pulse (e.g., Paus et al., 2001a, see the following text), to more long-term changes in ongoing oscillations after repetitive TMS, the latter presumably driven by TMS-induced changes in synaptic efficacy (for review see Thut and Pascual-Leone, 2010). We here focus on one particular interaction of TMS with brain oscillations, their entrainment. TMS can in principle be used to stimulate neural elements in a rhythmic fashion, at frequencies similar to those characterizing brain oscillations (Thut and Miniussi, 2009). This raises the intriguing possibility that TMS could be used to transiently entrain brain oscillations (see Thut et al., 2011a for a general overview on entrainment by transcranial stimulation techniques).

Entrainment is based on the self-sustained and dynamic nature of oscillating elements and perturbation of these oscillators by an external force (Glass, 2001; Pikovsky et al., 2003). If this force is periodic, the natural oscillation may then become synchronized to the periodic event (for models see Glass, 2001; Pikovsky et al., 2003). Synchronization here means that the oscillating element starts to cycle with the same period as the external force; that is, it becomes entrained or locked to the external event. At the neuronal population level, this leads to phase adjustment and amplitude increase in the underlying oscillation (see Thut et al., 2011a for a model). Entrainment is therefore about driving brain oscillations. Could the periodic electromagnetic force produced by rhythmic TMS be used for such entrainment of brain oscillations?

Early work combining EEG with TMS suggests that it could. For example, single TMS pulses can evoke a brain response whose frequency matches the natural oscillation of the targeted cortex (Paus et al., 2001a; Rosanova et al., 2009). Over the motor cortex, a single TMS pulse evokes EEG responses that cycle at beta- (to alpha-) frequency, which has been interpreted to reflect a TMS-induced phase reset of the natural oscillation of the stimulated cortex (Paus et al., 2001a; Fuggetta et al., 2005; Van Der Werf and Paus, 2006). Rosanova et al. (2009) have further validated this idea for other cortical areas. This suggests that the frequency of the TMS-evoked response echoes spectral features of the ongoing oscillations at the specific cortical stimulation site targeted with TMS. Importantly, this implies that the external, pulsed electromagnetic force leads to a phase locking of ongoing brain oscillations to the TMS pulse and, by extension, that entrainment is conceivable. With the application of further periodic pulses (and according to entrainment models, see the preceding text), ongoing oscillations should then become more and more entrained, especially if the repeated TMS pulse applications match the frequency of the intrinsic oscillation of a brain region (i.e., if the TMS frequency is tuned to the natural oscillatory activity).

This entrainment hypothesis has recently been tested in humans, by concurrently combining EEG with rhythmic TMS (Thut et al., 2011b). This initial work focused on parietal EEG activity in the alpha-frequency band, which represents a thalamo-cortical network rhythm (Lorincz et al., 2009). Thut et al. (2011b) showed that rhythmic TMS to parietal cortex tuned to its specific alpha frequency progressively enhances parietal alpha activity. By contrast, such effects do not occur when the same number of pulses is applied over the same time interval, but in an arrhythmic mode, or when sham stimulation is used. In addition, entrainment of parietal alpha depended on the phase at which the TMS train "caught" the ongoing oscillatory activity (Thut et al., 2011b).

Taken together, recent developments suggest that TMS may be used to entrain brain oscillations, possibly by phase resetting these correlates of network interactions. Dependence of entrainment on ongoing phase further illustrates the state dependence of TMS effects and suggests that a natural oscillation had been entrained, as opposed to creating artificial rhythms with TMS. It remains to be shown to what extent entrainment also affects communication within the targeted networks by means of synchronization (or de-synchronization) of their elements.

12.8 BEHAVIORAL CONSEQUENCES OF FREQUENCY-TUNED RHYTHMIC TMS: THE "ENTRAINMENT" APPROACH

How could such entrainment advance our understanding of brain–behavior relationships? Recent accounts have proposed that brain oscillations are causally responsible for routing of information between brain regions during perception and cognition (cf. Buzsáki and Draguhn, 2004; Schnitzler and Gross, 2005), as opposed to representing mere epiphenomena. Such claims are difficult to test. However, by inducing brain oscillations via entrainment, one can directly assess the resulting perceptual and behavioral consequences and thereby start to probe for causal roles of brain oscillations in the healthy human brain. For example, a key prediction of the hypothesis that a particular brain oscillation is causal for a specific behavior is that frequency-tuned rhythmic TMS, which targets that specific oscillation, will change perception or behavior in a systematic way. Importantly, there is already an existing EEG/MEG literature that has established correlative relationships between brain oscillations and behavior and can be used to guide predictions for TMS probes into brain oscillations.

For example, there is EEG/MEG evidence that brain rhythms associated with attention/perception dissociate at the level of spatial locations versus visual features (location-based versus feature-based selection). With respect to location-based selection, EEG and MEG research has identified oscillatory signatures in the posterior alpha band (8–14 Hz) as a possible key player. Occipito-parietal alpha oscillations predict whether and/or where in space an upcoming stimulus is perceived (Ergenoglu et al., 2004; Thut et al., 2006; Hanslmayr et al., 2007; van Dijk et al., 2008; Romei et al., 2008; Yamagishi et al., 2008) and are modulated by the deployment of visual spatial attention (Worden et al., 2000; Sauseng et al., 2005; Kelly et al., 2006; Thut et al., 2006; Rihs et al., 2007, 2009; Yamagishi et al., 2008; Siegel et al., 2009; review by Foxe and Snyder, 2011). With respect to selection of visual features, an intriguing EEG finding indicates differential involvement of parietal beta and theta oscillations in local versus global feature extraction (Smith et al., 2006). Smith et al. (2006) showed that these two rhythms are differentially associated with conscious experience of local versus global aspects of a visual scene. This suggests that visual spatial attention and feature selection operate in distinct frequency bands and leads to distinct predictions as to impact of rhythmic TMS on location-based versus feature-based visual task performance and electrophysiological correlates.

These differential predictions were tested in two recent TMS studies (Romei et al., 2010, 2011). If oscillations are causally implicated in these attentional functions, then frequency-tuned TMS over parietal cortex should bias perception depending on the stimulation frequency that is applied. In accordance with the EEG/MEG work outlined earlier, TMS at alpha frequency suppressed visual perception in the visual field contralateral to TMS, but enhanced perception ipsilaterally. Specifically, TMS biased perception toward/away from specific spatial locations (Romei et al., 2010). By contrast, TMS at beta and theta frequencies had no such effect (Romei et al., 2010). However, TMS at beta and theta frequencies differentially affected the processing of local and global stimulus information (Romei et al., 2011). Similarly, parietal TMS at alpha frequency (but not at control frequencies) suppressed visual information contralateral to TMS in working memory tasks (Sauseng et al., 2009). The behavioral consequences of TMS therefore depend on the specific oscillatory activity with which the stimulation is synchronized.

Finally, while we here review evidence for entrainment of brain oscillations by rhythmic TMS, and its impact on behavior, parallel research using transcranial direct or alternating current stimulation in oscillatory patterns (e.g., tACS) has drawn very similar conclusions. These studies have also confirmed that brain oscillations can be entrained in the human brain by noninvasive transcranial rhythmic brain stimulation (Marshall et al., 2006; Pogosyan et al., 2009; Ozen et al., 2010; Zaehle et al., 2011) and, furthermore, that this can lead to specific behavioral changes that causally confirm prior EEG/MEG evidence on the functional role of these oscillations (Marshall et al., 2006, Pogosyan et al., 2009; for a recent review of studies on perceptual outcomes, see Thut et al., 2011a).

In summary, combined TMS/EEG evidence suggests that, when stimulation frequency is matched to the intrinsic oscillatory signature of a given cortical territory, this can alter both ongoing neural activity and specific aspects of perception and behavior in a predictable manner. This kind of work promises to further refine our hypotheses on TMS action and thereby allow TMS to be used in a more hypothesis-driven approach, offering potential benefits for both fundamental brain research and possibly also for clinical applications.

12.9 SUMMARY AND CONCLUSIONS

A key ongoing contribution of combined TMS/neuroimaging is to provide us with an ever-expanding wealth of data on the widespread and distributed nature of TMS-induced neural effects. As discussed, this distributed character of TMS effects had been suggested by the earliest studies, but it is only in the last 10 years, since the combination of TMS with PET, fMRI, and EEG became technically feasible, that these hints could be investigated experimentally. Combined TMS/neuroimaging studies have demonstrated that these distributed neural effects are not merely epiphenomenal, but that they can have functional impact, whether or not they issue in induced behavioral changes. The capacity to observe whole-brain activity in conjunction with TMS offers us a more complex picture of TMS actions. Just as for real lesions caused by brain damage, TMS-induced "virtual lesions" can alter the functioning of distributed neural networks and can provoke rapid compensatory adjustments in network functioning. In some cases, the absence of TMS-induced behavioral effects can provide the starting point for interesting questions, which could not have been investigated prior to the advent of combination methods. With combined TMS/neuroimaging, we can now characterize how the stimulated region, together with other brain regions, distal from but connected to the stimulated site, shows adjustments in activity and at which point in information processing such adjustments occur. In some cases, this may help to finesse current models about the causal role of a specific brain region in a given behavior. For example, with concurrent TMS/EEG, we can now, in principle, ask whether behavioral perturbations arise solely because the stimulated region is at its highest activity at the time of stimulation or, alternatively, whether it may instead have the strongest interactions with other crucial parts of a network at the critical time. Given the complexity of brain structure and functioning, and the ever-burgeoning means of noninvasively stimulating the brain, there is unlikely to emerge a single account of "how TMS works" or of "how TMS affects the brain." Rather, combined TMS/neuroimaging methods offer us an expanded set of tools and a rich vein of information with which to interrogate and reconceptualize how complex neural network interactions can give rise to behavior.

REFERENCES

Alexander, G.E., DeLong, M.R., and Strick, P.L. 1986. Parallel organization of functionally segregated circuits linking basal ganglia and cortex. *Annu Rev Neurosci* 9:357–381.

Amassian, V.E., Cracco, R.Q., Maccabee, P.J., Cracco, J.B., Rudell, A., and Eberle, L. 1989. Suppression of visual perception by magnetic coil stimulation of human occipital cortex. *Electroencephalogr Clin Neurophysiol* 74:458–462.

Ashbridge, E., Walsh, V., and Cowey, A. 1997. Temporal aspects of visual search studied by transcranial magnetic stimulation. *Neuropsychologia* 35:1121–1131.

Barker, A.T., Jalinous, R., and Freeston, I.L. 1985. Non-invasive magnetic stimulation of human motor cortex. *Lancet* 1:1106–1107.

Baudewig, J., Nitsche, M.A., Paulus, W., and Frahm, J. 2001. Regional modulation of BOLD MRI responses to human sensorimotor activation by transcranial direct current stimulation. *Magn Reson Med* 45:196–201.

Bestmann, S., Baudewig, J., and Frahm, J. 2003. On the synchronization of transcranial magnetic stimulation and functional echo-planar imaging. *J Magn Reson Imaging* 17:309–316.

Bestmann, S., Baudewig, J., Siebner, H.R., Rothwell, J.C., and Frahm, J. 2004. Functional MRI of the immediate impact of transcranial magnetic stimulation on cortical and subcortical motor circuits. *Eur J Neurosci* 19:1950–1962.

Bestmann, S., Baudewig, J., Siebner, H.R., Rothwell, J.C., and Frahm, J. 2005. BOLD MRI responses to repetitive TMS over human dorsal premotor cortex. *Neuroimage* 28:22–29.

Bestmann, S., Oliviero, A., Voss, M. et al. 2006. Cortical correlates of TMS-induced phantom hand movements revealed with concurrent TMS-fMRI. *Neuropsychologia* 44:2959–2971.

Bestmann, S., Ruff, C.C., Blakemore, C., Driver, J., and Thilo, K.V. 2007. Spatial attention changes excitability of human visual cortex to direct stimulation. *Curr Biol* 17:134–139.

Bestmann, S., Ruff, C.C., Blankenburg, F., Weiskopf, N., Driver, J., and Rothwell, J.C. 2008a. Mapping causal interregional influences with concurrent TMS-fMRI. *Exp Brain Res* 191:383–402.

Bestmann, S., Ruff, C.C., Driver, J., and Blankenburg, F. 2008b. Concurrent TMS and functional magnetic resonance imaging: Methods and current advances. In: *Oxford Handbook of Transcranial Stimulation* (Wasserman, E.A., Epstein, C.M., Ziemann, U., Walsh, V., Paus, T., and Lisanby, S.H. eds.). Oxford, U.K.: Oxford University Press.

Bestmann, S., Swayne, O., Blankenburg, F., et al. 2008c. Dorsal premotor cortex exerts state-dependent causal influences on activity in contralateral primary motor and dorsal premotor cortex. *Cereb Cortex* 18(6):1281–1291.

Bestmann, S., Swayne, O., Blankenburg, F. et al. 2010. The role of contralesional dorsal premotor cortex after stroke as studied with concurrent TMS-fMRI. *J Neurosci* 30(36):11926–11937.

Blankenburg, F., Ruff, C.C., Bestmann, S. et al. 2008. Interhemispheric effect of parietal TMS on somatosensory response confirmed directly with concurrent TMS-fMRI. *J Neurosci* 28:13202–13208.

Blankenburg, F., Ruff, C.C., Bestmann, S. et al. 2010. Studying the role of human parietal cortex in visuospatial attention with concurrent TMS-fMRI. *Cereb Cortex* 20:2702–2711.

Bohning, D.E., Denslow, S., Bohning, P.A., Lomarev, M.P., and George, M.S. 2003a. Interleaving fMRI and rTMS. *Suppl Clin Neurophysiol* 56:42–54.

Bohning, D.E., Pecheny, A.P., Epstein, C.M. et al. 1997. Mapping transcranial magnetic stimulation (TMS) fields in vivo with MRI. *Neuroreport* 8:2535–2538.

Bohning, D.E., Shastri, A., Lomarev, M.P., Lorberbaum, J.P., Nahas, Z., and George, M.S. 2003b. BOLD-fMRI response vs. transcranial magnetic stimulation (TMS) pulse-train length: Testing for linearity. *J Magn Reson Imaging* 17:279–290.

Bohning, D.E., Shastri, A., McConnell, K.A. et al. 1999. A combined TMS/fMRI study of intensity-dependent TMS over motor cortex. *Biol Psychiatry* 45:385–394.

Bohning, D.E., Shastri, A., Nahas, Z. et al. 1998. Echoplanar BOLD fMRI of brain activation induced by concurrent transcranial magnetic stimulation. *Invest Radiol* 33:336–340.

Bohning, D.E., Shastri, A., Wassermann, E.M. et al. 2000. BOLD-f MRI response to single-pulse transcranial magnetic stimulation (TMS). *J Magn Reson Imaging* 11:569–574.

Bonato, C., Zanette, G., Manganotti, P. et al. 1996. 'Direct' and 'crossed' modulation of human motor cortex excitability following exercise. *Neurosci Lett* 216:97–100.

Bonato, C., Zanette, G., Polo, A. et al. 1994. Cortical output modulation after rapid repetitive movements. *Ital J Neurol Sci* 15:489–494.

Bonnard, M., Spieser, L., Meziane, H.B., de Graaf, J.B., and Pailhous, J. 2009. Prior intention can locally tune inhibitory processes in the primary motor cortex: Direct evidence from combined TMS-EEG. *Eur J Neurosci* 30:913–923.

Boorman, E.D., O'Shea, J., Sebastian, C., Rushworth, M.F., and Johansen-Berg, H. 2007. Individual differences in white-matter microstructure reflect variation in functional connectivity during choice. *Curr Biol* 17(16):1426–1431.

Boy, F., Evans, C.J., Edden, R.A., Singh, K.D., Husain, M., and Sumner, P. 2010. Individual differences in subconscious motor control predicted by GABA concentration in SMA. *Curr Biol* 20(19):1779–1785.

Bungert, A., Chambers, C.D., Phillips, M., and Evans, C.J. 2012. Reducing image artefacts in concurrent TMS/fMRI by passive shimming. *Neuroimage* 59(3):2167–2174.

Buzsáki, G. and Draguhn, A. 2004. Neuronal oscillations in cortical networks. *Science* 304:1926–1929.

Cheeran, B., Talelli, P., Mori, F. et al. 2008. A common polymorphism in the brain-derived neurotrophic factor gene (BDNF) modulates human cortical plasticity and the response to rTMS. *J Physiol* 586(Pt 23):5717–5725.

Chouinard, P.A., Van Der Werf, Y.D., Leonard, G., and Paus, T. 2003. Modulating neural networks with transcranial magnetic stimulation applied over the dorsal premotor and primary motor cortices. *J Neurophysiol* 90:1071–1083.

Cohen, L.G., Celnik, P., Pascual-Leone, A. et al. 1997. Functional relevance of cross-modal plasticity in blind humans. *Nature* 389(6647):180–183.

Cowey, A. and Walsh, V. 2001. Tickling the brain: Studying visual sensation, perception and cognition by transcranial magnetic stimulation. *Prog Brain Res* 134:411–425.

Cracco, R.Q. 1987. Evaluation of conduction in central motor pathways: techniques, pathophysiology, and clinical interpretation. *Neurosurgery* 20:199–203.

Cracco, J.B. and Cracco, R.Q. 1999. The physiological basis of transcranial magnetic stimulation. *Electroencephalogr Clin Neurophysiol Suppl* 49:217–221.

Denslow, S., Lomarev, M., George, M.S., and Bohning, D.E. 2005. Cortical and subcortical brain effects of transcranial magnetic stimulation (TMS)-induced movement: An interleaved TMS/functional magnetic resonance imaging study. *Biol Psychiatry* 57:752–760.

van Dijk, H., Schoffelen, J.M., Oostenveld, R., and Jensen, O. 2008. Prestimulus oscillatory activity in the alpha band predicts visual discrimination ability. *J Neurosci* 28:1816–1823.

Ergenoglu, T., Demiralp, T., Bayraktaroglu, Z., Ergen, M., Beydagi, H., and Uresin, Y. 2004. Alpha rhythm of the EEG modulates visual detection performance in humans. *Brain Res* 20:376–383.

Ferbert, A., Priori, A., Rothwell, J.C., Day, B.L., Colebatch, J.G., and Marsden, C.D. 1992. Interhemispheric inhibition of the human motor cortex. *J Physiol.* 453:525–546.

Flöel, A., Meinzer, M., Kirstein, R. et al. 2011. Short-term anomia training and electrical brain stimulation. *Stroke* 42(7):2065–2067.

Fox, P., Ingham, R., George, M.S. et al. 1997. Imaging human intra-cerebral connectivity by PET during TMS. *Neuroreport* 8:2787–2791.

Foxe, J.J. and Snyder, A.C. 2011. The role of alpha-band brain oscillations as a sensory suppression mechanism during selective attention. *Front Psychol* 2:154.

Fregni, F., Boggio, P.S., Valle, A.C. et al. 2006. A sham-controlled trial of a 5-day course of repetitive transcranial magnetic stimulation of the unaffected hemisphere in stroke patients. *Stroke* 37:2115–2122.

Fregni, F. and Pascual-Leone A. 2007. Technology insight: Noninvasive brain stimulation in neurology-perspectives on the therapeutic potential of rTMS and tDCS. *Nat Clin Pract Neurol.* 3(7):383–393.

Fritsch, B., Reis, J., Martinowich, K. et al. 2010. Direct current stimulation promotes BDNF-dependent synaptic plasticity: Potential implications for motor learning. *Neuron* 66(2):198–204.

Fuggetta, G., Fiaschi, A., and Manganotti, P. 2005. Modulation of cortical oscillatory activities induced by varying single-pulse transcranial magnetic stimulation intensity over the left primary motor area: A combined EEG and TMS study. *Neuroimage* 27:896–908.

Fuggetta, G., Pavone, E.F., Walsh, V., Kiss, M., and Eimer, M. 2006. Cortico-cortical interactions in spatial attention: A combined ERP/TMS study. *J Neurophysiol* 95:3277–3280.

Fujiwara, T. and Rothwell, J.C. 2004. The after effects of motor cortex rTMS depend on the state of contraction when rTMS is applied. *Clin Neurophysiol* 115:1514–1518.

Glass, L. 2001. Synchronization and rhythmic processes in physiology. *Nature* 410:277–284.

Groppa, S., Schlaak, B.H., Münchau, A. et al. 2012. The human dorsal premotor cortex facilitates the excitability of ipsilateral primary motor cortex via a short latency cortico-cortical route. *Hum Brain Mapp* 33(2):419–430.

Hamidi, M., Slagter, H.A., Tononi, G., and Postle, B.R. 2010. Brain responses evoked by high-frequency repetitive TMS: An ERP study. *Brain Stimul* 3:2–17.

Hanslmayr, S., Aslan, A., Staudigl, T., Klimesch, W., Herrmann, C.S., and Bäuml, K.H. 2007. Prestimulus oscillations predict visual perception performance between and within subjects. *Neuroimage* 37:1465–1473.

Hufnagel, A., Jaeger, M., and Elger, C.E. 1990. Transcranial magnetic stimulation: Specific and non-specific facilitation of magnetic motor evoked potentials. *J Neurol* 237:416–419.

Hummel, F., Celnik, P., Giraux, P. et al. 2005. Effects of non-invasive cortical stimulation on skilled motor function in chronic stroke. *Brain* 128(Pt 3):490–499.

Ilmoniemi, R.J. and Karhu, J. 2008. TMS and electroencephalography: Methods and current advances. In: *Oxford Handbook of Transcranial Stimulation* (Wasserman, E.M., Epstein, C.M., Ziemann, U., Walsh, V., Paus T., and Lisanby S.H., eds.), pp. 593–608. Oxford, U.K.: Oxford University Press.

Ilmoniemi, R.J. and Kicić, D. 2010. Methodology for combined TMS and EEG. *Brain Topogr* 22:233–248.

Ilmoniemi, R.J., Virtanen, J., Ruohonen, J. et al. 1997. Neuronal responses to magnetic stimulation reveal cortical reactivity and connectivity. *Neuroreport* 8:3537–3540.

Ives, J.R., Rotenberg, A., Poma, R., Thut, G., and Pascual-Leone, A. 2006. Electroencephalographic recording during transcranial magnetic stimulation in humans and animals. *Clin Neurophysiol* 117:1870–1875.

Iwahashi, M., Arimatsu, T., Ueno, S., and Iramina, K. 2008. Differences in evoked EEG by transcranial magnetic stimulation at various stimulus points on the head. *Conf Proc IEEE Eng Med Biol Soc* 2008:2570–2573.

Johansen-Berg, H., Rushworth, M.F., Bogdanovic, M.D., Kischka, U., Wimalaratna, S., and Matthews, P.M. 2002. The role of ipsilateral premotor cortex in hand movement after stroke. *Proc Natl Acad Sci USA* 99(22):14518–14523.

Johnson, J.S., Hamidi, M., and Postle, B.R. 2010. Using EEG to explore how rTMS produces its effects on behavior. *Brain Topogr* 22:281–293.

Julkunen, P., Paakkonen, A., Hukkanen, T. et al. 2008. Efficient reduction of stimulus artefact in TMS-EEG by epithelial short-circuiting by mini-punctures. *Clin Neurophysiol* 119:475–481.

Kahkonen, S., Komssi, S., Wilenius, J., and Ilmoniemi, R.J. 2005. Prefrontal TMS produces smaller EEG responses than motor-cortex TMS: Implications for rTMS treatment in depression. *Psychopharmacology (Berl)* 181:16–20.

Kelly, S.P., Lalor, E.C., Reilly, R.B., and Foxe, J.J. 2006. Increases in alpha oscillatory power reflect an active retinotopic mechanism for distracter suppression during sustained visuospatial attention. *J Neurophysiol* 95:3844–3851.

Khedr, E.M., Ahmed, M.A., Fathy, N., and Rothwell, J. 2005. Therapeutic trial of repetitive transcranial magnetic stimulation after acute ischemic stroke. *Neurology* 65:466–468.

Kimbrell, T.A., Dunn, R.T., George, M.S. et al. 2002. Left prefrontal-repetitive transcranial magnetic stimulation (rTMS) and regional cerebral glucose metabolism in normal volunteers. *Psychiatry Res* 115:101–113.

Kleim, J.A., Chan, S., Pringle, E. et al. 2006. BDNF val66met polymorphism is associated with modified experience-dependent plasticity in human motor cortex. *Nat Neurosci* 9(6):735–737.

Klöppel, S., Bäumer, T., Kroeger, J. et al. 2008. The cortical motor threshold reflects microstructural properties of cerebral white matter. *Neuroimage* 40(4):1782–1791.

Ko, J.H., Monchi, O., Ptito, A., Bloomfield, P., Houle, S., and Strafella, A.P. 2008. Theta burst stimulation-induced inhibition of dorsolateral prefrontal cortex reveals hemispheric asymmetry in striatal dopamine release during a set-shifting task: A TMS-[(11)C]raclopride PET study. *Eur J Neurosci* 28:2147–2155.

Ko, J.H. and Strafella, A.P. 2012. Dopaminergic neuro transmission in the human brain: New lessons from perturbation and imaging. *Neuroscientist* 18(2):149–168.

Koch, G., Oliveri, M., Cheeran, B. et al. 2008. Hyperexcitability of parietal-motor functional connections in the intact left-hemisphere of patients with neglect. *Brain* 131 (Pt 12):3147–3155.

Komssi, S., Aronen, H.J., Huttunen, J. et al. 2002. Ipsi- and contralateral EEG reactions to transcranial magnetic stimulation. *Clin Neurophysiol* 113:175–184.

Komssi, S. and Kahkonen, S. (2006) The novelty value of the combined use of electroencephalography and transcranial magnetic stimulation for neuroscience research. *Brain Res Rev* 52:183–192.

Lindenberg, R., Renga, V., Zhu, L.L., Nair, D., and Schlaug, G. 2010. Bihemispheric brain stimulation facilitates motor recovery in chronic stroke patients. *Neurology* 75(24):2176–2184.

Litvak, V., Komssi, S., Scherg, M. et al. 2007. Artifact correction and source analysis of early electroencephalographic responses evoked by transcranial magnetic stimulation over primary motor cortex. *Neuroimage* 37:56–70.

Lorincz, M.L., Kékesi, K.A., Juhász, G., Crunelli, V., and Hughes, S.W. 2009. Temporal framing of thalamic relay-mode firing by phasic inhibition during the alpha rhythm. *Neuron* 63:683–696.

Lotze, M., Markert, J., Sauseng, P., Hoppe, J., Plewnia, C., and Gerloff, C. 2006. The role of multiple contralesional motor areas for complex hand movements after internal capsular lesion. *J Neurosci* 26:6096–6102.

Maeda, F., Gangitano, M., Thall, M., and Pascual-Leone, A. 2002. Inter- and intra-individual variability of paired-pulse curves with transcranial magnetic stimulation (TMS). *Clin Neurophysiol* 113:376–382.

Marshall, L., Helgadottir, H., Molle, M., and Born, J. 2006. Boosting slow oscillations during sleep potentiates memory. *Nature* 444:610–613.

Massimini, M., Ferrarelli, F., Esser, S.K. et al. 2007. Triggering sleep slow waves by transcranial magnetic stimulation. *Proc Natl Acad Sci USA* 104:8496–8501.

Massimini, M., Ferrarelli, F., Huber, R., Esser, S.K., Singh, H., and Tononi, G. 2005. Breakdown of cortical effective connectivity during sleep. *Science* 309:2228–2232.

Mazzocchio, R., Rothwell, J.C., Day, B.L., and Thompson, P.D. 1994. Effect of tonic voluntary activity on the excitability of human motor cortex. *J Physiol* 474:261–267.

McHughen, S.A., Rodriguez, P.F., Kleim, J.A. et al. 2010. BDNF val66met polymorphism influences motor system function in the human brain. *Cereb Cortex* 20(5):1254–1262.

Meyer, B.U., Roricht, S., Grafin, v.E., Kruggel, F., and Weindl, A. 1995. Inhibitory and excitatory interhemispheric transfers between motor cortical areas in normal humans and patients with abnormalities of the corpus callosum. *Brain* 118:429–440.

Mills, K.R., Murray, N.M., and Hess, C.W. 1987. Magnetic and electrical transcranial brain stimulation: Physiological mechanisms and clinical applications. *Neurosurgery* 20:164–168.

Miniussi, C., Ruzzoli, M., and Walsh, V. 2010. The mechanism of transcranial magnetic stimulation in cognition. *Cortex* 46:128–130.

Moisa, M., Pohmann, R., Ewald, L., and Thielscher, A. 2009. New coil positioning method for interleaved transcranial magnetic stimulation (TMS)/functional MRI (fMRI) and its validation in a motor cortex study. *J Magn Reson Imaging* 29:189–197.

Moisa, M., Pohmann, R., Uludag, K., and Thielscher, A. 2010. Interleaved TMS/CASL: Comparison of different rTMS protocols. *Neuroimage* 49:612–620.

Morbidi, F., Garulli, A., Prattichizzo, D., Rizzo, C., Manganotti, P., and Rossi, S. 2007. Off-line removal of TMS-induced artifacts on human electroencephalography by Kalman filter. *J Neurosci Methods* 162:293–302.

Morishima, Y., Akaishi, R., Yamada, Y., Okuda, J., Toma, K., and Sakai, K. 2009. Task-specific signal transmission from prefrontal cortex in visual selective attention. *Nat Neurosci* 12:85–91.

Muthukumaraswamy, S.D., Edden, R.A., Jones, D.K., Swettenham, J.B., and Singh, K.D. 2009. Resting GABA concentration predicts peak gamma frequency and fMRI amplitude in response to visual stimulation in humans. *Proc Natl Acad Sci USA* 9;106(20):8356–8361.

Mumford, D. 1992. On the computational architecture of the neocortex. II. The role of cortico-cortical loops. *Biol Cybern* 66:241–251.

Nahas, Z., Lomarev, M., Roberts, D.R. et al. 2001. Unilateral left prefrontal transcranial magnetic stimulation (TMS) produces intensity-dependent bilateral effects as measured by interleaved BOLD fMRI. *Biol Psychiatry* 50:712–720.

O'Shea, J., Johansen-Berg, H., Trief, D., Göbel, S., and Rushworth, M.F. 2007. Functionally specific reorganization in human premotor cortex. *Neuron* 54(3):479–490.

Opitz, A., Windhoff, M., Heidemann, R.M., Turner, R., and Thielscher, A. 2011. How the brain tissue shapes the electric field induced by transcranial magnetic stimulation. *Neuroimage* 58(3):849–859.

Ozen, S., Sirota, A., Belluscio, M.A. et al. 2010. Transcranial electric stimulation entrains cortical neuronal populations in rats. *J Neurosci* 30:11476–11485.

Pascual-Leone, A., Grafman, J., and Hallett, M. 1994. Modulation of cortical motor output maps during development of implicit and explicit knowledge. *Science* 263:1287–1289.

Pascual-Leone, A., Nguyet, D., Cohen, L.G., Brasil-Neto, J.P., Cammarota, A., and Hallett, M. 1995. Modulation of muscle responses evoked by transcranial magnetic stimulation during the acquisition of new fine motor skills. *J Neurophysiol* 74:1037–1045.

Pascual-Leone, A. and Walsh, V. 2001. Fast backprojections from the motion to the primary visual area necessary for visual awareness. *Science* 292:510–512.

Pascual-Leone, A., Walsh, V., and Rothwell, J. 2000. Transcranial magnetic stimulation in cognitive neuroscience—Virtual lesion, chronometry, and functional connectivity. *Curr Opin Neurobiol* 10:232–237.

Paus, T. 2005. Inferring causality in brain images: A perturbation approach. *Philos Trans R Soc Lond B Biol Sci* 360:1109–1114.

Paus, T., Castro-Alamancos, M.A., and Petrides, M. 2001b. Cortico-cortical connectivity of the human mid-dorsolateral frontal cortex and its modulation by repetitive transcranial magnetic stimulation. *Eur J Neurosci* 14:1405–1411.

Paus, T., Jech, R., Thompson, C.J., Comeau, R., Peters, T., and Evans, A.C. 1997. Transcranial magnetic stimulation during positron emission tomography: A new method for studying connectivity of the human cerebral cortex. *J Neurosci* 17:3178–3184.

Paus, T., Jech, R., Thompson, C.J., Comeau, R., Peters, T., and Evans, A.C. 1998. Dose-dependent reduction of cerebral blood flow during rapid-rate transcranial magnetic stimulation of the human sensorimotor cortex. *J Neurophysiol* 79:1102–1107.

Paus, T., Sipila, P.K., and Strafella, A.P. 2001a. Synchronization of neuronal activity in the human primary motor cortex by transcranial magnetic stimulation: An EEG study. *J Neurophysiol* 86:1983–1990.

Paus, T. and Wolforth, M. 1998. Transcranial magnetic stimulation during PET: Reaching and verifying the target site. *Hum Brain Mapp* 6:399–402.

Pikovsky, A., Rosenblum, M., and Kurths, J. 2003. *Synchronization: A Universal Concept in Nonlinear Sciences*. 1st edn. Cambridge, U.K.: Cambridge University Press.

Pogosyan, A., Gaynor, L.D., Eusebio, A., and Brown, P. 2009. Boosting cortical activity at Beta-band frequencies slows movement in humans. *Curr Biol* 19:1637–1641.

Reichenbach, A., Whittingstall, K., and Thielscher, A. 2011. Effects of transcranial magnetic stimulation on visual evoked potentials in a visual suppression task. *Neuroimage* 54:1375–1384.

Reithler, J., Peters, J.C., and Sack, A.T. 2011. Multimodal transcranial magnetic stimulation: Using concurrent neuroimaging to reveal the neural network dynamics of noninvasive brain stimulation. *Prog Neurobiol* 94:149–165.

Ridding, M.C., Taylor, J.L., and Rothwell, J.C. 1995. The effect of voluntary contraction on cortico-cortical inhibition in human motor cortex. *J Physiol* 487:541–548.

Rihs, T.A., Michel, C.M., and Thut, G. 2007. Mechanisms of selective inhibition in visual spatial attention are indexed by alpha-band EEG synchronization. *Eur J Neurosci* 25:603–610.

Rihs, T.A., Michel, C.M., and Thut, G. 2009. A bias for posterior alpha-band power suppression versus enhancement during shifting versus maintenance of spatial attention. *Neuroimage* 44:190–199.

Romei, V., Brodbeck, V., Michel, C., Amedi, A., Pascual-Leone, A., and Thut, G. 2008. Spontaneous fluctuations in posterior alpha-band EEG activity reflect variability in excitability of human visual areas. *Cereb Cortex* 18:2010–2018.

Romei, V., Driver, J., Schyns, P.G., and Thut, G. 2011. Rhythmic TMS over parietal cortex links distinct brain frequencies to global versus local visual processing. *Curr Biol* 21:334–337.

Romei, V., Gross, J., and Thut, G. 2010. On the role of prestimulus alpha rhythms over occipito-parietal areas in visual input regulation: Correlation or causation? *J Neurosci* 30:8692–8697.

Rosanova, M., Casali, A., Bellina, V., Resta, F., Mariotti, M., and Massimini, M. 2009. Natural frequencies of human corticothalamic circuits. *J Neurosci* 29:7679–7685.

Rothwell, J.C., Day, B.L., Thompson, P.D., Dick, J.P., and Marsden, C.D. 1987a. Some experiences of techniques for stimulation of the human cerebral motor cortex through the scalp. *Neurosurgery* 20:156–163.

Rothwell, J.C., Thompson, P.D., Day, B.L. et al. 1987b. Motor cortex stimulation in intact man. 1. General characteristics of EMG responses in different muscles. *Brain* 110:1173–1190.

Ruff, C.C., Bestmann, S., Blankenburg, F. et al. 2008. Distinct causal influences of parietal versus frontal areas on human visual cortex: Evidence from concurrent TMS fMRI. *Cereb Cortex* 18:817–827.

Ruff, C.C., Blankenburg, F., Bjoertomt, O. et al. 2006. Concurrent TMS-fMRI and psychophysics reveal frontal influences on human retinotopic visual cortex. *Curr Biol* 16:1479–1488.

Ruff, C.C., Blankenburg, F., Bjoertomt, O., Bestmann, S., Weiskopf, N., and Driver, J. 2009a. Hemispheric differences in frontal and parietal influences on the human occipital cortex: Direct confirmation with concurrent TMS-fMRI. *J Cogn Neurosci* 21:1146–1161.

Ruff, C.C., Driver, J., and Bestmann, S. 2009b. Combining TMS and fMRI: From 'virtual lesions' to functional-network accounts of cognition. *Cortex* 45:1043–1049.

Sack, A.T., Kohler, A., Bestmann, S. et al. 2007. Imaging the brain activity changes underlying impaired visuospatial judgments: Simultaneous FMRI, TMS, and behavioral studies. *Cereb Cortex* 17:2841–2852.

Sauseng, P., Klimesch, W., Heise, K.F. et al. 2009. Brain oscillatory substrates of visual short-term memory capacity. *Curr Biol* 19:1846–1852.

Sauseng, P., Klimesch, W., Stadler, W. et al. 2005. A shift of visual spatial attention is selectively associated with human EEG alpha activity. *Eur J Neurosci* 22:2917–2926.

Schnitzler, A. and Gross, J. 2005. Normal and pathological oscillatory communication in the brain. *Nat Rev Neurosci* 6:285–296.

Siebner, H.R., Peller, M., Willoch, F. et al. 1999. Imaging functional activation of the auditory cortex during focal repetitive transcranial magnetic stimulation of the primary motor cortex in normal subjects. *Neurosci Lett* 270:37–40.

Siebner, H.R., Peller, M., Willoch, F. et al. 2000. Lasting cortical activation after repetitive TMS of the motor cortex: A glucose metabolic study. *Neurology* 54:956–963.

Siebner, H.R., Willoch, F., Peller, M. et al. 1998. Imaging brain activation induced by long trains of repetitive transcranial magnetic stimulation. *Neuroreport* 9:943–948.

Siegel, M., Donner, T.H., Oostenveld, R., Fries, P., and Engel, A.K. 2009. Neuronal synchronization along the dorsal visual pathway reflects the focus of spatial attention. *Neuron* 60:709–719.

Silvanto, J., Lavie, N., and Walsh, V. 2005. Double dissociation of V1 and V5/MT activity in visual awareness. *Cereb Cortex* 15:1736–1741.

Silvanto, J., Muggleton, N., and Walsh, V. 2008. State-dependency in brain stimulation studies of perception and cognition. *Trends Cogn Sci* 12:447–454.

Smith, M.L., Gosselin, F., and Schyns, P.G. 2006. Perceptual moments of conscious visual experience inferred from oscillatory brain activity. *Proc Natl Acad Sci USA* 103:5626–5631.

Speer, A.M., Willis, M.W., Herscovitch, P. et al. 2003. Intensity-dependent regional cerebral blood flow during 1-Hz repetitive transcranial magnetic stimulation (rTMS) in healthy volunteers studied with H215O positron emission tomography: II. Effects of prefrontal cortex rTMS. *Biol Psychiatry* 54:826–832.

Stagg, C.J., Bachtiar, V., and Johansen-Berg, H. 2011a. The role of GABA in human motor learning. *Curr Biol* 21(6):480–484.

Stagg, C.J., Bachtiar, V., O'Shea, J. et al. 2012. Cortical activation changes underlying stimulation-induced behavioural gains in chronic stroke. *Brain* 135(pt1):276–284.

Stagg, C.J., Bestmann, S., Constantinescu, A.O. et al. 2011b. Relationship between physiological measures of excitability and levels of glutamate and GABA in the human motor cortex. *J Physiol.* 589(Pt 23):5845–5855.

Stagg, C.J., Wylezinska, M., Matthews, P.M. et al. 2009. Neurochemical effects of theta burst stimulation as assessed by magnetic resonance spectroscopy. *J Neurophysiol* 101(6):2872–2877.

Stinear, C.M., Barber, P.A., Smale, P.R., Coxon, J.P., Fleming, M.K., and Byblow, W.D. 2007. Functional potential in chronic stroke patients depends on corticospinal tract integrity. *Brain* 130(Pt 1):170–180.

Stokes, M.G., Chambers, C.D., Gould, I.C. et al. 2007. Distance-adjusted motor threshold for transcranial magnetic stimulation. *Clin Neurophysiol* 118(7):1617–1625.

Strafella, A.P., Ko, J.H., Grant, J., Fraraccio, M., and Monchi, O. 2005. Corticostriatal functional interactions in Parkinson's disease: arTMS/[11C] raclopride PET study. *Eur J Neurosci* 22:2946–2952.

Strafella, A.P. and Paus, T. 2001. Cerebral blood-flow changes induced by paired-pulse transcranial magnetic stimulation of the primary motor cortex. *J Neurophysiol* 85:2624–2629.

Strafella, A.P., Paus, T., Barrett, J., and Dagher, A. 2001. Repetitive transcranial magnetic stimulation of the human prefrontal cortex induces dopamine release in the caudate nucleus. *J Neurosci* 21:RC157.

Strafella, A.P., Paus, T., Fraraccio, M., and Dagher, A. 2003. Striatal dopamine release induced by repetitive transcranial magnetic stimulation of the human motor cortex. *Brain* 126:2609–2615.

Sumner, P., Edden, R.A., Bompas, A., Evans, C.J., and Singh, K.D. 2010. More GABA, less distraction: A neurochemical predictor of motor decision speed. *Nat Neurosci* 13(7):825–827.

Swayne, O.B., Rothwell, J.C., Ward, N.S., and Greenwood, R.J. 2008. Stages of motor output reorganization after hemispheric stroke suggested by longitudinal studies of cortical physiology. *Cereb Cortex* 18(8):1909–1922.

Tachibana, Y., Nambu, A., Hatanaka, N., Miyachi, S., and Takada, M. 2004. Input-output organization of the rostral part of the dorsal premotor cortex, with special reference to its corticostriatal projection. *Neurosci Res* 48:45–57.

Takada, M., Tokuno, H., Nambu, A., and Inase, M. 1998a. Corticostriatal projections from the somatic motor areas of the frontal cortex in the macaque monkey: Segregation versus overlap of input zones from the primary motor cortex, the supplementary motor area, and the premotor cortex. *Exp Brain Res* 120:114–128.

Takada, M., Tokuno, H., Nambu, A., and Inase, M. 1998b. Corticostriatal input zones from the supplementary motor area overlap those from the contra – rather than ipsilateral primary motor cortex. *Brain Res* 791:335–340.

Taylor, P.C., Nobre, A.C., and Rushworth, M.F. 2007a. FEFTMS affects visual cortical activity. *Cereb Cortex* 17:391–399.

Taylor, P.C., Nobre, A.C., and Rushworth, M.F. 2007b. Subsecond changes in top down control exerted by human medial frontal cortex during conflict and action selection: A combined transcranial magnetic stimulation electroencephalography study. *J Neurosci* 27:11343–11353.

Taylor, P.C., Walsh, V., and Eimer, M. 2008. Combining TMS and EEG to study cognitive function and cortico-cortico interactions. *Behav Brain Res* 191:141–147.

Taylor, P.C., Walsh, V., and Eimer, M. 2010. The neural signature of phosphene perception. *Hum Brain Mapp* 31:1408–1417.

Tegenthoff, M., Ragert, P., Pleger, B. et al. 2005. Improvement of tactile discrimination performance and enlargement of cortical somatosensory maps after 5 Hz rTMS. *PLoS Biol* 3(11):e362.

Thielscher, A., Opitz, A., and Windhoff, M. 2011. Impact of the gyral geometry on the electric field induced by transcranial magnetic stimulation. *Neuroimage* 54(1):234–243.

Thut, G., Ives, J.R., Kampmann, F., Pastor, M.A., and Pascual-Leone, A. 2005. A new device and protocol for combining TMS and online recordings of EEG and evoked potentials. *J Neurosci Methods* 141:207–217.

Thut, G. and Miniussi, C. 2009. New insights into rhythmic brain activity from TMS-EEG studies. *Trends Cogn Sci* 13:182–189.

Thut, G., Nietzel, A., Brandt, S.A., and Pascual-Leone, A. 2006. Alpha-band electroencephalographic activity over occipital cortex indexes visuospatial attention bias and predicts visual target detection. *J Neurosci* 26:9494–9502.

Thut, G. and Pascual-Leone, A. 2010. A review of combined TMS-EEG studies to characterize lasting effects of repetitive TMS and assess their usefulness in cognitive and clinical neuroscience. *Brain Topogr* 22:219–232.

Thut, G., Schyns, P.G., and Gross, J. 2011a. Entrainment of perceptually relevant brain oscillations by non-invasive rhythmic stimulation of the human brain. *Front Psychol* 2:170.

Thut, G., Veniero, D., Romei, V., Miniussi, C., Schyns, P., and Gross, J. 2011b. Rhythmic TMS causes local entrainment of natural oscillatory signatures. *Curr Biol* 21:1176–1185.

Van Der Werf, Y.D., and Paus, T. 2006. The neural response to transcranial magnetic stimulation of the human motor cortex. I. Intracortical and cortico-cortical contributions. *Exp Brain Res* 175:231–245.

Veniero, D., Bortoletto, M., and Miniussi, C. 2009. TMS-EEG co-registration: On TMS-induced artifact. *Clin Neurophysiol* 120:1392–1399.

Virtanen, J., Ruohonen, J., Naatanen, R., and Ilmoniemi, R.J. 1999. Instrumentation for the measurement of electric brain responses to transcranial magnetic stimulation. *Med Biol Eng Comput* 37:322–326.

Wahl, M., Lauterbach-Soon, B., Hattingen, E. et al. 2007. Human motor corpus callosum: Topography, somatotopy, and link between microstructure and function. *J Neurosci* 27(45):12132–12138.

Worden, M.S., Foxe, J.J., Wang, N., and Simpson, G.V. 2000. Anticipatory biasing of visuospatial attention indexed by retinotopically specific alpha-band electroencephalography increases over occipital cortex. *J Neurosci* 20:RC63.

Weiskopf, N., Josephs, O., Ruff, C.C. et al. 2009. Image artifacts in concurrent transcranial magnetic stimulation (TMS) and fMRI caused by leakage currents: Modeling and compensation. *J Magn Reson Imaging* 29:1211–1217.

Wig, G.S., Grafton, S.T., Demos, K.E., and Kelley, W.M. 2005. Reductions in neural activity underlie behavioral components of repetition priming. *Nat Neurosci* 8(9):1228–1233.

Wilson, S.A., Thickbroom, G.W., and Mastaglia, F.L. 1995. Comparison of the magnetically mapped corticomotor representation of a muscle at rest and during low-level voluntary contraction. *Electroencephalogr Clin Neurophysiol* 97:246–250.

Yamagishi, N., Callan, D.E., Anderson, S.J., and Kawato, M. 2008. Attentional changes in pre-stimulus oscillatory activity within early visual cortex are predictive of human visual performance. *Brain Res* 1197:115–122.

Yeterian, E.H. and Pandya, D.N. 1991. Prefrontostriatal connections in relation to cortical architectonic organization in rhesus monkeys. *J Comp Neurol* 312:43–67.

Zaehle, T., Rach, S., and Herrmann, C.S. 2010. Transcranial alternating current stimulation enhances individual alpha activity in human EEG. *PLoS One* 5:e13766.

Part IV

Perception and Cognition

Part IV

Perception and Cognition

13 Transcranial Magnetic and Electric Stimulation in Perception and Cognition Research

Carlo Miniussi, Géza Gergely Ambrus,
Maria Concetta Pellicciari,
Vincent Walsh, and Andrea Antal

CONTENTS

13.1 INTRODUCTION

In recent years, we have witnessed the emergence of new techniques for studying the mechanisms that underlie perceptual and cognitive function in the human brain. An important contribution has come from the introduction of non-invasive brain stimulation (NIBS). The development of NIBS techniques to study perception and cognition constitutes a significant breakthrough in our understanding of the changes in the brain that may account for behavioral plasticity. NIBS approaches aim to induce changes in the activity of the brain, which can lead to alterations in the performance of a wide range of behavioral tasks (Sandrini et al. 2011). NIBS techniques that are used to modulate cortical activity include transcranial magnetic stimulation (TMS) (see Chapter 1) and transcranial electric stimulation (tES) (see Chapter 4). TMS and tES can transiently influence behavior by altering neuronal activity through different mechanisms, which may have facilitative or inhibitory effects. The relevance that NIBS has recently gained in the field of cognitive neuroscience is mainly derived from its ability to transiently probe the functions of the stimulated cortical area/network by changing behavior. These behavioral changes can sometimes be related to its effects on modulating cortical excitability, but the explanatory route is not always direct. This opportunity to probe and modulate functional brain mechanisms opens up new possibilities for basic cognitive neuroscience and in the field of cognitive rehabilitation in directing adaptive cognitive plasticity in pathological conditions.

13.2   TRANSCRANIAL MAGNETIC STIMULATION

TMS is a technique that can be used to investigate brain-behavior relationships and to explore the state of different regions of the brain (see Chapter 1 for an overview on technical aspects). Since its discovery (Barker et al. 1985), TMS has been used to investigate the state of cortical excitability, and the excitability of cortico-cortical and cortico-spinal pathways (Rothwell et al. 1987). Moreover, this technique has been used in cognitive neuroscience to investigate the role of a given brain region in a particular cognitive function (Robertson et al. 2003) and to examine the timing of its activity (Walsh and Cowey 1998, 2000). TMS is a tool that involves the induction of a brief electric current in the cortical surface under a coil, which causes a depolarization of a population of cortical neurons. The spatial and temporal resolution of this technique enables the investigation of two important questions in cognitive neuroscience: what information is processed in a given brain structure, and when this processing occurs (Sandrini et al. 2011, Walsh and Cowey 2000, Walsh et al. 1998). Accordingly, TMS has been used in many different cognitive domains to establish causality in brain-behavior relationships.

The results of functional neuroimaging (PET, fMRI) and high-resolution electroencephalography (EEG, MEG) experiments have revealed important correlative evidence for the involvement of a number of brain regions in perception and cognition. Neuroimaging and EEG techniques based on in vivo measurements of local changes in activity provide the best spatial and temporal resolution available. Functional neuroimaging is helpful in identifying brain regions involved in a given task; however, it cannot distinguish between the areas that play a critical role in the task (Price and Friston 1999).

Numerous lesion studies have reported a putative role in brain areas dedicated to the execution of cognitive tasks, and this approach is still very productive. Neuropsychology is, of course, a valuable bedrock of information about brain organization and function. Nevertheless, studies that attempt to infer normal function from a single patient with brain damage are susceptible to criticism because, among other issues, such cases provide evidence about the brain organization of a single individual and may not be generalizable to the population as a whole. A second, more important criticism leveled at these studies is that chronic brain lesions can often lead to plastic changes that affect the damaged region and may cause undamaged subsystems to be used in new ways. Therefore, the behavioral changes that are observed could reflect the functional reorganization of the intact systems rather than the loss of the damaged system. Thus, results from single cases, while extremely valuable, must always be interpreted with some caution, and it is important to obtain converging evidence using a variety of methods.

The use of TMS has the advantage of combining lesion and neuroimaging approaches, which allows for more information to be obtained in functionally relevant areas. Therefore, TMS is an excellent tool for directly investigating the functional participation of a brain area in an ongoing cognitive process (Walsh and Cowey 1998). This approach does not depend on the measurement of electrophysiological or hemodynamic responses to cognitive challenges; therefore, it complements traditional neuroimaging techniques by offering the unique opportunity to directly interfere with and investigate cortical area functions and the related neuronal circuitry during the execution of a task.

13.3 GENERAL ASPECTS IN THE USE OF TMS

We now know more about some of the basic properties of TMS effects. These properties depend on several technical parameters, the intensity (% of maximum stimulator output or % of motor threshold determined by the stimulation intensity necessary to produce a response of least 50 μV in amplitude in a relaxed muscle in at least 5 out of 10 consecutive stimulations [Rossi et al. 2009]), the number of stimulator discharges (frequency), the coil orientation, coil shape and dimension (focality, with a circular coil being less focal than a figure-eight-shaped coil), and the depth of stimulation as well as the possible interactions between these factors. The effects also depend on a number of variables related to the stimulated subject, including age, gender, eventual pharmacological treatments, and the activity state of the subject (Landi and Rossini 2010, Miniussi et al. 2010, Silvanto et al. 2008). This basic knowledge and selection of opportune parameters are essential when planning TMS studies.

A direct demonstration of the intimate mechanisms of TMS in the field of cognitive neuroscience is still lacking, although it is reasonable to believe that two possible mechanisms are at play. One mechanism implies the interruption of neural processes (a reduction in signal strength (Harris et al. 2008b). This might reflect an alteration in membrane permeability directly induced by TMS or an enhancement of inhibitory GABAergic activity, which is more likely the case (Mantovani et al. 2006, Moliadze et al. 2003). The other mechanism that may explain the TMS effect on cognition has been attributed to an introduction of random activity into the system (neural noise) (Harris et al. 2008b, Miniussi et al. 2010, Ruzzoli et al. 2010, 2011). Both of these mechanisms are consistent with the impact of TMS on depolarizing neurons.

In the field of cognitive neuroscience, TMS has been mainly used for the stimulation of cortical areas with the aim of interfering with cognitive processing at a precise time during task execution (Sack and Linden 2003). This type of application is called online TMS, and the functional impact is due to the ability to impinge on neuronal function temporarily, which modifying information processing that is dependent on the activity of the involved neurons (Ruzzoli et al. 2010, Silvanto et al. 2007).

In this respect, the specificity of TMS is remarkable in space and time. TMS combines good spatial and temporal resolution, and the rapid rise-time and short duration of the magnetic pulses offer millisecond precision. It is difficult to determine the exact spatial extent of TMS but some strong inferences can be made. For example, one can produce phosphenes in different regions of the visual field with an accuracy of 1°–2° of visual angle. In the motor cortex, one can selectively activate cortical representations of finger muscles without affecting other finger muscles or facial representations. The distinction between these areas is of the order or 1–2 cm across the cortex. This does not mean that the induced field only affects that 1–2 cm of cortex. Rather, it is a functional way of determining the physiological efficacy of the spread of the stimulation. When one induces a phosphene in a given part of the visual space the extent of that phosphene is a good measure of the amount of stimulation that is above the threshold for neuronal activation.

Temporal resolution is related to the duration of a single TMS pulse and its physiological effects over the area. The physical duration of the TMS pulse is very brief and is on the order of microseconds (less than 1 ms), whereas physiologically induced effects are more complex and last for approximately several hundred milliseconds (Moliadze et al. 2005). It is clear, however, that not all of these physiological effects are functionally effective. This is a temporal analogue of the spatial resolution issue. Although there are measurable effects of a TMS pulse that last for several hundred milliseconds, it is evident that single pulses can have effects with a resolution of tens or even one or two milli seconds (Amassian et al. 1989, Corthout et al. 2003, Pascual-Leone and Walsh 2001), which means that the excess, recordable activity is not functionally effective. In this respect, TMS can be applied at different time points during the execution of a perceptual or cognitive task to provide valuable information about when a brain region is involved in that task.

The initial application of TMS involved the delivery of single magnetic pulses. More recent technological advances allow for the delivery of rhythmic trains of magnetic pulses in sequences with a repetition rate as fast as 100 Hz, a technique that is called repetitive TMS (rTMS). rTMS can be used to map the flow of information across different brain regions during the execution of a task with a large temporal window. For example, Harris and Miniussi (2003) investigated whether neural activity in the parietal cortex is essential for successful mental rotation by observing the effects of disrupting this activity during the execution of a mental rotation task. rTMS was applied at 200–400, 400–600, or 600–800 ms after the onset of a mental rotation trial on the left or right parietal cortex. Only stimulation of the left parietal cortex at 400–600 ms affected the performance reliably, which provided information on the brain area involved and when it functioned during the task.

In summary, rTMS delivered during the execution of a cognitive task triggers synchronous activity in a subpopulation of neurons located under the stimulating coil, which results in a disruption in the pattern of activity that occurs at the same time as the stimulation (Jahanshahi and Rothwell 2000) and enables the adequate execution of the task. This allows information to be obtained about the timing of the contribution of a given cortical region to a specific behavior, which enables the study of the mental chronometry of a cognitive process, using TMS with high temporal resolution (Pascual-Leone et al. 2000, Sack 2006).

There are two distinct approaches to the application of rTMS. Manipulating cognitive processing when rTMS is applied during the performance of a task, as just described, is called online TMS. In contrast, rTMS may be applied several minutes before the subject is tested on the task, which is defined as offline stimulation (see Rossi et al. [2009] for a classification of the TMS approaches).

Presumably, in the online application, the faster the rTMS frequency, the greater the disruption of the activity of the targeted brain region, and the greater the final behavioral effects will be. However, the potential risks are that greater and more prominent nonspecific behavioral and attentional effects will be observed, which can make the results more difficult to interpret (Rossi et al. 2009). Moreover, the effects induced by online stimulation are generally short-lived, lasting approximately a few hundred milliseconds to a few seconds.

The alternative offline approach, which has achieved some popularity over the last few years, is to stimulate the site of interest for several seconds, using theta burst stimulation (TBS) (Huang et al. 2005, Vallesi et al. 2007) or for minutes (5–30 min) at a given frequency (low or high) "before" beginning a cognitive task. In this case, rTMS affects the modulation of cortical excitability (increased vs. decreased) beyond the duration of the application itself, and the aim of rTMS is to alter cognitive performance. The division between high and low frequency is not arbitrary. The cut-off is empirically based on direct and indirect measurements of brain activity as well as on behavioral outputs. Therefore, treating low- and high-frequency rTMS as separate phenomena is essential because the application of these two types of stimulation for several minutes might produce distinct effects on brain activity. Converging evidence has indicated that continuous rTMS below 1 Hz (low frequency) causes a reduction in neuronal firing and decreases cortical excitability locally and in functionally related regions. By contrast, intermittent rTMS above 5 Hz (high frequency), which leads to increased neuronal firing, appears to have the opposite effect (Chen et al. 1997, Maeda et al. 2000, Pascual-Leone et al. 1994). However, studies have not always confirmed the strict and unequivocal association between behavioral improvement and excitation or between behavioral disruption and inhibition (Andoh et al. 2006, Drager et al. 2004, Hilgetag et al. 2001, Kim et al. 2005, Waterston and Pack 2010). Therefore, we need to separate the physiological effects from the behavioral effects (Miniussi et al. 2010).

Some evidence has suggested that the effects induced by different offline rTMS approaches were site specific; however, they were not site limited (Bestmann et al. 2008). Thus, the long-term consequences induced by sustained and repetitive brain stimulation were most likely due to activity changes in a given network of cortical and subcortical areas rather than local inhibition or excitation of an individual brain area (Selimbeyoglu and Parvizi 2010). This means that brain stimulation can modulate the ongoing properties of a neuronal network by amplifying or reducing its activity. Moreover, the stimulated area cannot be considered to be isolated from its own functions or the functional status induced by the state of the subject (Harris et al. 2008a, Pasley et al. 2009, Ruzzoli et al. 2010, Silvanto et al. 2008). These aspects suggest that the functional effects induced in one area could be co-opted into different functions in other areas depending on the mode of activation or which of its interconnected networks was activated (Harris et al. 2008a, Selimbeyoglu and Parvizi 2010, Silvanto et al. 2005).

Moreover, the effects of offline rTMS using specific protocols have been shown to outlast the stimulation period itself, and synaptic long-term potentiation and depression (LTP and LTD, respectively) have been suggested to account for these modifications (Cooke and Bliss 2006, Thickbroom 2007, Ziemann and Siebner 2008). In general, one of the advantages of TMS is that it can be used on a larger population of subjects, and the location of the coil can be precisely controlled using a neuronavigation approach. Manipulation with rTMS can be meaningful if the coil position can be accurately localized on an individual basis, especially in situations where inter-individual differences are particularly relevant (Manenti et al. 2010). Therefore, in some cases, it is very important to guide the positioning of the coil over the target area using a neuronavigation system with single subject functional magnetic resonance imaging (Sack et al. 2009). Nevertheless, it should also be mentioned that even though the location of the stimulation can be precisely controlled, the spatial resolution of the induced effects has not been completely determined (Bestmann et al. 2008). Therefore, sometimes the spatial resolution of rTMS effects hinders a precise interpretation of the observed functional effects in terms of anatomical localization. For example, the discharging coil produces a clicking sound that may induce arousal and disrupt task performance irrespective of the exact demands of the experimental design. While this issue may be addressed by giving the subject earplugs, this approach is not practical in all cases, such as in language experiments that require the subject to listen to voices or sounds. Therefore, a control condition must be used to try to ensure that changes in performance are specifically attributable to the effects of TMS on the brain. One of these controls is a sham (placebo) stimulation, which should be used to obtain a baseline measurement. In the sham condition, one should ensure that no effective magnetic stimulation reaches the brain (Rossi et al. 2007) while all other experimental parameters are identical. Another approach is the stimulation of contralateral homologous areas (homotopic) or vertex areas while the subject performs the same task under identical auditory and somatosensory perceptions. This allows for the comparison of the effects of rTMS at different sites where only one site has functional relevance. Finally, it is also possible to observe subject's behavior across a number of distinct tasks following stimulation at one site. Consequently, many studies have also taken the approach of observing behavior across several distinct tasks following stimulation at one site (Sandrini et al. 2011). Following stimulation at one site, only one task is functionally related to the stimulated site.

All of these technical controls are critically important in experiments that involve cognition because the functional effects that can be induced after stimulation of a cortical area can have different manifestations depending on which of the interconnected networks are engaged in a given task (Sack and Linden 2003).

13.4 TMS AND COGNITIVE NEUROSCIENCE

The use of TMS as an investigative tool in the study of specific cognitive functions has been previously established (Walsh and Cowey 2000). In the last few years, many TMS studies have significantly contributed to our understanding of the role of different cortical sites in various perceptual and cognitive functions. For example, the application of TMS over prefrontal sites (for a review, see Guse et al. 2010) has allowed for understanding the role of these sites in cognitive tasks involving working memory (Pascual-Leone and Hallett 1994), episodic memory (Rossi et al. 2001, 2004, Sandrini et al. 2003), and implicit learning (Pascual-Leone et al. 1996). Moreover, spatial attention (Ashbridge et al. 1997, Thut et al. 2005), somatosensation (Seyal et al. 1995), object recognition (Harris et al. 2008a), and numerical processing (Rusconi et al. 2005, Sandrini et al. 2004) are a subset of the functions that have been investigated through stimulation of the parietal cortex. TMS over the occipital cortex has been used to examine a number of aspects of visual processing as well as visual motion (Ruzzoli et al. 2010) and color perception (Maccabee et al. 1991). Additional studies have applied TMS over the temporal cortex to understand the functions related to language and semantic cognition (Pobric et al. 2010).

Although it is impossible to summarize all the studies that have used TMS in the field of cognitive neuroscience in an exhaustive manner, it is possible to report the principal questions that may be addressed with TMS in cognitive function studies. As highlighted by Jahanshahi and Rothwell (2000), TMS may be used as a tool to investigate and understand the role and timing of the involvement of a target area during a specific performance, the contribution of different sites to different aspects of a task, the relative timing of the contribution of two or more areas to task performance and the function of intracortical and transcallosal connectivity. In general, the possibility of understanding the location, timing and functional relevance of the neuronal activity underlying cortical functions makes TMS an essential technique in perception and cognitive research.

13.5 TRANSCRANIAL ELECTRIC STIMULATION

tES, like TMS, is a technique that can be used to investigate brain-behavior relationships and explore the state of different regions of the brain. The tES technique (see Chapter 4 for an overview of the technical aspects) involves applying weak electrical currents directly to the head for several minutes. These currents generate an electrical field that modulates neuronal activity according to the modality of the application, which can be direct (transcranial direct current stimulation, tDCS), alternating current (transcranial alternating current stimulation, tACS), or random noise (transcranial random noise stimulation, tRNS).

tDCS applied through the skull was shown to directly modulate the excitability of motor (Nitsche and Paulus 2000, 2001) and visual (Antal and Paulus 2008) cortices in human subjects. tDCS is applied for a much longer duration than TMS, and has been shown to modulate the resting membrane potential and related cortical activity and induces transient functional changes in the human brain (Nitsche and Paulus 2000, 2001).

From a methodological prospective, most of the general points made for TMS are valid for tES in cognitive neuroscience research; however, there are a few exceptions. There is a reduction in spatial and temporal resolution (Dmochowski et al. 2011), although there is an advantage in terms of applicability. tES does not produce the noise and discomfort produced by TMS; therefore, changes in performance are not attributable to nonspecific effects (tactile or auditory).

With regard to the mechanism and efficacy of tDCS, the vast majority of evidence comes from stimulation of the primary motor cortex (M1). It is still not clear to what extent these findings are transferable to other areas of the cortex, although it is likely that some of the mechanisms of action are similar. Cognitive neuroscience using tES can still be considered a nascent field; therefore, in this chapter, we describe the current knowledge of the physiological and behavioral effects of tES on perception and cognition.

13.6 tDCS AND VISUAL PERCEPTION

Several studies have investigated the effect of anodal and cathodal tDCS over the occipital cortex (for an overview, see Antal and Paulus 2008). In a study using large Gabor patch stimuli with a spatial frequency of 4 cycles/degree, only cathodal stimulation significantly decreased static and dynamic contrast sensitivity (Antal et al. 2001). However, a recent study (Kraft et al. 2010) demonstrated that anodal tDCS of the visual cortex could also cause a transient increase in contrast sensitivity for central positions at eccentricities smaller than 2°. In this study, cathodal stimulation of the visual cortex did not affect contrast sensitivity. This might result from the different stimulation durations used in the two studies; however, it also may be due to the specific visual stimuli used.

When applied over the visual cortex, tDCS is capable of modifying the perception threshold of phosphenes. Induction of phosphenes can be evoked by single pulses or by repetitive TMS of the visual cortex (V1). These phosphenes are commonly described as spots of light or stars that tend to persist during the time of the stimulation and disappear with its cessation. Cathodal stimulation increases, whereas anodal stimulation decreases phosphene thresholds (PT) (Antal et al. 2003a,b). Compared to PT measurements, the measurement of visual evoked potential (VEP), which characterizes occipital activation in response to visual activation, is a more objective and widely accepted method for evaluating visual-cortical function in humans. Using montages obtained from a number of stimulating electrode positions, only the occipital (Oz)-vertex (Cz) electrode position was effective in inducing after-effects, which shows that the stimulation efficacy of tDCS highly depends on the direction of the current flow (Antal et al. 2004a). This finding is similar to what is observed using tDCS in M1 (Nitsche and Paulus 2000). Cathodal tDCS over V1 decreased the amplitude of a negative waveform at 70 ms, the N70 component of the VEP, whereas anodal tDCS increased N70 amplitude. However, significant effects can only be observed when low-contrast visual stimuli were shown. High-contrast stimuli most likely activate the appropriate visual-cortical pathways and areas optimally; therefore, subthreshold excitability modulation induced by tDCS could not produce a clear change in the VEP amplitude. With regard to stimulation polarity, the opposite effect was also observed in another study, and anodal stimulation resulted in a reduction in the P100 amplitude, whereas cathodal stimulation increased the P100 amplitude (Accornero et al. 2007). This apparent discrepancy may be due to the different VEP modalities used. The first study used a sinusoidal onset pattern, whereas the second study used a checkerboard pattern-reversal stimulation. Furthermore, the position of the reference electrode (Cz in the first study, and the neck in the second study) could have a strong influence on DC-induced after-effects.

Results from several studies provide evidence that external modulation of visual neural excitability using tDCS goes beyond V1 and could influence complex, visual adaptation-related processes. Neural cells in the MT and medial-superior temporal (MST) areas (the human analogue is called V5 or MT+) are particularly sensitive to motion, and many cells in these areas selectively react to optical flow (Lappe et al. 1996). Direct current stimulation over MT+/V5 using tDCS affected the strength of the perceived motion after-effect (MAE), which supports the involvement of MT+/V5 in motion adaptation processes (Antal et al. 2004d). Interestingly, both cathodal and anodal stimulation over this area resulted in a significant reduction in the MAE duration. One possible explanation of this effect is that tDCS affects the interaction between the neural representations of different motion directions in MT+/V5. It has been suggested that adaptation results in an imbalance of mutual inhibition processes between different motion directions, which will lead to an illusory perception of motion. Modulation of neural excitability with anodal and cathodal tDCS might result in attenuated expression of the adaptation-induced imbalance in both cases, and consequently, weakened motion after-effects.

In addition, cathodal stimulation to the right temporo-parietal cortex reduced the magnitude of facial adaptation, whereas stimulation over V1 did not have a significant effect (Varga et al. 2007). These data imply that lateral temporo-parietal cortical areas play a major role in facial adaptation and in facial gender discrimination, which supports the idea that the observed after-effects are the result of high-level, configurational adaptation mechanisms. In agreement with previous studies, the inhibitory effect of cathodal tDCS on adaptation may be related to the focal diminishment of cortical excitability due to membrane hyperpolarization.

Recent results also demonstrated that anodal tDCS applied over the posterior parietal cortex (PPC) is a promising technique for enhancing visuo-spatial abilities when combined with a visual field exploration training task. When anodal tDCS was applied to the right PPC, it increased the training-induced behavioral improvement of visual exploration when compared to sham tDCS (Bolognini et al. 2010). In addition, stimulation of the right PPC enhanced covert visual orientation and stimulation of this area by itself, even without associated training or enhanced visual exploration. Unilateral stimulation of the PPC bidirectionally modulated the performance of healthy subjects in a visual dot-detection task depending on the side of stimulation and current polarity (Schweid et al. 2008, Sparing et al. 2009). Anodal tDCS improved the detection of contralateral stimuli, whereas cathodal tDCS ameliorated the detection of ipsilateral stimuli and worsened the detection of contralateral stimuli using bilateral stimulation conditions with an extinction-like pattern. These findings are encouraging for future interventions in brain-damaged patients with visuo-spatial disabilities.

In summary, the results from the studies mentioned above show that the effects of anodal and cathodal stimulation highly depend on the task and on the activity state of the visual system. To date, the definition of relevant stimulation parameters, such as task type, strength and duration of stimulation, size of electrodes, and position of electrodes, have been elaborated thoroughly for the motor system (for a review, see Nitsche et al. 2008); however, this still needs to be established for the visual system (Jacobson et al. 2011).

13.7 tDCS AND MOTOR/VISUO-MOTOR LEARNING

As learning requires functional changes in the cortical construction that involves modifications of excitability, the induction of neuroplastic changes using tDCS is an interesting tool with which to modulate these processes. tDCS has been investigated as a means of modulating visual perception and as a tool that can modulate cortical excitability to reveal causal relationships between brain regions, cognitive functions and facilitate skill acquisition and learning. Generally, combining tDCS with behavioral interventions could be a powerful method to enhance the response of the target system and increase behavioral performance. However, the effects of tDCS on learning and cognitive processing not only depend on the stimulation parameters applied, but also on the stimulated area and task used. For example, it has been shown that anodal stimulation of M1 specifically improves implicit motor learning in its acquisition phase, whereas stimulation of other motor-related areas, such as the premotor and prefrontal cortices, had no effect (Nitsche et al. 2003). However, studying the functional effects of tDCS over the visual areas revealed that the percentage of correct tracking movements in a visuo-motor task significantly increased during and immediately after cathodal tDCS of V5. In contrast, anodal stimulation had no effect when a previously learned manual visuo-motor tracking task was applied (Antal et al. 2004c). Indeed, the effect of cathodal tDCS is highly specific in reducing excitability in V5 and enhancing performance in this visually guided tracking task. This effect is most likely explained by the complexity of the perceptual information processing needed for the task. This task most likely produces a noisy activation state in its neuronal patterns in response to different directions of movement. In this activation state, cathodal stimulation may have a focusing effect by decreasing global excitation levels and diminishing the amount of activation of concurrent patterns below the threshold for eliciting a (Antal et al. 2004c). Therefore, cathodal stimulation increases the signal-to-noise ratio and improves performance. However, when tDCS was applied during the learning phase of the same visuo-motor coordination task in a different subject group, performance was significantly increased 5–10 min after the beginning of anodal stimulation of V5 or M1, whereas cathodal stimulation had no significant effect (Antal et al. 2004b). The positive effect of anodal tDCS was restricted to the learning phase, which suggests a highly specific temporal effect of stimulation (Stagg et al. 2011).

However, tDCS can also influence long-term motor skill learning. Reis et al. (2009) used a computerized motor skill task to evaluate the effects of anodal tDCS during the course of learning. Speed and accuracy were measured on the training day and on the days before and after training. The experimental training group received five sessions of anodal tDCS for 20 min over the left M1 region, whereas the two control groups received either sham or cathodal tDCS. Anodal tDCS showed greater effects on learning over the entire training period when compared to cathodal or sham stimulation. These effects were maintained in follow-up sessions, and the anodal group still performed better than the two control groups.

Interestingly, in some tasks, such as the Jebsen-Taylor hand function task that mimics daily activities and is often used in stroke research, the increase in performance was only observed when the brain area of the non-dominant hand was stimulated, which suggests a ceiling effect of the stimulation (Boggio et al. 2006).

Generally, compared to the M1, the after-effects are relatively short lasting in the visual areas using the same stimulation durations. However, visual and motor cortices vary with regard to factors influencing neuroplasticity and in their excitatory/inhibitory circuitry. Differences in cortical connections and neuronal membrane properties, including receptor expression, between primary motor and visual cortices may also account for the contrasting responses to the application of tDCS. Furthermore, the results indicate that gender differences exist within the visual cortex of humans and may be subject to the influences of modulatory neurotransmitters or gonadal hormones, which induce short-term neuroplasticity (Chaieb et al. 2008). Alternatively, it is possible that the threshold for stimulating M1 is lower than the visual areas.

13.8 tDCS AND COGNITION

tDCS can modulate many aspects of cognition. However, studies investigating the influence of tDCS on cognitive function show facilitatory as well as inhibitory effects. In general, considering the bipolar nature of tDCS, it would be too simplistic to assume that anodal stimulation has a beneficial effect and cathodal stimulation has a disrupting effect on cognition. For example, Kincses et al. (2004) demonstrated that anodal stimulation over the left dorsolateral prefrontal cortex (DLPFC) improved implicit classification learning, whereas cathodal tDCS had no effect. Anodal stimulation of the left DLPFC also improved the accuracy of performance during a sequential letter working-memory task in healthy subjects (Fregni et al. 2005). More recently, Zaehle et al. (2011) observed that 15 min of anodal tDCS to the left DLPFC resulted in significantly greater working memory performance on a 2-back task in healthy controls when compared with sham and cathodal tDCS stimulation. In another study (Ambrus et al. 2011), the effects of tDCS over the DLPFC on categorization using the *A-not-A* dot-pattern version of a prototype distortion task was investigated. Contrary to the expectations of the authors, it was determined that anodal tDCS of the DLPFC (linked to categorization processes in previous studies) did not improve categorization performance. Rather, the prototype effect present in the sham stimulation condition disappeared in all stimulation conditions (anodal tDCS of the right and left DLPFC, cathodal tDCS of the right DLPFC). In another study, Ferrucci et al. (2008) showed that anodal and cathodal tDCS over the cerebellum disrupted practice-dependent improvements during a modified Sternberg verbal working-memory task. In addition, intermittent frontal bilateral tDCS during a modified Sternberg task impaired response selection and preparation in this task (Marshall et al. 2005). The opposing results in these studies might be due to task characteristics, for example, the Sternberg paradigm demands more complex information processing than the three-back letter task; however, the results may have been caused by the different stimulation protocols that were employed.

The state of the cortex during stimulation might also be an important factor regarding the effect of tDCS. In a study by Andrews et al. (2011), the execution of a working memory task during anodal tDCS stimulation of the DLPFC increased performance on the digit span task compared to either stimulation without the digit span task or sham tDCS with this secondary task.

To date, there are very few studies showing that tDCS over temporal areas can modulate cognitive function in healthy subjects. Language learning and visual picture naming can be improved by anodal stimulation of the left perisylvian area (Fertonani et al. 2010, Sparing et al. 2008). A more recent study demonstrated that visual memory can be enhanced in healthy people using tDCS (Chi et al. 2010). In this study, 13 min of bilateral tDCS was applied to the anterior temporal lobes. Only participants who received left cathodal stimulation along with right anodal stimulation showed an improvement in visual memory. This 110% improvement was similar to the performance of individuals with autism (who are known to be more literal) compared to normal subjects in an identical visual task. Participants receiving stimulation of the opposite polarity (left anodal with right cathodal stimulation) failed to show any change in memory performance. The authors argued that this was the first brain stimulation study to employ an identical task that had been previously used in testing subjects with autism, which suggests a possible technique for temporarily inducing an autistic-like performance in healthy people.

A recent study explored whether tDCS could also be effective in modulating multisensory audiovisual interactions in the human brain (Bolognini et al. 2010). In different sessions, healthy participants performed the sound-induced flash illusion task (Shams et al. 2000) while receiving anodal, cathodal, or sham stimulation with an intensity of 2 mA for 8 min to the occipital, temporal, or posterior parietal cortices. The perception of a single flash combined with two beeps was enhanced after anodal tDCS of the temporal cortex and decreased after anodal stimulation of the occipital cortex. A reversal of these effects was induced by cathodal tDCS. However, the perceptual fusion of multiple flashes combined with a single beep was unaffected by tDCS. This result might open new possibilities for modulating multisensory perception in humans.

13.9 tACS AND VISUAL PERCEPTION

In a recent study, it was demonstrated that tACS could induce phosphenes and interact with ongoing rhythmic activities in V1 in a frequency-specific manner (Kanai et al. 2008). In this study, an oscillatory current was applied over V1 using different frequencies to observe interactions with ongoing cortical rhythms, and the effects of delivering tACS under conditions of light or darkness were compared. Stimulation induced phosphenes most effectively when the beta frequency range was applied in an illuminated room, whereas the most effective stimulation frequency shifted to the alpha frequency range during testing in darkness. Stimulation with theta or gamma frequencies did not produce any visual phenomena. These results show that tACS can induce perceptions more effectively if it is used to strengthen spontaneously occurring oscillations. These findings might lead to new implementations of rhythmic stimulation as tools in therapy and neurorehabilitation. Nevertheless it has been suggested that current spread from the occipital electrode might evokes phosphenes in the retina (Schwiedrzik 2009).

Several recent studies, in line with tACS results, have indicated that manipulating cortical activity using rhythmic TMS can positively influence cognitive performance in normal subjects (Thut et al. 2011) and in patients affected by neural disorders, such as unilateral neglect or dementia (Thut and Miniussi 2009). The modification of cortical activity with rhythmic electrical stimulation might regulate maladaptive patterns of brain oscillations and provide the possibility of inducing a new balance within an affected functional network. It is likely that by restoring sufficient synchronization, improvements in sensory function can be achieved.

13.10   tRNS AND LEARNING

tRNS consists of the application of a random electrical oscillation spectrum over the cortex. tRNS can be applied at different frequency band ranges over the entire spectrum from 0.1 to 640 Hz. Terney et al. (2008) have demonstrated that 10 min of tRNS at high frequency (101–640 Hz) on the motor cortex is able to modulate cortical excitability (an increase in averaged motor-evoked potential [MEP] amplitude) that persists after cessation of the stimulation. Moreover, the potential of tRNS at high frequency was also demonstrated by Fertonani et al. (2011). They applied tRNS to the visual cortices of healthy subjects and observed a significant improvement in the performance of healthy subjects in a visual perceptual learning task. This improvement was significantly higher than the improvement obtained with anodal tDCS or with low frequency (0.1–100 Hz) tRNS. Therefore, tRNS might potentiate the activity of neural populations involved in tasks that facilitate brain plasticity by strengthening synaptic transmission between neurons (Fertonani et al. 2011, Terney et al. 2008). Modulation of the efficacy of synaptic transmission can result in excitability and activity changes in specific cortical networks that are activated by the execution of the task, and these changes correlate with cognitive plasticity at the behavioral level. These data show the great potential of tRNS, especially considering the substantially reduced cutaneous sensations elicited by tRNS, which makes verum tRNS indistinguishable from placebo tRNS (Ambrus et al. 2010).

13.11   PROSPECTS FOR COGNITIVE REHABILITATION BY NIBS

TMS and tES can transiently influence behavior by producing an alteration of the stimulated cortical areas. Generally, this alteration interferes with task execution. Nevertheless, TMS and tES as reported in this chapter may also lead to enhanced performance (Antal et al. 2004a, Cappa et al. 2002, Cohen Kadosh et al. 2010, Harris et al. 2008a, Vallar and Bolognini 2011, Walsh et al. 1998).

For TMS, the facilitation effects could be due to nonspecific factors, such as general arousal due to stimulation. These effects can be directly controlled, and there are cases where TMS-induced improvements in accuracy or reaction times are not due to nonspecific factors but clearly depend on the site of stimulation or the type of task given to the participants (Miniussi et al. 2008). tES does not produce this type of nonspecific effect because the induced activation of esteroceptive somatosensory receptors is minimal, and there is no activation of the auditory system. In addition, framing the experimental hypothesis in a signal-to-noise context should also predict performance facilitation, although a reduction of signal strength would predict a performance decrease. In non-linear systems, such as the nervous system, noise can even enhance performance through a phenomenon called "stochastic resonance," which produces an "optimal" level of noise (Antal et al. 2004b, Miniussi et al. 2010, Ruzzoli et al. 2010, 2011, Stein et al. 2005).

These mechanisms have important implications for neurorehabilitation because, in principle, they may begin to allow for the specific enhancement of impaired functions as a component of therapeutic interventions. Most of these effects are transient (in the order of minutes) but their application, in concert with learning and plasticity processes, can perpetuate facilitatory effects beyond the end of the stimulation period, which provides important opportunities for progress.

In clinical populations, tES over the visual cortex is a promising technique for modulating residual visual capacities. In motor cognition and rehabilitation, tES has already shown its clinical usefulness, e.g., anodal tDCS improved motor performance in stroke patients with hemiparesis (Hummel et al. 2005, Tanaka et al. 2011). To our knowledge, there have been a limited number of studies where the excitability of the occipital cortex was modified for clinical rehabilitation. In vision rehabilitation, the idea of a visual prosthesis (Brindley and Lewin 1968) for blind subjects or patients with cortical visual field defects may be rejuvenated with tDCS. In cases where the excitability of the occipital cortex should be changed, such as in photosensitive epilepsy, stroke or migraine (Antal et al. 2011) the application of tES/rTMS could also be therapeutically beneficial. These are just a few examples of many recent studies that have been reported to improve cognitive abilities in patients with unilateral spatial neglect (Hesse et al. 2011), aphasia (Cotelli et al. 2011) or memory problems (Boggio et al. 2011, Cotelli et al. 2012).

In summary, NIBS methods are able to induce cortical plasticity modifications, which may outlast the stimulation period itself. Given this potential, there is currently a growing interest in therapeutically applying these methodologies to reduce cognitive deficits in patients with stroke or chronic neurodegenerative diseases. Although in its infancy, this approach is poised to deliver novel insight into the fundamental aspects of cognitive rehabilitation (Stuss 2011), thereby paving the way for more effective neuromodulatory therapeutic interventions (Miniussi and Rossini 2011, Miniussi and Vallar 2011, Miniussi et al. 2008, Paulus 2011).

13.12 CONCLUSIONS

There are a number of exciting prospects for the use of tES and TMS as tools to promote changes in brain activity that are paralleled by behavioral improvements. There is emerging evidence that multiple sessions may increase the duration of these behavioral effects to several weeks in healthy controls (Reis et al. 2009). According to previous studies and results from our own laboratories, tES and TMS appear to be promising methods for inducing acute, as well as prolonged, cortical excitability and activity modulation. Recently, significant efforts have been made to combine tDCS and TMS with other techniques, such as fMRI, EEG (see Chapter 12; Antal et al. 2011, Baudewig et al. 2001, Miniussi and Thut 2010, Miniussi et al. 2012, Siebner et al. 2004, 2009, Taylor et al. 2008). The combination of these techniques appears to be a very important approach in learning more about the location, time course and functional specifications of brain areas involved in visual and visuo-cognitive tasks. To make this tool relevant for basic research purposes and for clinical application, additional studies are necessary.

It is interesting to note that the effects of NIBS are proportional to the level of neuronal activation during the application of the stimulus (Epstein and Rothwell 2003). In the motor system, an increase in the amplitude of MEPs can be achieved by voluntary contraction of the target muscle (Rothwell et al. 1987). A similar effect was observed by Silvanto et al. (2007) using a neural adaptation paradigm over the visual area. These authors systematically manipulated the activity of distinct neural populations within V1 through perceptual adaptation induced before TMS application, which enabled them to study the interaction between the state of activation of the stimulated area (adapted vs. not adapted) and the neural activity induced by TMS. When TMS was applied over V1, the induced perceived phosphenes, which are generally colorless, were the same color as the adapting stimulus. Moreover, it has been showed by Antal et al. (2007) that the effect of tDCS differs if it is applied to an active, rested or fatigued cortical area (Antal et al. 2007). For example, also the induction of a phosphene by TMS is strictly related to the amount of alpha activity present in the occipital cortex (Romei et al. 2008). These findings suggest that NIBS effects are sensitive to changes in the cortical state and open the intriguing possibility that administration of NIBS while a subject is in a given condition, or as the subject performs a behavioral task, may permit the targeting of specific circuitry.

REFERENCES

Accornero, N., Li Voti, P., La Riccia, M., and Gregori, B. 2007. Visual evoked potentials modulation during direct current cortical polarization. *Exp Brain Res* 178: 261–266.

Amassian, V. E., Cracco, R. Q., Maccabee, P. J. et al. 1989. Suppression of visual perception by magnetic coil stimulation of human occipital cortex. *Electroencephalogr Clin Neurophysiol* 74: 458–462.

Ambrus, G. G., Paulus, W., and Antal, A. 2010. Cutaneous perception thresholds of electrical stimulation methods: Comparison of tDCS and tRNS. *Clin Neurophysiol* 121: 1908–1914.

Ambrus, G. G., Zimmer, M., Kincses, Z. T. et al. 2011. The enhancement of cortical excitability over the DLPFC before and during training impairs categorization in the prototype distortion task. *Neuropsychologia* 49: 1974–1980.

Andoh, J., Artiges, E., Pallier, C. et al. 2006. Modulation of language areas with functional MR image-guided magnetic stimulation. *Neuroimage* 29: 619–627.

Andrews, S. C., Hoy, K. E., Enticott, P. G. et al. 2011. Improving working memory: The effect of combining cognitive activity and anodal transcranial direct current stimulation to the left dorsolateral prefrontal cortex. *Brain Stimul* 4: 84–89.

Antal, A., Kincses, T. Z., Nitsche, M. A., Bartfai, O., and Paulus, W. 2004a. Excitability changes induced in the human primary visual cortex by transcranial direct current stimulation: Direct electrophysiological evidence. *Invest Ophthalmol Vis Sci* 45: 702–707.

Antal, A., Kincses, T. Z., Nitsche, M. A., and Paulus, W. 2003a. Manipulation of phosphene thresholds by transcranial direct current stimulation in man. *Exp Brain Res* 150: 375–378.

Antal, A., Kincses, T. Z., Nitsche, M. A., and Paulus, W. 2003b. Modulation of moving phosphene thresholds by transcranial direct current stimulation of V1 in human. *Neuropsychologia* 41: 1802–1807.

Antal, A., Nitsche, M. A., Kincses, T. Z. et al. 2004b. Facilitation of visuo-motor learning by transcranial direct current stimulation of the motor and extrastriate visual areas in humans. *Eur J Neurosci* 19: 2888–2892.

Antal, A., Nitsche, M. A., Kruse, W. et al. 2004c. Direct current stimulation over V5 enhances visuomotor coordination by improving motion perception in humans. *J Cogn Neurosci* 16: 521–527.

Antal, A., Nitsche, M. A., and Paulus, W. 2001. External modulation of visual perception in humans. *Neuroreport* 12: 3553–3555.

Antal, A. and Paulus, W. 2008. Transcranial direct current stimulation and visual perception. *Perception* 37: 367–374.

Antal, A., Polania, R., Schmidt-Samoa, C., Dechent, P., and Paulus, W. 2011. Transcranial direct current stimulation over the primary motor cortex during fMRI. *Neuroimage* 55: 590–596.

Antal, A., Terney, D., Poreisz, C., and Paulus, W. 2007. Towards unravelling task-related modulations of neuroplastic changes induced in the human motor cortex. *Eur J Neurosci* 26: 2687–2691.

Antal, A., Varga, E. T., Nitsche, M. A. et al. 2004d. Direct current stimulation over MT+/V5 modulates motion aftereffect in humans. *Neuroreport* 15: 2491–2494.

Ashbridge, E., Walsh, V., and Cowey, A. 1997. Temporal aspects of visual search studied by transcranial magnetic stimulation. *Neuropsychologia* 35: 1121–1131.

Barker, A. T., Jalinous, R., and Freeston, I. L. 1985. Non-invasive magnetic stimulation of human motor cortex. *Lancet* 1: 1106–1107.

Baudewig, J., Nitsche, M. A., Paulus, W., and Frahm, J. 2001. Regional modulation of BOLD MRI responses to human sensorimotor activation by transcranial direct current stimulation. *Magn Reson Med* 45: 196–201.

Bestmann, S., Ruff, C. C., Blankenburg, F. et al. 2008. Mapping causal interregional influences with concurrent TMS-fMRI. *Exp Brain Res* 191: 383–402.

Boggio, P. S., Castro, L. O., Savagim, E. A. et al. 2006. Enhancement of non-dominant hand motor function by anodal transcranial direct current stimulation. *Neurosci Lett* 404: 232–236.

Boggio, P. S., Valasek, C. A., Campanha, C. et al. 2011. Non-invasive brain stimulation to assess and modulate neuroplasticity in Alzheimer's disease. *Neuropsychol Rehabil* 21: 703–716.

Bolognini, N., Fregni, F., Casati, C., Olgiati, E., and Vallar, G. 2010. Brain polarization of parietal cortex augments training-induced improvement of visual exploratory and attentional skills. *Brain Res* 1349: 76–89.

Brindley, G. S. and Lewin, W. S. 1968. The visual sensations produced by electrical stimulation of the medial occipital cortex. *J Physiol* 194: 54–5P.

Cappa, S. F., Sandrini, M., Rossini, P. M., Sosta, K., and Miniussi, C. 2002. The role of the left frontal lobe in action naming: rTMS evidence. *Neurology* 59: 720–723.

Chaieb, L., Antal, A., and Paulus, W. 2008. Gender-specific modulation of short-term neuroplasticity in the visual cortex induced by transcranial direct current stimulation. *Vis Neurosci* 25: 77–81.

Chen, R., Classen, J., Gerloff, C. et al. 1997. Depression of motor cortex excitability by low-frequency transcranial magnetic stimulation. *Neurology* 48: 1398–1403.

Chi, R. P., Fregni, F., and Snyder, A. W. 2010. Visual memory improved by non-invasive brain stimulation. *Brain Res* 1353: 168–175.

Cohen Kadosh, R., Soskic, S., Iuculano, T., Kanai, R., and Walsh, V. 2010. Modulating neuronal activity produces specific and long-lasting changes in numerical competence. *Curr Biol* 20: 2016–2020.

Cooke, S. F. and Bliss, T. V. 2006. Plasticity in the human central nervous system. *Brain* 129: 1659–1673.

Corthout, E., Hallett, M., and Cowey, A. 2003. Interference with vision by TMS over the occipital pole: A fourth period. *Neuroreport* 14: 651–655.

Cotelli, M., Calabria, M., Manenti, R. et al. 2012. Brain stimulation improves associative memory in an individual with amnestic mild cognitive impairment. *Neurocase* 18: 217–223.

Cotelli, M., Fertonani, A., Miozzo, A. et al. 2011. Anomia training and brain stimulation in chronic aphasia. *Neuropsychol Rehabil* 21: 717–741.

Dmochowski, J. P., Datta, A., Bikson, M., Su, Y., and Parra, L. C. 2011. Optimized multi-electrode stimulation increases focality and intensity at target. *J Neural Eng* 8: 046011.

Drager, B., Breitenstein, C., Helmke, U., Kamping, S., and Knecht, S. 2004. Specific and nonspecific effects of transcranial magnetic stimulation on picture-word verification. *Eur J Neurosci* 20: 1681–1687.

Epstein, C. M. and Rothwell, J. C. 2003 In *Therapeutic Uses of rTMS*. Cambridge, U.K.: Cambridge University Press.

Ferrucci, R., Marceglia, S., Vergari, M. et al. 2008. Cerebellar transcranial direct current stimulation impairs the practice-dependent proficiency increase in working memory. *J Cogn Neurosci* 20: 1687–1697.

Fertonani, A., Pirulli, C., and Miniussi, C. 2011. Random noise stimulation improves neuroplasticity in perceptual learning. *J Neurosci* 31: 15416–15423.

Fertonani, A., Rosini, S., Cotelli, M., Rossini, P. M., and Miniussi, C. 2010. Naming facilitation induced by transcranial direct current stimulation. *Behav Brain Res* 208: 311–318.

Fregni, F., Boggio, P. S., Nitsche, M. et al. 2005. Anodal transcranial direct current stimulation of prefrontal cortex enhances working memory. *Exp Brain Res* 166: 23–30.

Guse, B., Falkai, P., and Wobrock, T. 2010. Cognitive effects of high-frequency repetitive transcranial magnetic stimulation: A systematic review. *J Neural Transm* 117: 105–122.

Harris, I. M., Benito, C. T., Ruzzoli, M., and Miniussi, C. 2008a. Effects of right parietal transcranial magnetic stimulation on object identification and orientation judgments. *J Cogn Neurosci* 20: 916–926.

Harris, I. M. and Miniussi, C. 2003. Parietal lobe contribution to mental rotation demonstrated with rTMS. *J Cogn Neurosci* 15: 315–323.

Harris, J. A., Clifford, C. W., and Miniussi, C. 2008b. The functional effect of transcranial magnetic stimulation: Signal suppression or neural noise generation? *J Cogn Neurosci* 20: 734–740.

Hesse, M. D., Sparing, R., and Fink, G. R. 2011. Ameliorating spatial neglect with non-invasive brain stimulation: From pathophysiological concepts to novel treatment strategies. *Neuropsychol Rehabil* 21: 676–702.

Hilgetag, C. C., Theoret, H., and Pascual-Leone, A. 2001. Enhanced visual spatial attention ipsilateral to rTMS-induced 'virtual lesions' of human parietal cortex. *Nat Neurosci* 4: 953–957.

Huang, Y. Z., Edwards, M. J., Rounis, E., Bhatia, K. P., and Rothwell, J. C. 2005. Theta burst stimulation of the human motor cortex. *Neuron* 45: 201–206.

Hummel, F., Celnik, P., Giraux, P. et al. 2005. Effects of non-invasive cortical stimulation on skilled motor function in chronic stroke. *Brain* 128: 490–499.

Jacobson, L., Koslowsky, M., and Lavidor, M. 2011. tDCS polarity effects in motor and cognitive domains: A meta-analytical review. *Exp Brain Res* 216: 1–10.

Jahanshahi, M. and Rothwell, J. 2000. Transcranial magnetic stimulation studies of cognition: An emerging field. *Exp Brain Res* 131: 1–9.

Kanai, R., Chaieb, L., Antal, A., Walsh, V., and Paulus, W. 2008. Frequency-dependent electrical stimulation of the visual cortex. *Curr Biol* 18: 1839–1843.

Kim, Y. H., Min, S. J., Ko, M. H. et al. 2005. Facilitating visuospatial attention for the contralateral hemifield by repetitive TMS on the posterior parietal cortex. *Neurosci Lett* 382: 280–285.

Kincses, T. Z., Antal, A., Nitsche, M. A., Bartfai, O., and Paulus, W. 2004. Facilitation of probabilistic classification learning by transcranial direct current stimulation of the prefrontal cortex in the human. *Neuropsychologia* 42: 113–117.

Kraft, A., Roehmel, J., Olma, M. C. et al. 2010. Transcranial direct current stimulation affects visual perception measured by threshold perimetry. *Exp Brain Res* 207: 283–290.

Landi, D. and Rossini, P. M. 2010. Cerebral restorative plasticity from normal ageing to brain diseases: A "never ending story". *Restor Neurol Neurosci* 28: 349–366.

Lappe, M., Bremmer, F., Pekel, M., Thiele, A., and Hoffmann, K. P. 1996. Optic flow processing in monkey STS: A theoretical and experimental approach. *J Neurosci* 16: 6265–6285.

Maccabee, P. J., Amassian, V. E., Cracco, R. Q. et al. 1991. Magnetic coil stimulation of human visual cortex: Studies of perception. *Electroencephalogr Clin Neurophysiol Suppl* 43: 111–120.

Maeda, F., Keenan, J. P., Tormos, J. M., Topka, H., and Pascual-Leone, A. 2000. Modulation of corticospinal excitability by repetitive transcranial magnetic stimulation. *Clin Neurophysiol* 111: 800–805.

Manenti, R., Cotelli, M., Calabria, M., Maioli, C., and Miniussi, C. 2010. The role of the dorsolateral prefrontal cortex in retrieval from long-term memory depends on strategies: A repetitive transcranial magnetic stimulation study. *Neuroscience* 166: 501–507.

Mantovani, M., Van Velthoven, V., Fuellgraf, H., Feuerstein, T. J., and Moser, A. 2006. Neuronal electrical high frequency stimulation enhances GABA outflow from human neocortical slices. *Neurochem Int* 49: 347–350.

Marshall, L., Molle, M., Siebner, H. R., and Born, J. 2005. Bifrontal transcranial direct current stimulation slows reaction time in a working memory task. *BMC Neurosci* 6: 23.

Miniussi, C., Brignani, D., and Pellicciari, M. C. 2012. Combining transcranial electrical stimulation with electroencephalography: A multimodal approach. *J Clin EEG Neurosci* 43: 184–191.

Miniussi, C., Cappa, S. F., Cohen, L. G. et al. 2008. Efficacy of repetitive transcranial magnetic stimulation/transcranial direct current stimulation in cognitive neurorehabilitation. *Brain Stimul* 1: 326–336.

Miniussi, C. and Rossini, P. M. 2011. Transcranial magnetic stimulation in cognitive rehabilitation. *Neuropsychol Rehabil* 21: 579–601.

Miniussi, C., Ruzzoli, M., and Walsh, V. 2010. The mechanism of transcranial magnetic stimulation in cognition. *Cortex* 46: 128–130.

Miniussi, C. and Thut, G. 2010. Combining TMS and EEG offers new prospects in cognitive neuroscience. *Brain Topogr* 22: 249–256.

Miniussi, C. and Vallar, G. 2011. Brain stimulation and behavioural cognitive rehabilitation: A new tool for neurorehabilitation? *Neuropsychol Rehabil* 21: 553–559.

Moliadze, V., Giannikopoulos, D., Eysel, U. T., and Funke, K. 2005. Paired-pulse transcranial magnetic stimulation protocol applied to visual cortex of anaesthetized cat: Effects on visually evoked single-unit activity. *J Physiol* 566: 955–965.

Moliadze, V., Zhao, Y., Eysel, U., and Funke, K. 2003. Effect of transcranial magnetic stimulation on single-unit activity in the cat primary visual cortex. *J Physiol* 553: 665–679.

Nitsche, M.A., Cohen, L.G., Wassermann, E.M. et al. 2008. Transcranial direct current stimulation: State of the art. *Brain Stimul* 1:206–223.

Nitsche, M. A. and Paulus, W. 2000. Excitability changes induced in the human motor cortex by weak transcranial direct current stimulation. *J Physiol* 527 (Pt 3): 633–639.

Nitsche, M. A. and Paulus, W. 2001. Sustained excitability elevations induced by transcranial DC motor cortex stimulation in humans. *Neurology* 57: 1899–1901.

Nitsche, M. A., Schauenburg, A., Lang, N. et al. 2003. Facilitation of implicit motor learning by weak transcranial direct current stimulation of the primary motor cortex in the human. *J Cogn Neurosci* 15: 619–626.

Pascual-Leone, A. and Hallett, M. 1994. Induction of errors in a delayed response task by repetitive transcranial magnetic stimulation of the dorsolateral prefrontal cortex. *Neuroreport* 5: 2517–2520.

Pascual-Leone, A., Valls-Sole, J., Wassermann, E. M., and Hallett, M. 1994. Responses to rapid-rate transcranial magnetic stimulation of the human motor cortex. *Brain* 117 (Pt 4): 847–858.

Pascual-Leone, A. and Walsh, V. 2001. Fast backprojections from the motion to the primary visual area necessary for visual awareness. *Science* 292: 510–512.

Pascual-Leone, A., Walsh, V., and Rothwell, J. 2000. Transcranial magnetic stimulation in cognitive neuroscience—Virtual lesion, chronometry, and functional connectivity. *Curr Opin Neurobiol* 10: 232–237.

Pascual-Leone, A., Wassermann, E. M., Grafman, J., and Hallett, M. 1996. The role of the dorsolateral prefrontal cortex in implicit procedural learning. *Exp Brain Res* 107: 479–485.

Pasley, B. N., Allen, E. A., and Freeman, R. D. 2009. State-dependent variability of neuronal responses to transcranial magnetic stimulation of the visual cortex. *Neuron* 62: 291–303.

Paulus, W. 2011. Transcranial electrical stimulation (tES - tDCS; tRNS, tACS) methods. *Neuropsychol Rehabil* 21: 602–617.

Pobric, G., Jefferies, E., and Lambon Ralph, M. A. 2010. Category-specific versus category-general semantic impairment induced by transcranial magnetic stimulation. *Curr Biol* 20: 964–968.

Price, C. J. and Friston, K. J. 1999. Scanning patients with tasks they can perform. *Hum Brain Mapp* 8: 102–108.

Reis, J., Schambra, H. M., Cohen, L. G. et al. 2009. Noninvasive cortical stimulation enhances motor skill acquisition over multiple days through an effect on consolidation. *Proc Natl Acad Sci USA* 106: 1590–1595.

Robertson, E. M., Theoret, H., and Pascual-Leone, A. 2003. Studies in cognition: The problems solved and created by transcranial magnetic stimulation. *J Cogn Neurosci* 15: 948–960.

Romei, V., Brodbeck, V., Michel, C. et al. 2008. Spontaneous fluctuations in posterior alpha-band EEG activity reflect variability in excitability of human visual areas. *Cereb Cortex* 18: 2010–2018.

Rossi, S., Cappa, S. F., Babiloni, C. et al. 2001. Prefrontal cortex in long-term memory: An "interference" approach using magnetic stimulation. *Nat Neurosci* 4: 948–952.

Rossi, S., Ferro, M., Cincotta, M. et al. 2007. A real electro-magnetic placebo (REMP) device for sham transcranial magnetic stimulation (TMS). *Clin Neurophysiol* 118: 709–716.

Rossi, S., Hallett, M., Rossini, P. M., and Pascual-Leone, A. 2009. Safety, ethical considerations, and application guidelines for the use of transcranial magnetic stimulation in clinical practice and research. *Clin Neurophysiol* 120: 2008–2039.

Rossi, S., Miniussi, C., Pasqualetti, P. et al. 2004. Age-related functional changes of prefrontal cortex in long-term memory: A repetitive transcranial magnetic stimulation study. *J Neurosci* 24: 7939–7944.

Rothwell, J. C., Day, B. L., Thompson, P. D., Dick, J. P., and Marsden, C. D. 1987. Some experiences of techniques for stimulation of the human cerebral motor cortex through the scalp. *Neurosurgery* 20: 156–163.

Rusconi, E., Walsh, V., and Butterworth, B. 2005. Dexterity with numbers: rTMS over left angular gyrus disrupts finger gnosis and number processing. *Neuropsychologia* 43: 1609–1624.

Ruzzoli, M., Abrahamyan, A., Clifford, C. W. et al. 2011. The effect of TMS on visual motion sensitivity: An increase in neural noise or a decrease in signal strength? *J Neurophysiol* 106: 138–143.

Ruzzoli, M., Marzi, C. A., and Miniussi, C. 2010. The neural mechanisms of the effects of transcranial magnetic stimulation on perception. *J Neurophysiol* 103: 2982–2989.

Sack, A. T. 2006. Transcranial magnetic stimulation, causal structure-function mapping and networks of functional relevance. *Curr Opin Neurobiol* 16: 593–599.

Sack, A. T., Cohen Kadosh, R., Schuhmann, T. et al. 2009. Optimizing functional accuracy of TMS in cognitive studies: A comparison of methods. *J Cogn Neurosci* 21: 207–221.

Sack, A. T. and Linden, D. E. 2003. Combining transcranial magnetic stimulation and functional imaging in cognitive brain research: Possibilities and limitations. *Brain Res Brain Res Rev* 43: 41–56.

Sandrini, M., Cappa, S. F., Rossi, S., Rossini, P. M., and Miniussi, C. 2003. The role of prefrontal cortex in verbal episodic memory: rTMS evidence. *J Cogn Neurosci* 15: 855–861.

Sandrini, M., Rossini, P. M., and Miniussi, C. 2004. The differential involvement of inferior parietal lobule in number comparison: A rTMS study. *Neuropsychologia* 42: 1902–1909.

Sandrini, M., Umilta, C., and Rusconi, E. 2011. The use of transcranial magnetic stimulation in cognitive neuroscience: A new synthesis of methodological issues. *Neurosci Biobehav Rev* 35: 516–536.

Schweid, L., Rushmore, R. J., and Valero-Cabre, A. 2008. Cathodal transcranial direct current stimulation on posterior parietal cortex disrupts visuo-spatial processing in the contralateral visual field. *Exp Brain Res* 186: 409–417.

Schwiedrzik, C. M. 2009. Retina or visual cortex? The site of phosphene induction by transcranial alternating current stimulation. *Front Integr Neurosci* 3: 6.

Selimbeyoglu, A. and Parvizi, J. 2010. Electrical stimulation of the human brain: Perceptual and behavioral phenomena reported in the old and new literature. *Front Hum Neurosci* 4: 46.

Seyal, M., Ro, T., and Rafal, R. 1995. Increased sensitivity to ipsilateral cutaneous stimuli following transcranial magnetic stimulation of the parietal lobe. *Ann Neurol* 38: 264–267.

Shams, L., Kamitani, Y., and Shimojo, S. 2000. Illusions. What you see is what you hear. *Nature* 408: 788.

Siebner, H. R., Bergmann, T. O., Bestmann, S. et al. 2009. Consensus paper: Combining transcranial stimulation with neuroimaging. *Brain Stimul* 2: 58–80.

Siebner, H. R., Lang, N., Rizzo, V. et al. 2004. Preconditioning of low-frequency repetitive transcranial magnetic stimulation with transcranial direct current stimulation: Evidence for homeostatic plasticity in the human motor cortex. *J Neurosci* 24: 3379–3385.

Silvanto, J., Lavie, N., and Walsh, V. 2005. Double dissociation of V1 and V5/MT activity in visual awareness. *Cereb Cortex* 15: 1736–1741.

Silvanto, J., Muggleton, N. G., Cowey, A., and Walsh, V. 2007. Neural adaptation reveals state-dependent effects of transcranial magnetic stimulation. *Eur J Neurosci* 25: 1874–1881.

Silvanto, J., Muggleton, N., and Walsh, V. 2008. State-dependency in brain stimulation studies of perception and cognition. *Trends Cogn Sci* 12: 447–454.

Sparing, R., Dafotakis, M., Meister, I. G., Thirugnanasambandam, N., and Fink, G. R. 2008. Enhancing language performance with non-invasive brain stimulation—A transcranial direct current stimulation study in healthy humans. *Neuropsychologia* 46: 261–268.

Sparing, R., Thimm, M., Hesse, M. D. et al. 2009. Bidirectional alterations of interhemispheric parietal balance by non-invasive cortical stimulation. *Brain* 132: 3011–3020.

Stagg, C. J., Jayaram, G., Pastor, D. et al. 2011. Polarity and timing-dependent effects of transcranial direct current stimulation in explicit motor learning. *Neuropsychologia* 49: 800–804.

Stein, R. B., Gossen, E. R., and Jones, K. E. 2005. Neuronal variability: Noise or part of the signal? *Nat Rev Neurosci* 6: 389–397.

Stuss, D. T. 2011. The future of cognitive neurorehabilitation. *Neuropsychol Rehabil* 21: 755–768.

Tanaka, S., Sandrini, M., and Cohen, L. G. 2011. Modulation of motor learning and memory formation by non-invasive cortical stimulation of the primary motor cortex. *Neuropsychol Rehabil* 21: 650–675.

Taylor, P. C., Walsh, V., and Eimer, M. 2008. Combining TMS and EEG to study cognitive function and cortico-cortico interactions. *Behav Brain Res* 191: 141–147.

Terney, D., Chaieb, L., Moliadze, V., Antal, A., and Paulus, W. 2008. Increasing human brain excitability by transcranial high-frequency random noise stimulation. *J Neurosci* 28: 14147–14155.

Thickbroom, G. W. 2007. Transcranial magnetic stimulation and synaptic plasticity: Experimental framework and human models. *Exp Brain Res* 180: 583–593.

Thut, G., Ives, J. R., Kampmann, F., Pastor, M. A., and Pascual-Leone, A. 2005. A new device and protocol for combining TMS and online recordings of EEG and evoked potentials. *J Neurosci Methods* 141: 207–217.

Thut, G. and Miniussi, C. 2009. New insights into rhythmic brain activity from TMS-EEG studies. *Trends Cogn Sci* 13: 182–189.

Thut, G., Veniero, D., Romei, V. et al. 2011. Rhythmic TMS causes local entrainment of natural oscillatory signatures *Curr Biol* 21:1176–1185.

Vallar, G. and Bolognini, N. 2011. Behavioural facilitation following brain stimulation: Implications for neurorehabilitation. *Neuropsychol Rehabil* 21: 618–649.

Vallesi, A., Shallice, T., and Walsh, V. 2007. Role of the prefrontal cortex in the foreperiod effect: TMS evidence for dual mechanisms in temporal preparation. *Cereb Cortex* 17: 466–474.

Varga, E. T., Elif, K., Antal, A. et al. 2007. Cathodal transcranial direct current stimulation over the parietal cortex modifies facial gender adaptation. *Ideggyogy Sz* 60: 474–479.

Walsh, V. and Cowey, A. 1998. Magnetic stimulation studies of visual cognition. *Trends Cogn Sci* 2: 103–110.

Walsh, V. and Cowey, A. 2000. Transcranial magnetic stimulation and cognitive neuroscience. *Nat Rev Neurosci* 1: 73–79.

Walsh, V., Ellison, A., Battelli, L., and Cowey, A. 1998. Task-specific impairments and enhancements induced by magnetic stimulation of human visual area V5. *Proc Biol Sci* 265: 537–543.

Waterston, M. L. and Pack, C. C. 2010. Improved discrimination of visual stimuli following repetitive transcranial magnetic stimulation. *PLoS One* 5: e10354.

Zaehle, T., Sandmann, P., Thorne, J. D., Jancke, L., and Herrmann, C. S. 2011. Transcranial direct current stimulation of the prefrontal cortex modulates working memory performance: Combined behavioural and electrophysiological evidence. *BMC Neurosci* 12: 2.

Ziemann, U. and Siebner, H. R. 2008. Modifying motor learning through gating and homeostatic metaplasticity. *Brain Stimul* 1: 60–66.

Part V

Therapeutic Applications

Therapeutic Applications

14 Therapeutic Applications of Transcranial Magnetic Stimulation/Transcranial Direct Current Stimulation in Neurology

Anli Liu, Felipe Fregni, Friedhelm C. Hummel, and Alvaro Pascual-Leone

CONTENTS

14.1 INTRODUCTION

Studies in both animals and humans have demonstrated that transcranial magnetic stimulation (TMS) and transcranial direct current stimulation (tDCS) can modulate brain activity in a noninvasive manner. Both techniques can induce changes in cortical excitability outlasting the duration of the stimulation itself (Chen et al. 1997; Gangitano et al. 2002; Hummel and Cohen 2005; Romero et al. 2002). Depending on stimulation parameters, activity in the targeted brain region can be facilitated or suppressed, with variable behavioral consequences.

Because of their noninvasive nature, there has been significant interest in exploring the diagnostic and therapeutic applications of TMS and tDCS in neurology.

Noninvasive brain stimulation offers a complementary therapeutic modality to pharmacological treatments. While most psychoactive drug therapies work by substitution of endogenous neurotransmitters (e.g., replacement of dopamine in Parkinson's disease), brain stimulation is thought to utilize mechanisms of plasticity to promote changes in neural circuitry. The ultimate goal in neurostimulation is to tune neural network activity and brain plasticity toward adaptive behavior.

In TMS, a brief pulse of current passing through a coil of wire held over the subject's head generates a rapidly changing magnetic field. This alternating field penetrates through skin, scalp, and skull with minimal distortion or attenuation to generate a secondary electric current in the underlying cortex. This current can be of sufficient magnitude to depolarize neural elements.

Applied in trains of repetitive TMS (rTMS), the pulse frequency and pattern can be regulated to either enhance or suppress neural activity. In most subjects, a continuous train of low-frequency (≤ 1 Hz) pulses results in suppression, while bursting, intermittent trains of high-frequency (≥ 5 Hz) pulses results in facilitation of excitability in the targeted cortical region. Variations in coil geometry can generate a more restricted field of stimulation. For example, a figure-of-8 coil delivers a spatially more precise impulse than a circular coil (Hallett 2000).

More recently, newer protocols using theta burst stimulation (TBS), which deliver continuous or intermittent asynchronous trains with high- or low-frequency components, have demonstrated more durable and potent effect sizes. In most instances, when applied to healthy subjects, continuous TBS (cTBS) depresses cortical excitability, while intermittent (iTBS) enhances cortical excitability via the induction of long-term depression (LTD)- or long-term potentiation (LTP)-like mechanisms, respectively (Huang et al. 2005). However, prolonged theta burst paradigms have been demonstrated to induce a contrary effect. For example, when applied for twice the duration or number of pulses per session compared to standard protocols, normally facilitatory iTBS has been demonstrated to be inhibitory, while normally inhibitory cTBS becomes excitatory.

By contrast, tDCS is a method of applying a low-intensity (1–2 mA) direct current to the scalp to influence underlying cortical excitability. TDCS may depolarize or hyperpolarize neurons to modulate spontaneous firing rates (SFRs). Thus, tDCS can be thought of as a purely "neuromodulatory" intervention. Generally, both animal and human studies demonstrate increased spontaneous neuronal firing activity under the anode and decreased activity under the cathode (Bindman et al. 1964; Creutzfeldt et al. 1962; Nitsche and Paulus 2000; Purpura and McMurtry 1965). The direction of effect is also influenced by variables such as duration of stimulation, electrode montage, and concurrent cognitive activities (Brunoni et al. 2011b).

tDCS is an older technique than TMS and has some recognized advantages and disadvantages. From a pragmatic perspective, some benefits of tDCS include its low cost, portability, and ease of use. Because it induces less scalp sensation than TMS, tDCS has a more reliable sham condition, which allows for improved double-blinding in controlled clinical trials (Gandiga et al. 2006). Adequate blinding is particularly significant when testing subjective outcomes, such as mood and cognitive improvement, which are highly vulnerable to placebo effect.

The major limitation of tDCS is the delivery of a less focused stimulation than TMS. Direct current is delivered over relatively large electrodes (20–35 cm^2), which makes precise stimulation and cortical mapping more difficult. Some studies have reduced electrode size to produce a more restricted current, comparable to TMS-delivered stimulation (Nitsche et al. 2007). On the other hand, the inherently wider surface of stimulation delivered with tDCS may be beneficial when detailed knowledge about the cortical topography is unavailable. Furthermore, tDCS can easily be combined with other interventions such as mental imagery, computerized cognitive interventions, or robot-assisted motor activity. Much of the appeal of tDCS lays in its potential for use in these multimodal synergistic approaches and its practicality.

One reassuring aspect of continued use of TMS and tDCS in future clinical research is their safety and tolerability, now well established when practitioners follow consensus guidelines (Nitsche et al. 2008; Rossi et al. 2009). In TMS, reported adverse events have been infrequent and generally mild. The most common side effects include headache (23%) and neck pain (12%); rare events include nausea, tinnitus, mood fluctuations, and psychosis (Machii et al. 2006).

The induction of a seizure is the most serious possible complication associated with rTMS application. The risk of seizures seems greatest for high-frequency rTMS with short interval periods, during or immediately after stimulation. Up to 2008, there were 16 TMS-induced seizures reported, an exceptionally low figure given the number of subjects who have undergone TMS (Rossi et al. 2009). Most seizures induced from rTMS occurred before safety parameters were introduced and in the setting of subjects taking medications known to lower the seizure threshold (Rossi et al. 2009). Even in patients with epilepsy, the crude per-subject risk of developing a seizure has been reported to be 1.4% (Bae et al. 2007) with all but one case consisting of a typical seizure and no instances of status epilepticus reported. Likewise, there have been no reported seizures from published series of stroke patients, thought to have a lowered event threshold, who have received various forms of stimulation (50 Hz epidural stimulation, anodal tDCS, high-frequency rTMS) (Brown et al. 2006; Hummel et al. 2005; Khedr et al. 2005a; Kim et al. 2006).

To avoid adverse events, current guidelines recommend careful consideration of patient variables that may affect the seizure threshold, such as use of pro-epileptogenic medications (i.e., antidepressants, neuroleptics), timing in the menstrual cycle, age, level of anxiety, and sleep deprivation. Further safety parameters for stimulation intensity, frequency, train duration, and intertrain interval, as well as appropriate monitoring methods have been published (Rossi et al. 2009).

Because tDCS delivers only weak electric currents to modulate cortical excitability, it is widely considered to be even safer and more tolerable than TMS. Indeed, the most commonly reported adverse events have been itching, tingling, headache, burning sensation, nausea, fatigue, and insomnia. Seizures have never been reported (Brunoni et al. 2011a; Poreisz et al. 2007). Instead, the major possible complication is a heat-induced skin lesion (Palm et al. 2008). However, one meta-analysis has identified a selective bias for underreporting, as fewer than half of tDCS studies have systematically queried subjects for experienced side effects (Brunoni et al. 2011a).

14.2 MECHANISMS OF ACTION

The preliminary success of neurostimulation in diverse neurological conditions contrasts with the limited understanding of the underlying neurobiological effects. Neurophysiologic and neuroimaging studies in humans and animals reveal that TMS and tDCS induce both local and distant effects and that the effects are state dependent. However, further studies are needed to provide greater mechanistic insights. Specifically, animal and in vitro experiments promise to further clarify the mechanisms of action of tDCS and TMS and enable more specific therapeutic approaches.

There are several lines of evidence that demonstrate the local effect of TMS in altering concentrations of neurotransmitters, cerebral perfusion, and cortical excitability. In a landmark study using cat visual cortex, Allen et al. (2007) applied rTMS to induce a brief increase in spontaneous neural spiking, with augmented neural activity and hemodynamic changes at higher stimulation frequencies, intensity, and duration. When TMS was applied during visual activity, a longer-lasting depression in evoked baseline neural activity was induced, coupled with a more sustained decrease in hemoglobin concentration and tissue oxygenation (Allen et al. 2007). Together, these findings provide a hemodynamic explanation of how TMS can induce local effects outlasting TMS stimulation and underscore the state-dependent nature of the response (Bestmann 2008).

Glutamate appears to play a key neurotransmitter role in mediating cortical excitability. Luborzewski et al. performed a magnetic resonance spectroscopy (MRS) study investigating high-frequency (20 Hz) TMS applied to the dorsolateral prefrontal cortex (DLPFC) in patients with depression. Responders had decreased baseline concentrations of glutamate in the prefrontal cortex, which increased after stimulation in an intensity-dependent manner (Luborzewski et al. 2007). Similarly, a significant relationship between MRS-assessed glutamate levels and TMS measures of cortical excitability has been found, whereas no relationship was found for GABA activity (Stagg et al. 2011).

Newer protocols combining rTMS with neuroimaging have begun to elucidate the widespread interconnected networks affected beyond local stimulation (Bestmann et al. 2003, 2004, 2008). For example, rTMS applied to the sensorimotor cortex induces MRI BOLD signal changes in a network of primary and secondary motor regions including M1/S1, supplementary motor area, dorsal premotor cortex, cingulate, putamen, and thalamus (Bestmann et al. 2004; Denslow et al. 2005), comparable to the regions activated in volitional movement.

The neuromodulatory effects of tDCS have been broadly attributed to LTP- and LTD-like mechanisms of synaptic plasticity (Hattori et al. 1990; Islam et al. 1995; Moriwaki 1991). Liebetanz et al. conducted a series of pharmacologic experiments demonstrating the role of specific membrane elements in tDCS-induced neuroplasticity. Dextromethorphan, an NMDA (N-methyl-D-aspartic acid) antagonist, decreased the post-tDCS effects of both anodal and cathodal stimulation. Carbamazepine, which blocks sodium channels, selectively suppressed the effect of anodal stimulation, suggesting that LTP requires the depolarization of action potentials (Liebetanz et al. 2002). In animal models, Kabakov et al. (2012) recently showed that tDCS induces changes in excitability and synaptic plasticity that are dependent on the precise spatial relation of the direction of induced current and the orientation of affected neural pathways.

Investigations combining functional neuroimaging with tDCS have begun to define the local and distant effects of neuromodulation. A recent MR spectroscopy study demonstrated a local reduction of GABA in response to anodal stimulation, while cathodal stimulation resulted in a local decrease in glutamatergic activity (Stagg et al. 2009). An MR arterial spin labeling (MR-ASL) study applying tDCS to the primary motor cortex in healthy subjects demonstrated that anodal stimulation induced a significant increase in resting-state cerebral blood flow (rCBF) both during and after stimulation, in a linear relationship to current intensity. However, cathodal tDCS induced a smaller increase in perfusion during stimulation but a sustained decrease in the poststimulation period (Zheng et al. 2011). Of interest, the magnitude of observed increase in rCBF in anodal tDCS is comparable to the range of changes seen in PET or ASL studies of high-frequency (10 Hz) TMS (Moisa et al. 2010), while the lower increases in rCBF in cathodal stimulations is comparable to the changes seen with low-frequency (<2 Hz) TMS (Fox et al. 1997, 2006; Moisa et al. 2010).

Anodal tDCS applied to primary motor cortex has elicited more widespread perfusion changes in functionally related but distant regions such as the ipsilateral premotor cortex, with a lesser influence on the contralateral motor and premotor regions (Zheng et al. 2011). Distant effects of tDCS have also been demonstrated in PET (Lang et al. 2005) and fMRI studies (Kwon et al. 2008; Stagg et al. 2011) and resembled the patterns discovered in TMS studies (Moisa et al. 2010). Similarly, anodal tDCS applied to the primary motor cortex, with cathodal tDCS applied to the contralateral frontopolar cortex, has been demonstrated to increase functional connectivity patterns within premotor, motor, and sensorimotor areas within the stimulated hemisphere (Polania et al. 2011a,b).

A recent fMRI study applying tDCS to the DLPFC illustrates how stimulation can alter resting-state network connectivity (Keeser et al. 2011). Real tDCS caused significant changes in regional brain connectivity in the default mode network (DMN) and the fronto-parietal networks (FPNs), both local to the stimulation site and in associated brain regions. Similar insights have been obtained in studies in patients with Parkinson's disease (Pereira et al. 2012). Strengthened connectivity within the DMN has been associated with improved working memory (Hampson et al. 2006) and semantic memory (Wirth et al. 2011). Similarly, increased coactivation in the frontal and parietal regions has been implicated in attention and working memory. In particular, increased connectivity within the left FPN has been observed after cognitive training (Lewis et al. 2009; Mazoyer et al. 2009).

14.3 NEUROTHERAPEUTIC EXPERIENCE WITH TMS/tDCS TO DATE

This review summarizes recent research on therapeutic applications of TMS and tDCS in neurology, concentrating on developments in the field over the last 6 years. We conducted a literature search for articles published from 2005 to 2011, using the search terms "transcranial magnetic stimulation," "transcranial direct current stimulation," with "stroke rehabilitation," "epilepsy," "pain," "Parkinson's disease," "tinnitus," and "ataxia." We included mostly randomized controlled trials in our review, although occasionally smaller case series are included when prospective or controlled studies were sparse or unavailable.

14.3.1 FOCAL EPILEPSY

Seizure foci are characterized by an increase in focal irritability, caused by a pathological increase in excitatory (glutaminergic) terminals with a decrease in inhibitory (GABA-ergic) activity (Lowenstein 1996). It seems reasonable to assume that inhibitory neuromodulation (slow repetitive TMS or cathodal tDCS) might be able to induce LTD or normalization of a hyperexcitable territory, and that this may translate into a therapeutic advantage. Low-frequency stimulation in hippocampal and neocortical rat slices has already been demonstrated to decrease interictal discharge and seizure frequency, in a manner that outlasts the stimulation (Albensi et al. 2004; Schiller and Bankirer 2007). By reducing cortical irritability, targeted suppression could potentially treat medication-refractory focal seizure activity.

Since 2005, there have been several randomized prospective rTMS studies published, which demonstrate mixed efficacy in decreasing seizure frequency (Table 14.1). Fregni et al. (2006f) and Santiago et al. (2008) both performed low-frequency rTMS directly over the epileptogenic cortex and demonstrated a significant reduction in seizure frequency, with clinical benefit persisting at a 2 month follow-up interval. However, Joo et al. (2007) and Cantello et al. (2007) conducted similar randomized prospective studies of low-frequency rTMS, which did not result in a reduction in seizure frequency.

The heterogeneity among the study designs precludes a definitive explanation of the discrepancy in findings. However, one plausible explanation could be the inclusion of patients with multifocal-, indeterminate-, or even generalized-onset epilepsy in the studies performed by Joo et al. (2007) and Cantello et al. (2007). Neurostimulation may have more pronounced effects in patients with well-circumscribed and superficial seizure foci. Of note, while these latter studies did not produce a clinical benefit, they did show a significant decrease in the frequency of interictal spikes, raising the possibility of a subthreshold clinical effect.

Furthermore, there exist a few case reports and case series describing the use of active rTMS for interrupting epilepsy partialis continua (EPC). These small studies demonstrate mixed efficacy, but good safety. The largest series by Rotenberg included seven cases of refractory EPC lasting longer than 1 day. TMS stopped seizures in two patients, caused a 20–30 min seizure cessation in three cases, and had no effect in two cases. However, there were no cases of seizure exacerbation and the reported side effects were mild (Rotenberg et al. 2009).

There have been two prospective controlled studies investigating the use of tDCS to control partial-onset epilepsy. In a randomized controlled trial, Fregni et al. (2006g) found that one 20 min session of 1 mA cathodal tDCS applied over the site of cortical malformation produced a significant reduction in the frequency of interictal discharges and a trend toward decreased seizure frequency (−44%, 95% CI −95% to 7.1%). Varga et al. (2011) also applied one 20 min session of 1 mA cathodal tDCS to the epileptogenic zone in five patients with focal, refractory continuous spike and slow wave during sleep, but did not see a reduction in the spike index. They hypothesize

TABLE 14.1
Comparison of Neurostimulation Protocols for Epilepsy

References	No. of Subjects	Diagnosis	Coil Position	rTMS Frequency	No. of Stimuli	Intensity	Coil	Duration	Session Schedule	Effect
Kinoshita et al. (2005)	7	Medically refractory extratemporal lobe epilepsy	Region of most prominent epileptic activity seen in EEG	0.9 Hz	810 pulses (15 min ON–5 min OFF)	90% MT	Round coil	15 min	Two sessions daily for 5 days	Nonsignificant decrease in seizure frequency over 2 weeks follow-up
Fregni et al. (2005c)	8	Cortical malformations	Cortical malformation	0.5 Hz	600 pulses	65% MT	Figure 8	20 min	One session	Significant decrease in seizure frequency in 4 week follow-up, with mean reduction in seizure frequency of 51.2%. Decrease in epileptiform discharges at 2 and 4 weeks.

(continued)

TABLE 14.1 (continued)
Comparison of Neurostimulation Protocols for Epilepsy

References	No. of Subjects	Diagnosis	Coil Position	rTMS Frequency	No. of Stimuli	Intensity	Coil	Duration	Session Schedule	Effect
Fregni et al. (2006f)	21	Cortical malformations	Cortical malformation	1 Hz	1200 pulses	70% motor output (MO)	Figure 8	20 min	5 days	Significant decrease in seizure frequency, effect lasting ≥2 months; significant decrease in interictal discharges immediately after and at 4 weeks
Joo et al. (2007)	35	18 focal onset, 17 multifocal or indeterminate onset	Cz (n = 17), temporal (n = 12), L frontal (n = 3), R parietal (n = 3)	0.5 Hz	3000 pulses/train (n = 19), 1500 pulses/train (n = 16)	100% MT	Figure 8 Or round	100 min (n = 19), 50 min (n = 16)	5 days	No significant reduction in seizure frequency; over 8 weeks follow-ups decreased frequency of interictal discharges (p < 0.05), with disappearance in 17.1% of patients

Cantello et al. (2007)	43	Drug-resistant epilepsy. Mixed focal, multifocal, and diffuse onset	Vertex	0.3 Hz	500 pulses/train	100% MT (n = 34), 65% MO (n = 9)	Round	30 min	Two trains daily × 5 days	No significant reduction of seizure frequency; decreased interictal EEG discharges in one-third of patients ($p < 0.05$)
Santiago-Rodriguez et al. (2008)	12	Focal epilepsy (frontal and frontotemporal)	Seizure focus	0.5 Hz	900 pulses	110% MT	Figure 8	30 min	10 days	Significant reduction in seizure frequency by 71% during intervention period; significant reduction in seizure frequency by 50% during 8 week follow-up period
Brodbeck et al. (2010)	5	Focal epilepsy	Seizure focus	6 Hz priming/ 1 Hz stim	1200 pulses	90% MT/ 110% MT	Figure 8	10/10 min	One session	Variable reduction in spike frequency

(continued)

TABLE 14.1 (continued)
Comparison of Neurostimulation Protocols for Epilepsy

References	No. of Subjects	Diagnosis	Electrode Position	tDCS Polarity	Intensity	Duration	Session Schedule	Effect
Fregni et al. (2006f)	19	Cortical malformation	Cortical malformation	Cathodal tDCS	1 mA	20 min	One session	Significant reduction in interictal discharge frequency; nonsignificant trend toward reducing seizure frequency
Varga et al. (2011)	5	Focal-refractory continuous spike and waves during slow-wave sleep	Focus of epileptiform discharges	Cathodal tDCS	1 mA	20 min	One session	No significant reduction of spike index
San-Juan et al. (2011)	2	Rasmussen's encephalitis	Focus of epileptiform discharges	Cathodal tDCS	1 and 2 mA	60 min	Four sessions	One patient with significant reduction in seizure frequency; second patient seizure free at 6 and 12 month follow-up

that their use of a 25 cm² cathodal electrode may have been too small to prevent propagation of the epileptiform discharges. Finally, San-Juan et al. (2011) reported on two patients with atypical (i.e., adult onset) dominant hemisphere Rasmussen's encephalitis, treated with 60 min of 1 or 2 mA tDCS over four sessions in 2 months (days 0, 7, 30, and 60). At the 6 and 12 month follow-up, one patient was completely seizure free and the other had a significant reduction in seizure frequency.

Given these preliminary results, more studies investigating the use of longer and repeated sessions of cathodal tDCS applied to seizure foci are needed. Thus far, studies demonstrate mixed efficacy in decreasing seizure frequency and terminating focal status epilepticus. Future research might employ stricter inclusion criteria, i.e., patients with definitive focal-onset superficial epileptogenic foci such as seen with cortical malformations. Once these superficial malformations are identified, more precise methodologies targeting the irritable cortex are needed, e.g., real-time EEG monitoring and image guidance (Rotenberg 2010).

As described earlier, studies have varied by stimulation parameters, which may partly explain the variability in efficacy. Systematic investigations of variables such as coil size and position, stimulus frequency and intensity, and number of sessions are needed to determine optimal efficacy and effect size. Some novel patterns of stimulation such as TBS seem to offer a more durable depression and deserve further trials. Special coil designs that allow deeper penetration into the brain, reaching structures such as the insula or cingulate cortex, may potentially modulate deeper epileptogenic foci (Roth et al. 2007).

14.3.2 CHRONIC PAIN

Chronic pain has been attributed to maladaptive changes in both the central and the peripheral nervous system. Increased activity in peripheral nerve endings leads to oversensitization followed by central changes. Stimulation of cortical targets may normalize the activity of the corticothalamic network. Imaging research demonstrates that stimulation of the motor cortex via high-frequency rTMS or anodal tDCS alters distant activity in the thalamic and subthalamic nuclei, which may explain how pain perception is altered (Garcia-Larrea et al. 1997, 1999; Peyron et al. 1995).

The majority of rTMS and tDCS studies stimulating the M1 motor cortex demonstrate a significant effect on both subjective and objective pain perception (Table 14.2). Clinical benefit has been seen in patients with chronic neuropathic pain (Fregni et al. 2006a; Leung et al. 2009; Antal et al. 2010; Soler et al. 2010), visceral pain (Fregni et al. 2005a) as well as fibromyalgia (Fregni et al. 2006d; Passard et al. 2007; Antal et al. 2010). Response rates are high, ranging between 40% and 80% among patients with central pain refractory to treatment (Brown and Barbaro 2003; Khedr et al. 2005b; Lefaucheur et al. 2004; Nuti et al. 2005b; Pleger et al. 2004; Tsubokawa et al. 1991). Moreover, positive results are clinically meaningful, with rTMS offering a 20%–45% reduction in pain perception (Andre-Obadia et al. 2006; Khedr et al. 2005b; Lefaucheur et al. 2004; Pleger et al. 2004), with perhaps a larger effect with tDCS, up to 58% reduction (Fregni et al. 2006a). There is some evidence that rTMS has a greater effect on centrally over peripherally mediated neuropathic pain states (Leung et al. 2009).

TABLE 14.2

Comparison of Neurostimulation Studies for Pain

References	No. of Subjects	Pain Etiology	Coil Position	Method of Localization	rTMS Frequency	No. of Stimuli	Intensity	Type of Coil	Duration	Session Schedule	Effect
Fregni et al. (2005a)	5	Visceral pain from chronic pancreatitis	R/L secondary somatosensory area	MRI–stereotactic guidance	1/20 Hz	1600 pulses	1/20 Hz	Figure 8	Variable	One session weekly × 6 weeks	Only 1 Hz rTMS to R side with significant pain reduction, mean decrease 62%
Khedr et al. (2005b)	48	Chronic unilateral neuropathic pain (trigeminal neuralgia or post-stroke)	M1 hand area contralateral to pain	10/20 EEG system	20 Hz	2000 pulses	80% MT	Figure 8	10 min	5 days	Significant decrease in trigeminal neuralgia and post-stroke pain, lasted 2 weeks after treatment
Borckardt et al. (2006)	20	Postsurgical (gastric bypass) pain	L DLPFC	5 cm anterior to motor cortex	10 Hz	4000 pulses (10 s ON–20 s OFF)	100% MT	Figure 8	20 min	One session immediately after surgery	Significant pain reduction in real TMS group, with 40% less morphine usage

Study	N	Condition	Target	Location	Frequency	Pulses	Intensity	Coil	Duration	Sessions	Outcome
Passard et al. (2007)	30	Fibromyalgia	L M1	10/20 EEG system (C3)	10 Hz	2000 pulses (8 s ON–52 s OFF)	80% MT	Figure 8	25 min	10 sessions	Significant reduction of pain after fifth treatment session, persisted up to 2 weeks
Borckardt et al. (2008)	20	Postsurgical (gastric bypass) pain	L DLPFC	5 cm anterior to motor cortex	10 Hz	4000 pulses (10 s ON–20 s OFF)	100% MT	Figure 8	20 m	One session immediately after surgery	Significant pain reduction in real TMS group, with 35% less morphine usage

(continued)

TABLE 14.2 (continued)
Comparison of Neurostimulation Studies for Pain

References	No. of Subjects	Pain Etiology	Electrode Position	Method of Localization	tDCS Polarity	Intensity	Duration	Session Schedule	Effect
Fregni et al. (2006a)	17	Central pain, traumatic spinal cord injury	M1 contralateral to side of pain or dominant hemisphere M1	10/20 EEG system (C3 or C4)	Anodal tDCS	2 mA	20 min	5 days	Significant pain reduction in 3 days, maximum pain reduction in 5 days during treatment
Fregni et al. (2006d)	32	Fibromyalgia	L M1 or DLPFC	10/20 EEG System (C3 or F3)	Anodal tDCS	2 mA	20 min	5 days	Anodal tDCS of M1 with greater pain reduction than sham or stimulation of DLPFC, still significant within 3 weeks after treatment

Study	N	Condition	Target	Location	Electrode	Type	Intensity	Duration	Days	Outcome
Soler et al. (2010)	39	Neuropathic pain from spinal cord injury	M1	contralateral to side of pain	10/20 EEG System (C3 or C4)	Anodal tDCS ± visual illusion	2 mA	20 min	10 days	Real tDCS with visual illusion reduced pain more significantly than any other subgroup, with significant improvement within 12 weeks after treatment
Antal et al. 2010	21	Chronic pain (trigeminal neuralgia, poststroke pain position)	M1	Hand area determined by TMS	Anodal	1 MA	20 min	5 days	Anodal tDCS decreased pain rating greater than sham tDCS, lasting 3-4 weeks after treatment
Mori et al. 2010	19	Chronic neuropathic pain in MS	M1	10/20 EEG system (C3/C4)	Anodal	2 mA	20 min	5 days	Significant decrease in pain scores in patients w real anodal tDCS, lasting 3 weeks

For example, in a blinded randomized, sham-controlled trial, Passard et al. (2007) demonstrated that 10 days of high-frequency rTMS (10 Hz, 8 s stimulation trains, 52 s intertrain interval, 2000 pulses per day, 80% MT) applied to the primary motor cortex in patients with fibromyalgia showed a significant decrease in subjective pain perception and improved quality of life. The effects were seen at 5 days into stimulation and lasted up to 2 weeks after treatment. There were no significant changes in mood or anxiety. Other studies with slightly different stimulation parameters have shown a more durable effect, up to 15 days after 3 days of consecutive TMS stimulation. There are some reports of longer-lasting effects with tDCS, up to 3 months after treatment (Fregni et al. 2006d; Gabis et al. 2009; Soler et al. 2010)

Antal et al. (2011) applied cathodal versus sham tDCS over the visual cortex in a group of migraineurs using a crossover design (1 mA for 15 min each session, three sessions a week, 3 weeks per treatment arm). Real cathodal tDCS significantly reduced the intensity and duration of headache pain compared to baseline, suggesting that tDCS could be an effective prophylactic therapy in migraine.

In addition, there exists preliminary evidence of synergistic effects with other modalities of pain therapy. Soler et al. (2010) performed a sham-controlled double-blinded parallel group design of tDCS applied to the premotor cortex in patients with neuropathic pain following spinal cord injury. The combination of tDCS and visual illusion reduced the intensity of neuropathic pain significantly more than either treatment alone, with a benefit at 12 weeks after treatment. Furthermore, given that mechanisms of noninvasive brain stimulation are associated with top-down modulation, association with bottom-up approaches such as transcutaneous electrical nerve stimulation or diffuse noxious inhibitory control may augment analgesic effects as recently demonstrated (Boggio et al. 2009; Fregni 2010).

Recently, the modulatory role of the DLPFC in pain sensation has been explored. Some preliminary research on stimulating the DLPFC shows a modest effect size, apparently less than the stimulation of the primary motor cortex. Borckardt et al. (2009b) demonstrated that high-frequency rTMS applied to the left DLPFC exhibited a modest improvement in daily pain ratings in patients with chronic pain (19% decrease lasting 2 weeks), as well as an increase in thermal and mechanical pain thresholds. In the postoperative setting, rTMS applied to the left prefrontal region increased thermal pain thresholds, whereas sham TMS did not (Borckardt et al. 2009a). In the clinical setting, a single 20 min session of rTMS (10 Hz, 10 s stimulation trains, 20 s intertrain intervals, 4000 pulses total, 100% MT) applied immediately after surgery reduced patient-controlled morphine use by approximately 40%, independent of mood effects (Borckardt et al. 2006, 2008). Given the limitations of current pain management and the clinically meaningful results demonstrated thus far, further investigation is warranted. Notably, there have been no studies of tDCS in the postoperative setting to date (Borckardt et al. 2009a).

The role of neurostimulation in managing medication-refractory pain syndromes seems especially promising, with consistent results replicated across a number of studies. However, there are a number of questions still unresolved. For example, further interaction studies are needed to demonstrate the synergistic effect of different therapeutic modalities, i.e., with concurrent administration of analgesics or biofeedback mechanisms. While there has been a suggestion of superior efficacy of stimulation of the primary motor cortex compared to the DLPFC, more direct comparison studies are needed. Third, quality of life variables should be included in research outcome measures to determine what numerical threshold of pain reduction is clinically significant. Finally, given the suggestion that tDCS may induce longer-lasting effects, more investigations incorporating this technology, especially in the postoperative setting, are needed.

14.3.3 STROKE REHABILITATION: MOTOR FUNCTION

Functionality after stroke appears to reflect how the undamaged brain, including both adjacent and contralateral territories, adapts to injury. Severe functional impairment and structural damage to primary motor pathways after stroke result in widespread activation of bilaterally distributed primary and secondary motor regions during a hand grip exercise (Ward 2006, 2011; Ward et al. 2006). Conversely, patients with a better stroke outcome had a more similar activation pattern on fMRI compared to normal subjects (Ward et al. 2003). Together, these observations suggest that disruption of corticospinal tracts after stroke leads to greater reliance on a more distributed secondary motor network (Ward 2011).

However, the relationship between the healthy and lesioned hemispheres during stroke recovery is unclear. Human and animal models of stroke recovery demonstrate an altered interhemispheric dynamic (Dijkhuizen et al. 2003; Marshall et al. 2000). Some perturbations in this dynamic appear maladaptive and may limit recovery. The predominant model proposes that relative overactivity of the unaffected hemisphere may cause excessive transcallosal inhibition of the unaffected hemisphere (Grefkes et al. 2008; Murase et al. 2004). The concept of interhemispheric rivalry has been supported by empirical demonstrations with normal subjects (Kobayashi et al. 2004; Plewnia et al. 2003; Schambra et al. 2003; Williams et al. 2010), where down-regulation of one hemisphere improved function in the ipsilateral hand. A competitive dynamic has also been inferred from functional neuroimaging studies showing increased contralateral activity following stroke affecting motor function, attention, memory, and language (Belin et al. 1996; Duque et al. 2005; Kinsbourne 1997; Murase et al. 2004; Najib and Pascual-Leone 2011; Sparing et al. 2009). Furthermore, some neuroimaging evidence suggests that optimal stroke rehabilitation is associated with the development of alternative pathways from the affected hemisphere, without contribution from the unaffected hemisphere to the paretic limb (Hallett 2001).

Neurostimulation approaches based on this model of interhemispheric competition may work by suppressing activity in the healthy hemisphere or increasing activity in the lesioned hemisphere. Theoretically, restoration of interhemispheric balance can favor a more adaptive plasticity (Fregni and Pascual-Leone 2006; Hummel and Cohen 2006). Indeed, there have been a number of proof-of-principle demonstrations utilizing high-frequency rTMS (Chang et al. 2010; Khedr et al. 2005a) and anodal tDCS (Hesse et al. 2007; Hummel and Cohen 2005) over adjacent cortex in the lesioned hemisphere, as well as low-frequency rTMS over the contralesional motor cortex (Fregni et al. 2006c; Liepert et al. 2007; Mansur et al. 2005; Nowak et al. 2008), which have all shown significant improvement in motor function (Table 14.3). One of the largest studies involved 52 patients who received either real or sham rTMS

TABLE 14.3
Comparison of Neurostimulation Protocols for Motor Rehabilitation

References	No. of Subjects	Time Post-Stroke	Coil Position	Methodology	Frequency	No. of Stimuli	Intensity	Coil	Duration	Session Schedule	Effect
Khedr et al. (2005a)	52	Acute 5–10 days	Ipsilesional M1	Real vs. sham rTMS stimulation to ipsilesional M1 with standard physical therapy	3 Hz	300 pulses (10 s ON–50 s OFF)	120% MT	Figure 8	10 min	10 days	Real rTMS with higher independence scores and milder disability at 10 days posttreatment than sham group
Mansur et al. (2005)	10	Chronic <12 months	Contralesional M1 and premotor	Crossover design (healthy M1, premotor, sham)	1 Hz	600 pulses	100% MT	Figure 8	10 min	One session per arm, 1 h between arms	Decrease in reaction time and improved Purdue Pegboard test in paretic hand with real rTMS to healthy M1

Study	N										Outcome
Fregni et al. (2006c)	15	Chronic >1 year	Contralesional M1	Inhibitory rTMS to unaffected M1	1 Hz	1200 pulses	100% MT	Figure 8	20 min	5 days	Significant improvement in motor function of paretic hand lasting for 2 weeks
Liepert et al. (2007)	12	Acute (mean 7.3 ± 4.5 days)	Contralesional M1	Inhibitory rTMS to unaffected M1 in crossover design	1 Hz	1200 pulses	90% MT	Figure 8	20 min	One session	Significant improvement in dexterity of paretic hand
Malcolm et al. (2007)	19	Chronic >1 year (mean 3.8 ± 3.3 years)	Ipsilesional M1	Constraint-induced therapy (CIT) ± rTMS	20 Hz	2000 pulses (2 s ON–28 s OFF)	90% MT	Figure 8	25 min	10 days	No additional effect of rTMS to CIT
Nowak et al. (2008)	15	Subacute 4 weeks–4 months	Contralesional M1	Unaffected M1, crossover, sham-controlled design	1 Hz	600 pulses	100% MT	Figure 8	10 min	One session	Significant improvement in kinematics and finger and grasp movement of paretic hand

(continued)

TABLE 14.3 (continued)
Comparison of Neurostimulation Protocols for Motor Rehabilitation

References	No. of Subjects	Time Post-Stroke	Coil Position	Methodology	Frequency	No. of Stimuli	Intensity	Coil	Duration	Session Schedule	Effect
Takeuchi et al. (2009)	30	Chronic >6 months (mean 26.1 ± 28.0 months)	Contralesional M1 (1 Hz) or ipsilesional M1 (10 Hz), or both	rTMS with motor training	1 Hz/10 Hz	1000 pulses/ hemisphere	90% MT	Figure 8	Variable	20 sessions	Bilateral rTMS with motor training improved pinch force more than 1 Hz rTMS for 1 week; no effect of 10 Hz rTMS on motor function
Khedr et al. (2009a)	36	Acute 7–20 days (mean 17.1 ± 3.6 days)	Ipsilesional M1 (3 Hz) or contralesional M1 (1 Hz)	Real TMS with standard physical therapy	3 Hz/1 Hz	900 pulses (10 s ON–2 s OFF)	130% MT (3 Hz)/ 100% MT (1 Hz)	Figure 8	15 min	5 days	Both real 3 and 1 Hz improved in hand function at 3 months, with 1 Hz over contralesional M1 performing better than 3 Hz group

Study	N	Time since stroke	Site	Intervention	Frequency	Pulses	Intensity	Coil	Duration	Sessions	Outcome
Chang et al. (2010)	28	Acute <1 months (mean 12.9 days ± 5.2)	Ipsilesional M1	rTMS with motor training	10 Hz	1000 pulses (5 s ON–55 s OFF with motor training)	90% MT	Figure 8	20 min	10 days	rTMS improved motor function greater than motor training alone, with lasting effects 3 months after stroke
Emara et al. (2010)	60	Subacute >1 month (2–14 months)	Ipsilesional M1 (5 Hz)/contralesional M1 (1 Hz)	rTMS and standard physical therapy	5 Hz/1 Hz	750 pulses 150 pulses	80%–90% MT/110%–120% MT	Figure 8	2.5 min	10 days	Patients with real 5 or 1 Hz rTMS showed improvement on finger tapping, functional status, and disability score at 2 and 12 weeks

(continued)

TABLE 14.3 (continued)
Comparison of Neurostimulation Protocols for Motor Rehabilitation

References	No. of Subjects	Time Post-Stroke	Electrode Position	tDCS Polarity	Intensity	Duration	Session Schedule	Effect
Hummel et al. (2005)	6	Chronic >1 year (3.7 ± 1.1 years)	Ipsilesional hand area	Anodal tDCS	1 mA	20 min	One session, crossover design	Hand function in paretic side improved with real tDCS, outlasting stimulation
Hesse et al. (2007)	10	Subacute 4–8 weeks	Ipsilesional hand area	Anodal tDCS with robot-assisted motor training	1.5 mA	20 min	30 sessions	Arm function of three patients (two with subcortical stroke improved significantly
Lindenberg et al. (2010)	20	Subacute (mean 40.3 ± 23.4)	Bilateral hemisphere	Anodal tDCS to ipsilesional hemisphere and cathodal tDCS to contralesional hemisphere	1.5 mA	30 min	Five sessions	Improved motor function for 1 week post-intervention

over the M1 of the affected hemisphere in the acute phase (5–10 days) after their stroke, in conjunction with standard rehabilitation therapy. Patients who received real rTMS demonstrated a clinical improvement in motor function, without significant side effects (Khedr et al. 2005a). There have also been several studies in laboratory animals demonstrating that electrical stimulation of the motor cortex around the lesion improved performance in the paretic limb, especially when combined with a training regimen (Adkins et al. 2006, 2008; Adkins-Muir and Jones 2003; Kleim et al. 2003).

On the other hand, there is some evidence that the contralateral premotor cortex facilitates the function of the lesioned motor cortex. Disruption of ipsilesional (Fridman et al. 2004) or contralateral (Johansen-Berg et al. 2002) dorsal premotor cortex using online TMS worsens motor performance in patients with chronic stroke, suggesting the supportive role of these regions after injury. Bestmann et al. (2010) used a paired coil TMS technique to probe the influence of the contralateral dorsal premotor (cPMd) cortex over the ipsilateral M1 (iM1) in a group of 12 patients with subcortical stroke. For patients with poor recovery, the influence of cPMd over iM1 appeared more excitatory at rest (Bestmann et al. 2010). This observation appears to contradict the prior finding of excessive inhibition of cM1 over iM1 (Murase et al. 2004). One proposed explanation is that the inhibitory influence of cM1 over iM1, normally regulated by inputs adjacent to iM1, becomes pathologically inhibitory when these neighboring inputs are damaged by stroke. In a similar fashion, the facilitation that cPMd normally provides iM1 becomes overly excitatory after stroke (Ward 2011).

Furthermore, some critics argue that a model of interhemispheric rivalry is too simplistic for purposes of clinical rehabilitation. Most proof-of-principle studies conducted to date include either healthy subjects or small samples of stroke patients (Hummel et al. 2008). Patient variables such as stroke location (cortical/subcortical, motor/premotor), size (i.e., degree of disrupted corticospinal tract integrity), and age (Hummel et al. 2008) will likely influence the interhemispheric balance and should theoretically be considered in the therapeutic approach.

More recent research has explored simultaneous bihemispheric stimulation, i.e., enhancement of lesioned hemisphere, suppression of healthy hemisphere, compared to either enhancement or suppression of either hemisphere alone (Table 14.3). There is limited evidence for favorable outcomes with bihemispheric stimulation. For example, Takeuchi et al. (2009) showed that in the chronic phase of stroke recovery, bilateral rTMS was superior to inhibitory rTMS to the unaffected hemisphere, with the affect lasting 1 week after the stroke. Stimulating the lesioned hemisphere did not improve motor function at all. Lindenberg et al. (2010) also demonstrated that bihemispheric tDCS was superior to sham stimulation when combined with motor training in the chronic phase of recovery, with the effects persisting up to 1 week after treatment.

Moreover, newer protocols combining neuromodulation with more traditional rehabilitation methods can result in a synergistic effect, although the evidence is certainly mixed (Table 14.3). Both suppression of activity in the healthy hemisphere (using low-frequency rTMS) and potentiation of activity in the lesioned hemisphere (using high-frequency rTMS) have demonstrated positive results, with a suggestion of more consistent success with targeting of the healthy hemisphere. In the subacute phase of stroke recovery, Chang et al. (2010) found that rTMS stimulation of the affected primary motor cortex with simultaneous motor training provided additional improvement over motor training alone, persisting for 3 months after stroke. Lindenberg et al. (2010) found that bihemispheric tDCS modulation in the chronic phase of recovery with simultaneous physical and occupational therapy showed a significant improvement in hand function, with the effects lasting 5 days. However, Malcolm et al. (2007) did not show any additional benefit of stimulatory rTMS to the lesioned cortex tested with constraint-induced therapy in the chronic phase of stroke recovery.

More investigation on the optimal timing of intervention is needed. Most studies occur in the chronic stage (>12 months) after the infarct has remodeled into scar tissue. However, given the enhanced metabolic activity after a stroke and active process of cortical reorganization, earlier intervention might produce larger and more durable benefits. There have been a few studies conducted during the acute and subacute stages, but no studies comparing efficacy at different intervention times during recovery. Studies applying high-frequency rTMS or anodal tDCS stimulation to the affected hemisphere during the acute and subacute phases of recovery showed a significant benefit in performance, with no reports of seizures or other negative outcomes (Chang et al. 2010; Hesse et al. 2007; Khedr et al. 2005a).

14.3.4 STROKE REHABILITATION: LANGUAGE AND COGNITION

Some beneficial therapeutic effect of neurostimulation in cognitive rehabilitation has been demonstrated, in the realms of aphasia and hemispatial neglect (Table 14.4). Most investigations have been conducted with TMS, using either an online or an off-line stimulation paradigm. Online stimulation uses TMS to temporarily interfere with cognitive tasks. Because the effects last only from milliseconds to seconds, online TMS is useful for proof-of-principle experiments to establish that any given area of cortex is responsible for a certain behavior, rather than clinical effect (Miniussi et al. 2008). Miniussi et al. (2008) propose that transient response to online TMS may be useful to demonstrate residual and compensatory function and therefore potentially useful for patient selection purposes.

There have been several studies showing longer-lasting effects from off-line rTMS stimulation or repeated stimulation before task performance. For example, Naeser et al.'s (2005) study of off-line TMS used to suppress the right homologue of Broca area in four patients with chronic aphasia demonstrated a significant improvement in picture naming at 2 months poststimulation, with effects lasting up to 8 months. Another study investigated the effect of anodal tDCS targeted to the perilesional brain area with greatest fMRI activation during a picture-naming task. In a double-blinded crossover design including only patients with fluent aphasia, subjects who received real anodal tDCS had higher picture-naming speed compared to those who received sham treatment (Fridriksson et al. 2011).

There has also been some promising work in patients with hemispatial neglect, usually following right parietal injuries. Low-frequency rTMS over the unlesioned parietal cortex has shown improvement in attention to the ipsilateral space, which was seen between 2 and 4 weeks after treatment, with lingering effects to 6 weeks (Shindo et al. 2006). A study utilizing tDCS for stroke patients with left hemifield neglect found that real anodal stimulation to the lesioned posterior parietal cortex

TABLE 14.4
Comparison of Neurostimulation Protocols for Cognitive Rehabilitation

References	No. of Subjects	Diagnosis	Coil Position	Method of Localization	Stimulation	Frequency	No of Stimuli	Intensity	Coil	Duration	Session Schedule	Effect
Cotelli et al. (2006)	24	AD	L/R DLPFC	MRI guidance	Online rTMS	20 Hz	10 pulses	90% MT	Figure 8	500 ms	One session	Stimulation of L/R DLPFC improved accuracy of action naming
Naeser et al. (2005)	4	Chronic aphasia (5–11 years)	R Broca's homologue	MRI guidance	Off-line rTMS	1 Hz	1200 pulses	90% MT	Figure 8	20 min	10 days	Improved picture naming at 2 months posttreatment
Finocchiaro et al. (2006)	1	Primary progressive aphasia	L PFC	6 cm anterior and 1 cm ventral to M1	Off-line rTMS	20 Hz	400 pulses (2 s ON–30 s OFF)	90%	Figure 8	5 min	5 days	Improved sentence completion at 45 days posttreatment
Shindo et al. (2006)	2	Chronic R parietal stroke	Unaffected L posterior parietal cortex	10/20 EEG system (P5)	Off-line rTMS	0.9 Hz	900 pulses	95%	Figure 8	~17 m	6 days	Improved attention to ipsilateral space lasting up to 2–6 weeks

(continued)

TABLE 14.4 (continued)
Comparison of Neurostimulation Protocols for Cognitive Rehabilitation

References	No. of Subjects	Diagnosis	Coil Position	Method of Localization	Stimulation	Frequency	No of Stimuli	Intensity	Coil	Duration	Session Schedule	Effect
Cotelli et al. (2008)	24	Alzheimer's disease	L/R DLPFC	Template MRI Guidance	Online rTMS	20 Hz	10 pulses	90% MT	Figure 8	500 ms	One session	Improved action naming in mild AD group; improved action and object naming in mod/severe AD group

References	No. of Subjects	Diagnosis	Electrode Position	Method of Localization	tDCS Polarity	Intensity	Duration	Session Schedule	Effect
Sparing et al. (2009)	10	Subacute stroke-induced L visuospatial neglect	L/R posterior parietal cortex (PPC)	10/20 EEG system (P3/P4)	Anodal tDCS (ipsi PPC), Cathodal tDCS (contra PPC)	1 mA	10 min	One session, 1 h between treatment arms	Both inhibition of unlesioned PPC and stimulation of lesioned PPC reduced symptoms of visuospatial neglect
Baker et al. (2010)	10	Chronic stroke-induced aphasia	L frontal perilesional	fMRI guidance during picture-naming task	Anodal tDCS	1 mA	20 min	5 days	Improved accuracy of picture naming, lasting up to 1 week
Fridriksson (2011)	8	Chronic stroke-induced aphasia	L frontal Perilesional	fMRI guidance to active perilesional region on naming tasks	Anodal tDCS	1 mA	20 min	5 days	Decreased processing time in picture-naming task, lasting up to 3 weeks

and real cathodal stimulation over the unlesioned posterior parietal cortex both reduced symptoms of visuospatial neglect (Sparing et al. 2009)

These positive results in language and neglect rehabilitation should be considered preliminary because of the low numbers of patients, lack of rigorous controls, and testing of restricted functions. Larger studies with tighter controls on patients with stroke and neurodegenerative conditions, with measures of functional improvement, are still needed for clinical validation (Miniussi et al. 2008).

Thus far, neurostimulation protocols in rehabilitation to improve motor function, language, and spatial neglect have shown some promising results. There is some preliminary evidence that stimulation of both hemispheres and concurrent application of traditional rehabilitation practices may provide additional benefits. Further studies combining neurostimulation with more traditional methods of rehabilitation as well as determination of efficacy at various time points during stroke recovery are needed.

14.3.5 PARKINSON'S DISEASE

Parkinson's disease results from a degeneration of dopaminergic neurons in the substantia nigra pars compacta, resulting in more widespread network dysfunction in the basal ganglia and motor cortex. Treatment generally consists of dopamine supplementation, although existing medications are limited by motor fluctuations and peak-dose dyskinesias, which may be as debilitating as disease itself. While invasive methods of neurostimulation such as deep brain stimulation are widely used, there is significant clinical interest in noninvasive technologies.

One apparent challenge to noninvasive stimulation is the ability to stimulate deeper structures of the basal ganglia. More recently, the H-coil has been used to modulate the activity of deeper neural circuits to maximum depth of 6 cm (Harel et al. 2011, 2012; Bersani 2012).

An alternative method to reach deeper structures may be to use cortical targets as "windows" to influence more distributed neural networks. The primary and supplementary motor regions may qualify as candidate sites for neurostimulation because of their extensive glutaminergic projections to the basal ganglia. Some functional neuroimaging studies have suggested that early, milder PD is associated with a hypoactivity of the primary motor cortex. On the other hand, more advanced PD is associated with hyperactivity, which may represent medication-induced cortical reorganization in response to deficient subcortical motor pathways and may partly explain why advanced PD patients experience dyskinesias (Haslinger et al. 2001; Sabatini et al. 2000). Stimulating motor and premotor cortices may treat hypokinetic features such as bradykinesia or gait freezing, whereas suppressing these regions may treat hyperkinetic features such as dyskinesias.

A number of excitatory neurostimulation protocols have demonstrated a modest benefit on bradykinesia and gait instability (Hamada et al. 2008, 2009, Lomarev et al. 2006, Fregni et al. 2006b). In a metanalysis, Fregni et al. (2005b) pooled 12 studies of TMS and electroconvulsive therapy (ECT) efficacy on motor symptoms of patients with PD and found a small but significant effect size for TMS. ECT carried a larger effect size; however, this finding was supported by a smaller subset of studies.

In another meta-analysis of controlled clinical trials of TMS in PD, Elahi et al. (2009) concluded that high-frequency rTMS studies produced a significant improvement on motor symptoms as reflected by lower UPDRS scores, whereas low-frequency rTMS studies demonstrated significant variability and did not.

Arguably the most robust and durable clinical effect was seen in the study performed by Lomarev et al. (2006). In this study, four cortical targets (left and right M1, left and right DLPFC) received high-frequency (25 Hz) stimulation during each session, with several sessions over a 4 week period. The real rTMS sessions had a significant effect in reducing bradykinesia as well as improving gait speed, which remained significant even 1 month after the end of treatment. Lomarev et al.'s protocol differs from others in several ways, including the extremely high frequency of stimulation, the stimulation of multiple cortical targets, as well as the number of sessions occurring over a prolonged period of time. Each of these variables deserves further exploration in controlled comparative studies.

Fregni et al. (2006b) conducted a double-blind sham-controlled tDCS study of 17 patients, comparing anodal stimulation of M1, cathodal stimulation of M1, anodal stimulation of the DLPFC, and sham stimulation. Only active anodal stimulation of M1 produced a significant motor improvement as measured by UPDRS scores and increased the amplitude and area of the associated motor-evoked potential.

Dyskinesias are another debilitating consequence of long-term dopamine use with some modest benefit demonstrated by pilot neurostimulation studies (Table 14.5). Filipovic et al. (2009) and Wagle-Shukla et al. (2007) both demonstrated that low-frequency rTMS applied to the motor cortex produced a significant decrease in dyskinesias compared to baseline. Koch et al. (2009) performed two studies applying cTBS (inhibitory) to the cerebellum demonstrating significant decrease in dyskinesia. Of interest, their stimulation of both cerebellar lobes for 10 sessions in patients with advanced PD resulted in significant decrease in dyskinesias lasting at least 4 weeks after treatment (Koch et al. 2009). While there has been some evidence that TBS temporarily increases cortical excitability in PD patients (Zamir et al. 2011), none of the studies stimulating the primary motor cortex seems to have demonstrated any motor improvement (Benninger et al. 2011; Eggers et al. 2010; Rothkegel et al. 2009).

In summary, the available research suggests that high-frequency repetitive stimulation and anodal tDCS of the primary motor cortex seem to improve the bradykinetic features of parkinsonism. An ongoing, multisite study supported by the Michael J. Fox Foundation (MASTER-PD) is seeking to provide more definite insights. There is also some evidence that low-frequency repetitive stimulation over the primary motor cortex and TBS over the cerebellum may reduce levodopa-induced dyskinesias, with follow-up studies warranted.

Several other questions merit further investigation, such as the optimal focus for stimulation. Comparative trials of low-, high-, and theta burst frequency stimulation over the primary motor cortex, supplementary motor cortex, and cerebellum are needed to determine which are the optimal parameters for treating bradykinesia, gait disturbance, and levodopa-induced dyskinesias. While most protocols have stimulated only the side contralateral to the more severely affected limb, more studies comparing bilateral to unilateral stimulation are needed as symptoms typically affect both sides. More functional neuroimaging studies are

TABLE 14.5

Comparison of Neurostimulation Protocols for Parkinson's Disease

References	No. of Subjects	Diagnosis	Coil Position	Stimulation	Frequency	No. of Stimuli	Intensity	Coil	Duration	Session Schedule	Effect
Koch et al. (2005)	8	Advanced PD with dyskinesia	Supplementary motor area, 3 cm anterior to Cz of 10/20 EEG system	rTMS	1 Hz/ 5 Hz	900 pulses (10 s ON–40 s OFF)	90% MT 110% MT	Figure 8	15 min	One session, crossover design	rTMS at 1 Hz significantly decreased drug-induced dyskinesias
Koch et al. (2009)	10	Advanced PD with dyskinesia	Lateral cerebellum ipsilateral to greatest dyskinesia B cerebellum	cTBS	50 Hz	600 pulses (3 pulses every 200 ms) to each cerebellar lobe	80% MT	Figure 8	40 s	One session 10 days	Transient decrease in dyskinesia Significant decrease in dyskinesia lasting up to 4 weeks after treatment
Eggers et al. (2010)	8	PD	M1 contralateral to more significant bradykinesia	cTBS	50 Hz	600 pulses	80% MT	Figure 8	40 s	Two sessions, crossover study with 1 week between real and sham treatments	No effect on motor performance or cortical excitability

Study	N	Condition	Target	Type	Frequency	Pulses	Intensity	Coil	Duration	Schedule	Outcome
Benninger et al. (2011)	26	Mild–moderate PD	Bilateral M1 + DLPFC	iTBS	50 Hz	600 pulses	80% MT	Round	200 s	8 days	Safe and improved mood. No effect on gait, UE bradykinesia
Rektorova et al. (2008)	6	PD with off-related freezing of gait	L DLPFC or M1 leg	rTMS	10 Hz	1350 pulses	90% MT	Figure 8	Variable	Delivered during ON-state × five sessions	No effect on OFF-state freezing of gait
Hamada et al. (2008, 2009)	98	PD	Supplementary motor cortex	rTMS	5 Hz	1000 pulses (10 s ON–50 s OFF)	110% MT	Figure 8	20 min	Once weekly × 8 weeks	Modest improvement in bradykinesia
Lomarev et al. (2006)	18	PD	L/R M1 L/R DLPFC	rTMS	25 Hz	300 pulses/target	100% MT	Figure 8	12 s/target	Eight sessions over 4 weeks	Significant improvement in gait and UE bradykinesia, with effect persisting 1 month past treatment

(continued)

TABLE 14.5 (continued)
Comparison of Neurostimulation Protocols for Parkinson's Disease

References	No. of Subjects	Diagnosis	Coil Position	Stimulation	Frequency	No. of Stimuli	Intensity	Coil	Duration	Session Schedule	Effect
Filipovic et al. (2010)	10	PD	M1 contralateral to more severely affected side	rTMS	1 Hz	1800 pulses (10 min ON–1 min OFF)	90% MT	Figure 8	32 min	Crossover study with four sessions (real or sham) per arm, with 2 week period between arms	No significant improvement in motor performance by UPDRS total score or subscore
Filipovic et al. (2009)	10	PD	M1 contralateral to more severely affected side	rTMS	1 Hz	1800 pulses	90% MT	Figure 8	32 min	Crossover study with four sessions (real or sham) per arm, with 2 week period between arms	Real rTMS with small but significant reduction in dyskinesias compared to baseline. Major effect on dystonia subscore

Wagle-shukla et al. (2007)	6	PD	M1 contralateral to more affected side	rTMS	1 Hz	900 pulses	90% MT	Figure 8	15 min	10 days	Significant decrease in dyskinesias immediately after stimulation (day 15), but no difference in other motor features
Minks et al. (2011)	20	Early PD	R cerebellum	rTMS	1 Hz	600 pulses	100% MT	Round	10 min	One session. Crossover design with 3 month interval	Significantly faster response on gross motor test, slower response on fine motor task
Arias et al. (2010)	18	PD	Vertex	rTMS	1 Hz	100 pulses	90% MT	Round	N/A	10 sessions	No motor improvement

(continued)

TABLE 14.5 (continued)
Comparison of Neurostimulation Protocols for Parkinson's Disease

References	No. of Subjects	Diagnosis	Electrode Position	tDCS Polarity	Intensity	Duration	Session Schedule	Effect
Fregni et al. (2006b)	17	PD	M1 DLPFC	Anodal tDCS, Cathodal tDCS	1 mA	20 min	One session, crossover design with 48 h washout period between stimulations	Anodal stimulation of M1 associated with significant motor improvement, also increased MEP amplitude and area; no effect with anodal stimulation of DLPFC or cathodal stimulation of M1

needed to confirm the more distant, subcortical effects caused by superficial stimulation. Finally, more studies are needed to understand stimulation effects during ON–OFF states, as related to fluctuations in medication level.

14.3.6 TINNITUS

Tinnitus is the subjective perception of sound in the absence of an external stimulus. About 5%–15% of the population in Western societies experience tinnitus, with a higher prevalence among the elderly and after hearing loss (Lockwood et al. 2002). This phantom auditory phenomenon can be very distressing. Tinnitus adversely affects quality of life, with high rates of concurrent depression (50%) and insomnia (40%) (Meyeroff and Cooper 1991; Phoon et al. 1993). Despite these significant comorbidities, no reliably effective treatments exist (Elgoyhen and Langguth 2010).

Part of the challenge in developing effective therapies for tinnitus is the absence of a clear pathophysiological model. However, there are several lines of evidence suggesting the importance of distorted central processing of sound, involving the auditory cortex, higher-order association areas, and limbic structures (De Ridder et al. 2006; Kaltenbach 2000; Muhlnickel et al. 1998; Salvi et al. 2000). Deprivation of primary afferent input caused by hearing loss may lead to hyperactivity in the central auditory system, reflected in an increase in the spontaneous firing rate (SFR) in cortical and subcortical structures (Eggermont and Roberts 2004; Kaltenbach 2006). While tinnitus has been associated with increased activity in the primary auditory and temporoparietal auditory association cortices, the manner of reorganization is controversial. For example, the laterality of abnormal auditory processing is unclear (Plewnia 2011).

Specific challenges in combining functional neuroimaging with tinnitus include subjects' sensitivity to external noise and the potential for study contamination with noise artifact. Therefore, quieter magnetoencephalogram (MEG) and PET protocols have been preferred over fMRI protocols, ideally including normal subjects as controls (Plewnia 2011). MEG studies on human subjects have demonstrated a reorganization of the auditory cerebral cortex, such that the tonotopic map may be shifted to the contralateral hemisphere (Muhlnickel et al. 1998). Other PET research suggests a higher level of spontaneous activity on the left-sided auditory cortex, regardless of the laterality of perceived tinnitus (Arnold et al. 1996; Kleinjung et al. 2005; Langguth et al. 2006). Moreover, the severity of tinnitus has been correlated with altered network connectivity (Schlee et al. 2008, 2009) and reduced volume of distant sensory and limbic structures (Landgrebe et al. 2009; Muhlau et al. 2006; Schneider et al. 2009).

Animal studies have demonstrated abnormal synchronized firing in the deprived frequency regions after noise trauma (Norena and Eggermont 2003) and enhanced burst firing in the inferior colliculus and primary and secondary auditory cortices after chemically induced tinnitus (Chen and Jastreboff 1995; Kenmochi and Eggermont 1997; Norena and Eggermont 2003). However, it is unclear what roles these other structures play in modulating sound processing.

Most neurostimulation studies have targeted the temporoparietal cortex to attempt to normalize maladaptive activity (Moller 2003). High-frequency rTMS has been demonstrated to create a "virtual lesion" to transiently suppress tinnitus.

TABLE 14.6
Comparison of Neurostimulation Protocols for Tinnitus

References	No. of Subjects	Tinnitus Laterality	Coil Position	Method of Localization	Stimulation	Frequency (Hz)	No of Stimuli	Intensity	Coil	Duration	Session Schedule	Effect
Fregni et al. (2006e)	7	Bi	L temporoparietal and mesial parietal areas	10/20 EEG system	rTMS (10 Hz); anodal and cathodal tDCS	10 Hz	30 pulses	120% MT; 1 mA	Figure 8	3 s/train × 3 trains; 3 min	One session	Three responders for active rTMS and anodal tDCS stimulation of left temporoparietal target
Folmer et al. (2006)	15	8 R 7 L	L and R temporal cortices	10/20 EEG system	rTMS	10 Hz (3 s ON–57 s OFF)	150 pulses	100% MT	Figure 8	5 min	One session	Six responders for active stimulation (five left temporal cortex, one right temporal cortex), two responders for sham

Plewnia et al. (2007a)	9	8 Bi 1 R	Area of maximum tinnitus	PET neuronavigation system	rTMS	1 Hz	300, 900, 1800 pulses with intertrain intervals of 30 min	120% MT	Figure 8	Three trains of 5, 15, 30 min	One session	Six responders for active rTMS with reduced tinnitus perception. Higher number of pulses with greater suppression. Shorter tinnitus direction with more effect
Plewnia et al. (2007b)	6	Bi	Area of maximum tinnitus	PET activation, neuronavigation system	rTMS	1 Hz	1800 pulses	120% MT	Figure 8	30 min	10 days real, 10 days sham crossover	Five responders for active rTMS, lasting for 2 weeks
Kleinjung et al. (2007)	45	30 Bi, 8 L, 7 R	L primary auditory cortex	Neuronavigation system	rTMS	1 Hz	2000 pulses	110% MT	Figure 8	33 min	10 days	18 responders, characterized by shorter-duration tinnitus and less hearing impairment

(continued)

TABLE 14.6 (continued)
Comparison of Neurostimulation Protocols for Tinnitus

References	No. of Subjects	Tinnitus Laterality	Coil Position	Method of Localization	Stimulation	Frequency (Hz)	No of Stimuli	Intensity	Coil	Duration	Session Schedule	Effect
De Ridder et al. (2007)	46	Uni, white noise	Contralateral to tinnitus side	Neuronavigation system	Tonic and burst stimulation	5/10/20 Hz	200 pulses	90% MT	Figure 8	Variable	One session	14 responders (5 with maximal suppression with theta burst, 2 with alpha burst, and 7 with beta burst); Burst rTMS suppressed narrow-band/white tinnitus better than tonic rTMS

Study												Results
Mennemeier et al. (2008)	1	Bi	R posterior superior lateral temporal gyrus	PET guidance	rTMS	1 Hz	1800 pulses	110% MT	N/A	30 min	5 days real, 5 days sham crossover	Tinnitus lowest after active rTMS than at baseline, post-sham rTMS, 3 and 6 month follow-up periods. Follow-up PET showed decreased metabolism in R temporal region compared to baseline
Poreisz et al. (2009)	20	N/A	L inferior temporal cortex	N/A	cTBS/iTBS/imTBS	Variable	600 pulses each arm	80% MT	Figure 8	40–190 s	One session	Only cTBS resulted in short-term improvement of symptoms
Khedr et al. (2009b)	66	N/A	L temporoparietal cortex	10/20 EEG system	rTMS	1/10/25 Hz	1500 pulses/ day	100% MT	Figure 8	Variable	10 days	Some patients treated with real rTMS with lasting benefit at 1 year, suggesting that 10/25 Hz superior to 1 Hz

(continued)

TABLE 14.6 (continued)
Comparison of Neurostimulation Protocols for Tinnitus

References	No. of Subjects	Tinnitus Laterality	Coil Position	Method of Localization	Stimulation	Frequency (Hz)	No of Stimuli	Intensity	Coil	Duration	Session Schedule	Effect
Khedr et al. (2010)	62	Uni	Temporoparietal cortex ipsilateral or contralateral to tinnitus side	10/20 EEG system	rTMS	1/25 Hz	2000 pulses/day	100% MT	Figure 8	Variable	2 weeks	Contralateral stimulation has greater effect than ipsilateral stimulation; twenty patients without tinnitus after 3 months; patients with shortest clinical history of tinnitus responded better
Marcondes et al. (2010)	20	Bi	L temporoparietal cortex	10/20 EEG system	rTMS	1 Hz	1020 pulses	110% MT Figure 8	Figure 8	17 min	5 days	Improvement in tinnitus up to 6 months after stimulation

References	No. of Subjects	Electrode Position	Method of Localization	tDCS Polarity	Intensity	Duration	Session Schedule	Effect
Frank et al. (2012)	32	R/L DLPFC	10/20 EEG system	Anodal/ Cathodal	1.5 mA	30 min	2 days/ week, 3 weeks	Improvement in tinnitus loudness, unpleasantness, and discomfort, but not in tinnitus or depression scales
Garin et al. (2011)	20	L temporoparietal area	10/20 EEG system	Anodal, cathodal tDCS	1 mA	20 min	One session	Anodal tDCS reduced tinnitus intensity after stimulation; no effect of cathodal tDCS

These initial studies have been conceptualized as proof-of-principle studies, to help define the functional neuroanatomy of this epiphenomenon (Plewnia 2011).

The largest study by De Ridder et al. involved 114 patients with unilateral tinnitus and investigated the effect of variable frequency of rTMS stimulation (1, 3, 5, 10, 20 Hz). Stimulation resulted in good effect (80%–100% suppression) in about 25% of patients, but no effect (0%–19% improvement) in 47% of patients (Table 14.6). Duration of tinnitus varied inversely with the effectiveness of treatment and also with the frequency of stimulation most likely to induce benefit (i.e., longer duration more likely helped by lower-frequency stimulation) (De Ridder et al. 2005). Other high-frequency rTMS stimulation protocols have also demonstrated significant interindividual variability of response (Folmer 2006; Fregni et al. 2006e; Khedr et al. 2010). Another comparative study found that high-frequency application of rTMS applied to the left temporoparietal cortex, but not real stimulation to the mesial parietal cortex or sham stimulation, resulted in a significant reduction of tinnitus. Likewise, anodal tDCS applied to the left temporoparietal cortex, but not cathodal tDCS or sham tDCS, induced a significant effect (Fregni et al. 2006e).

While high-frequency stimulation may elicit a temporary effect, low-frequency rTMS (≤1 Hz) could result in a more durable suppression of tinnitus. One of the largest studies included 45 patients with bilateral or unilateral (left or right) tinnitus. This group targeted the left primary auditory cortex and found that about one-third of their patients responded (Kleinjung et al. 2007). In a proof-of-principle study by Plewnia et al, nine patients with chronic tinnitus underwent low-frequency rTMS (1 Hz, 120% MT) applied to hyperactive cortical regions as individually determined by PET. Tinnitus loudness was significantly reduced for a brief period, and inversely correlation with prior duration of tinnitus (Plewnia et al. 2007a). Generally, the duration of improvement following serial applications of low-frequency rTMS has been reported to last several weeks, although there are some studies reporting benefit over 6 months (Kleinjung et al. 2005; Marcondes et al. 2010) and even 1 year (Khedr et al. 2009b).

There has been some preliminary investigation of newer stimulation protocols, e.g., TBS (Table 14.6). One study of 46 patients with narrow-band/white noise unilateral tinnitus compared continuous to intermittent TBS applied contralaterally to the side of perceived tinnitus (De Ridder et al. 2007). About one-third of patients responded, with intermittent stimulation more effective than continuous stimulation, at all frequencies. However, another study investigating a single session of different types of TBS (intermittent, continuous, intermediate) applied over the left inferior temporal cortex found a slight attenuation of tinnitus in about 50% of subjects, with no main effect of stimulation type (Poreisz et al. 2009).

In summary, neurostimulation protocols for tinnitus have demonstrated a high interindividual variability of responses, likely a product of the wide range of patient and study attributes. Studies to date have demonstrated a moderate effect size for a subset of patients (Plewnia 2011). There is some evidence that longer duration of tinnitus predicts a poorer response to neurostimulation (De Ridder et al. 2005; Plewnia et al. 2007a). However, further subgroup analyses are needed to better predict who will respond, by variables such as degree and frequency of hearing loss, tinnitus duration, and age (Langguth et al. 2008).

The major source of study heterogeneity has been the choice of stimulation target, reflecting an unclear understanding of underlying pathophysiology. Some protocols stimulated only left-sided temporoparietal cortex, others over the side of perceived tinnitus, others contralateral to the side of perceived stimulus, and still others over hyperactive regions determined by functional neuroimaging. Future studies may compare stimulation sites and protocols to achieve more clinically meaningful results.

14.3.7 OTHER CLINICAL INDICATIONS

There has been an explosion of investigatory applications for neurostimulation within neurology, in fields as diverse as ataxia, dystonia, migraine, tremor, and traumatic brain injury. Overall, such investigations are very preliminary and have to be considered, at best, proof-of-principle trials. While a review of the emerging research in each of these applications is beyond the scope of this chapter, therapeutic applications via TMS stimulation of the cerebellum deserve some consideration.

The cerebellum maintains numerous direct and indirect connections to nearly the entire nervous system. As an important regulatory center within the motor system, cerebellar lesions may result in movement disorders as diverse as limb, truncal, or gait ataxia; speech dysregulation; or extraocular dysmetria (Daskalakis et al. 2004). However, as its efferent connections to the frontal, parietal, temporal, and occipital cortices are being revealed by functional neuroimaging and electrophysiological studies, the role of cerebellum in cognition (including attention, learning, memory, and emotion) is beginning to be appreciated (Minks et al. 2010).

While most TMS studies on the cerebellum have been performed on healthy subjects, there have been small preliminary studies on patients with neurological disease. Shimizu et al. applied rTMS over the, left, right, and middle cerebellum in four patients with spinocerebellar ataxia daily for 21 days and found significant improvements in walking and a balance in all patients (Shimizu et al. 1999). Brighina et al. have used TMS as a diagnostic tool to probe the relationship between the cerebellum and the motor cortex in patients with dystonia, concluding that there is a reduction in cerebellar modulation of motor cortex excitability in these patients (Brighina et al. 2009).

14.4 CONCLUDING REMARKS

The last several years have witnessed an exponential increase in the number of well-designed prospective studies in neurostimulation. While studies are still relatively small and limited in number, there exists some convincing evidence of a therapeutic effect in focal-onset epilepsy, chronic pain, stroke rehabilitation, Parkinson's disease, and tinnitus. Some of the variability in results may be a result of diverse patient case mix and variable stimulation parameters. However, the positive effect already demonstrated is encouraging as many of these conditions have no alternative medical therapies or existing therapies are suboptimal. The importance of these results can also be underscored due to the fact that neurostimulation induces therapeutic effects via neuroplasticity, a fundamentally different mechanism than pharmacological treatments.

Here it must be emphasized that positive proof-of-principle evidence does not equal clinical therapeutic utility. Establishing therapeutic utility requires appropriately powered, controlled clinical trials. While there are a number of randomized, double-blinded, sham-controlled studies, the majority of these studies rely on small sample sizes. Small positive studies may overestimate effect estimates due to the variability and instability of data. On the other hand, small studies that yield negative results carry a higher risk of a type II error and are less likely to be published. Indeed, appropriately powered studies are needed, as are phase IV translational trials. Further studies using neuronavigational devices to specify cortical targets are needed (Lefaucheur 2010).

A host of unanswered questions merit further investigation, most notably effective and optimal stimulation parameters. Drawing helpful distinctions between those studies with and without clinical benefit is nearly impossible because of the heterogeneity of stimulation protocols. For example, little is known about the duration of stimulation required to induce a long-term clinical effect. Yet length of treatment is perhaps the single treatment variable that would most affect patient compliance. Previous studies demonstrating clinical efficacy have generally administered stimulation over a longer duration. While studies that administer a single session of stimulation play a role in establishing functional neuroanatomy and initial clinical effects, future studies hoping to demonstrate clinically meaningful improvement should administer repeated sessions of stimulation. However, such study designs will also require careful longitudinal assessment of safety, and thus controlled trials are warranted prior to broader clinical adoption. Lastly, combining neurostimulation with functional and structural neuroimaging techniques, e.g., fMRI, MR perfusion, PET, DTI, EEG, and voxel-based morphometry, can help elucidate the local and remote cortical and subcortical network changes caused by superficial focal stimulation.

REFERENCES

Adkins, D. L., P. Campos, D. Quach et al. 2006. Epidural cortical stimulation enhances motor function after sensorimotor cortical infarcts in rats. *Exp Neurol* 200 (2):356–370.

Adkins, D. L., J. E. Hsu, and T. A. Jones. 2008. Motor cortical stimulation promotes synaptic plasticity and behavioral improvements following sensorimotor cortex lesions. *Exp Neurol* 212 (1):14–28.

Adkins-Muir, D. L. and T. A. Jones. 2003. Cortical electrical stimulation combined with rehabilitative training: Enhanced functional recovery and dendritic plasticity following focal cortical ischemia in rats. *Neurol Res* 25 (8):780–788.

Albensi, B. C., G. Ata, E. Schmidt, J. D. Waterman, and D. Janigro. 2004. Activation of long-term synaptic plasticity causes suppression of epileptiform activity in rat hippocampal slices. *Brain Res* 998 (1):56–64.

Allen, E. A., B. N. Pasley, T. Duong, and R. D. Freeman. 2007. Transcranial magnetic stimulation elicits coupled neural and hemodynamic consequences. *Science* 317 (5846):1918–1921.

Andre-Obadia, N., R. Peyron, P. Mertens et al. 2006. Transcranial magnetic stimulation for pain control. Double-blind study of different frequencies against placebo, and correlation with motor cortex stimulation efficacy. *Clin Neurophysiol* 117 (7):1536–1544.

Antal, A., N. Kriener, N. Lang, K. Boros, and W. Paulus. 2011. Cathodal transcranial direct current stimulation of the visual cortex in the prophylactic treatment of migraine. *Cephalalgia* 31 (7):820–828.

Antal, A., D. Terney, S. Kuhnl, and W. Paulus. 2010. Anodal transcranial direct current stimulation of the motor cortex ameliorates chronic pain and reduces short intracortical inhibition. *J Pain Symptom Manage* 39 (5):890–903.

Arias, P., J. Vivas, K. L. Grieve, and J. Cudeiro. 2010. Controlled trial on the effect of 10 days low-frequency repetitive transcranial magnetic stimulation (rTMS) on motor signs in Parkinson's disease. *Mov Disord* 25 (12):1830–1838.

Arnold, W., P. Bartenstein, E. Oestreicher, W. Romer, and M. Schwaiger. 1996. Focal metabolic activation in the predominant left auditory cortex in patients suffering from tinnitus: A PET study with [18F]deoxyglucose. *ORL J Otorhinolaryngol Relat Spec* 58 (4):195–199.

Bae, E. H., L. M. Schrader, K. Machii et al. 2007. Safety and tolerability of repetitive transcranial magnetic stimulation in patients with epilepsy: A review of the literature. *Epilepsy Behav* 10 (4):521–528.

Baker, J. M., C. Rorden, and J. Fridriksson. 2010. Using transcranial direct-current stimulation to treat stroke patients with aphasia. *Stroke* 41 (6):1229–1236.

Belin, P., P. Van Eeckhout, M. Zilbovicius et al. 1996. Recovery from nonfluent aphasia after melodic intonation therapy: A PET study. *Neurology* 47 (6):1504–1511.

Benninger, D. H., B. D. Berman, E. Houdayer et al. 2011. Intermittent theta-burst transcranial magnetic stimulation for treatment of Parkinson disease. *Neurology* 76 (7):601–609.

Bersani, F. S., A. Minichino, P. G. Enticott et al. 2012. Deep transcranial magnetic stimulation as a treatment for psychiatric disorders: A comprehensive review. *Eur Psychiatry* (in press).

Bestmann, S. 2008. The physiological basis of transcranial magnetic stimulation. *Trends Cogn Sci* 12 (3):81–83.

Bestmann, S., J. Baudewig, H. R. Siebner, J. C. Rothwell, and J. Frahm. 2003. Subthreshold high-frequency TMS of human primary motor cortex modulates interconnected frontal motor areas as detected by interleaved fMRI-TMS. *Neuroimage* 20 (3):1685–1696.

Bestmann, S., J. Baudewig, H. R. Siebner, J. C. Rothwell, and J. Frahm. 2004. Functional MRI of the immediate impact of transcranial magnetic stimulation on cortical and subcortical motor circuits. *Eur J Neurosci* 19 (7):1950–1962.

Bestmann, S., C. C. Ruff, F. Blankenburg et al. 2008. Mapping causal interregional influences with concurrent TMS-fMRI. *Exp Brain Res* 191 (4):383–402.

Bestmann, S., O. Swayne, F. Blankenburg et al. 2010. The role of contralesional dorsal premotor cortex after stroke as studied with concurrent TMS-fMRI. *J Neurosci* 30 (36):11926–11937.

Bindman, L. J., O. C. Lippold, and J. W. Redfearn. 1964. The action of brief polarizing currents on the cerebral cortex of the rat (1) during current flow and (2) in the production of long-lasting after-effects. *J Physiol* 172:369–382.

Boggio, P. S., E. J. Amancio, C. F. Correa et al. 2009. Transcranial DC stimulation coupled with TENS for the treatment of chronic pain: A preliminary study. *Clin J Pain* 25 (8):691–695.

Borckardt, J. J., S. Reeves, and M. S. George. 2009a. The potential role of brain stimulation in the management of postoperative pain. *J Pain Manag* 2 (3):295–300.

Borckardt, J. J., S. T. Reeves, M. Weinstein et al. 2008. Significant analgesic effects of one session of postoperative left prefrontal cortex repetitive transcranial magnetic stimulation: A replication study. *Brain Stimul* 1 (2):122–127.

Borckardt, J. J., A. R. Smith, S. T. Reeves et al. 2009b. A pilot study investigating the effects of fast left prefrontal rTMS on chronic neuropathic pain. *Pain Med* 10 (5):840–849.

Borckardt, J. J., M. Weinstein, S. T. Reeves et al. 2006. Postoperative left prefrontal repetitive transcranial magnetic stimulation reduces patient-controlled analgesia use. *Anesthesiology* 105 (3):557–562.

Brighina, F., M. Romano, G. Giglia et al. 2009. Effects of cerebellar TMS on motor cortex of patients with focal dystonia: A preliminary report. *Exp Brain Res* 192 (4): 651–656.

Brodbeck, V., G. Thut, L. Spinelli et al. 2010. Effects of repetitive transcranial magnetic stimulation on spike pattern and topography in patients with focal epilepsy. *Brain Topogr* 22 (4):267–280.

Brown, J. A. and N. M. Barbaro. 2003. Motor cortex stimulation for central and neuropathic pain: Current status. *Pain* 104 (3):431–435.

Brown, J. A., H. L. Lutsep, M. Weinand, and S. C. Cramer. 2006. Motor cortex stimulation for the enhancement of recovery from stroke: A prospective, multicenter safety study. *Neurosurgery* 58 (3):464–473.

Brunoni, A. R., J. Amadera, B. Berbel et al. 2011a. A systematic review on reporting and assessment of adverse effects associated with transcranial direct current stimulation. *Int J Neuropsychopharmacol* 14 (3):1133–1145.

Brunoni, A. R., M. A. Nitsche, N. Bolognini, M. Bikson, T. Wagner, L. Merabet, D. J. Edwards et al. 2011b. Clinical research with transcranial direct current stimulation (tDCS): Challenges and future directions. *Brain Stimul* 5 (3):175–195.

Cantello, R., S. Rossi, C. Varrasi et al. 2007. Slow repetitive TMS for drug-resistant epilepsy: Clinical and EEG findings of a placebo-controlled trial. *Epilepsia* 48 (2):366–374.

Chang, W. H., Y. Kim, O. Y. Bang, S. T. Kim, Y. H. Park, and P. Lee 2010. Long-term effects of rTMS on motor recovery in patients after subacute stroke. *J Rehabil Med* 42:758–764.

Chen, R., J. Classen, C. Gerloff et al. 1997. Depression of motor cortex excitability by low-frequency transcranial magnetic stimulation. *Neurology* 48 (5):1398–1403.

Chen, G. D. and P. J. Jastreboff. 1995. Salicylate-induced abnormal activity in the inferior colliculus of rats. *Hear Res* 82 (2):158–179.

Cotelli, M., R. Manenti, S. F. Cappa et al. 2006. Effect of transcranial magnetic stimulation on action naming in patients with Alzheimer disease. *Arch Neurol* 63 (11):1602–1604.

Cotelli, M., R. Manenti, S. F. Cappa, O. Zanetti, and C. Miniussi. 2008. Transcranial magnetic stimulation improves naming in Alzheimer disease patients at different stages of cognitive decline. *Eur J Neurol* 15 (12):1286–1292.

Creutzfeldt, O. D., G. H. Fromm, and H. Kapp. 1962. Influence of transcortical d-c currents on cortical neuronal activity. *Exp Neurol* 5:436–452.

Daskalakis, Z. J., G. O. Paradiso, B. K. Christensen et al. 2004. Exploring the connectivity between the cerebellum and motor cortex in humans. *J Physiol* 557 (Pt 2):689–700.

De Ridder, D., G. De Mulder, E. Verstraeten et al. 2006. Primary and secondary auditory cortex stimulation for intractable tinnitus. *ORL J Otorhinolaryngol Relat Spec* 68 (1): 48–54; discussion 54–55.

De Ridder, D., E. van der Loo, K. Van der Kelen et al. 2007. Theta, alpha and beta burst transcranial magnetic stimulation: Brain modulation in tinnitus. *Int J Med Sci* 4 (5):237–241.

De Ridder, D., E. Verstraeten, K. Van der Kelen et al. 2005. Transcranial magnetic stimulation for tinnitus: Influence of tinnitus duration on stimulation parameter choice and maximal tinnitus suppression. *Otol Neurotol* 26 (4):616–619.

Denslow, S., M. Lomarev, M. S. George, and D. E. Bohning. 2005. Cortical and subcortical brain effects of transcranial magnetic stimulation (TMS)-induced movement: An interleaved TMS/functional magnetic resonance imaging study. *Biol Psychiatry* 57 (7):752–760.

Dijkhuizen, R. M., A. B. Singhal, J. B. Mandeville et al. 2003. Correlation between brain reorganization, ischemic damage, and neurologic status after transient focal cerebral ischemia in rats: A functional magnetic resonance imaging study. *J Neurosci* 23 (2):510–517.

Duque, J., F. Hummel, P. Celnik et al. 2005. Transcallosal inhibition in chronic subcortical stroke. *Neuroimage* 28 (4):940–946.

Eggermont, J. J. and L. E. Roberts. 2004. The neuroscience of tinnitus. *Trends Neurosci* 27 (11):676–682.

Eggers, C., G. R. Fink, and D. A. Nowak. 2010. Theta burst stimulation over the primary motor cortex does not induce cortical plasticity in Parkinson's disease. *J Neurol* 257 (10):1669–1674.

Elahi, B., B. Elahi, and R. Chen. 2009. Effect of transcranial magnetic stimulation on Parkinson motor function—Systematic review of controlled clinical trials. *Mov Disord* 24 (3):357–363.

Elgoyhen, A. B. and B. Langguth. 2010. Pharmacological approaches to the treatment of tinnitus. *Drug Discov Today* 15 (7–8):300–305.

Emara, T. H., R. R. Moustafa, N. M. Elnahas et al. 2010. Repetitive transcranial magnetic stimulation at 1 Hz and 5 Hz produces sustained improvement in motor function and disability after ischaemic stroke. *Eur J Neurol* 17 (9):1203–1209.

Filipovic, S. R., J. C. Rothwell, and K. Bhatia. 2010. Low-frequency repetitive transcranial magnetic stimulation and off-phase motor symptoms in Parkinson's disease. *J Neurol Sci* 291 (1–2):1–4.

Filipovic, S. R., J. C. Rothwell, B. P. van de Warrenburg, and K. Bhatia. 2009. Repetitive transcranial magnetic stimulation for levodopa-induced dyskinesias in Parkinson's disease. *Mov Disord* 24 (2):246–253.

Finocchiaro, C., M. Maimone, F. Brighina, F. Piccoli, G. Giglia, and B. Fierro. 2006. A case study of Primary Progressive Aphasia: Improvement on verbs after rTMS treatment. *Neurocase* 12 (6):317–321.

Folmer, R. L., J. R. Carroll, A. Rahim, Y. Shi, and W. Hal Martin. 2006. Effects of repetitive transcranial magnetic stimulation (rTMS) for chronic tinnitus. *Acta Otolaryngol Suppl* (556):96–101.

Fox, P., R. Ingham, M. S. George et al. 1997. Imaging human intra-cerebral connectivity by PET during TMS. *Neuroreport* 8 (12):2787–2791.

Fox, P. T., S. Narayana, N. Tandon et al. 2006. Intensity modulation of TMS-induced cortical excitation: Primary motor cortex. *Hum Brain Mapp* 27 (6):478–487.

Frank, E., M. Schecklmann, M. Landgrebe et al. 2012. Treatment of chronic tinnitus with repeated sessions of prefrontal transcranial direct current stimulation: Outcomes from an open-label pilot study. *J Neurol* 259 (2):327–333.

Fregni, F. 2010. Analgesia with noninvasive electrical cortical stimulation: Challenges to find optimal parameters of stimulation. *Anesth Analg* 111 (5):1083–1085.

Fregni, F., P. S. Boggio, M. C. Lima et al. 2006a. A sham-controlled, phase II trial of transcranial direct current stimulation for the treatment of central pain in traumatic spinal cord injury. *Pain* 122 (1–2):197–209.

Fregni, F., P. S. Boggio, M. C. Santos et al. 2006b. Noninvasive cortical stimulation with transcranial direct current stimulation in Parkinson's disease. *Mov Disord* 21 (10): 1693–1702.

Fregni, F., P. S. Boggio, A. C. Valle et al. 2006c. A sham-controlled trial of a 5-day course of repetitive transcranial magnetic stimulation of the unaffected hemisphere in stroke patients. *Stroke* 37 (8):2115–2122.

Fregni, F., D. DaSilva, K. Potvin et al. 2005a. Treatment of chronic visceral pain with brain stimulation. *Ann Neurol* 58 (6):971–972.

Fregni, F., R. Gimenes, A. C. Valle et al. 2006d. A randomized, sham-controlled, proof of principle study of transcranial direct current stimulation for the treatment of pain in fibromyalgia. *Arthritis Rheum* 54 (12):3988–3998.

Fregni, F., R. Marcondes, P. S. Boggio et al. 2006e. Transient tinnitus suppression induced by repetitive transcranial magnetic stimulation and transcranial direct current stimulation. *Eur J Neurol* 13 (9):996–1001.

Fregni, F., P. T. Otachi, A. Do Valle et al. 2006f. A randomized clinical trial of repetitive transcranial magnetic stimulation in patients with refractory epilepsy. *Ann Neurol* 60 (4):447–455.

Fregni, F. and A. Pascual-Leone. 2006. Hand motor recovery after stroke: Tuning the orchestra to improve hand motor function. *Cogn Behav Neurol* 19 (1):21–33.

Fregni, F., D. K. Simon, A. Wu, and A. Pascual-Leone. 2005b. Non-invasive brain stimulation for Parkinson's disease: A systematic review and meta-analysis of the literature. *J Neurol Neurosurg Psychiatry* 76 (12):1614–1623.

Fregni, F., S. Thome-Souza, F. Bermpohl et al. 2005c. Antiepileptic effects of repetitive transcranial magnetic stimulation in patients with cortical malformations: An EEG and clinical study. *Stereotact Funct Neurosurg* 83 (2–3):57–62.

Fregni, F., S. Thome-Souza, M. A. Nitsche et al. 2006g. A controlled clinical trial of cathodal DC polarization in patients with refractory epilepsy. *Epilepsia* 47 (2):335–342.

Fridman, E. A., T. Hanakawa, M. Chung et al. 2004. Reorganization of the human ipsilesional premotor cortex after stroke. *Brain* 127 (Pt 4):747–758.

Fridriksson, J., J. D. Richardson, J. M. Baker, and C. Rorden. 2011. Transcranial direct current stimulation improves naming reaction time in fluent aphasia: A double-blind, sham-controlled study. *Stroke* 42 (3):819–821.

Gabis, L., B. Shklar, Y. K. Baruch et al. 2009. Pain reduction using transcranial electrostimulation: A double blind "active placebo" controlled trial. *J Rehabil Med* 41 (4):256–261.

Gandiga, P. C., F. C. Hummel, and L. G. Cohen. 2006. Transcranial DC stimulation (tDCS): A tool for double-blind sham-controlled clinical studies in brain stimulation. *Clin Neurophysiol* 117 (4):845–850.

Gangitano, M., A. Valero-Cabre, J. M. Tormos et al. 2002. Modulation of input-output curves by low and high frequency repetitive transcranial magnetic stimulation of the motor cortex. *Clin Neurophysiol* 113 (8):1249–1257.

Garcia-Larrea, L., R. Peyron, P. Mertens et al. 1997. Positron emission tomography during motor cortex stimulation for pain control. *Stereotact Funct Neurosurg* 68 (1–4 Pt 1):141–148.

Garcia-Larrea, L., R. Peyron, P. Mertens et al. 1999. Electrical stimulation of motor cortex for pain control: A combined PET-scan and electrophysiological study. *Pain* 83 (2):259–273.

Garin, P., C. Gilain, J. P. Van Damme et al. 2011. Short- and long-lasting tinnitus relief induced by transcranial direct current stimulation. *J Neurol* 258 (11):1940–1948.

Grefkes, C., D. A. Nowak, S. B. Eickhoff et al. 2008. Cortical connectivity after subcortical stroke assessed with functional magnetic resonance imaging. *Ann Neurol* 63 (2): 236–246.

Hallett, M. 2000. Transcranial magnetic stimulation and the human brain. *Nature* 406 (6792):147–150.

Hallett, M. 2001. Functional reorganization after lesions of the human brain: Studies with transcranial magnetic stimulation. *Rev Neurol (Paris)* 157 (8–9 Pt 1):822–826.

Hamada, M., Y. Ugawa, and S. Tsuji. 2008. High-frequency rTMS over the supplementary motor area for treatment of Parkinson's disease. *Mov Disord* 23 (11):1524–1531.

Hamada, M., Y. Ugawa, and S. Tsuji. 2009. High-frequency rTMS over the supplementary motor area improves bradykinesia in Parkinson's disease: Subanalysis of double-blind sham-controlled study. *J Neurol Sci* 287 (1–2):143–146.

Hampson, M., N. R. Driesen, P. Skudlarski, J. C. Gore, and R. T. Constable. 2006. Brain connectivity related to working memory performance. *J Neurosci* 26 (51):13338–13343.

Harel, E. V., L. Rabany, L. Deutsch et al. 2012. H-coil repetitive transcranial magnetic stimulation for treatment resistant major depressive disorder: An 18-week continuation safety and feasibility study. *World J Biol Psychiatry* (Epub. ahead of print).

Harel, E. V., I. Zangen, Y. Roth et al. 2011. H-coil repetitive transcranial magnetic stimulation for the treatment of bipolar depression: An add-on, safety and feasibility study. *World J Biol Psychiatry* 12 (2):119–126.

Haslinger, B., P. Erhard, N. Kampfe et al. 2001. Event-related functional magnetic resonance imaging in Parkinson's disease before and after levodopa. *Brain* 124 (Pt 3):558–570.

Hattori, Y., A. Moriwaki, and Y. Hori. 1990. Biphasic effects of polarizing current on adenosine-sensitive generation of cyclic AMP in rat cerebral cortex. *Neurosci Lett* 116 (3): 320–324.

Hesse, S., C. Werner, E. M. Schonhardt et al. 2007. Combined transcranial direct current stimulation and robot-assisted arm training in subacute stroke patients: A pilot study. *Restor Neurol Neurosci* 25 (1):9–15.

Huang, Y. Z., M. J. Edwards, E. Rounis, K. P. Bhatia, and J. C. Rothwell. 2005. Theta burst stimulation of the human motor cortex. *Neuron* 45 (2):201–206.

Hummel, F., P. Celnik, P. Giraux et al. 2005. Effects of non-invasive cortical stimulation on skilled motor function in chronic stroke. *Brain* 128 (Pt 3):490–499.

Hummel, F. C., P. Celnik, A. Pascual-Leone et al. 2008. Controversy: Noninvasive and invasive cortical stimulation show efficacy in treating stroke patients. *Brain Stimul* 1 (4):370–382.

Hummel, F. C. and L. G. Cohen. 2005. Drivers of brain plasticity. *Curr Opin Neurol* 18 (6): 667–674.

Hummel, F. C. and L. G. Cohen. 2006. Non-invasive brain stimulation: A new strategy to improve neurorehabilitation after stroke? *Lancet Neurol* 5 (8):708–712.

Islam, N., M. Aftabuddin, A. Moriwaki, Y. Hattori, and Y. Hori. 1995. Increase in the calcium level following anodal polarization in the rat brain. *Brain Res* 684 (2):206–208.

Johansen-Berg, H., M. F. Rushworth, M. D. Bogdanovic et al. 2002. The role of ipsilateral premotor cortex in hand movement after stroke. *Proc Natl Acad Sci USA* 99 (22): 14518–14523.

Joo, E. Y., S. J. Han, S. H. Chung et al. 2007. Antiepileptic effects of low-frequency repetitive transcranial magnetic stimulation by different stimulation durations and locations. *Clin Neurophysiol* 118 (3):702–708.

Kabakov, A. Y., P. A. Muller, A. Pascual-Leone, F. E. Jensen, and A. Rotenberg. 2012. Contribution of axonal orientation to pathway-dependent modulation of excitatory transmission by direct current stimulation in isolated rat hippocampus. *J Neurophysiol* 107 (7):1881–1889.

Kaltenbach, J. A. 2000. Neurophysiologic mechanisms of tinnitus. *J Am Acad Audiol* 11 (3):125–137.

Kaltenbach, J. A. 2006. Summary of evidence pointing to a role of the dorsal cochlear nucleus in the etiology of tinnitus. *Acta Otolaryngol Suppl* (556):20–26.

Keeser, D., T. Meindl, J. Bor et al. 2011. Prefrontal transcranial direct current stimulation changes connectivity of resting-state networks during fMRI. *J Neurosci* 31 (43): 15284–15293.

Kenmochi, M. and J. J. Eggermont. 1997. Salicylate and quinine affect the central nervous system. *Hear Res* 113 (1–2):110–116.

Khedr, E. M., M. R. Abdel-Fadeil, A. Farghali, and M. Qaid. 2009a. Role of 1 and 3 Hz repetitive transcranial magnetic stimulation on motor function recovery after acute ischaemic stroke. *Eur J Neurol* 16 (12):1323–1330.

Khedr, E. M., N. Abo-Elfetoh, J. C. Rothwell et al. 2010. Contralateral versus ipsilateral rTMS of temporoparietal cortex for the treatment of chronic unilateral tinnitus: Comparative study. *Eur J Neurol* 17 (7):976–983.

Khedr, E. M., M. A. Ahmed, N. Fathy, and J. C. Rothwell. 2005a. Therapeutic trial of repetitive transcranial magnetic stimulation after acute ischemic stroke. *Neurology* 65 (3):466–468.

Khedr, E. M., H. Kotb, N. F. Kamel et al. 2005b. Longlasting antalgic effects of daily sessions of repetitive transcranial magnetic stimulation in central and peripheral neuropathic pain. *J Neurol Neurosurg Psychiatry* 76 (6):833–838.

Khedr, E. M., J. C. Rothwell, and A. El-Atar. 2009b. One-year follow up of patients with chronic tinnitus treated with left temporoparietal rTMS. *Eur J Neurol* 16 (3):404–408.

Kim, Y. H., S. H. You, M. H. Ko et al. 2006. Repetitive transcranial magnetic stimulation-induced corticomotor excitability and associated motor skill acquisition in chronic stroke. *Stroke* 37 (6):1471–1476.

Kinoshita, M., A. Ikeda, T. Begum et al. 2005. Low-frequency repetitive transcranial magnetic stimulation for seizure suppression in patients with extratemporal lobe epilepsy-a pilot study. *Seizure* 14 (6):387–392.

Kinsbourne, M. 1997. Hemineglect and hemispheric rivalry. *Adv Neurol* 18:41–49.

Kleim, J. A., R. Bruneau, P. VandenBerg et al. 2003. Motor cortex stimulation enhances motor recovery and reduces peri-infarct dysfunction following ischemic insult. *Neurol Res* 25 (8):789–793.

Kleinjung, T., P. Eichhammer, B. Langguth et al. 2005. Long-term effects of repetitive transcranial magnetic stimulation (rTMS) in patients with chronic tinnitus. *Otolaryngol Head Neck Surg* 132 (4):566–569.

Kleinjung, T., T. Steffens, P. Sand et al. 2007. Which tinnitus patients benefit from transcranial magnetic stimulation? *Otolaryngol Head Neck Surg* 137 (4):589–595.

Kobayashi, M., S. Hutchinson, H. Theoret, G. Schlaug, and A. Pascual-Leone. 2004. Repetitive TMS of the motor cortex improves ipsilateral sequential simple finger movements. *Neurology* 62 (1):91–98.

Koch, G., L. Brusa, C. Caltagirone et al. 2005. rTMS of supplementary motor area modulates therapy-induced dyskinesias in Parkinson disease. *Neurology* 65 (4):623–625.

Koch, G., L. Brusa, F. Carrillo et al. 2009. Cerebellar magnetic stimulation decreases levodopa-induced dyskinesias in Parkinson disease. *Neurology* 73 (2):113–119.

Kwon, Y. H., M. H. Ko, S. H. Ahn et al. 2008. Primary motor cortex activation by transcranial direct current stimulation in the human brain. *Neurosci Lett* 435 (1):56–59.

Landgrebe, M., B. Langguth, K. Rosengarth et al. 2009. Structural brain changes in tinnitus: Grey matter decrease in auditory and non-auditory brain areas. *Neuroimage* 46 (1):213–218.

Lang, N., H. R. Siebner, N. S. Ward et al. 2005. How does transcranial DC stimulation of the primary motor cortex alter regional neuronal activity in the human brain? *Eur J Neurosci* 22 (2):495–504.

Langguth, B., P. Eichhammer, A. Kreutzer et al. 2006. The impact of auditory cortex activity on characterizing and treating patients with chronic tinnitus—First results from a PET study. *Acta Otolaryngol Suppl* 556:84–88.

Langguth, B., D. de Ridder, J. L. Dornhoffer et al. 2008. Controversy: Does repetitive transcranial magnetic stimulation/transcranial direct current stimulation show efficacy in treating tinnitus patients? *Brain Stimul* 1 (3):192–205.

Lefaucheur, J. P. 2010. Why image-guided navigation becomes essential in the practice of transcranial magnetic stimulation. *Neurophysiol Clin* 40 (1):1–5.

Lefaucheur, J. P., X. Drouot, I. Menard-Lefaucheur et al. 2004. Neurogenic pain relief by repetitive transcranial magnetic cortical stimulation depends on the origin and the site of pain. *J Neurol Neurosurg Psychiatry* 75 (4):612–616.

Leung, A., M. Donohue, R. Xu, et al. 2009. rTMS for suppressing neuropathic pain: A meta-analysis. *J Pain* 10 (12):1205–1216.

Lewis, C. M., A. Baldassarre, G. Committeri, G. L. Romani, and M. Corbetta. 2009. Learning sculpts the spontaneous activity of the resting human brain. *Proc Natl Acad Sci USA* 106 (41):17558–17563.

Liebetanz, D., M. A. Nitsche, F. Tergau, and W. Paulus. 2002. Pharmacological approach to the mechanisms of transcranial DC-stimulation-induced after-effects of human motor cortex excitability. *Brain* 125 (Pt 10):2238–2247.

Liepert, J., S. Zittel, and C. Weiller. 2007. Improvement of dexterity by single session low-frequency repetitive transcranial magnetic stimulation over the contralesional motor cortex in acute stroke: A double-blind placebo-controlled crossover trial. *Restor Neurol Neurosci* 25 (5–6):461–465.

Lindenberg, R., V. Renga, L. L. Zhu, D. Nair, and G. Schlaug. 2010. Bihemispheric brain stimulation facilitates motor recovery in chronic stroke patients. *Neurology* 75 (24):2176–2184.

Lockwood, A. H., R. J. Salvi, and R. F. Burkard. 2002. Tinnitus. *N Engl J Med* 347 (12): 904–910.

Lomarev, M. P., S. Kanchana, W. Bara-Jimenez et al. 2006. Placebo-controlled study of rTMS for the treatment of Parkinson's disease. *Mov Disord* 21 (3):325–331.

Lowenstein, D. H. 1996. Recent advances related to basic mechanisms of epileptogenesis. *Epilepsy Res Suppl* 11:45–60.

Luborzewski, A., F. Schubert, F. Seifert et al. 2007. Metabolic alterations in the dorsolateral prefrontal cortex after treatment with high-frequency repetitive transcranial magnetic stimulation in patients with unipolar major depression. *J Psychiatr Res* 41 (7):606–615.

Machii, K., D. Cohen, C. Ramos-Estebanez, and A. Pascual-Leone. 2006. Safety of rTMS to nonmotor cortical areas in healthy participants and patients. *Clin Neurophysiol* 117 (2):455–471.

Malcolm, M. P., W. J. Triggs, K. E. Light, L. J. Gonzalez Rothi, S. Wu, K. Reid, and S. E. Nadeau. 2007. Repetitive transcranial magnetic stimulation as an adjunct to constraint-induced therapy: An exploratory randomized controlled trial. *Am J Phys Med Rehabil* 86 (9):707–715.

Mansur, C. G., F. Fregni, P. S. Boggio et al. 2005. A sham stimulation-controlled trial of rTMS of the unaffected hemisphere in stroke patients. *Neurology* 64 (10):1802–1804.

Marcondes, R. A., T. G. Sanchez, M. A. Kii et al. 2010. Repetitive transcranial magnetic stimulation improve tinnitus in normal hearing patients: A double-blind controlled, clinical and neuroimaging outcome study. *Eur J Neurol* 17 (1):38–44.

Marshall, R. S., G. M. Perera, R. M. Lazar et al. 2000. Evolution of cortical activation during recovery from corticospinal tract infarction. *Stroke* 31 (3):656–661.

Mazoyer, B., O. Houde, M. Joliot, E. Mellet, and N. Tzourio-Mazoyer. 2009. Regional cerebral blood flow increases during wakeful rest following cognitive training. *Brain Res Bull* 80 (3):133–138.

Mennemeier, M., K. C. Chelette, J. Kyhill, P. Taylor-Cooke, T. Bartel, W. Triggs, T. Kimbrell, and J. Dornhoffer. 2008. Maintenance repetitive transcranial magnetic stimulation can inhibit the return of tinnitus. *Laryngoscope* 118 (7):1228–1232.

Meyeroff, W. L. and J. C. Cooper. 1991. Tinnitus. In *Otolaryngology* 3rd ed., eds M. M. Paparella, D. A. Shumrick, J. L. Gluckman, W. L. Meyeroff, pp. 1669–1675. Philadelphia, PA: Saunders.

Miniussi, C., S. F. Cappa, L. G. Cohen et al. 2008. Efficacy of repetitive transcranial magnetic stimulation/transcranial direct current stimulation in cognitive neurorehabilitation. *Brain Stimul* 1 (4):326–336.

Minks, E., M. Kopickova, R. Marecek, H. Streitova, and M. Bares. 2010. Transcranial magnetic stimulation of the cerebellum. *Biomed Pap Med Fac Univ Palacky Olomouc Czech Repub* 154 (2):133–139.

Minks, E., R. Marecek, T. Pavlik, P. Ovesna, and M. Bares. 2011. Is the cerebellum a potential target for stimulation in Parkinson's disease? results of 1-Hz rTMS on upper limb motor tasks. *Cerebellum* 10 (4):804–811.

Moisa, M., R. Pohmann, K. Uludag, and A. Thielscher. 2010. Interleaved TMS/CASL: Comparison of different rTMS protocols. *Neuroimage* 49 (1):612–620.

Moller, A. R. 2003. Pathophysiology of tinnitus. *Otolaryngol Clin North Am* 36 (2):249–266, v–vi.

Mori, F., C. Codecà, H. Kusayanagi, et al. 2010. Effects of anodal transcranial direct current stimulation on chronic neuropathic pain in patients with multiple sclerosis. *J Pain* 11 (5):436–442.

Moriwaki, A. 1991. Polarizing currents increase noradrenaline-elicited accumulation of cyclic AMP in rat cerebral cortex. *Brain Res* 544 (2):248–252.

Muhlau, M., J. P. Rauschecker, E. Oestreicher et al. 2006. Structural brain changes in tinnitus. *Cereb Cortex* 16 (9):1283–1288.

Muhlnickel, W., T. Elbert, E. Taub, and H. Flor. 1998. Reorganization of auditory cortex in tinnitus. *Proc Natl Acad Sci USA* 95 (17):10340–10343.

Murase, N., J. Duque, R. Mazzocchio, and L. G. Cohen. 2004. Influence of interhemispheric interactions on motor function in chronic stroke. *Ann Neurol* 55 (3):400–409.

Naeser, M. A., P. I. Martin, M. Nicholas et al. 2005. Improved picture naming in chronic aphasia after TMS to part of right Broca's area: An open-protocol study. *Brain Lang* 93 (1):95–105.

Najib, U. and Pascual-Leone, A. 2011. Paradoxical functional facilitation with noninvasive brain stimulation. In *The Paradoxical Brain*, eds. A. Pascual-Leone, N. Kapur, V. Ramachandran, J. Cole, S. Della Sala, T. Manly, and A. Mayes, pp. 234–260. New York: Cambridge University Press.

Nitsche, M. A., L. G. Cohen, E. M. Wassermann et al. 2008. Transcranial direct current stimulation: State of the art 2008. *Brain Stimul* 1 (3):206–223.

Nitsche, M. A., S. Doemkes, T. Karakose et al. 2007. Shaping the effects of transcranial direct current stimulation of the human motor cortex. *J Neurophysiol* 97 (4):3109–3117.

Nitsche, M. A. and W. Paulus. 2000. Excitability changes induced in the human motor cortex by weak transcranial direct current stimulation. *J Physiol* 527 (Pt 3):633–639.

Norena, A. J. and J. J. Eggermont. 2003. Changes in spontaneous neural activity immediately after an acoustic trauma: Implications for neural correlates of tinnitus. *Hear Res* 183 (1–2):137–153.

Nowak, D. A., C. Grefkes, M. Dafotakis et al. 2008. Effects of low-frequency repetitive transcranial magnetic stimulation of the contralesional primary motor cortex on movement kinematics and neural activity in subcortical stroke. *Arch Neurol* 65 (6):741–747.

Nuti, C., R. Peyron, L. Garcia-Larrea et al. 2005. Motor cortex stimulation for refractory neuropathic pain: Four year outcome and predictors of efficacy. *Pain* 118 (1–2): 43–52.

Palm, U., D. Keeser, C. Schiller, Z. Fintescu et al. 2008. Skin lesions after treatment with transcranial direct current stimulation (tDCS). *Brain Stimul* 1:386–387.

Passard, A., N. Attal, R. Benadhira et al. 2007. Effects of unilateral repetitive transcranial magnetic stimulation of the motor cortex on chronic widespread pain in fibromyalgia. *Brain* 130 (Pt 10):2661–2670.

Pereira, J. B., C. Junqué, D. Bartrés-Faz, M. J. Martí, R. Sala-Llonch, and Y. Compta, C. Falcón, P. Vendrell, A. Pascual-Leone, J. Valls-Solé, and E. Tolosa. 2012. Modulation of verbal fluency networks using transcranial direct current stimulation in Parkinson's disease. *Brain Stimul.* [Epub. ahead of print.]

Peyron, R., L. Garcia-Larrea, M. P. Deiber et al. 1995. Electrical stimulation of precentral cortical area in the treatment of central pain: Electrophysiological and PET study. *Pain* 62 (3):275–286.

Phoon, W. H., H. S. Lee, and S. E. Chia. 1993. Tinnitus in noise-exposed workers. *Occup Med (Lond)* 43 (1):35–38.

Pleger, B., F. Janssen, P. Schwenkreis et al. 2004. Repetitive transcranial magnetic stimulation of the motor cortex attenuates pain perception in complex regional pain syndrome type I. *Neurosci Lett* 356 (2):87–90.

Plewnia, C. 2011. Brain stimulation: New vistas for the exploration and treatment of tinnitus. *CNS Neurosci Ther* 17 (5):449–461.

Plewnia, C., M. Lotze, and C. Gerloff. 2003. Disinhibition of the contralateral motor cortex by low-frequency rTMS. *Neuroreport* 14 (4):609–612.

Plewnia, C., M. Reimold, A. Najib et al. 2007a. Dose-dependent attenuation of auditory phantom perception (tinnitus) by PET-guided repetitive transcranial magnetic stimulation. *Hum Brain Mapp* 28 (3):238–246.

Plewnia, C., M. Reimold, A. Najib et al. 2007b. Moderate therapeutic efficacy of positron emission tomography-navigated repetitive transcranial magnetic stimulation for chronic tinnitus: A randomised, controlled pilot study. *J Neurol Neurosurg Psychiatry* 78 (2):152–156.

Polania, R., M. A. Nitsche, and W. Paulus. 2011a. Modulating functional connectivity patterns and topological functional organization of the human brain with transcranial direct current stimulation. *Hum Brain Mapp* 32 (8):1236–1249.

Polania, R., W. Paulus, and M. A. Nitsche. 2011b. Modulating cortico-striatal and thalamo-cortical functional connectivity with transcranial direct current stimulation. *Hum Brain Mapp*. [Epub. ahead of print.]

Poreisz, C., K. Boros, A. Antal, and W. Paulus. 2007. Safety aspects of transcranial direct current stimulation concerning healthy subjects and patients. *Brain Res Bull* 72 (4–6):208–214.

Poreisz, C., W. Paulus, T. Moser, and N. Lang. 2009. Does a single session of theta-burst transcranial magnetic stimulation of inferior temporal cortex affect tinnitus perception? *BMC Neurosci* 10:54.

Purpura, D. P. and J. G. McMurtry. 1965. Intracellular activities and evoked potential changes during polarization of motor cortex. *J Neurophysiol* 28:166–185.

Rektorova, I., S. Sedlackova, S. Telecka, A. Hlubocky, and I. Rektor. 2008. Dorsolateral prefrontal cortex: A possible target for modulating dyskinesias in Parkinson's disease by repetitive transcranial magnetic stimulation. *Int J Biomed Imaging* 2008:372125.

Romero, J. R., D. Anschel, R. Sparing, M. Gangitano, and A. Pascual-Leone. 2002. Subthreshold low frequency repetitive transcranial magnetic stimulation selectively decreases facilitation in the motor cortex. *Clin Neurophysiol* 113 (1):101–107.

Rossi, S., M. Hallett, P. M. Rossini, and A. Pascual-Leone. 2009. Safety, ethical considerations, and application guidelines for the use of transcranial magnetic stimulation in clinical practice and research. *Clin Neurophysiol* 120 (12):2008–2039.

Rotenberg, A. 2010. Prospects for clinical applications of transcranial magnetic stimulation and real-time EEG in epilepsy. *Brain Topogr* 22 (4):257–266.

Rotenberg, A., E. H. Bae, M. Takeoka et al. 2009. Repetitive transcranial magnetic stimulation in the treatment of epilepsia partialis continua. *Epilepsy Behav* 14 (1):253–257.

Roth, Y., A. Amir, Y. Levkovitz, and A. Zangen. 2007. Three-dimensional distribution of the electric field induced in the brain by transcranial magnetic stimulation using figure-8 and deep H-coils. *J Clin Neurophysiol* 24 (1):31–38.

Rothkegel, H., M. Sommer, T. Rammsayer, C. Trenkwalder, and W. Paulus. 2009. Training effects outweigh effects of single-session conventional rTMS and theta burst stimulation in PD patients. *Neurorehabil Neural Repair* 23 (4):373–381.

Sabatini, U., K. Boulanouar, N. Fabre et al. 2000. Cortical motor reorganization in akinetic patients with Parkinson's disease: A functional MRI study. *Brain* 123 (Pt 2):394–403.

Salvi, R. J., J. Wang, and D. Ding. 2000. Auditory plasticity and hyperactivity following cochlear damage. *Hear Res* 147 (1–2):261–274.

San-Juan, D., D. Calcaneo Jde, M. F. Gonzalez-Aragon et al. 2011. Transcranial direct current stimulation in adolescent and adult Rasmussen's encephalitis. *Epilepsy Behav* 20 (1):126–131.

Santiago-Rodriguez, E., L. Cardenas-Morales, T. Harmony et al. 2008. Repetitive transcranial magnetic stimulation decreases the number of seizures in patients with focal neocortical epilepsy. *Seizure* 17 (8):677–683.

Schambra, H. M., L. Sawaki, and L. G. Cohen. 2003. Modulation of excitability of human motor cortex (M1) by 1 Hz transcranial magnetic stimulation of the contralateral M1. *Clin Neurophysiol* 114 (1):130–133.

Schiller, Y. and Y. Bankirer. 2007. Cellular mechanisms underlying antiepileptic effects of low- and high-frequency electrical stimulation in acute epilepsy in neocortical brain slices in vitro. *J Neurophysiol* 97 (3):1887–1902.

Schlee, W., T. Hartmann, B. Langguth, and N. Weisz. 2009. Abnormal resting-state cortical coupling in chronic tinnitus. *BMC Neurosci* 10:11.

Schlee, W., N. Weisz, O. Bertrand, T. Hartmann, and T. Elbert. 2008. Using auditory steady state responses to outline the functional connectivity in the tinnitus brain. *PLoS One* 3 (11):e3720.

Schneider, P., M. Andermann, M. Wengenroth et al. 2009. Reduced volume of Heschl's gyrus in tinnitus. *Neuroimage* 45 (3):927–939.

Shimizu, H., T. Tsuda, Y. Shiga et al. 1999. Therapeutic efficacy of transcranial magnetic stimulation for hereditary spinocerebellar degeneration. *Tohoku J Exp Med* 189 (3):203–211.

Shindo, K., K. Sugiyama, L. Huabao et al. 2006. Long-term effect of low-frequency repetitive transcranial magnetic stimulation over the unaffected posterior parietal cortex in patients with unilateral spatial neglect. *J Rehabil Med* 38 (1):65–67.

Soler, M. D., H. Kumru, R. Pelayo et al. 2010. Effectiveness of transcranial direct current stimulation and visual illusion on neuropathic pain in spinal cord injury. *Brain* 133 (9): 2565–2577.

Sparing, R., M. Thimm, M. D. Hesse et al. 2009. Bidirectional alterations of interhemispheric parietal balance by non-invasive cortical stimulation. *Brain* 132 (Pt 11):3011–3020.

Stagg, C. J., S. Bestmann, A. O. Constantinescu et al. 2011. Relationship between physiological measures of excitability and levels of glutamate and GABA in the human motor cortex. *J Physiol* 589 (Pt 23):5845–5855.

Stagg, C. J., J. O'Shea, Z. T. Kincses et al. 2009. Modulation of movement-associated cortical activation by transcranial direct current stimulation. *Eur J Neurosci* 30 (7):1412–1423.

Takeuchi, N., T. Tada, M. Toshima, Y. Matsuo, K. Ikoma. 2009. Repetitive transcranial magnetic stimulation over bilateral hemispheres enhances motor function and training effect of paretic hand in patients after stroke. *J Rehabil Med* 41:1049–1054.

Tsubokawa, T., Y. Katayama, T. Yamamoto, T. Hirayama, and S. Koyama. 1991. Chronic motor cortex stimulation for the treatment of central pain. *Acta Neurochir Suppl (Wien)* 52:137–139.

Varga, E. T., D. Terney, M. D. Atkins et al. 2011. Transcranial direct current stimulation in refractory continuous spikes and waves during slow sleep: A controlled study. *Epilepsy Res* 97 (1–2):142–145.

Wagle-Shukla, A., M. J. Angel, C. Zadikoff et al. 2007. Low-frequency repetitive transcranial magnetic stimulation for treatment of levodopa-induced dyskinesias. *Neurology* 68 (9):704–705.

Ward, N. 2011. Assessment of cortical reorganisation for hand function after stroke. *J Physiol* 589 (Pt 23):5625–5632.

Ward, N. S. 2006. The neural substrates of motor recovery after focal damage to the central nervous system. *Arch Phys Med Rehabil* 87 (12 Suppl 2):S30–S35.

Ward, N. S., M. M. Brown, A. J. Thompson, and R. S. Frackowiak. 2003. Neural correlates of motor recovery after stroke: A longitudinal fMRI study. *Brain* 126 (Pt 11):2476–2496.

Ward, N. S., J. M. Newton, O. B. Swayne et al. 2006. Motor system activation after subcortical stroke depends on corticospinal system integrity. *Brain* 129 (Pt 3):809–819.

Williams, J. A., A. Pascual-Leone, and F. Fregni. 2010. Interhemispheric modulation induced by cortical stimulation and motor training. *Phys Ther* 90 (3):398–410.

Wirth, M., K. Jann, T. Dierks et al. 2011. Semantic memory involvement in the default mode network: A functional neuroimaging study using independent component analysis. *Neuroimage* 54 (4):3057–3066.

Zamir, O., C. Gunraj, Z. Ni, F. Mazzella, and R. Chen. 2011. Effects of theta burst stimulation on motor cortex excitability in Parkinson's disease. *Clin Neurophysiol* 123 (4):815–821.

Zheng, X., D. C. Alsop, and G. Schlaug. 2011. Effects of transcranial direct current stimulation (tDCS) on human regional cerebral blood flow. *Neuroimage* 58 (1):26–33.

Part VI

Safety

15 Safety of Transcranial Magnetic Stimulation
With a Note on Regulatory Aspects

Simone Rossi

CONTENTS

15.1 INTRODUCTION

After about 10 years from the publication of the first safety guidelines for the use of transcranial magnetic stimulation (TMS) (Wassermann et al. 1998), a large group of experts, including neurologists, neurophysiologists, psychiatrists, experimental psychologists, cognitive neuroscientists, physicists, engineers, representatives of TMS equipment manufacturers, representatives from various worldwide regulatory agencies, as well as basic and applied scientists, have met in Siena in march 2008, on behalf of the International Federation of Clinical Neurophysiology (IFCN), with the aim to revise all the available material regarding the safety of TMS appeared in the literature since 1996 to the meeting date.

A consensus had been reached for most of the treated items regarding not only safety, but also ethical issues and recommendations for the use of TMS in research and clinical settings, and a new document appeared in the late 2009 (Rossi et al. 2009). Thanks to the contribution of the several leading scientists in their respective fields, that article rapidly has had a great diffusion, witnessed by the fact that it has been the most downloaded article in 2011 from the website of *Clinical Neurophysiology* (i.e., the official Journal of the IFCN) and that it is the third most cited article of that journal in the last 5 years, despite being published in the late 2009. It is downloadable as an open-access article from many websites, including that of the IFCN and that of *Clinical Neurophysiology.*

Since the consensus meeting time, a few new aspects regarding safety have emerged, suggesting that safety tables, introduced in 1996 and partly updated in 2009, were basically successful in preventing major adverse events of the procedure. Therefore, the present chapter will not be a copy-and-paste duplicate of the latest version of safety guidelines (Rossi et al. 2009). All safety tables published in that article are still valid and not formally updated yet. Rather, it will highlight those few new aspects that have emerged from 2009 up to the end of 2011, as well as some topics representing still an open discussion since a full consensus was not reached in the 2008 meeting.

For example, the conventional excitatory aftereffect of 5 Hz rTMS applied in blocks seems to be reversed when the same rTMS is given continuously (Rothkegel et al. 2010). These data suggest that introducing intervals in high-frequency rTMS is not preventive in terms of the relation between cortical excitability and assumed seizure risk; on the contrary, intervals are responsible for the switch from inhibitory 5 Hz rTMS effects to excitation.

15.2   SAFETY ASPECTS RELATED TO TECHNICAL AND METHODOLOGICAL ADVANCEMENTS

15.2.1   COILS FOR DEEP STIMULATION

The use of single-pulse or double-pulse TMS in routine clinical practice, either with round or eight-shaped coils, does not pose particular safety concerns (Groppa et al. 2012), provided that patients are screened with the revised version of the appropriate questionnaire (Rossi et al. 2011).

A recent technological advance in TMS is the H-shaped coils, which are claimed to stimulate deeper than conventional coils. At the time of the consensus meeting in 2008, such coils were just introduced, so that safety data were very preliminary. Recent studies have begun to address the safety of such coils in small groups of patients with bipolar depression treated daily (20 Hz, 2 s ON, 20 s OFF, a total of 1680 stimuli/day) for 4 weeks (Harel et al. 2011) and after rTMS performed as continuation therapy for 18 weeks (Harel et al. 2012). Although no significant adverse effects were found, such promising results should be replicated in placebo-controlled clinical trials in different labs, preferably performed by independent researchers without commercial relationships with the manufacturers of coils.

15.2.2 THETA BURST STIMULATION

Theta burst stimulation (TBS) is an increasingly common method of administering repetitive TMS, since comparable after effects to those induced by conventional rTMS interventions can be obtained following shorter sessions of stimulation, due to the higher inner frequencies of TBS. This latter aspect, however, carries the intrinsic risk of TBS of conferring an even higher risk of seizure than other rTMS protocols. At the time of the consensus meeting of 2008, there was not so much specific safety literature, and recommendations for the maximum duration or intensity of stimulation when applying patterned trains of TMS were lacking. A recent paper (Oberman et al. 2011) filled this gap: authors performed a meta-analysis of 67 studies (1040 subjects/patients for a total of about 4500 TMS sessions). While most of the subjects were healthy controls, 225 were clinical patients with a variety of diagnoses including autism spectrum disorders (n = 27), chronic pain (n = 6), stroke (n = 42), tinnitus (n = 67), Parkinson's disease (PD) (n = 37), dystonia (n = 14), amyotrophic lateral sclerosis (n = 20), fragile X syndrome (n = 2), and multiple sclerosis (n = 10). Not only that most of the subjects (632) received TBS over the primary motor cortex (M1), but also their other brain regions were stimulated: prefrontal cortex (PFC), including supplementary motor area (150), frontal eye fields (20), dorsolateral PFC (97), as well as primary sensory cortex (98), additional parietal sites (56), temporal cortex (67), occipital cortex (102), cerebellum (44). In the 10 patients with multiple sclerosis, more than one site was stimulated.

Overall, the crude risk for mild adverse events during TBS was estimated to be 1.1% (Oberman et al. 2011), while that of seizure was 0.02%: these reported adverse events were comparable to those most commonly described for conventional rTMS, in terms of both severity and rough incidence (see Table 1 of Rossi et al. 2009). These were (1) seizure in 1 healthy control subject during continuous TBS, a case fully described in the following paragraph; (2) mild headache in 24 subjects (20 healthy controls, 2 patients with tinnitus, and 2 PD patients; (3) nonspecific discomfort in 5 patients with tinnitus; (4) mild discomfort due to cutaneous sensation and neck muscle contraction in 5 healthy subjects; (5) transient worsening tinnitus in 3 patients; (6) nausea in 1 PD patient with; (7) light headedness or vagal responses in 11 healthy controls; (8) unilateral eye pain and lacrimation in 1 healthy subject (which ceased upon cessation of the treatment session). The most common reported adverse event during TBS was also the most common in other rTMS protocols, namely a transient headache and neck pain. This adverse event has been reported in up to 40% of patients undergoing high-frequency rTMS (Rossi et al. 2009) and was experienced by less than 3% of the subjects receiving TBS. This is probably because the intensity for TBS is usually lower (80% of active motor threshold) than that used for rTMS applications (usually ranging from 90% to 120% of resting motor threshold [RMT]), for investigational or treatment paradigms. The next warranted step for the neuroscientific community is to provide safety tables summarizing safe combinations of intensity/duration of TBS according to the targeted brain regions, also in view of the recent findings indicating a reversal of TBS aftereffects with prolonged stimulation (Gamboa et al. 2010).

15.2.3 rTMS with Implanted Brain Electrodes

When applying rTMS in patients with an implanted deep brain stimulation (DBS) system, one potential safety hazard is the induction of significant voltages in the subcutaneous leads in the scalp, which could result in unintended electric currents in the electrode contacts used for DBS. This can be the case either when the internal pulse generator (IPG) is turned ON or OFF. The situation may be exacerbated by the coiling of the electrode lead into several loops near the electrode insertion point in the skull (Rossi et al. 2009). Voltages as high as 100 V, resulting in currents as high as 83 mA, can be induced in the DBS leads by a TMS pulse, irrespective of IPG modes. These currents are an order of magnitude higher than the normal DBS pulses and can expose patients to potential tissue damage. When the IPG is turned OFF, electrode currents flow only if the TMS-induced voltage exceeds 5 V (Deng et al. 2011). It is clear that caution is required when doing TMS in patients with DBS electrodes. It remains that TMS, if justified, should eventually be applied far away from the IPG (Rossi et al. 2009).

15.2.4 An Emerging Field of Research: TMS Interleaved with Transcranial Current Application

In recent studies, TMS has been used in combination with other forms of transcranial stimulation based on the simultaneous delivery of weak electric currents on the scalp surface. This has been done, e.g., to explore whether transcranial alternating current stimulation (tACS) may alter cortical excitability in a frequency-dependent manner, both in the visual and in the motor systems (Kanai et al. 2010, Feurra et al. 2011). Such a new field of neuroscience poses new theoretically relevant safety aspects: does the biophysical interaction between tACS and TMS increase the risk of altering cortical excitability, thereby exposing subjects to a higher risk of seizures or other side effects? The current, albeit partial, response seems negative, although ad hoc studies are warranted, especially regarding the time length of tACS application. An even more important aspect is that, up to now, only single-pulse TMS has been coupled with concurrent tACS; there are no studies yet addressing potential effects of rTMS during tACS. It is advisable that in case of studies like those, strict monitoring of subjects will be required.

15.3 rTMS AND SEIZURES

The incidence of seizures during rTMS remained definitely low from 2009 up to now, despite thousands of subjects and patients who have undergone TMS. The following are three single, but highly instructive, case reports that show (1) how safety guidelines, if strictly followed, are able to prevent seizure occurrence; (2) how also TBS may potentially trigger seizures; (3) how seizures can occur at a distance from the TMS site of application.

15.3.1 When Safety Guidelines Are Not Followed

The first case concerns a healthy subject who had a partial, secondarily generalized seizure with 20 Hz rTMS to the right motor cortex during muscle contractions

(Edwardson et al. 2011). Published safety tables consider the combination of frequency/intensity/duration of rTMS while subjects are at rest. Therefore, it remains unclear whether contracting a muscle sufficiently to generate motor-evoked potentials (MEPs) during rTMS trains may affect the risk for a seizure. The paradigm of stimulation was quite complicated: briefly, the experimental TMS sessions consisted of 40 min of 0.5–20 Hz (average 8.7 Hz) variable-frequency TMS, at an intensity ranging from 50% of the maximal stimulator output to 90% of the RMT for the hand muscle, in which single TMS pulses were triggered immediately when the electromyographic (EMG) amplitude exceeded a threshold of 30% of maximum voluntary contraction, but were rate limited to no more than 20 pulses/s. Stimulus train durations varied according to an algorithm based on established safety guidelines with short durations (as short as 1.5 s) for high-frequency stimulation and long durations (as long as 270 s) for low-frequency stimulation. Additional control sessions consisted of 20 Hz rTMS delivered in 1.5 s trains. The inter-train interval (ITI) for both the experimental and control sessions was 30 s. The crucial point here was that during all sessions, the subject was asked to contract the muscle of study (either a hand or a forearm muscle) isometrically, at 30% of maximum amplitude during the stimulus trains and to rest during the ITIs. This protocol, intermingled with a TMS mapping to assess cortical plasticity, was repeated for 4 weeks, for a total of five sessions (including the mapping session). As stated by authors (Edwardson et al. 2011), "during the first few minutes of the last session, rTMS evoked motor potentials in the left ECR and in some intrinsic hand muscles during the stimulus trains. Five minutes into the session the subject began to experience MEPs in more proximal muscles, including the left biceps and deltoid. Eight minutes into the session he started to develop flexor posturing of the left upper extremity during a stimulus train." This was finally followed by a tonic seizure lasting 60 s, accompanied by tongue laceration, but not bladder incontinence, and an overt post-ictal state. Of course, the seizure could be easily avoided by stopping the rTMS as soon as muscle twitches in the biceps appeared, since TMS was targeted for hand muscles. This is a clear violation of the suggested safety guidelines (Rossi et al. 2009). This case highlights the importance of EMG monitoring during new paradigms of rTMS and suggests that muscle contraction, through the lowering of the excitability threshold of the corticospinal system, should be performed with extreme caution during rTMS and in the presence of medically qualified personnel.

15.3.2   TBS AND SEIZURES

The second case concerns a single case of seizure occurrence induced by TBS, described by Oberman and Pascual-Leone (2009) in a 33 year old healthy man with no risk factors for epilepsy. The subject had a flight from Europe to Boston in the previous days, but he did not complain jet lag or sleep deprivation in the night preceding the TMS experiment. The seizure occurred following approximately 50 trains (10 s) of cTBS to the M1 at an intensity of 100% of RMT. This is an important factor (and a red flag for future protocols), since TBS is usually given at intensities below 80% of the active motor threshold. Given this one incident of a seizure among the 4500 TBS sessions in more than 1000 subjects described in the most recent meta-analysis about safety of TBS (Oberman et al. 2011), the resulting crude risk per subject of seizure as a result of TBS is estimated as 0.1% the crude risk of seizure per session of TBS was approximately 0.02%. However, this finding highlights the need to evaluate motor threshold with EMG monitoring rather than that with visible muscle twitches, a topic that did not reach a full consensus in the 2008 meeting (Rossi et al. 2009).

15.3.3 Seizure Occurring at a Distance from the TMS Site

A still "in press" paper (Vernet et al. 2012) suggests that TMS-induced seizures may also originate from a cortical focus, which is different from the stimulation site. A 22 year old male epileptic patient with a periventricular nodular heterotopia, while undergoing paired-pulse TMS over a cortical region distant from his seizure focus, had a spontaneous seizure of his usual semiology (tonic head version without clonic jerks), with demonstrated ictal EEG onset from his expected epileptogenic focus (Vernet et al. 2012). The patient had a spontaneous occurrence of such seizures every 2–6 weeks, with no precipitating factors identified.

Initially, he received 91 single biphasic pulses with intensity between 30% and 58% of maximum stimulator output (MSO) over the M1, with an RMT of 46%. Then, 35 single biphasic pulses (55% MSO–120% RMT) were delivered over the posterior end of the right inferior frontal gyrus. Subsequently, 38 monophasic pulses (intensity 63%–84% MSO) were given over M1 to establish the intensity needed to obtain stable MEPs. The following monophasic pulses were then delivered over the programmed target: (a) 22 single pulses (76% MSO); (b) 22 paired pulses with an interstimulus interval (ISI) of 3 ms (50% and 76% MSO); (c) 22 paired pulses (ISI 12 ms, 50% and 76% MSO); and (d) 6 paired pulses (ISI 100 ms, 76% MSO). All stimulations were separated by at least 5 s; in total, 286 pulses were delivered over approximately 45 min, with on average a 3.5 min break between each condition. The seizure occurred after the (d) session.

Although the most probable explanation is seizure occurrence not related to an at-distance effect of TMS, this case report again emphasizes the need of neurophysiological monitoring in patients at risk undergoing TMS.

15.4 NEED FOR MONITORING SUBJECTS UNDERGOING rTMS

It should be stressed that neurophysiological monitoring is strongly recommended for all those studies that are based on TMS protocols that are not sufficiently tested yet by a safety point of view, as well as for   all those protocols that use a combination of parameters of stimulation (including new coils) that are close to upper safety limits of the published tables. This applies to both healthy subjects for research use and, even more, patient populations.

Two measures have been proposed to detect potential early signs of increasing brain excitability that might lead to a seizure: spread of excitation to neighboring cortical areas and possible manifestations of EEG afterdischarges. The methods to achieve these requirements have been extensively described (Rossi et al. 2009) and summarized here:

In studies where rTMS is not expected to elicit stimulus-locked muscular responses (called motor-evoked potentials, MEPs) (e.g., stimulation of the motor cortex below threshold, or of a scalp site outside it at any intensity), the EMG can be monitored continuously from a contralateral hand muscle, since these muscles have the lowest threshold for the production of MEPs: indeed, the appearance of MEPs under these circumstances may indicate a lowering of threshold in the subthreshold stimulated motor cortex or the spread of excitation from neighboring areas to the motor cortex.

In studies where the stimulation is expected to produce MEPs in a distal muscle (i.e., the hand), an additional muscle at a proximal segment of the same limb can be monitored. The appearance of "proximal" MEPs in a forearm muscle (such as the extensor carpi radialis), or in an arm muscle (as the deltoid), would indicate the intra-cortical spread of excitation or lowering of the excitability threshold.

Visual monitoring of subjects/patients (by qualified personnel) during rTMS is mandatory. Muscle twitching time-locked to the stimulus provides a potentially important indication of the spread of evoked motor activity. More work is probably needed to explore new monitoring methods based on EEG co-registered to TMS.

15.5 SAFETY ISSUES OF rTMS AS REHABILITATIVE/TREATMENT PROCEDURE IN DIFFERENT PATIENT POPULATIONS

15.5.1 DEPRESSION AND OTHER PSYCHIATRIC DISORDERS

A comprehensive description of most recent published papers on rTMS and depression, with relative safety reports, can be found in a recent review (Lefaucheur et al. 2011). There is sufficient evidence to consider rTMS as definitely beneficial in schizophrenic patients to treat auditory hallucinations. A very recent double-blind trial, again with inhibitory 1-Hz rTMS, has confirmed the potential utility of this treatment (Mantovani et al. 2006) in the reduction of pharmacoresistant obsessive-compulsive symptoms (Mantovani et al. 2010). Once adhered to general precautions of TMS use, no major safety hazards have been described under these circumstances (i.e., when low-frequency rTMS is applied).

In depression, both high- and low-frequency rTMS treatments have been tested. The author will comment here, as an example of safety despite the most aggressive rTMS treatment ever published, about the study of Hadley et al. (2011): they enrolled 19 patients who were in a current major depressive episode with treatment-resistant unipolar or bipolar depression, and treated them in their acute episode and in a maintenance fashion for 18 months. The patients received daily left prefrontal rTMS at 120% RMT, 10 Hz, 5 s ON, and 10 s OFF and for a mean of 6,800 stimuli per session (34,000 stimuli/week), more than twice the dose delivered in the pivotal Food and Drug Administration (FDA) trial. Moreover, patients were taking one to eight concomitant psychotropic medications, including stimulants, benzodiazepines, heterocyclics, selective serotonin reuptake inhibitors, monoamine oxidase inhibitors, lithium, other antidepressants, and sleep aids. Also three subjects had vagus nerve stimulation (VNS) implants. Even with the aggressive stimulation parameters given to patients taking medications that might lower the seizure threshold (see Rossi et al. 2009, Lefaucheur et al. 2011 for complete lists of these drugs), there were no serious adverse events. Specifically, there were no syncopal spells or seizures. Two subjects decided to discontinue treatment after the first session for a variety of reasons, some being concerned with the painfulness of the procedure, the most common adverse effect of rTMS (Rossi et al. 2009). There were no other major adverse effects and no problems to the pulse generator for VNS were reported, despite one VNS patient received the most aggressive weekly treatment: 9,000 pulses twice a day for 2 weeks (excluding weekends), meaning a total of 90,000 pulses in a week for 2 weeks, more than doubling what was previously shown to be safe (Anderson et al. 2011). Interestingly, one subject received rTMS treatment for more than 2 years. Her daily stimuli were 5100 (25 treatments), 6800 (90 treatments), 6000 (22 treatments), and 8000 (48 treatments and ongoing). Over this period, she received a total of 1,243,500 stimuli. Despite receiving more than a million stimuli in 2 years, this subject reported no adverse effects and is reported still doing well, having graduated from college. She is continuing to receive approximately 8000 stimuli per weekly session and has experienced relief from her depressive symptoms (Hadley et al. 2011).

15.5.2 Patients with "Positive Sensory Phenomena"

A recent review addressed safety of rTMS in patients with pathologic positive sensory phenomena (Muller et al. 2011), a topic that was not covered specifically in the last available guidelines (Rossi et al. 2009). Among these are tinnitus, auditory and visual hallucinations, and pain syndromes, all conditions that are potentially linked to a pathological increase in cortical excitability in sensory or associative brain regions. On a sample of 1815 subjects, adverse events were generally mild and occurred in 16.7% of subjects. Seizure was the most serious adverse event, and occurred in three patients, with a 0.16% crude per-subject risk. The second most severe adverse event involved aggravation of sensory phenomena, occurring in 1.54%, mainly in patients with tinnitus, a transient finding that had been already reported (Rossi et al. 2007). Results of the meta-analysis suggest that rTMS in these patients appears to be a reasonably safe and well-tolerated procedure (Muller et al. 2011), with a risk of developing side effects not different from studies done with rTMS on other neuropsychiatric disorders.

15.5.3 Posttraumatic Brain Injury

It has been also proposed that neuromodulation via rTMS may be a fruitful avenue for alleviating symptoms in posttraumatic brain injury patients (Demirtas-Tatlidede et al. 2011), another topic not covered in the previous guidelines. Some possible concerns related to the use of rTMS in these patients are the higher risk of seizures following head trauma, as well as factors related to a distorted conductance and magnitude of the electric current being induced in cortical regions due to the skull damage and fractures. Indeed, skull injuries significantly modify the distribution of the induced currents, which may become concentrated toward the edges of large skull defects, depending on the combination of type of stimulation and nature of the defect (Madhavan et al. 2011). Of course, besides the presence of skull defects, also craniotomy with placement of skull plates might add another potential risk for application of rTMS in traumatic brain injury patients (Rotenberg et al. 2007, Rotenberg and Pascual-Leone 2009). Nowadays, available safety data of rTMS on these patients are still very limited: there is only one case study, in which a detailed safety assessment reported a lack of adverse events in a patient with severe traumatic brain injury following application of a specific rTMS protocol over five consecutive days through 6 weeks (Louise-Bender Pape et al. 2009).

15.6 CONCLUSIONS AND REGULATORY ASPECTS

The field of therapeutic/rehabilitative application of rTMS and TBS is steadily growing worldwide. Thousands of patients have undergone TMS in the setting of clinical trials and increasingly in clinical programs in a variety of settings (from academic medical centers to private practice clinics). The expanded use of TMS sometimes goes even beyond solid scientific evidence, especially in private structures, and careful assessment of the existing knowledge is important. An updated review of all current applications of rTMS in neurology, neurorehabilitation, psychiatry, and otolaryngology, together with an attempt to classify the level of evidence, can be found in the recently published French version of guidelines for the use of TMS (Lefaucheur et al. 2011).

Several TMS protocols and devices are officially approved in a few countries for specific clinical/therapeutic uses: in the United States, the FDA in 2008 cleared the Neurostar treatment using the Neuronetics device for a specific indication (i.e., in patients with major depression who have failed one type of medical treatment). In the following years, rTMS treatment for depression was approved in Canada and Israel. In October 2011, the Federal Council of Medicine in Brazil approved the use of rTMS for the treatment of major depression—both unipolar and bipolar—and also for the treatment of hallucinations in schizophrenia and for neurosurgical mapping (Felipe Fregni, personal communication, see the official document at http://www.portalmedico.org.br/pareceres/CFM/2011/37_2011.htm). In the approval document, it is also specified that therapeutic TMS needs to be performed by physicians and in an environment that can give rapid medical support, if needed. Regarding presurgical motor mapping, in the United States, the FDA cleared the eXimia Nexstim device.

To the best of my knowledge, there is still no official document regulating the clinical/therapeutic use of rTMS in Europe, Asia, and Australia.

Therefore, maybe it is now time that Scientific Societies (mostly Clinical Neurophysiology, Psychiatry and Neurology) might consider the possibility to promote the official therapeutic use of rTMS for those diseases where a satisfactory level of efficacy, based on independent scientific evidences, has been demonstrated (major depression, auditory hallucinations in schizophrenia, chronic pain, tinnitus), as well as to advise against the use of rTMS when beneficial effects have not been reproduced in controlled trials (at least with current protocols). For example, the American Psychiatric Association is considering endorsement of TMS for depression and auditory hallucinations (Alvaro Pascual-Leone, personal communication).

Following the line of the last consensus conference (Rossi et al. 2009) and of the current work done by an "ad hoc commission" in France (Lefaucheur et al. 2011), a first step toward this unitary goal could be done through the formation of an international panel (or, at least, national panels) of experts, with the aim to provide public health administrations of each country with accreditation requirements for TMS operators, as well as with all the scientific instruments to definitely judge the fields where the use of therapeutic/rehabilitative applications of rTMS could be formally approved.

REFERENCES

Anderson B, Mishory A, Nahas Z et al. 2011. Tolerability and safety of high daily doses of repetitive transcranial magnetic stimulation in healthy young men. *J ECT* 22:49Y53.

Demirtas-Tatlidede A, Vahabzadeh-Hagh AM, Bernabeu M, Tormos JM, and Pascual-Leone A. 2011. Noninvasive brain stimulation in traumatic brain injury. *J Head Trauma Rehabil* 27:274–292.

Deng ZD, Lisanby SH, and Peterchev AV. 2011. Transcranial magnetic stimulation in the presence of deep brain stimulation implants: Induced electrode currents. *Conf Proc IEEE Eng Med Biol Soc* 2011:5473–5476.

Edwardson M, Fetz EE, and Avery DH. 2011. Seizure produced by 20 Hz transcranial magnetic stimulation during isometric muscle contraction in a healthy subject. *Clin Neurophysiol* 122:2324–2327.

Feurra M, Bianco G, Santarnecchi E, Del Testa M, Rossi A, and Rossi S. 2011. Frequency-dependent tuning of the human motor system induced by transcranial oscillatory potentials. *J Neurosci* 31:12165–12170.

Gamboa OL, Antal A, Moliadze V, and Paulus W. 2010. Simply longer is not better: Reversal of theta-burst after-effect with prolonged stimulation. *Exp Brain Res* 204:181–187.

Groppa S, Oliviero A, Eisen A, Quartarone A, Cohen LG, Mall V, Kaelin-Lang A et al. 2012. A practical guide to diagnostic transcranial magnetic stimulation: Report of an IFCN committee. *Clin Neurophysiol* 123:858–882.

Hadley D, Anderson BS, Borckardt JJ, Arana A, Li X, Nahas Z, and George MS. 2011. Safety, tolerability, and effectiveness of high doses of adjunctive daily left prefrontal repetitive transcranial magnetic stimulation for treatment-resistant depression in a clinical setting. *J ECT* 27:18–25.

Harel EV, Rabany L, Deutsch L, Bloch Y, Zangen A, and Levkovitz Y. 2012. H-coil repetitive transcranial magnetic stimulation for treatment resistant major depressive disorder: An 18-week continuation safety and feasibility study. *World J Biol Psychiatry* [Epub. ahead of print].

Harel EV, Zangen A, Roth Y, Reti I, Braw Y, and Levkovitz Y. 2011. H-coil repetitive transcranial magnetic stimulation for the treatment of bipolar depression: An add-on, safety and feasibility study. *World J Biol Psychiatry* 12:119–126.

Kanai R, Paulus W, and Walsh V. 2010. Transcranial alternating current stimulation (tACS) modulates cortical excitability as assessed by TMS-induced phosphene thresholds. *Clin Neurophysiol* 121:1551–1554.

Lefaucheur JP, André-Obadia N, Poulet E, Devanne H, Haffen E, Londero A, Cretin B et al. 2011. French guidelines on the use of repetitive transcranial magnetic stimulation (rTMS): Safety and therapeutic indications. *Neurophysiol Clin* 41:221–295 (French).

Louise-Bender Pape T, Rosenow J, Lewis G, Ahmed G, Walker M, Guernon A, Roth H, and Patil V. 2009. Repetitive TMS- associated neurobehavioral gains during coma recovery. *Brain Stimul* 2:22–35.

Madhavan S, Weber KA IInd, and Stinear JW. 2011. Non-invasive brain stimulation enhances fine motor control of the hemiparetic ankle: Implications for rehabilitation. *Exp Brain Res* 209:9–17.

Mantovani A, Lisanby SH, Pieraccini F, Ulivelli M, Castrogiovanni P, and Rossi S. 2006. Repetitive transcranial magnetic stimulation (rTMS) in the treatment of obsessive-compulsive disorder (OCD) and Tourette's syndrome (TS). *Int J Neuropsychopharmacol* 9:95–100.

Mantovani A, Simpson HB, Fallon BA, Rossi S, and Lisanby SH. 2010. Randomized sham-controlled trial of repetitive transcranial magnetic stimulation in treatment-resistant obsessive-compulsive disorder. *Int J Neuropsychopharmacol* 13:217–227.

Muller A, Pascual-Leone A, and Rotenberg A. 2011. Safety and tolerability of repetitive transcranial magnetic stimulation inpatients with pathologic positive sensory phenomena: A review of literature. *Brain Stimul* [Epub. ahead of print].

Oberman L, Edwards D, Eldaief M, and Pascual-Leone A. 2011. Safety of theta burst transcranial magnetic stimulation: A systematic review of the literature. *Clin Neurophysiol* 28:67–74.

Oberman L and Pascual-Leone A. 2009. Report of seizure induced by continuous theta burst stimulation. *Brain Stimul* 2:246–247.

Rossi S, De Capua A, Ulivelli M, Bartalini S, Falzarano V, Filippone G, and Passero S. 2007. Effects of repetitive transcranial magnetic stimulation on chronic tinnitus: A randomised, crossover, doubleblind, placebo controlled study. *J Neurol Neurosurg Psychiatry* 78:857–863.

Rossi S, Hallett M, Rossini PM, and Pascual-Leone A. 2011. Screening questionnaire before TMS: An update. *Clin Neurophysiol* 122:1866.

Rossi S, Hallett M, Rossini PM, Pascual-Leone A, and Safety of TMS Consensus Group. 2009. Safety, ethical considerations, and application guidelines for the use of transcranial magnetic stimulation in clinical practice and research. *Clin Neurophysiol* 120:2008–2039.

Rotenberg A, Harrington MG, Birnbaum DS et al. 2007. Minimal heating of titanium skull plates during 1 Hz repetitive transcranial magnetic stimulation. *Clin Neurophysiol* 118: 2536–2538.

Rotenberg A and Pascual-Leone A. 2009. Safety of 1 Hz repetitive transcranial magnetic stimulation (rTMS) in patients with titanium skull plates. *Clin Neurophysiol* 120:1417.

Rothkegel H, Sommer M, and Paulus W. 2010. Breaks during 5 Hz rTMS are essential for facilitatory after effects. *Clin Neurophysiol* 121:426–430.

Vernet M, Walker L, Yoo WK, Pascual-Leone A, and Chang BS. 2012. EEG onset of a seizure during TMS from a focus independent of the stimulation site. *Clin Neurophysiol* (Epub. ahead of print).

Wassermannn EM. 1998. Risk and safety of repetitive transcranial magnetic stimulation: report and suggested guidelines from the International Workshop on the Safety of Repetitive Transcranial Magnetic Stimulation. *Electroencephalogr Clin Neurophysio* 108:1–16.

Index

T - #0066 - 071024 - C8 - 234/156/25 - PB - 9780367380571 - Gloss Lamination